WATER DISTRIBUTION SYSTEMS HANDBOOK

WATER DISTRIBUTION SYSTEMS HANDBOOK

Larry W. Mays, Editor in Chief
Department of Civil and Environmental Engineering
Arizona State University
Tempe, Arizona

McGraw-Hill
New York • San Francisco • Washington, D.C. • Auckland • Bogotá • Caracas • Lisbon
• London • Madrid • Mexico City • Milan • Montreal • New Delhi • San Juan
• Singapore • Sydney • Tokyo • Toronto

Library of Congress Cataloging-in-Publication Data
Water distribution systems handbook/Larry W. Mays, ed.
  p. cm.
  Includes bibliographical references.
  ISBN 0-07-134213-3
  1. Water—Distributions Hanbooks, manuals, etc.
  2. Water—supply engineering Handbooks, manuals, etc. I. Mays, Larry W.
  TD481.W375   1999
  628. 1'44—dc21   99–16987
                            CIP

McGraw-Hill

A Division of The **McGraw·Hill** Companies

Copyright © 2000 by The McGraw-Hill Companies, Inc. All rights reserved. Printed in the United States of America. Except as permitted under the United States Copyright Act of 1976, no part of this publication may be reproduced or distributed in any form or by any means, or stored in a data base or retrieval system, without the prior written permission of the publisher.

    4 5 6 7 8 9 0 DOC/DOC   0 4 3 2 1 0

ISBN 0-07-134213-3
The sponsoring editor for this book was Larry Hager and the production supervisor was Sherri Souffrance. It was set in Times Roman by Compuvision.

Printed and bound by R. R. Donnelley & Sons Company.

This book was printed on acid-free paper.

McGraw-Hill books are available at special quantity discounts to use as premiums and sales promotions, or for use in corporate training programs. For more information, please write to the Director of Special Sales, McGraw-Hill, Inc., Two Penn Plaza, New York, NY, 10121-2298. Or contact your local bookstore.

Information contained in this work has been obtained by The McGraw-Hill Companies, Inc. ("McGraw-Hill") from sources believed to be reliable. However, neither McGraw-Hill nor its authors guarantee the accuracy or completeness of any information published herein, and neither McGraw-Hill nor its authors shall be responsible for any errors, omissions, or damages arising out of use of this information. This work is published with the understanding that McGraw-Hill and its authors are supplying information but are not attempting to render engineering or other professional services. If such services are required, the assistance of an appropriate professional should be sought.

CONTENTS

Contributors xxv
Preface xxvii
Acknowledgments xxix

CHAPTER 1 INTRODUCTION

1.1 BACKGROUND 1.1
1.2 HISTORICAL ASPECTS OF WATER DISTRIBUTION 1.3
   1.2.1 Ancient Urban Water Supplies 1.3
   1.2.2 Status of Water Distribution Systems in the 19th Century 1.9
   1.2.3 Perspectives on Water Distribution Mains in the United States 1.10
   1.2.4 Early Pipe Flow Computational Methods 1.16
1.3 MODERN WATER DISTRIBUTION SYSTEMS 1.16
   1.3.1 The Overall Systems 1.16
   1.3.2 System Components 1.20
   1.3.3 System Operation 1.26
   1.3.4 The Future 1.29
REFERENCES 1.30

CHAPTER 2 HYDRAULICS OF PRESSURIZED FLOW

2.1 INTRODUCTION 2.1
2.2 IMPORTANCE OF PIPELINE SYSTEMS 2.2
2.3 NUMERICAL MODELS: BASIS FOR PIPELINE ANALYSIS 2.3
2.4 MODELING APPROACH 2.4
   2.4.1 Properties of Matter (What?) 2.5
   2.4.2 Laws of Conservation (How?) 2.6
   2.4.3 Conservation of Mass 2.7
      2.4.3.1 Law of conservation of chemical species. 2.7
      2.4.3.2 Steady flow 2.8
   2.4.4 Newton's Second Law 2.9
2.5 SYSTEM CAPACITY: PROBLEMS IN TIME AND SPACE 2.10
2.6 STEADY FLOW 2.13
   2.6.1 Turbulent Flow 2.15
   2.6.2 Headloss Caused by Friction 2.16
   2.6.3 Comparison of Loss Relations 2.18
   2.6.4 Local Losses 2.21
   2.6.5 Tractive Force 2.22
   2.6.6 Conveyance System Calculations: Steady Uniform Flow 2.23
   2.6.7 Pumps: Adding Energy to the Flow 2.26
   2.6.8 Sample Application Including Pumps 2.28
   2.6.9 Networks—Linking Demand and Supply 2.30
2.7 QUASI-STEADY FLOW: SYSTEM OPERATION 2.30

2.8 UNSTEADY FLOW: INTRODUCTION OF FLUID TRANSIENTS 2.32
  2.8.1 Importance of Waterhammer 2.32
  2.8.2 Cause of Transients 2.34
  2.8.3 Physical Nature of Transient Flow 2.35
    2.8.3.1 Implication 1. Water has a high density 2.35
    2.8.3.2 Implication 2. Water is only slightly compressible 2.35
    2.8.3.3 Implication 3. Local action and control of valves 2.36
  2.8.4 Equation of State-Wavespeed Relations 2.37
  2.8.5 Increment of Head-Change Relation 2.38
  2.8.6 Transient Conditions in Valves 2.39
    2.8.6.1 Gate discharge equation 2.40
    2.8.6.2 Alternate valve representation 2.41
    2.8.6.3 Pressure regulating valves 2.42
  2.8.7 Conclusion 2.42
REFERENCES 2.42

CHAPTER 3 SYSTEM DESIGN: AN OVERVIEW

3.1 INTRODUCTION 3.1
  3.1.1 Overview 3.1
  3.1.2 Definitions 3.2
3.2 DISTRIBUTION SYSTEM PLANNING 3.2
  3.2.1 Water Demands 3.2
  3.2.2 Planning and Design Criteria 3.7
    3.2.2.1 Supply 3.7
    3.2.2.2 Storage 3.7
    3.2.2.3 Fire demands 3.8
    3.2.2.4 Distribution system analysis 3.8
    3.2.2.5 Service pressures 3.8
  3.2.3 Peaking Coefficients 3.9
  3.2.4 Computer Models and System Modeling 3.9
    3.2.4.1 History of computer models 3.10
    3.2.4.2 Software packages 3.10
    3.2.4.3 Development of a system model 3.11
3.3 PIPELINE PRELIMINARY DESIGN 3.11
  3.3.1 Alignment 3.11
  3.3.2 Subsurface Conflicts 3.13
  3.3.3 Rights-of-Way 3.13
3.4 PIPING MATERIALS 3.13
  3.4.1 Ductile Iron Pipe (DIP) 3.14
    3.4.1.1 Materials 3.14
    3.4.1.2 Available sizes and thicknesses 3.14
    3.4.1.3 Joints 3.14
    3.4.1.4 Gaskets 3.14
    3.4.1.5 Fittings 3.14
    3.4.1.6 Linings 3.16
    3.4.1.7 Coatings 3.17
  3.4.2 Polyvinyl Chloride (PVC) Pipe 3.18
    3.4.2.1 Materials 3.18
    3.4.2.2 Available sizes and thicknesses 3.19
    3.4.2.3 Joints 3.19

3.4.2.4 Gaskets	3.20
3.4.2.5 Fittings	3.20
3.4.2.6 Linings and Coatings	3.20
3.4.3 Steel Pipe	3.21
3.4.3.1 Materials	3.21
3.4.3.2 Available sizes and thicknesses	3.21
3.4.3.3 Joints	3.22
3.4.3.4 Gaskets	3.22
3.4.3.5 Fittings	3.22
3.4.3.6 Linings and Coatings	3.23
3.4.4 Reinforced Concrete Pressure Pipe (RCPP)	3.25
3.4.4.1 Steel cylinder pipe, AWWA C300	3.26
3.4.4.2 Prestressed steel cylinder pipe, AWWA C301	3.26
3.4.4.3 Noncylinder pipe, AWWA C302	3.28
3.4.4.4 Pretensioned steel cylinder, AWWA C300	3.28
3.4.5 High-Density Polyethylene (HDPE) Pipe	3.29
3.4.5.1 Materials	3.29
3.4.5.2 Available sizes and thicknesses	3.30
3.4.5.3 Joints	3.30
3.4.5.4 Gaskets	3.31
3.4.5.5 Fittings	3.31
3.4.5.6 Linings and coatings	3.31
3.4.6 Asbestos-Cement Pipe (ACP)	3.31
3.4.6.1 Available sizes and thicknesses	3.31
3.4.6.2 Joints and fittings	3.32
3.4.7 Pipe Material Selection	3.32
3.5 PIPELINE DESIGN	3.34
3.5.1 Internal Pressures	3.34
3.5.2 Loads on Buried Pipe	3.34
3.5.2.1 Earth loads	3.35
3.5.2.2 Rigid pipe	3.36
3.5.2.3 Flexible pipe	3.37
3.5.3 Thrust Restraint	3.38
3.5.3.1 Thrust blocks	3.39
3.5.3.2 Restrained joints	3.41
3.6 DISTRIBUTION AND TRANSMISSION SYSTEM VALVES	3.44
3.6.1 Isolation Valves	3.44
3.6.1.1 Gate valves	3.45
3.6.1.2 Butterfly valves	3.45
3.6.2 Control Valves	3.46
3.6.2.1 Pressure-reducing valve	3.46
3.6.2.2 Pressure-sustaining valves	3.47
3.6.2.3 Flow-control valves	3.47
3.6.2.4 Altitude valves	3.47
3.6.2.5 Pressure-relief valves	3.47
3.6.3 Blow-offs	3.47
3.6.4 Air Release and Vacuum-Relief Valves	3.48
REFERENCES	3.48

viii Contents

CHAPTER 4 HYDRAULICS OF WATER DISTRIBUTION SYSTEMS

4.1 INTRODUCTION 4.1
   4.1.1 Configuration and Components of Water Distribution Systems 4.1
   4.1.2 Conservation Equations for Pipe Systems 4.3
   4.1.3 Network Components 4.3
4.2 STEADY-STATE HYDRAULIC ANALYSIS 4.5
   4.2.1 Series and Parallel Pipe Systems 4.5
   4.2.2 Branching Pipe Systems 4.7
   4.2.3 Pipe Networks 4.11
      4.2.3.1 Hardy Cross method 4.11
      4.2.3.2 Linear theory method 4.17
      4.2.3.3 Newton-Raphson method and the node equations 4.18
      4.2.3.4 Gradient algorithm 4.20
      4.2.3.5 Comparison of solution methods 4.22
      4.2.3.6 Extended-period simulation 4.23
4.3 UNSTEADY FLOW IN PIPE NETWORK ANALYSIS 4.24
   4.3.1 Governing Equations 4.24
   4.3.2 Solution Methods 4.25
      4.3.2.1 Loop formulation 4.25
      4.3.2.2 Pipe formulation with gradient algorithm 4.26
4.4 COMPUTER MODELING OF WATER DISTRIBUTION SYSTEMS 4.26
   4.4.1 Applications of Models 4.27
   4.4.2 Model Calibration 4.27
REFERENCES 4.28

CHAPTER 5 PUMP SYSTEM HYDRAULIC DESIGN

5.1 PUMP TYPES AND DEFINITIONS 5.1
   5.1.1 Pump Standards 5.1
   5.1.2 Pump Definitions and Terminology 5.2
   5.1.3 Types of Centrifugal Pumps 5.6
5.2 PUMP HYDRAULICS 5.8
   5.2.1 Pump Performance Curves 5.8
   5.2.2 Pipeline Hydraulics and System Curves 5.8
      5.2.2.1 Hazen-Williams equation 5.8
      5.2.2.2 Manning's equation 5.11
      5.2.2.3 Darcy-Weisbach equation 5.11
      5.2.2.4 Comparisons of f, C, and n 5.12
   5.2.3 Hydraulics of Valves 5.12
   5.2.4 Determination of Pump Operating Points-Single Pump 5.13
   5.2.5 Pumps Operating in Parallel 5.13
   5.2.6 Variable-Speed Pumps 5.13
5.3 CONCEPT OF SPECIFIC SPEED 5.18
   5.3.1 Introduction: Discharge-Specific Speed 5.18
   5.3.2 Suction-Specific Speed 5.19
5.4 NET POSITIVE SUCTION HEAD 5.19
   5.4.1 Net Positive Suction Head Available 5.19
   5.4.2 Net Positive Suction Head Required by a Pump 5.20
   5.4.3 NPSH Margin or Safety Factor Considerations 5.22
   5.4.4 Cavitation 5.22
5.5 CORRECTED PUMP CURVES 5.22

5.6 HYDRAULIC CONSIDERATIONS IN PUMP SELECTION 5.27
　5.6.1 Flow Range of Centrifugal Pumps 5.27
　5.6.2 Causes and Effects of Centrifugal Pumps Operating Outside
　　　　Allowable Flow Ranges 5.28
　5.6.3 Summary of Pump Selection 5.28
5.7 APPLICATION OF PUMP HYDRAULIC ANALYSIS TO DESIGN
　　OF PUMPING STATION COMPONENTS 5.30
　5.7.1 Pump Hydraulic Selections and Specifications 5.30
　　　5.7.1.1 Pump operating ranges 5.30
　　　5.7.1.2 Specific pump hydraulic operating problems 5.32
　5.7.2 Piping 5.32
　　　5.7.2.1 Pump suction and discharge piping installation guidelines 5.33
　　　5.7.2.2 Fluid velocity 5.33
　　　5.7.2.3 Design of pipe wall thickness (pressure design) 5.33
　　　5.7.2.4 Design of pipe wall thickness (vacuum conditions) 5.34
　　　5.7.2.5 Summary of pipe design criteria 5.35
5.8 IMPLICATIONS OF HYDRAULIC TRANSIENTS IN PUMPING
　　STATION DESIGN 5.35
　5.8.1 Effect of Surge on Valve Selection 5.35
　5.8.2 Effect of Surge on Pipe Material Selection 5.36
REFERENCES 5.36
APPENDIX 5.37

**CHAPTER 6 HYDRAULIC TRANSIENT DESIGN FOR
PIPELINE SYSTEMS**

6.1 INTRODUCTION TO WATERHAMMER AND SURGING 6.1
6.2 FUNDAMENTALS OF WATERHAMMER AND SURGE 6.2
　6.2.1 Definitions 6.2
　6.2.2 Acoustic Velocity 6.2
　6.2.3 Joukowsky (Waterhammer) Equation 6.3
6.3 HYDRAULIC CHARACTERISTICS OF VALVES 6.4
　6.3.1 Descriptions of Various Types of Valves 6.5
　6.3.2 Definition of Geometric Characteristics of Valves 6.6
　6.3.3 Definition of Hydraulic Performance of Valves 6.6
　6.3.4 Typical Geometric and Hydraulic Valve Characteristics 6.8
　6.3.5 Valve Operation 6.9
6.4 HYDRAULIC CHARACTERISTICS OF PUMPS 6.9
　6.4.1 Definition of Pump Characteristics 6.10
　6.4.2 Homologous (Affinity) Laws 6.10
　6.4.3 Abnormal Pump (Four–Quadrant) Characteristics 6.12
　6.4.4 Representation of Pump Data for Numerical Analysis 6.15
　6.4.5 Critical Data Required for Hydraulic Analysis
　　　　of Systems with Pumps 6.16
6.5 SURGE PROTECTION AND SURGE CONTROL DEVICES 6.18
　6.5.1 Critical Parameters for Transients 6.18
　6.5.2 Critique of Surge Protection 6.20
　6.5.3 Surge Protection Control and Devices 6.22
6.6 DESIGN CONSIDERATIONS 6.24
6.7 NEGATIVE PRESSURES AND WATER COLUMN SEPARATION IN
　　NETWORKS 6.26
6.8 TIME CONSTANTS FOR HYDRAULIC SYSTEMS 6.27

6.9 CASE STUDIES 6.27
   6.9.1 Case Study with One-way and Simple Surge Tanks 6.27
   6.9.2 Case Study with Air chamber 6.28
   6.9.3 Case Study with Air-vacuum Breaker 6.31
REFERENCES 6.32

CHAPTER 7 OPTIMAL DESIGN OF WATER DISTRIBUTION SYSTEMS

7.1 OVERVIEW 7.1
7.2 PROBLEM DEFINITION 7.1
7.3 MATHEMATICAL FORMULATION 7.3
7.4 OPTIMIZATION METHODS 7.4
   7.4.1 Branched Systems 7.4
   7.4.2 Looped Pipe Systems via Linearization 7.5
   7.4.3 General System Design via Nonlinear Programming 7.7
   7.4.4 Stochastic Search Techniques 7.8
7.5 APPLICATIONS 7.9
7.6 SUMMARY 7.12
REFERENCES 7.13

CHAPTER 8 WATER-QUALITY ASPECTS OF CONSTRUCTION AND OPERATIONS

8.1 INTRODUCTION 8.1
8.2 DISINFECTION OF NEW WATER MAINS 8.1
   8.2.1 Need for Disinfection 8.2
   8.2.2 Disinfection Chemicals 8.2
   8.2.3 Disinfection Procedures 8.2
      8.2.3.1 The tablet method 8.2
      8.2.3.2 The continous feed method 8.3
      8.2.3.3 The slug method 8.3
   8.2.4 Testing New Mains 8.3
   8.2.5 Main Repairs 8.3
   8.2.6 Disposal of Highly Chlorinated Water 8.3
8.3 DISINFECTION OF STORAGE TANKS 8.4
   8.3.1 Disinfection Procedures for Filling Tanks 8.4
      8.3.1.1 Method 1 ....8.4
      8.3.1.2 Method 2 ....8.4
      8.3.1.3 Method 3 ....8.5
   8.3.2 Underwater Inspection 8.5
8.4 CROSS-CONNECTION CONTROL 8.5
   8.4.1 Definitions 8.5
   8.4.2 Cross-Connection Control Programs 8.5
   8.4.3 Backflow Prevention 8.6
      8.4.3.1 Air Gap 8.6
      8.4.3.2 Reduced-pressure backflow preventers and
             double-check valve assemblies 8.6
      8.4.3.3 Atmospheric and pressure vacuum
             breakers, and barometric loops 8.6
      8.4.3.4 Single and dual check valves 8.7
   8.4.4 Application of Backflow Preventers 8.7

8.5 FLUSHING OF DISTRIBUTION SYSTEMS 8.8
  8.5.1 Background 8.8
  8.5.2 Flushing Procedures 8.8
  8.5.3 Directional Flushing 8.9
  8.5.4 Alternating of Disinfectants 8.9
REFERENCES 8.10

CHAPTER 9 WATER QUALITY

9.1 INTRODUCTION 9.1
  9.1.1 Overview 9.1
  9.1.2 Definitions 9.2
9.2 WATER-QUALITY PROCESSES 9.3
  9.2.1 Loss of Disinfectant Residual 9.3
    9.2.1.1 Disinfection methods 9.4
    9.2.1.2 Rates of disinfectant loss 9.5
    9.2.1.3 Mitigation of disinfectant loss 9.5
  9.2.2 Growth of Disinfection By-products 9.6
  9.2.3 Internal Corrosion 9.6
    9.2.3.1 Types of corrosion 9.7
    9.2.3.2 Factors affecting corrosion 9.7
    9.2.3.3 Indicators of corrosion 9.8
    9.2.3.4 Control of corrosion 9.8
  9.2.4 Biofilms 9.9
    9.2.4.1 Origins 9.9
    9.2.4.2 Composition 9.9
    9.2.4.3 Significance 9.10
    9.2.4.4 Treatment and control 9.10
9.3 WATER-QUALITY MONITORING 9.11
  9.3.1 Routine Monitoring 9.11
    9.3.1.1 Regulatory requirements 9.11
    9.3.1.2 Sampling methods 9.11
    9.3.1.3 Sampling parameters 9.11
  9.3.2 Synoptic Monitoring 9.11
9.4 WATER-QUALITY MODELING 9.15
  9.4.1 History 9.16
  9.4.2 Governing Equations 9.16
    9.4.2.1 Advective transport in pipes 9.17
    9.4.2.2 Mixing at pipe junctions 9.17
    9.4.2.3 Mixing in storage facilities 9.17
    9.4.2 4 Bulk flow reactions 9.17
    9.4.2.5 Pipe wall reactions 9.18
    9.4.2.6 System of equations 9.18
  9.4.3 Solution Methods 9.18
    9.4.3.1 Steady-state models 9.18
    9.4.3.2 Dynamic models 9.19
  9.4.4 Data Requirements 9.20
    9.4.4.1 Hydraulic data 9.20
    9.4.4.2 Water-quality data 9.20
    9.4.4.3 Reaction-rate data 9.20
  9.4.5 Model Calibration 9.21

xii Contents

    9.4.5.1 Calibration of conservative substances 9.21
    9.4.5.2 Calibration of nonconservative substances 9.21
    9.4.5.3 Uses for hydraulic calibration. 9.21
REFERENCES 9.22

CHAPTER 10 HYDRAULIC DESIGN OF WATER DISTRIBUTION STORAGE TANKS

10.1 INTRODUCTION 10.1
10.2 BASIC CONCEPTS 10.1
   10.2.1 Equalization 10.2
   10.2.2 Pressure Maintenance 10.2
   10.2.3 Fire Storage 10.2
   10.2.4 Emergency Storage 10.2
   10.2.5 Energy Consumption 10.3
   10.2.6 Water Quality 10.3
   10.2.7 Hydraulic Transient Control 10.3
   10.2.8 Aesthetics 10.4
10.3 DESIGN ISSUES 10.4
   10.3.1 Floating Versus Pumped Storage 10.4
   10.3.2 Ground Versus Elevated Tank 10.5
   10.3.3 Effective Versus Total Storage 10.6
   10.3.4 Private Versus Utility Owned Tanks 10.6
   10.3.5 Pressurized Tanks 10.6
10.4 LOCATION 10.7
   10.4.1 Clearwell Storage 10.7
   10.4.2 Tanks Downstream of the Demand Center 10.8
   10.4.3 Multiple Tanks in the Pressure Zone 10.8
   10.4.4 Multiple Pressure-Zone Systems 10.9
   10.4.5 Other Siting Considerations 10.9
10.5 TANK LEVELS 10.9
   10.5.1 Setting Tank Overflow Levels 10.9
   10.5.2 Identifying Tank Service Areas 10.10
   10.5.3 Identifying Pressure Zones 10.10
10.6 TANK VOLUME 10.11
   10.6.1 Trade-offs in Tank Volume Design 10.11
   10.6.2 Standards-Driven Sizing 10.12
   10.6.3 Functional Design 10.12
      10.6.3.1 Equalization Storage. 10.12
      10.6.3.2 Fire Storage. 10.14
      10.6.3.3 Emergency Storage. 10.16
      10.6.3.4 Combination Equalization, Fire and Emergency Storage. 10.16
      10.6.3.5 Summary of Functional Sizing. 10.16
   10.6.4 Staging Requirements 10.16
   10.6.5 Useful Dead Storage 10.17
10.7 OTHER DESIGN CONSIDERATIONS 10.18
   10.7.1 Altitude Valves 10.18
   10.7.2 Cathodic Protection and Coatings 10.18
   10.7.3 Overflows and Vents 10.18
REFERENCES 10.19

CHAPTER 11 QUALITY OF WATER IN STORAGE

11.1 INTRODUCTION 11.1
   11.1.1 Overview 11.1
   11.1.2 Definitions 11.2
11.2 WATER QUALITY PROBLEMS 11.2
   11.2.1 Chemical Problems 11.2
      11.2.1.1 Loss of disinfectant residual 11.2
      11.2.1.2 Formation of disinfection by-products 11.3
      11.2.1.3 Development of taste and odor 11.3
      11.2.1.4 Increase in pH 11.4
      11.2.1.5 Corrosion 11.4
      11.2.1.6 Buildup of iron and manganese 11.4
      11.2.1.7 Occurrence of hydrogen sulfide 11.5
      11.2.1.8 Leachate from internal coatings 11.5
   11.2.2 Microbiological Problems 11.5
      11.2.2.1 Bacterial regrowth 11.5
      11.2.2.2 Nitrification 11.6
      11.2.2.3 Worms and Insects 11.6
   11.2.3 Physical Problems 11.7
      11.2.3.1 Sediment buildup 11.7
      11.2.3.2 Entry of contaminants 11.7
      11.2.3.3 Temperature 11.8
11.3 MIXING AND AGING IN STORAGE FACILITIES 11.8
   11.3.1 Ideal Flow Regimes 11.8
   11.3.2 Jet Mixing 11.9
   11.3.3 Mixing Times 11.9
   11.3.4 Stratification 11.10
   11.3.5 Aging 11.11
11.4 MONITORING AND SAMPLING 11.12
   11.4.1 Routine Monitoring 11.12
      11.4.1.1 Typical parameters of water quality 11.13
      11.4.1.2 Parameters of nitrification monitoring 11.13
      11.4.1.3 Parameters of sediment monitoring 11.13
      11.4.1.4 Parameters of biofilm monitoring 11.17
   11.4.2 Sampling Methods and Equipment 11.17
   11.4.3 Monitoring Frequency and Location of Samples 11.18
   11.4.4 Special Studies 11.20
      11.4.4.1 Intensive studies of water quality and tracers 11.20
      11.4.4.1 Temperature monitoring 11.20
11.5 MODELING 11.22
   11.5.1 Scale Models 11.22
      11.5.1.1 Principles of similitude 11.22
      11.5.1.2 Construction of a model 11.23
      11.5.1.3 Types of tracers 11.24
      11.5.1.4 Temperature modeling 11.25
   11.5.2 Computational Fluid Dynamics 11.25
      11.5.2.1 Mathematical formulations of CFD models 11.26
      11.5.2.2 Application of CFD Models 11.27
   11.5.3 Systems Models 11.28

xiv Contents

    11.5.3.1 Background 11.28
    11.5.3.2 Elemental systems models 11.28
    11.5.3.3 Compartment models 11.28
    11.5.3.4 Application of systems models 11.28
11.6 DESIGN AND OPERATIONAL ISSUES 11.30
  11.6.1 Water-Quality Design Objectives 11.30
  11.6.2 Modes of Operation: Simultaneous Inflow-Outflow Versus Fill
  and Draw 11.30
  11.6.3 Flow Regimes: Complete Mix Versus Plug Flow 11.30
    11.6.3.1 Effects os flow regime on loss of disinfectant in reservoirs 11.31
    11.6.3.2 Mixed flow 11.31
    11.6.3.3 Plug flow 11.32
    11.6.3.4 Recommendations 11.33
  11.6.4 Stratification in Reservoirs 11.33
11.7 INSPECTION AND MAINTENANCE ISSUES 11.34
  11.7.1 Inspections 11.34
  11.7.2 Maintenance 11.36
REFERENCES 11.36

CHAPTER 12 COMPUTER MODELS/EPANET

12.1 INTRODUCTION 12.1
  12.1.1 Need for Computer Models 12.1
  12.1.2 Uses of Computer Models 12.2
  12.1.3 History of Computer Models 12.2
12.2 USE OF A COMPUTER MODEL 12.3
  12.2.1 Network Representation 12.3
    12.2.1.1 Network components 12.3
    12.2.1.2 Network skeletonization 12.4
  12.2.2 Compilation of Data 12.4
    12.2.2.1 ID labels 12.5
    12.2.2.2 Nodal elevations 12.5
    12.2.2.3 Pipe diameters 12.5
    12.2.2.4 Pipe roughness 12.6
    12.2.2.5 Pump curves 12.6
  12.2.3 Estimation of Demand 12.6
  12.2.4 Operating Characteristics 12.7
  12.2.5 Reaction–Rate Information 12.7
  12.2.6 Model Calibration 12.8
12.3 COMPUTER MODEL INTERNALS 12.8
  12.3.1 Input Processing 12.9
  12.3.2 Topological Processing 12.9
  12.3.3 Hydraulic Solution Algorithms 12.9
  12.3.4 Linear-Equation Solver 12.11
  12.3.5 Extended-Period Solver 12.11
  12.3.6 Water-Quality Algorithms 12.12
  12.3.7 Output Processing 12.12
12.4 EPANET PROGRAM 12.13
  12.4.1 Background 12.13
  12.4.2 Program Features 12.14

12.4.3 User Interface 12.15
12.4.4 Solver Module 12.17
12.4.5 Programmer's Toolkit 12.20
12.5 CONCLUSION 12.20
REFERENCES 12.21

CHAPTER 13 WATER QUALITY MODELING-CASE STUDIES

13.1 INTRODUCTION 13.1
13.2 DESIGN OF DISTRIBUTION SYSTEMS IN THE
UNITED STATES 13.2
13.3 WATER QUALITY IN NETWORKS 13.3
13.4 HYDRAULIC AND WATER-QUALITY MODELS 13.4
  13.4.1 Steady-State-Water Quality Models 13.5
  13.4.2 Dynamic Water-Quality Models 13.5
13.5 EARLY APPLICATIONS OF WATER-QUALITY MODELING 13.6
  13.5.1 North Penn Study 13.6
    13.5.1.1 Network modeling 13.7
    13.5.1.2 Variations in water quality data 13.8
    13.5.1.3 Development of dynamic water-quality algoritm 13.8
  13.5.2 South Central Connecticut Regional Water Authority 13.9
    13.5.2.1 System modeling 13.13
    13.5.2.2 Design of the field study 13.13
    13.5.2.3 Results from the field study 13.13
    13.5.2.4 Verification study 13.17
    13.5.2.5 Presampling procedures 13.17
    13.5.2.6 Analysis of sampling results 13.17
    13.5.2.7 Modeling of chlorine residual 13.21
  13.5.3 Case Study of Cabool, Missouri 13.22
13.6 EVOLUTION OF WATER QUALITY MODELING 13.22
13.7 MODELING PROPAGATION OF CONTAMINANTS 13.23
  13.7.1 Case Study of the North Marin Water District 13.24
    13.7.1.1 Water-quality study 13.29
    13.7.1.2 Modeling of total trihalomethane formations 13.30
    13.7.1.3 Chlorine demand 13.34
    13.7.1.4 Effect of system demand 13.34
  13.7.2 Complement to the North Marin study 13.34
  13.7.3 Waterborne Outbreak in Gideon, Missouri 13.36
    13.7.3.1 Description of the system 13.38
    13.7.3.2 Identification of the outbreak 13.39
    13.7.3.3 Possible causes 13.40
    13.7.3.4 Evaluation of the System 13.41
    13.7.3.5 Performance of the System 13.41
    13.7.3.6 Propagation of the contaminant 13.43
13.8 CURRENT TRENDS IN WATER-QUALITY MODELING 13.44
  13.8.1 Study in Cholet, France 13.44
  13.8.2 Case Study in Southington, Connecticut 13.44
  13.8.3 Mixing in Storage Tanks 13.45
13.9 SUMMARY AND CONCLUSIONS 13.45
REFERENCES 13.46

CHAPTER 14 CALIBRATION OF HYDRAULIC NETWORK MODELS

14.1 INTRODUCTION 14.1
   14.1.1 Network Characterization 14.1
   14.1.2 Network Data Requirements 14.1
   14.1.3 Model Parameters 14.3
14.2 IDENTIFY THE INTENDED USE OF THE MODEL 14.3
14.3 DETERMINE ESTIMATES OF THE MODEL PARAMETERS 14.3
   14.3.1 Pipe Roughness Values 14.4
      14.3.1.1 Chart the pipe roughness 14.4
      14.3.1.2 Field test the pipe roughness 14.6
   14.3.2 Distribution of Nodal Demands 14.9
      14.3.2.1 Spatial distribution of demands 14.10
      14.3.2.2 Temporal distribution of demands 14.12
14.4 COLLECT CALIBRATION DATA 14.12
   14.4.1 Fire-Flow Tests 14.12
   14.4.2 Telemetric Data 14.13
   14.4.3 Water-Quality Data 14.14
14.5 EVALUATE THE RESULTS OF THE MODEL 14.14
14.6 PERFORM A MACRO-LEVEL CALIBRATION OF THE MODEL 14.15
14.7 PERFORM A SENSITIVITY ANALYSIS 14.16
14.8 PERFORM A MACRO-LEVEL CALIBRATION OF THE MODEL 14.16
   14.8.1 Analytical Approaches 14.17
   14.8.2 Simulation Approaches 14.17
   14.8.3 Optimization Approaches 14.17
14.9 FUTURE TRENDS 14.21
14.10 SUMMARY AND CONCLUSION 14.21
REFERENCES 14.21

CHAPTER 15 OPERATION OF WATER DISTRIBUTION SYSTEMS

15.1 INTRODUCTION 15.1
15.2 HOW SYSTEMS ARE OPERATED 15.2
   15.2.1 Typical Operating Indexes 15.2
   15.2.2 Operating Criteria 15.3
   15.2.3 Water Quality and Operations 15.4
   15.2.4 Emergency Operations 15.4
15.3 MONITORING OF SYSTEM PERFORMANCE WITH SCADA
SYSTEMS 15.5
   15.3.1 Anatomy of a SCADA System 15.6
   15.3.2 Data Archiving 15.9
15.4 CONTROL OF WATER DISTRIBUTION SYSTEM 15.9
   15.4.1 Control Strategies 15.10
      15.4.1.1 Supervisory control 15.10
      15.4.1.2 Automatic control 15.10
      15.4.1.3 Advanced control 15.10
   15.4.2 Centralized Versus Local Control 15.11
15.5 LINKING OF SCADA SYSTEMS WITH ANALYSIS AND CONTROL
MODELS 15.11
   15.5.1 Data Requirements of Analysis and Control Models 15.12
   15.5.2 Establishment of the Link 15.13

15.6 USE OF CENTRAL DATABASES IN SYSTEM CONTROL	15.15
15.7 WHAT THE FUTURE HOLDS	15.16
REFERENCES	15.16

CHAPTER 16 OPTIMIZATION MODELS FOR OPERATIONS

16.1 INTRODUCTION	16.1
16.2 FORMULATIONS FOR MINIMIZING ENERGY COST MINIMIZATION	16.3
16.2.1 Energy Management	16.3
16.2.2 Management Strategies	16.3
16.2.3 Management Models	16.5
16.2.3.1 Hydraulic network models	16.5
16.2.3.2 Demand forecast models	16.7
16.2.3.3 Control models	16.8
16.2.4 Optimization Models	16.9
16.2.4.1 Problem formulation	16.9
16.2.4.2 System classification	16.10
16.2.5 Summary and Conclusions	16.14
16.3 FORMULATIONS TO SATISFY WATER QUALITY	16.16
16.4 SOLUTION METHODS AND APPLICATIONS FOR WATER-QUALITY PURPOSES	16.19
16.4.1 Mathematical Programming Approach	16.19
16.4.2 Simulated Annealing Approach	16.22
16.4.3 Development of Cost Function	16.24
16.4.4 Sample Application	16.26
16.4.5 Advantages and Disadvantages of the Two Methods	16.28
16.5 OPTIMAL SCHEDULING OF BOOSTER DISINFECTION	16.28
16.5.1 Background 1: Linear Superposition	16.33
16.5.2 Background 2: Dynamic Network Water-Quality Models in a Planning Context	16.34
16.5.3 Optimal Scheduling of Booster-Station Dosages as Linear Pogramming Problem	16.36
16.5.4 Optimal Location and Scheduling of Booster-Station Dosage as a Mixed-Integer Linear Programming Problem	16.36
16.5.5 Optimal Location of Booster Stations as a Maximum Set-Covering Problem	16.38
16.5.6 Solution of the Optimization Models	16.40
16.5.7 Available Software	16.41
16.5.8 Summary	16.42
REFERENCES	16.43

CHAPTER 17 MAINTENANCE AND REHABILITATION/REPLACEMENT

17.1 INTRODUCTION	17.1
17.1.1 Maintenance and Rehabilitation Problems	17.1
17.1.1.1 Normal wear	17.1
17.1.1.2 Corrosion	17.2
17.1.1.3 Unforeseen loads	17.2
17.1.1.4 Poor manufacture and installation	17.2
17.1.2. Preview of the Chapter	17.2

17.2 UNACCOUNTED-FOR WATER 17.2
　17.2.1 Indicators for Unaccounted-for Water 17.3
　17.2.2. Understanding the Causes of Unaccounted-for Water 17.3
　17.2.3 Components of Unaccounted-for Water 17.5
　　17.2.3.1 Water main leakage 17.5
　　17.2.3.2 Service pipe leakage 17.7
　　17.2.3.3 System pressure 17.7
　　17.2.3.4 Fire fighting 17.7
　　17.2.3.5 Main flushing 17.8
　　17.2.3.6 Blowoffs 17.8
　　17.2.3.7 Flat rate customers 17.8
　　17.2.3.8 Authorized unmetered uses 17.8
　　17.2.3.9 Meter under registration 17.8
　　17.2.3.10 Theft of water 17.9
　17.2.4. Summary 17.9
17.3 PIPE BREAKS 17.10
　17.3.1 Corrosion 17.10
　　17.3.1.1 External soil corrosion 17.10
　　17.3.1.2 Internal corrosion 17.11
　　17.3.1.3 Stray current corrosion 17.11
　　17.3.1.4 Bimetallic connections 17.11
　17.3.2. External Loads 17.11
　17.3.3 Poor Tapping 17.13
　17.3.4. Pressure-Related Breaks 17.13
　17.3.5. Repair Versus Replacement 17.14
17.4 HYDRAULIC CARRYING CAPACITY 17.16
　17.4.1 Diagnosis of Pressure Problems 17.16
　　17.4.1.1 Pressure gauges 17.16
　　17.4.1.2 Hydraulic modeling 17.16
　17.4.2 Correction of Pressure Problems 17.17
　　17.4.2.1 Closed isolating valves 17.17
　　17.4.2.2 Elevation and pressure zone issues 17.18
　　17.4.2.3 Carrying capacity 17.18
　　17.4.2.4 Inadequate capacity 17.19
　17.4.3 Pipe Rehabilitation Technology 17.20
　17.4.4. Evaluation of Pipe Rehabilitation 17.21
17.5 MAINTENANCE INFORMATION SYSTEMS 17.21
　17.5.1 System Mapping 17.22
　17.5.2 System Database 17.22
　17.5.3 Geographic Information Systems 17.22
　17.5.4 Maintenance Management Systems 17.23
　17.5.5 SCADA Systems 17.23
REFERENCES 17.24

CHAPTER 18 RELIABILITY ANALYSIS FOR DESIGN

18.1 FAILURE MODES FOR WATER DISTRIBUTION SYSTEMS 18.1
　18.1.1 Need and Justification 18.1
　18.1.2 Definitions of Distribution System Repairs 18.3
　18.1.3 Failure Modes 18.4
　　18.1.3.1 Performance failure 18.4
　　18.1.3.2 Component (mechanical) failure 18.5

18.1.4 Reliability: Indexes and Approaches 18.5
18.2 PRACTICAL ASPECTS OF PROVIDING RELIABILITY 18.6
   18.2.1 Improving the Reliability of Water Distribution Systems 18.6
      18.2.1.1 Piping materials 18.6
      18.2.1.2 Construction methods 18.7
      18.2.1.3 Pipe sizing 18.7
      18.2.1.4 Looped water distribution system 18.7
      18.2.1.5 Emergency Storage 18.8
      18.2.1.6 Backup pumping and control valves 18.8
      18.2.1.7 Standby power 18.8
      18.2.1.8 Emergency controls 18.8
      18.2.1.9 Emergency interconnections 18.9
      18.2.1.10 Water distribution system modeling 18.9
      18.2.1.11 Transient analysis 18.9
      18.2.1.12 Operational considerations 18.9
      18.2.1.13 Maintenance considerations 18.10
   18.2.2 Analyzing the Effect of Valving on System Reliability 18.10
      18.2.2.1 Background 18.10
      18.2.2.2 Diagrams of distribution segments 18.10
      18.2.2.3 Loops served from transmission mains 18.11
      18.2.2.4 Emergency interconnections 18.11
      18.2.2.5 Transmission lines connected to old systems 18.13
      18.2.2.6 Typical cross-intersections 18.13
      18.2.2.7 Application of segments in valve locations
                and reliability evaluation 18.14
18.3 COMPONENT RELIABILITY ANALYSIS 18.15
   18.3.1 Failure Density, Failure Rate, and Mean Time To Failure 18.15
   18.3.2 Availability and Unavailability 18.19
18.4 REVIEW OF MODELS FORE RELIABILITY OF WATER DISTRIBUTION
SYSTEMS 18.21
   18.4.1 Reliability of a System Failure 18.21
   18.4.2 Failure Modes 18.22
   18.4.3 Approaches to the Assessment of Reliability 18.25
   18.4.4 Models and Techniques for Assessing Network Reliability 18.29
      18.4.4.1 Simulation models 18.29
      18.4.4.2 Analytical approaches 18.33
      18.4.4.3 Heuristic techniques 18.39
      18.4.4.4 Redundancy based measures 18.39
   18.4.5 Overview of Reliability Measures 18.40
   18.4.6 Observations 18.42
18.5 MEASURE OF LINK IMPORTANCE 18.43
REFERENCES 18.49

Index follows Chapter 18

CONTRIBUTORS

Bayard Bosserman II *Boyle Engineering Corporation* (CHAPTER 5)
Francious Bouchart *Heriot-Watt University* (CHAPTER 18)
Donald V. Chase *University of Dayton* (CHAPTER 15)
Robert Clark *U. S. Environmental Protection Agency* (CHAPTER 13)
Edwin E. Geldreich *Consulting Microbiologist* (CHAPTER 9)
Fred E. Goldman *Goldman, Toy, and Associates* (CHAPTER 16)
Ian Goulter *Swinburne University of Technology* (CHAPTER 18)
Walter M. Grayman *Consulting Engineer* (CHAPTER 9, 11)
Bryan W. Karney *University of Toronto* (CHAPTER 2)
Gregory J. Kirmeyer *Economic and Engineering Services, Inc.* (CHAPTER 11)
Kevin Lansey *University of Arizona* (CHAPTER 4, 7)
Srinivasa Lingireddy *University of Kentucky* (CHAPTER 14)
James W. Male *University of Portland* (CHAPTER 17)
C. Samuel Martin *Georgia Institute of Technology* (CHAPTER 6)
Larry W. Mays *Arizona State University* (CHAPTER 1, 4, 16, 18)
Lindell E. Ormsbee *University of Kentucky* (CHAPTER 14, 16)
Lewis A. Rossman *U. S. Environmental Protection Agency* (CHAPTER 9, 12)
A. Burcu Altan Sakarya *Middle East Technical University* (CHAPTER 16, 18)
Yeou-Koung Tung *Hong Kong University of Science and Technology* (CHAPTER 18)
Jim Uber *University of Cincinnati* (CHAPTER 16)
Thomas M. Walski *Pennsylvania American Water Co.* (CHAPTER 8, 10, 17, 18)
Mark Ysusi *Montgomery Watson* (CHAPTER 3)

PREFACE

At the beginning of year 2000 this is an exciting time to be involved in writing about the delivery of safe drinking water. Today's increased awareness and concern for safe drinking water on a national and international basis, coupled with limited budgets of not only developing countries but also of developed countries, has generated an exponential increase in interest in the future of water distribution.

In the U. S. and in other developed countries the populations take the ability to have safe drinking water at any time and place for granted. According to the World Health Organization and the United Nations, however, the needs for urban water and rural water supply are tremendous. The urban population in developing countries in 1990 without access to safe drinking water was approximately 243 million people, and in rural areas in developing countries was approximately 989 million people, for a total of 1,232 million people without access to safe drinking water. The expected population increase in urban areas in developing countries from year 1990 to 2000 is expected to be 570 million people, making the total in urban areas requiring the service of safe drinking water to be 813 million people. The expected population increase in rural areas in developing countries from 1990 to 2000 is expected to be 312 million people, making the total in rural areas requiring the service of safe drinking water to be 1,301 million people. The total population needs in developing countries requiring safe drinking water by year 2000 is 2,114 million people.

The *Water Distribution Systems Handbook*, referred to herein as the *Handbook*, has been an extensive effort to develop a comprehensive reference book on water distribution systems. A substantial amount of new knowledge concerning the design, operation, and analysis of water distribution systems has accumulated over the past decade. In particular, many new developments have taken place on the subjects of water quality of storage, modeling of water quality, optimal operation, reliability of water distribution systems, and many other subjects. Some of this information is dispersed in professional and scientific journals and reports. Within the *Handbook* the various authors have synthesized this accumulated knowledge and presented it in a concise and accessible form.

There are obviously many other topics that could have been covered, making the *Handbook* even more comprehensive; however, I had to make choices on the coverage. These choices obviously reflect my vision of what is needed most in the *Handbook*. The topics covered are the ones that I feel are the most important for state-of-the art design, analysis, modeling, and operation of water distribution systems. The detail of each topic is fairly well balanced among the chapters and among the topics in each chapter. There is also a reflection of my perspective on the subject, with the constraint that all the material fits within one handbook. Hopefully, the readers will have an understanding and appreciation of what is being accomplished in this *Handbook*.

First and foremost this handbook is intended to be a reference for those wishing to expand their knowledge of water distribution systems. The *Handbook* will be of value to engineers, managers, operators, and analysts involved with the design, analysis, operation, maintenance, and rehabilitation of water distribution systems. This handbook can also be a valuable reference, if not the text in both undergraduate and graduate courses for teaching the design and analysis of water distribution systems.

Each of the authors is a leading expert in the field of water distribution systems. They have published extensively in the literature on water distribution systems, and many of them have had extensive experience in the design, operation, and analysis of distributions systems. Each of the authors was chosen because of their proven knowledge in the specific area of contribution.

As editor in chief of the *Handbook,* I felt that it was important to provide a brief historical perspective (Chapter 1) of the knowledge of water distribution, starting from the ancient times to the present. This historical perspective begins with the pressurized water distribution systems at Knossos (circa 2000 BC) and provides examples of other ancient water systems. The developments during the 19th and 20th centuries are particularly important to understand our present status at the start of year 2000 with this handbook in place. To better understand where we are and where we may be going, it is wise to look at where we have been.

In 1952 Albert Einstein was offered the presidency of Israel but declined because he thought he was too naïve in politics. Perhaps his real reason, according to Stephen W. Hawking (*A Brief History of Time*), was different. To quote Einstein, "Equations are more important to me, because politics is for the present, but an equation is something for eternity." Hopefully, this handbook is not only for the present, but also will be a contribution for the future.

Each book that I have worked on has been a part of my lifelong journey in water resources. The *Handbook* certainly is no exception. I have gained more from this experience than can ever be measured in words.

I dedicate this handbook to humanity and human welfare.

Larry W. Mays
Scottsdale, Arizona

Acknowledgments

I must first acknowledge the authors who made this handbook possible. It has been a sincere privilege to have worked with such an excellent group of dedicated people. They are all experienced professionals who are among the leading experts in their fields. References to material in this handbook should be attributed to the respective chapter authors.

During the past twenty-three years of my academic career as a professor, I have received help and encouragement from so many people that it is not possible to name them all. These people represent a wide range of universities, research institutions, government agencies, and professions. To all of you I express my deepest thanks.

I would like to acknowledge Arizona State University, especially the time afforded me to pursue this handbook.

I sincerely appreciate the advice and encouragement of Larry Hager of McGraw-Hill throughout this project. Larry has always been a great guy to work with on the three handbooks that we have done together. He is always a joy to talk to, as he's one of the few that is willing to listen to my fly fishing and snow skiing experiences.

This handbook has been a part of a personal journey that began years ago when I was a young boy with a love of water. Books are companions along the journey of learning. I hope that you will be able to use this handbook in your own journey of learning about water. Have a happy and wonderful journey.

Larry W. Mays

ABOUT THE EDITOR

Larry W. Mays is professor of civil and environmental engineering at Arizona State University and former chair of the department. He was formerly director of the Center for Research in Water Resources at the University of Texas at Austin, where he also held an Engineering Foundation Endowed Profesorship. A registered professional engineer in seven states and a registered professional hydrologist, he has served as a consultant to many organizations. A widely published expert on water resources, he wrote *Optimal Control of Hydrosystems* (Marcel Dekker) and was editor in chief of both *Water Resources Handbook* (McGraw-Hill) and *Hydraulic Design Handbook* (McGraw-Hill). Co-author of both *Applied Hydrology* and *Hydrosystems Engineering and Management* published by McGraw-Hill and was the editor in chief of *Reliability Analysis of Water Distribution Systems* (ASCE), and co-editor of *Computer Modeling of Free-Surface and Pressurized Flows* (Kluwer Academic Publishers). He has published extensively on his research in water resources management.

CHAPTER 1
INTRODUCTION

Larry W. Mays
*Department of Civil and Environmental Engineering
Arizona State University
Tempe, AZ*

1.1 BACKGROUND

The cornerstone of any healthy population is access to safe drinking water. The goal of the United Nations International Drinking Water Supply and Sanitation Decade from 1981 to 1990 was safe drinking water for all. A substantial effort was made by the United Nations to provide drinking water and sanitation services to populations lacking those services. Unfortunately, the population growth in developing countries almost entirely wiped out the gains. In fact, nearly as many people lack those services today as they did at the beginning of the 1980s (Gleick, 1993). Table 1.1 lists the developing countries needs for urban and rural water supplies and sanitation. Four-fifths of the world's population and approximately 100 percent of the population of developing countries are covered by this table. Also refer to Gleick (1998).

Because of the importance of safe drinking water for the needs of society and for industrial growth, considerable emphasis recently has been given to the condition of the infrastructure. Large capital expenditures will be needed to bring the concerned systems to higher levels of serviceability and to lend vigor to U.S. industry and help it remain competitive in the world economy. One of the most vital services to industrial growth is an adequate water supply system—without it, industry cannot survive.

The lack of adequate water supply systems is due to both the deterioration of aging water supplies in older urbanized areas and to the nonexistence of water supply systems in many areas that are undergoing rapid urbanization, such as in the southwestern United States. In other words, methods for evaluation of the nation's water supply services need to consider not only rehabilitation of existing urban water supply systems but also the future development of new water supply systems to serve expanding population centers. Both the adaptation of existing technologies and the development of new innovative technologies will be required to improve the efficiency and cost-effectiveness of future and existing water supply systems and facilities necessary for industrial growth.

An Environmental Protection Agency (EPA) survey (Clark et al., 1982) of previous water supply projects concluded that the distribution facilities in water supply

TABLE 1.1 Developing Country Needs for Urban and Rural Water Supply and Sanitation, 1990 and 2000

	Population Not Served in 1990 (10^6)	Expected Population Increase 1990–2000 (10^6)	Total Additional Population Requiring Service by 2000 (10^6)
Water supply			
Urban	243	570	813
Rural	989	312	1301
Total	1232	882	2114
Sanitation			
Urban	377	570	947
Rural	1364	312	1676
Total	1741	882	2623

Source: From Gleick (1993).
These data present the drinking water and sanitation service needs in developing countries only and use United Nations population estimates for 2000. The level of service is typically defined by the World Meteorological Organization. As used here by the World Health Organization (WHO), safe drinking water includes treated surface water and untreated water from protected springs, boreholes, and wells. The WHO defines access to safe drinking water in urban areas as piped water to housing units or to public standpipes within 200 m. In rural areas, reasonable access implies that fetching water does not take up a disproportionate part of the day.

systems will account for the largest cost item in future maintenance budgets. The aging, deteriorating systems in many areas raise tremendous maintenance decision-making problems, which are further complicated by the expansion of existing systems. Deterioration of the water distribution systems in many areas has translated into a high proportion of unaccounted-for water caused by leakage. Not only does this amount to loss of a valuable resource; it also raises concerns about safe drinking water because of possible contamination from cracked pipes.

The reliability of the existing aging systems is continually decreasing (Mays, 1989). Only recently have municipalities been willing or able to finance rehabilitation of deteriorating pipelines, and needed maintenance and replacement of system components is still being deferred until a catastrophe occurs or the magnitude of leakage justifies the expense of repair. Water main failures have been extensive in many cities.

As a result of governmental regulations and consumer-oriented expectations, a major concern now is the transport and fate of dissolved substances in water distribution systems. The passage of the Safe Drinking Water Act in 1974 and its Amendments in 1986 (SDWAA) changed the manner in which water is treated and delivered in the United States. The EPA is required to establish maximum contaminant level (MCL) goals for each contaminant that may have an adverse effect on the health of persons. These goals are set to the values at which no known or expected adverse effects on health can occur. By allowing a margin of safety (Clark, 1987), previous regulatory concerns were focused on water as it left the treatment plant before entering the distribution system (Clark, 1987), disregarding the variations in water quality which occurred in the water distribution systems.

To understand better where we are and where we may be going, it is sometimes wise to look at where we have been. This is particularly true in water management, where understanding the lessons of the history of water management may provide clues to solving some of the present-day and future problems. The next section is devoted to the aspects of the historical development of water distribution systems.

1.2 HISTORICAL ASPECTS OF WATER DISTRIBUTION

1.2.1 Ancient Urban Water Supplies

Humans have spent most of their history as hunters and food gatherers. Only in the last 9000–10,000 years have human beings discovered how to raise crops and tame animals. This agricultural revolution probably took place first in the hills to the north of present-day Iraq and Syria. From there, the agricultural revolution spread to the Nile and Indus Valleys. During the time of this agricultural breakthrough, people began to live in permanent villages instead of leading a wandering existence. About 6000–7000 years ago, farming villages of the Near and Middle East became cities. The first successful efforts to control the flow of water were made in Mesopotamia and Egypt. Remains of these prehistoric irrigation canals still exist. Table 1.2 from Crouch (1993) presents a chronology of water knowledge. Crouch (1993) pointed out, traditional water knowledge relied on geological and meteorological observation plus social consensus and administrative organization, particularly among the ancient Greeks.

Knossos, approximately 5 km from Herakleion, the modern capital of Crete, was one of the most ancient and unique cities of the Aegean Sea area and of Europe. Knossos was first inhabited shortly after 6000 B.C., and within 3000 years it had became the largest Neolithic (Neolithic Age, ca. 5700–28 B.C.) settlement in the Aegean. During the Bronze Age (ca. 2800–1100 B.C.), the Minoan civilization developed and reached its culmination as the first Greek cultural miracle of the Aegean world. During the neopalatial period (1700–1400

TABLE 1.2 Chronology of Water Knowledge

Prehistorical period	Springs
3rd–2nd millennium B.C.	Cisterns
*3rd millennium B.C.	Dams
3rd millennium B.C.	Wells
? Probably very early	Reuse of excrement as fertilizer
*2 millennium B.C.	Gravity flow supply, pipes or channels and drains, pressure pipes (subsequently forgotten)
8th–6th c B.C.	Long-distance water supply lines with tunnels and bridges as well as intervention in and harnessing of karstic water systems
6th c. B.C. at latest	Public as well as private bathing facilities consisting of: bathtubs or showers, footbaths, washbasins, latrines or toilets, laundry and dishwashing facilities
6th c. B.C. at latest	Utilization of definitely two and probably three qualities of water: potable, subpotable, and nonpotable, including irrigation using storm runoff, probably combined with waste waters
6th–3rd c. B.C.	Pressure pipes and siphon systems

Source: Crouch (1993).
* Indicates an element discovered, probably forgotten, and rediscovered later.
? Indicates an educated guess.

B.C.), Knossos was at the height of its splendor. The city occupied an area of 75,000–125,000 m² and had an estimated population on the order of tens of thousands of inhabitants. The water supply system at Knossos was most interesting. An aqueduct supplied water through tubular conduits from the Knunavoi and Archanes regions and branched out to supply the city and the palace. Figure 1.1 shows the type of pressure conduits used within the palace for water distribution. Unfortunately, around 1450 B.C. the Mycenean palace was destroyed by an earthquake and fire, as were all the palatial cities of Crete.

The Acropolis in Athens, Greece, has been a focus of settlement starting in the earliest times. Not only its defensive capabilities, but also its water supply made it the logical location for groups who dominated the region. The location of the Acropolis on an outcropping of rock, the naturally occurring water, and the ability of the location to save the rain and spring water resulted in a number of diverse water sources, including cisterns, wells, and springs. Figure 1.2 shows the shaft of one archaic water holder at the site of the Acropolis.

Anatolia, also called Asia Minor, which is part of the present-day Republic of Turkey, has been the crossroads of many civilizations during the last 10,000 years. In this region, there are many remains of ancient water supply systems dating back to the Hittite period (2000–200 B.C.), including pipes, canals, tunnels, inverted siphons, aqueducts, reservoirs, cisterns, and dams.

An example of one ancient city with a well-developed water supply system is Ephesus in Anatolia, Turkey, which was founded during the 10th century B.C. as an Ionian city surrounding the Artemis temple. During the 6th century B.C., Ephesus was reestablished at the present site, where it further developed during the Roman period. Water for the great fountain, built during 4–14 A.D., was diverted by a small dam at Marnss and was conveyed to the city by a 6-km-long system consisting of one larger and two smaller clay pipe lines. Figure 1.3 shows the types of clay pipes used at Ephesus for water distribution purposes.

FIGURE 1.1 Water distribution pipe at Knossos, Crete. (Photograph by L.W. Mays).

FIGURE 1.2 Shaft of water holder at the Acropolis at Athens, Greece. (Photograph by L. W. Mays).

FIGURE 1.3 (A, B) Water distribution pipe in Ephesus, Turkey. (Photographs by L. W. Mays).

Baths were unique in ancient cities, such as the Skolacctica baths in Ephesus that had a salon and central heating. These baths had a hot bath (*caldarium*), a warm bath (*tepidarium*), a cold bath (*frigidarium*), and a dressing room (*apodyterium*). The first building of this bath, which was constructed in the 2nd century A.D., had three floors. A woman named Skolacticia modified the bath in the 4th century A.D., making it appealing to hundreds of people. There were public rooms and private rooms and those who wished could stay for many days. Hot water was provided using a furnace and a large boiler to heat the water.

Perge, located in Anatolia, is another ancient city that had a unique urban water infrastructure. Figure 1.4 illustrates the majestic fountain (Nymphaion), which consisted of a wide basin and a richly decorated architectural facade. Because of the architecture and statues of this fountain, it was one of Perge's most magnificent edifices. A water channel (shown in Fig. 1.4) ran along the middle, dividing each street and bringing life and coolness to the city. The baths of Perge were magnificent. As in other ancient cities in Anatolia, three separate baths existed (*caldarium, tepidarium,* and *frigidarium*).

The early Romans devoted much of their time to useful public works projects, building boats, harbor works, aqueducts, temples, forums, town halls, arenas, baths, and sewers. The prosperous bourgeois of early Rome typically had a dozen-room house, with a square hole in the roof to let rain in and a cistern beneath the roof to store the water. The Romans built many aqueducts; however, they were not the first. King Sennacherio built aqueducts, as did both the Phoenicians and the Hellenes. The Romans and Hellenes needed extensive aqueduct systems for their fountains, baths, and gardens. They also realized that water transported from springs was better for their health than river water and did not need to be lifted to street level as did river water. Roman aqueducts were built on elevated structures to provide the needed slope for

FIGURE 1.4 Majestic fountain (Nymphaion) at Perge, Anatolia, Turkey. (Photograph by L. W. Mays).

water flow. Knowledge of pipe making—using bronze, lead, wood, tile, and concrete—was in its infancy, and the difficulty of making pipes was a hindrance. Most Roman piping was made of lead, and even the Romans recognized that water transported by lead pipes was a health hazard.

The water source for a typical water supply system of a Roman city was a spring or a dug well, usually with a bucket elevator to raise the water. If the well water was clear and of sufficient quantity, it was conveyed to the city by aqueduct. Also, water from several sources was collected in a reservoir, then conveyed by aqueduct or pressure conduit to a distributing reservoir (castellum). Three pipes conveyed the water—one to pools and fountains, the second to the public baths, and the third to private houses for revenue to maintain the aqueducts (Rouse and Ince, 1957). Figure 1.5 illustrates the major aqueducts of ancient Rome. Figure 1.6 shows the Roman aqueduct at Segovia, Spain, which is probably one of the most interesting Roman remains in the world. This aqueduct, built during the second half of the 1st century A.D. or the early years of the 2nd century A.D., has a maximum height of 28.9 m.

Water flow in the Roman aqueducts was basically by gravity. Water flowed through an enclosed conduit (*specus* or *rivus*), which was typically underground, from the source to a terminus or distribution tanks (*castellum*). Aqueducts above ground were built on a raised embankment (*substructio*) or on an arcade or bridge. Settling tanks (*piscinae*) were located along the aqueducts to remove sediments and foreign matter. Subsidiary lines (*vamus*) were built at some locations along the aqueduct to supply additional water. Also, subsidiary or branch lines (*ramus*) were used. At distribution points, water was delivered through pipes (*fistulae*) made of either tile or lead. These pipes were connected to the castellum by a fitting or nozzle (*calix*) and were usually placed below the ground level along major streets. Refer to Evans (1994), Frontius (1973), Garbrecht (1982), Robbins (1946), and Van Deman (1934) for additional reading on the water supply of the city of Rome and other locations in the Roman Empire.

The following quote from Vitruvius's treatise on architecture, as translated by Morgan (1914), describes how the aqueduct castellern worked (as presented in Evans, 1994):

> When it [the water] has reached the city, build a reservoir with a distribution tank in three compartments connected with a reservoir to receive the water, and let the reservoir have three pipes, one for each of the connecting tanks, so that when the water runs over from the tanks at the ends, it may run into the one between them. From this central tank, pipes will be laid to all the basins and fountains; from the second tank, to baths, so that they yield an annual income to the state; and from the third, to private houses, so that water for public use will not run short; for people will be unable to divert it if they have only their own supplies from headquarters. This is the reason why I have made these divisions, and in order that individuals who take water into their houses may by their taxes help to maintain the conducting of the water by the contractors.

It is interesting that Vitrivius's treatise is frequently in conflict with what the actual practice was in the Roman world (Evans, 1994).

According to Evans (1994), the remains of distribution tanks (*castella*) that survive at Pompeii and Nines indicate that the tanks distributed water according to geography as opposed to use. The pipes from the *castellum*, located along the main streets, carried water to designated neighborhoods, with branched pipes supplying both public basins and private homes, (Richardson, 1988).

The Greco-Roman city of Pompeii is located on the Bay of Naples, south-southeast of Mt. Vesuvius in Italy. Sources of water for Pompeii included wells, cisterns, and other reservoirs, and a long-distance water supply line (Crouch, 1993). According to Richardson

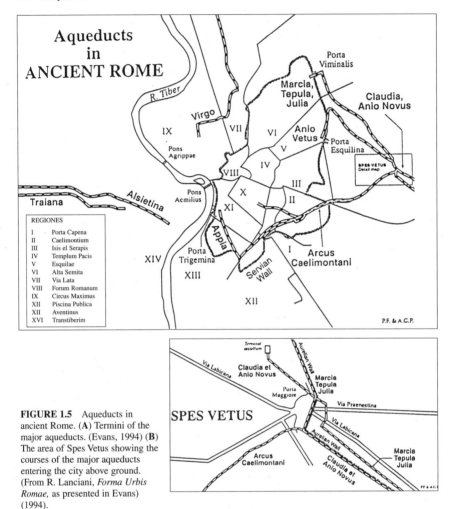

FIGURE 1.5 Aqueducts in ancient Rome. (**A**) Termini of the major aqueducts. (Evans, 1994) (**B**) The area of Spes Vetus showing the courses of the major aqueducts entering the city above ground. (From R. Lanciani, *Forma Urbis Romae,* as presented in Evans) (1994).

(1988), there were no springs within the city of Pompeii. The water table was tapped within Pompeii using wells as deep as 38 m below the surface (Maiuri, 1931). A long-distance water supply line from the hills to the east and northeast also supplied the city. Figure 1.7 illustrates the water distribution system of Pompeii (ca. 79 A.D.).

The fall of the Roman Empire extended over a 1000-year transition period called the Dark Ages. During this period, the concepts of science related to water resources probably retrogressed. After the fall of the Roman Empire, water sanitation and public health declined in Europe. Historical accounts tell of incredibly unsanitary conditions—polluted water, human and animal wastes in the streets, and water thrown out of windows onto passersby. Various epidemics ravaged Europe. During the same period, Islamic cultures, on the periphery of Europe, had religiously mandated high levels of personal hygiene, along with highly developed water supplies and adequate sanitation systems.

FIGURE 1.6 Roman aqueduct in Segovia, Spain. (Photograph by L.W. Mays).

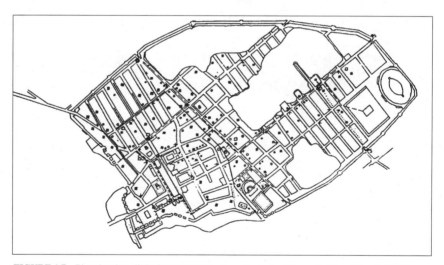

FIGURE 1.7 Plan showing all the known water system elements of Pompeii. (From Crouch 1993).

1.2.2 Status of Water Distribution Systems in the 19th Century

In J. T. Fannings's work, *A Practical Treatise on Hydraulic and Water-Supply Engineering* (1890), the following quote is presented in the preface:

> There is at present no sanitary subject of more general interest, or attracting more general attention, than that relating to the abundance and wholesomeness of domestic water supplies.
>
> Each citizen of a densely populated municipality must of necessity be personally interested in either its physiological or its financial bearing, or in both. Each closely settled town and city must give the subject earnest consideration early in its existence.
>
> At the close of the year 1875, fifty of the chief cities of the American Union had provided themselves with public water supplies at an aggregate cost of not less than ninety-five million dollars, and two hundred and fifty lesser cities and towns were also provided with liberal public water supplies at an aggregate cost of not less than fifty-five million dollars.
>
> The amount of capital annually invested in newly inaugurated water-works is already a large sum, and is increasing, yet the entire American literature relating to water-supply engineering exists, as yet, almost wholly in reports upon individual works, usually few of those especially in pamphlet form, and accessible each to but comparatively few of those especially interested in the subject.

Fanning (1890): discussed the use of wood pipes, the bored and Wychoff's patent pipes also.

> Bored Pipes. The wooden pipes used to replace the leaden pipes, in London, that were destroyed by the great fire, three-quarters of a century ago, reached a total length exceeding four hundred miles. These pipes were bored with a peculiar core-auger, that cut them out in nests, so that small pipes were made from cores of larger pipes.
>
> The earliest water-mains laid in America were chiefly of bored logs, and recent excavations in the older towns and cities have often uncovered the old cedar, pitch-pine, or chestnut pipe-logs that had many years before been laid by a single, or a few associated citizens, for a neighborhood supply of water.
>
> Bored pine logs, with conical faucet and spigot ends, and with faucet ends strengthened by wrought bands, were laid in Philadelphia as early as 1797.
>
> Detroit had at one time one hundred and thirty miles of small wood water pipes in her streets.
>
> Wyckoff's Patent pipe. A patent wood pipe, manufactured at Bay City, Michigan, has recently been laid in several western towns and cities, and has developed an unusual strength for wood pipes. Its chief peculiarities are, a spiral banding of hoop-iron, to increase its resistance to pressure and water-ram; a coating of asphaltum, to preserve the exterior of the shell; and a special form of thimble-joint.

Figures 1.8 to 1.13 present some of the various water distribution components presented in Fanning (1890).

1.2.3 Perspectives on Water Distribution Mains in the United States

In the United States, the construction of water supply systems dates back to 1754, when the system for the Moravian settlement of Bethehem, Pennsylvania, was built (American Public Works Association, 1976). This system consisted of spring water forced by a pump through bored logs. Philadelphia was also developing a water supply system during this same period. The water supply system included horse-driven pumps, as this was before the steam engine.

FIGURE 1.8 Tank stand pipe, South Abington Water Works, Massachusetts. (From Fanning, 1890).

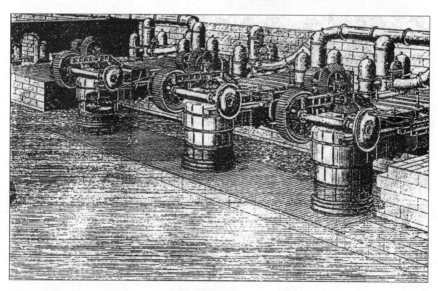

FIGURE 1.9 Fairmount pumping machinery, Philadelphia. (From Fanning, 1890).

1.12 Chapter One

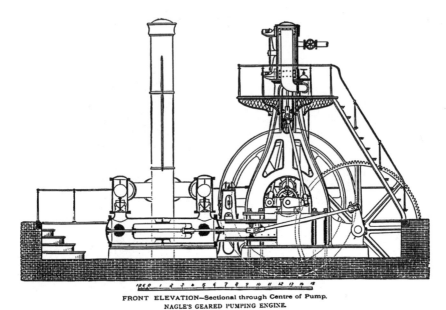

FRONT ELEVATION—Sectional through Centre of Pump.
NAGLE'S GEARED PUMPING ENGINE.

FIGURE 1.10 Nagle's geared pumping engine, front elevation—sectional through the center of pump. (From Fanning, 1890).

The following perspective on water mains is extracted from the Report to Congress of the Comptroller General of the United States (1980):

> Most water distribution mains in our older cites are made of cast iron, an extremely long-lasting material. Many American cities have cast iron mains over 100 years old which are still providing satisfactory service. No industry standard exists for replacing cast iron mains based on age alone. Ordinarily, breaks and leaks in mains are repaired, and large sections are replaced only if the mains are badly deteriorated or too small. A new form of cast, called ductile iron, has come into general use in recent years. This product has been almost failure free, a good sign for the future. Reduced carrying capacity caused by tuberculation—the products of internal corrosion—occurs in many older cast iron mains but can often be remedied by in-place cleaning and cement mortar lining, a less costly solution than replacement. Deterioration caused by external corrosion does not appear to be a major factor.
>
> *Water Distribution Mains in Older Cities*
>   Cast iron has been the material most used for water distribution mains in older cities since its introduction in the United States in the early 1800s. Current estimates of the total number of miles of distribution mains, or of cast iron mains, are not available. A survey done in the late 1960s by the Cast Iron Pipe Research Association (now called the Ductile Iron Pipe Research Association) reported that in the 100 largest cities, about 90 percent (87,000 miles) of water mains 4 inches and larger were cast iron. Twenty-eight of the cities reported having cast

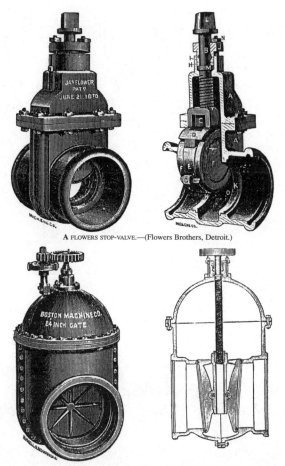

FIGURE 1.11 Stop-valves (**A**) Flowers stop-valve (Flowers Brothers, Detroit). (**B**) Coffin's stop-valve (Courtesy of Boston Machine Co., Boston) (From Fanning, 1890).

iron mains 100 years old or older. Based on this survey, this association estimated that the United States had over 400,000 miles of cast iron water mains in 1970. In Boston, 99 percent of the distribution system is cast and ductile iron; in Washington, D.C., 95 percent; and in New Orleans, 69 percent.

Early developments

America's first piped water supply was in Boston in 1652 when water was brought from springs and wells to near what is now the restored Quincy Market area. In about 1746, the first piped supply for an entire community was built in what is now Schaefferstown, Pennsylvania. In both instances, the water was stored in wooden tanks from which citizens filled buckets. Early systems used wooden

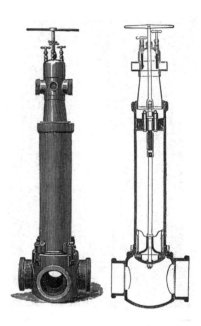

FIGURE 1.12 Lowry's flush hydrant (Courtesy of Boston Machine Co., Boston) (From Fanning, 1890).

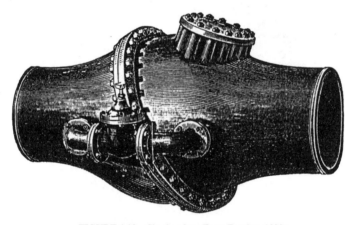

FIGURE 1.13 Check-valve. (From Fanning, 1890).

pipes and the force of gravity to move water from higher to lower elevations. Water systems as we know them today began when steam-driven pumps were first used in 1764 to move water uphill in Bethlehem, Pennsylvania.

Development of cast iron pipe

The first cast iron water main in the United States was laid in Philadelphia in 1817. Even that early in United States history, a cast iron main in Versailles, France, was already 153 years old. This main, laid in 1664, is still in use after more than three centuries.

Like most manufactured items, cast iron pipe has undergone a number of changes and improvements over the years. Early iron pipe was statically cast in horizontal sand molds. By the late 1800s, most pipe was cast vertically in static sand molds—often called pit casting. Some pipe made by both methods had portions of the pipe wall thinner than others because the mandrel around which the iron was poured to form the pipe bore shifted. While many cities have such "thick and thin" pipe still in use today, it does not withstand stress as well as more recently manufactured pipe.

In 1908, AWWA published the first standards for vertical pit casting. The formula for wall thickness considered internal pressures and included an arbitrary factor to provide for stresses which were unknown or could not be satisfactorily calculated.

Static casting continued until about 1921 when the centrifugal casting method came into use. This method, using either sand or metal molds, continues in use today. Centrifugal casting, combined with increased knowledge of metallurgy, produced a pipe with considerably more tensile strength than pit cast pipe. However, some of the early centrifugally cast pipe had very thin walls and broke easily.

In 1948, a metallurgically different cast iron pipe, having the favorable characteristics of both steel and cast iron, was invented. Called ductile iron, it is less brittle than its predecessors, collectively called gray iron, and has superior strength, flexibility, and impact resistance. National standards for this pipe were first published in 1965. In the last 3 or 4 years, virtually all cast iron pipe produced has been ductile iron.

Boston started using ductile iron in 1968 and has used it exclusively since 1970. At the end of 1979, at least 73 miles, or 7 percent, of the system was ductile iron. New Orleans and Washington have only small amounts of ductile iron mains.

Because cast iron has been so long lasting, older cities may have mains of each type. Louisville, for example, had some mains from every year since 1862 still in service at the end of 1976. Boston had some mains that were installed in 1853, and officials estimated that about 20 percent of the system was installed before 1900. About half of the cast iron mains in New Orleans were installed between 1904 and 1908 and most of the remainder from 1909 to 1950. Washington's present system went into service in the late 1800s, and most of the original mains remain.

Ways of joining cast iron pipes

Methods of joining pipes have also changed over the years. Until about 1935, the common joint for cast iron pipe was the "bell and spigot." The straight (spigot) was inserted in the larger (bell) end, and the space between was caulked with lead. If the pipe moved, the lead worked loose. In an 1851 report, the city of Boston noted an improved bell with a groove cast in it which would fill with lead to better hold the joint. From about 1920 until about 1955, some cities used a sulphur compound in place of lead. This material was cheaper and easier to use. Some water company officials stated, however, that it produced an extremely rigid joint which contributed to cast iron main breaks. Also about 1920, a bolted mechanical joint, developed for the natural gas industry, was first used for water mains. The next development was a rubber ring gasket which was used in place of the lead or sulpher caulking on bell and spigot pipe. Since 1955, new cast or ductile iron pipe has been installed with a rubber gasket that fits in a groove in the bell. This method produces a watertight joint with a good deal of flexibility.

1.2.4 Early Pipe Flow Computational Methods

In Fanning's *A Practical Treatise on Hydraulic and Water-Supply Engineering* (1980), the pipe flow formulas in Table 1.3 were compared. This book did not cover the flow in any type of pipe system, even in a simple branching system or a parallel pipe system. Le Conte (1926) and King et al. (1941) discussed branching pipes connecting three reservoirs and pipes in series and parallel.

The book *Water Supply Engineering*, by Babbitt and Doland (1939), stated, "A method of successive approximations has recently (1936) been developed by Prof. Hardy Cross which makes it possible to analyze rather complicated systems with the simple equipment of pencil, paper and slide rule." The authors then quoted the following method of solution from Cross (1936):

(a) Assume any distribution of flow.

(b) Compute in each pipe the loss of head, $h = rQ^n$. With due attention to sign (direction of potential drop), compute the total head loss around each elementary closed circuit, $\Sigma h = \Sigma rQ^n$.

(c) Compute also in each such closed circuit the sum of the quantities $R = nrQ^{n-1}$ without reference to sign.

(d) Set up in each circuit a counterbalancing flow to balance the head in that circuit (to make $\Sigma rQ^n = 0$) equal to

$$\Delta Q = \frac{\Sigma rQ^n \text{ (with due attention to direction of flow)}}{\Sigma rQ^{n-1} \text{ (with reference to direction of flow)}}$$

(e) Compute the revised flows and repeat the procedure. Continue to any desired precision. In applying the method, it is recommended that successive computations of the circuits be put on identical diagrams of the system. In office practice such diagrams will usually be white prints. Write in each elementary circuit the value ΣR, and outside the circuit write first (above) the value Σh for flow in a clockwise direction around the circuit. On the right of these figures put an arrow pointing to the large figure. This arrow will show correctly the direction of counter flow in the circuit.

1.3 MODERN WATER DISTRIBUTION SYSTEMS

1.3.1 The Overall Systems

Water utilities construct, operate, and maintain water supply systems. The basic function of these water utilities is to obtain water from a source, treat the water to an acceptable quality, and deliver the desired quantity of water to the appropriate place at the appropriate time. The analysis of a water utility is often devoted to the evaluation of one or more of the six major functional components of the utility: source development, raw water transmission, raw water storage, treatment, finished water storage, and finished water distribution as well as associated subcomponents. Because of their interaction, finished water storage is usually evaluated in conjunction with finished water distribution and raw water storage is usually evaluated in conjunction with the source. Figure 1.14 illustrates the six functional components of a water utility.

TABLE 1.3 Results Given by Various Formulas for Flow of Water in Smooth Pipes, under Pressure, Compared Data.- To find the velocity, given *Head*, $H = 100$ feet; *Diameter*, $d = 1$ foot; and *Lengths*, l, respectively as follows:

Authority	Equations	Lengths				
		5 ft Veloc.	50 ft Veloc.	100 ft Veloc.	1000 ft Veloc.	10,000 ft Veloc.
Equation (12)	$v = \left\{ \dfrac{2gH}{(1.5) + 4m\dfrac{l}{d}} \right\}^{\frac{1}{2}}$	63.463	51.111	43.111	17.386	5.392
Chezy	$v = \left\{ \dfrac{ghS}{\frac{1}{2}mlC} \right\}^{\frac{1}{2}}$	223.607	70.710	50.000	15.810	5.000
Du Buat	$v = \dfrac{88.5r^{\frac{1}{2}} - .03}{\left(\dfrac{l}{h}\right)^{\frac{1}{2}} - \text{hyp. Log.} \left(\dfrac{l}{h} + 1.6\right)^{\frac{1}{2}}} - .84(r^{\frac{1}{2}} - .03)$		102.918	81.510	13.662	3.9781
Prony (a)	$v = (9419.75ri + .00665)^{\frac{1}{2}} - .0816$	216.94	68.54	48.446	15.258	4.770
Prony (b)	$v = (9978.76ri + .02375)^{\frac{1}{2}} - .15412$	223.214	70.480	49.792	15.641	4.842
Eytelwein (a)	$v = (11703.95ri + .01698)^{\frac{1}{2}} - .1308$	241.778	76.367	53.960	16.975	5.280
Eytelwein (b)	$v = 50 \left\{ \dfrac{dh}{l + 50d} \right\}^{\frac{1}{2}}$	67.40	50.00	40.82	15.427	4.985
Saint Vennant	$v = 105.926(ri)^{\frac{11}{21}}$	246.171	73.682	51.247	15.232	4.592
D'Aubuisson (a)	$v = (9579ri + .00813)^{\frac{1}{2}} - .0902$	218.758	69.114	48.845	15.384	4.800
D'Aubuisson (b)	$v = 95.6 \sqrt{ri}$	213.761	67.589	47.804	15.114	4.780
Neville (a)	$v = \left\{ \dfrac{Hr}{.0234r + .0001085l} \right\}^{\frac{1}{2}}$	62.540	47.080	38.750	14.780	4.780
Neville (b)	$v = 140(ri)^{\frac{1}{2}} - 11\,(ri)^{\frac{1}{3}}$	294.650	90.263	63.070	18.917	5.507
Blackwell	$v = 47.913 \left\{ \dfrac{hd}{l} \right\}^{\frac{1}{2}}$	214.267	67.715	47.913	15.140	4.791
D'Arcy	$v = \left\{ \dfrac{ri}{.00007726 + \dfrac{.00000162}{r}} \right\}^{\frac{1}{2}}$	244.120	77.133	56.640	17.279	5.464
Leslie	$v = 100 \sqrt{ri}$	223.607	70.710	50.000	15.810	5.000
Jackson	$v = 50c(di)^{\frac{1}{2}}$	223.607	70.710	50.000	15.810	5.000
Hawksley	$v = 48.045 \left\{ \dfrac{dh}{l + 54d} \right\}^{\frac{1}{2}}$	62.555	47.084	38.724	14.797	4.804

In which
- C = contour of pipe, in feet;
- c = unity for smooth pipes, and is reduced for rough pipes.
- d = diameter, of pipe, in feet.
- H = entire head, in feet.
- h^{ll} = resistance head, in feet.
- l = length of pipe, in feet.
- m = coefficient of flow.
- r = hyd. mean radius, in feet, $= \dfrac{d}{4}$.
- S = sectional area of pipe, in square feet.
- i = sine of inclination, in feet, $= \dfrac{h^{ll}}{t}$
- v = velocity of flow, in feet per sec.

Source: Fanning (1890).

1.18 Chapter One

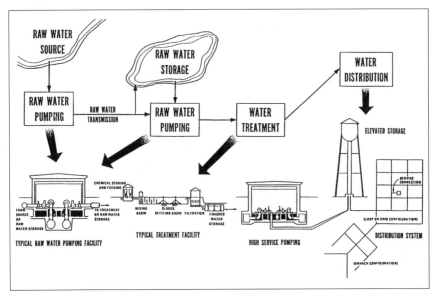

FIGURE 1.14 Functional components of a water utility. (Cullinane, 1989).

Urban water distribution is composed of three major components: distribution piping, distribution storage, and pumping stations. These components can be further divided into subcomponents, which can in turn be divided into sub-subcomponents. For example, the pumping station component consists of structural, electrical, piping, and pumping unit subcomponents. The pumping unit can be further divided into sub-subcomponents: pump, driver, controls, power transmission, and piping and valves. The exact definition of components, subcomponents, and sub-subcomponents is somewhat fluid and depends on the level of detail of the required analysis and, to a somewhat greater extent, the level of detail of available data. In fact, the concept component-subcomponent-subsubcomponent merely defines a hierarchy of building blocks used to construct the urban water distribution system. Figure 1.15 summarizes the relationship between components and subcomponents.

1. *Subsub-components.* Subsubcomponents represent the basic building blocks of systems. Individual sub-subcomponents may be common to a number of subcomponents within the water distribution system. Seven sub-subcomponents can be readily identified for analysis: pipes, valves, pumps, drivers, power transmission units, controls, and storage tanks.

2. *Subcomponents.* Subcomponents representing the basic building blocks for components are composed of one or more sub-subcomponents integrated into a common operational element. For example, the pumping unit subcomponent is composed of pipes, valves, pump, driver, power transmission, and control sub-subcomponents. Three subcomponents can be used to evaluate the reliability of the urban water distribution systems: pumping units, pipe links, and storage tanks.

3. *Components.* Components represent the largest functional elements in an urban water distribution system. Components are composed of one or more subcomponents.

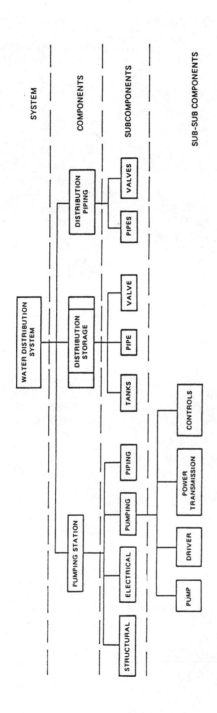

FIGURE 1.15 Hierarchical relationship of components, subcomponents, and sub-subcomponents for a water distribution system (Cullinane, 1989).

1.20 Chapter One

These include distribution piping, distribution storage, and pumping stations. Distribution piping is either branched as shown in Fig. 1.16, or looped, as shown in Fig. 1.17, or is a combination of branched, and looped. A typical pumping station is shown in Fig. 1.18. A typical elevated storage tank installation is shown in Fig. 1.19. A representation of distribution system in a pipe network model is illustrated in Fig. 1.17. A typical water distribution model display is illustrated in Fig. 1.20.

1.3.2 System Components

Pipe sections or links are the most abundant elements in the network. These sections are constant in diameter and may contain fittings and other appurtenances, such as valves, storage facilities, and pumps. Pipes are manufactured in different sizes and are composed of different materials, such as steel, cast or ductile iron, reinforced or prestressed concrete, asbestos cement, polyvinyl chloride, polyethylene, and fiberglass. The American Water Works Association publishes standards for pipe construction, installation, and performance in the C-series standards (continually updated). Pipes are the largest capital investment in a distribution system. Figure 1.21 shows a steel pipeline that is coated with polyethylene tape and lined by cement mortar once in place. Figure 1.22 shows a steel pipeline that is tape coated and epoxy lined. Figure 1.23 shows a prestressed concrete cylinder pipe (PCCP).

A node refers to either end of a pipe. Two categories of nodes are junction nodes and fixed-grade nodes. Nodes where the inflow or the outflow is known are referred to as junction nodes. These nodes have lumped demand, which may vary with time. Nodes to which a reservoir is attached are referred to as fixed-grade nodes. These nodes can take the form of tanks or large constant-pressure mains.

Control valves regulate the flow or pressure in water distribution systems. If conditions exist for flow reversal, the valve will close and no flow will pass. The most common type of control valve is the pressure-reducing (pressure-regulating) valve (PRV), which is placed at pressure zone boundaries to reduce pressure. The PRV maintains a constant pressure at the downstream side of the valve for all flows with a pressure lower than the

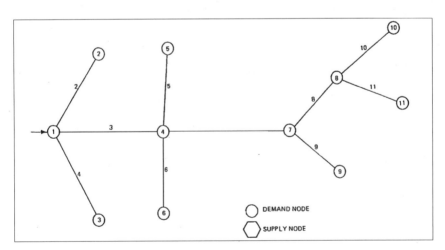

FIGURE 1.16 Typical branched distribution system.

Introduction **1.21**

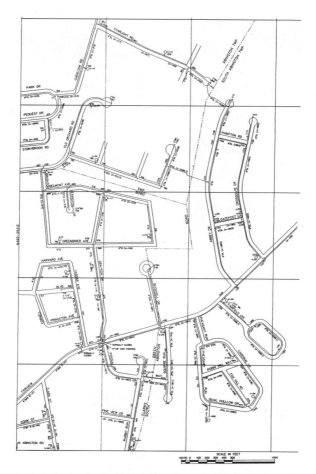

FIGURE 1.17 Typical water distribution map (from Pennsylvania American Water Company).

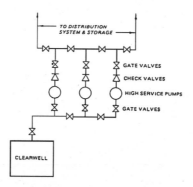

FIGURE 1.18 Schematic of a typical water distribution system pumping station.

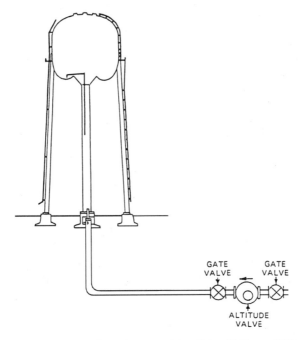

FIGURE 1.19 Typical elevated storage tank installation. (Cullinane, 1989).

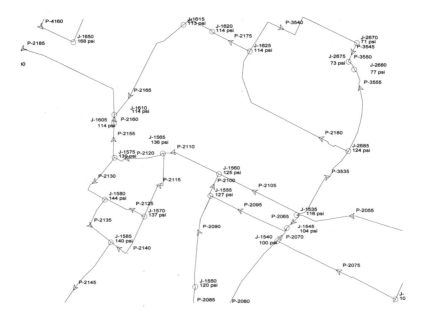

FIGURE 1.20 Typical water distribution model display (from T. Walski).

FIGURE 1.21 Steel pipeline, 81 in diameter, Seattle, Washington. Polyethylene tape coated, to be cement mortar lined in place. (Courtesy of Northwest Pipe Company).

FIGURE 1.22 Steel pipeline, 72 in diameter, tape coated and epoxy lined. (Courtesy of Northwest Pipe Company).

FIGURE 1.23 Prestressed concrete cylinder pipe, 160 in. in diameter, 20 ft long. Lake Garten Pipeline, City of Virginia Beach. (Courtesy of Price Brothers).

upstream head. When connecting high-pressure and low-pressure water distribution systems, the PRV permits flow from the high-pressure system if the pressure on the low side is not excessive.

The headloss through the valve varies, depending upon the downstream pressure and not on the flow in the pipe. If the downstream pressure is greater than the PRV setting, then the pressure in chamber A will close the valve. Another type of check valve, a horizontal swing valve, operates under similar principle. Pressure-sustaining valves operate similarly to PRVs monitoring pressure at the upstream side of the valve. Figure 1.24 illustrates a combined pressure-reducing and pressure-sustaining valve. Figure 1.25 illustrates the typical application of these valves.

There are many other types of valves, including isolation valves to shut down a segment of a distribution system; direction-control (check) valves to allow the flow of water in only one direction, such as swing check valves, rubber-flapper check valves, slanting check disk check valves, and double-door check valves; and air-release/vacuum-breaker valves to control flow in the main.

Distribution-system storage is needed to equalize pump discharge near an efficient operating point in spite of varying demands, to provide supply during outages of individual components, to provide water for fire fighting, and to dampen out hydraulic transients (Walski, 1996). Distribution storage in a water distribution network is closely associated with the water tank. Tanks are usually made of steel and can be built at ground level or be elevated at a certain height from the ground. The water tank is used to supply water to meet the requirements during high system demands or during emergency conditions when pumps cannot adequately satisfy the pressure requirements at the demand nodes. If a minimum volume of water is kept in the tank at all times, then unexpected high demands cannot be met during critical conditions. The higher the pump discharge, the lower the pump head becomes. Thus, during a period of peak demands, the amount of available pump head is low.

Pumps are used to increase the energy in a water distribution system. There are many different types of pumps (positive-displacement pumps, kinetic pumps, turbine pumps, horizontal centrifugal pumps, vertical pumps, and horizontal pumps). The most commonly used type of pump used in water distribution systems is the centrifugal pump. Figure 1.27 illustrates a pumping station with centrifugal pumps and Figure 1.26 illustrates vertical pumps. Pump stations house the pumps, motors, and the auxiliary equipment.

The metering (flow measurement) of water mains involves a wide array of metering devices. These include electromagnetic meters, ultrasonic meters, propeller or turbine

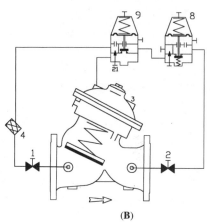

FIGURE 1.24 Pressure-reducing and pressure-sustaining valve. (**A**) Valve (**B**) Control diagram (Courtesy of Bermad).

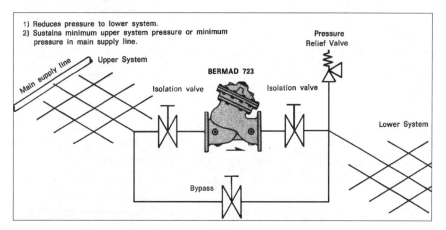

FIGURE 1.25 Typical application of a pressure-reducing and pressure-sustaining valve. (Courtesy of Bermad).

1.26 Chapter One

meters, displacement meters, multijet meters, proportional meters, and compound meters. Electromagnetic meters measure flow by means of a magnetic field generated around an insulated section of pipe. Ultrasonic meters utilize sound-generating and sound receiving sensors (transducers) attached to the sides of the pipe. Turbine meters (Fig. 1.28) have a measuring chamber that is turned by the flow of water. Multijet meters have a multiblade rotor mounted on a vertical spindle within a cylindrical measuring chamber. Proportional meters utilize restriction in the water line to divert a portion of water into a loop that holds a turbine or displacement meter, with the diverted flow being proportional to the flow in the main line. Compound meters connect different sized meters in parallel, as shown in Fig. 1.29. This meter has a turbine meter in parallel with a multijet meter. Fig. 1.30 illustrates the operation of these meters at low flows and at high flows. Fig. 1.31 illustrates a typical installation of a compound meter.

1.3.3 System Operation

The most fundamental decision that must be made in the operation of water distribution systems is which pumps should be operated at any given time. Walski (1996) pointed out three competing goals for water distribution system operation:

1. Maximize reliability, which is achieved by keeping the maximum amount of water in storage in case of emergencies, such as pipe breaks and fires.
2. Minimize energy costs, which is achieved by operating pumps against as low a head as possible (minimize water in storage) near the best efficiency point for the pump

FIGURE 1.26 Horizontal pumps at Gardner Creek Pump Station. Pennsylvania American Water Company (Photograph by T. Walski).

FIGURE 1.27 Vertical pumps at Gardner Creek Pump Station, Pennsylvania American Water Company. (Photograph by T. Walski).

FIGURE 1.28 Turbine meter with integral strainer. (Courtesy of Master Meter).

FIGURE 1.29 A dual-body (DB) compound meter, which combines a turbine meter on the main flow line and appropriately sized multijet meter on the low flow or bypass line. A differential pressure valve controls the flow of water through the appropriate measuring device. Piping sized to the bypass line connects the meters in a single assembly. Water flow through a bypass meter and usage is recorded on its register until the follow rate reaches approximately one-half the capacity of the bypass meter. At that point, the drop in pressure causes the differential pressure valve to open and water flows through both the main line and bypass meters. In its full open position, the valve allows unimpeded flow through both chambers, and registration is recorded on both meters. When flow is decreasing, the process is reversed, directing low flows through the multijet chamber. (Courtesy of Master Meter).

3. Meet water-quality standards, which involves minimizing the time the water is in the distribution system and storage tanks and is achieved by having storage-tank levels fluctuate as much as possible

The control of pumping operations can range from a simple manual operation an individual pump or valve to the use of a Supervisory Control and Data Acquisition (SCADA) system. Most utilities have some level of a SCADA system in place for use in operation of the system. Chapter 15 discusses SCADA systems in detail. The integration of hydraulic simulation models with SCADA systems is presently in its infancy.

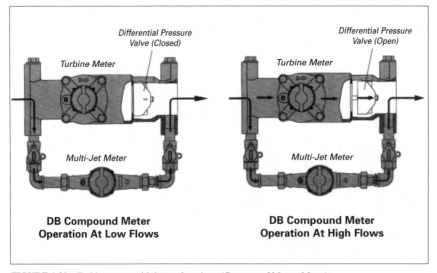

FIGURE 1.30 Turbine meter with integral strainer. (Courtesy of Master Meter).

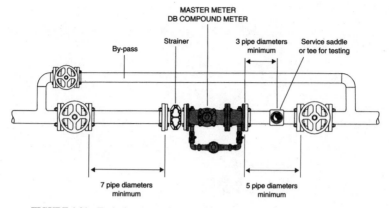

FIGURE 1.31 Typical compound meter installation. (Courtesy of Master Meter).

1.3.4 The Future

The future of water distribution system operation is illustrated in Fig. 1.32, by the optimal control system. The SCADA system would provide not only real-time hydraulic parameters but also real-time water quality information to the network model which solves both the hydraulics and the water quality. A demand-forecast model would be used in the real-time operation to forecast demands several hours in advance. The optimization model would be used to help make operation decisions based both the hydraulics and the water quality.

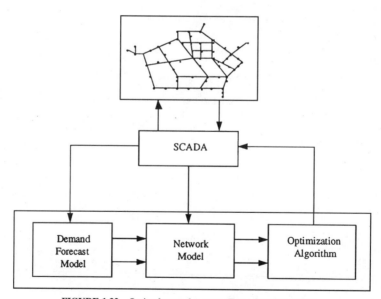

FIGURE 1.32 Optimal control system. (From Ormsbee, 1991).

REFERENCES

American Public Works Association, Historical Society, *History of Public Works in the United States*, American Public Works Association, Chicago, 1976.

Babbitt, H. E., and J. J. Doland, *Water Supply Engineering*, 3rd ed., McGraw-Hill, New York, 1939.

Clark, R. M., "Applying Water Quality Models," in M. H. Chaudhry and L. W. Mays, eds., *Computer Modeling of Free-Surface and Pressurized Flows*, Kluwer, Dordrecht Netherlands, pp. 581–612, 1987.

Clark, R. M., C. L. Stafford, and J. A. Goodrich, "Water Distribution Systems: A Spatial and Cost Evaluation," *Journal of the Water Resources Planning of the Water Resources Planning and Management Division*, ASCE, 108(8), pp. 243-256, October 1982.

Clark, R. M., Adams, J. Q., and Miltner, R. M., "Cost and Performance Modeling for Regulatory Decision Making," *Water*, 28(3); pp. 20–27, 1987.

Comptroller General of the United States, Report to the Congress, *Additional Federal Aid for Urban Water Distribution Systems Should Wait Until Needs Are Clearly Established*, CED-81-17, November 24, 1980.

Cross, H., "Analysis of Pipes in Networks of Conduits and Conductors," University of Illinois Experiment Station Bulletin 286, Urbana, IL, 1936.

Crouch, D. P. *Water Management in Ancient Greek Cities*, Oxford University Press, New York, 1993.

Cullinane, Jr., M. J., Methodologies for the evaluation of Water Distribution System Reliability/Availability, Ph. D. dissertation, University of Texas at Austin , May 1989

Evans, H. B., *Water Distribution in Ancient Rome*, University of Michigan Press, Ann Arbor 1994.

Fanning, J. T., *A Practical Treatise on Hydraulic and Water-Supply Engineering*, D. Van Nostrand, New York, 1890.

Frontius, Sextus Julius, *The Two Books on the Water Supply for the City of Rome*, AD 97 (translated by Clemens Hershel), republished by the New England Water Works Association, Boston, 1973.

Garbrecht, G., *Wasserversorgung Stechnik in Romischer Zeit*, R. Oldenbourg Verlag, Vienna, 1982.

Gleick, P. H., *Water in Crisis, A Guide to the World's Freshwater Resources*, Oxford University Press, New York, 1993.

Gleick, P. H., *The World's Water 1998-1999, The Biennial Report on Freshwater Resources*, Island Press, Washington, DC, 1998.

King, H. W., C. O. Wisler, and J. G. Woodburn, *Hydraulics*, 4th ed., John Wiley & Sons, New York, 1941.

Le Conte, J. N., *Hydraulics*, McGraw-Hill, New York, 1926.

Maiari, A., "Pompeii Pozzi e Condotture d'acqua . . ." and "Scoperta di grandi conditture in piombo dell'acquedoto urbano," *Notizie degli Scavi di Antichita*, 7: 546 –576, 1931.

Mays, L. W. (ed.), *Reliability Analysis of Water Distribution Systems*, American Society of Civil Engineers, New York, 1989.

Morgan, M. H., *Vitruvius, The Ten Books on Architecture*, Cambridge, MA., 1914.

Ormsbee, L. E. (ed.), "Energy Efficient Operation of Water Distribution Systems," *Report by the ASCE Task Committee on the Optimal Operation of Water Distribution Systems, Research Report No. UKCE 9104*, Department of Civil Engineering, University of Kentucky, Lexington, 1991.

Richardson, L. Jr., *Pompeii: An Architectural History*, Johns Hopkins University Press, Baltimore, MD, 1988.

Robbins, F. W., *The Story of Water Supply*, Oxford University, London, 1946.

Rouse, H. and S. Ince, *History of Hydraulics*, Dover Publication, Inc., New York, 1957.

Van Deman, E. B., *The Building of Roman Aqueducts,* Carnegie Institute of Washington, Washington, DC, 1934.

Walski, T. M., "Water Distribution," in L.W. Mays (ed.), *Water Resources Handbook,* McGraw-Hill, New York, 1996.

CHAPTER 2
HYDRAULICS OF PRESSURIZED FLOW

Bryan W. Karney
*Department of Civil Engineering
University of Toronto,
Toronto, Ontario,
Canada*

2.1 INTRODUCTION

The need to provide water to satisfy basic physical and domestic needs; use of maritime and fluvial routes for transportation and travel, crop irrigation, flood protection, development of stream power; all have forced humanity to face water from the beginning of time. It has not been an easy rapport. City dwellers who day after day see water flowing from faucets, docile to their needs, have no idea of its idiosyncrasy. They cannot imagine how much patience and cleverness are needed to handle our great friend-enemy; how much insight must be gained in understanding its arrogant nature in order to tame and subjugate it; how water must be "enticed" to agree to our will, respecting its own at the same time. That is why a hydraulician must first be something like a water psychologist, thoroughly knowledgeable of its nature. (Enzo Levi, *The Science of Water: The Foundations of Modern Hydraulics*, ASCE, 1995, p. xiii.)

Understanding the hydraulics of pipeline systems is essential to the rational design, analysis, implementation, and operation of many water resource projects. This chapter considers the physical and computational bases of hydraulic calculations in pressurized pipelines, whether the pipelines are applied to hydroelectric, water supply, or wastewater systems. The term *pressurized pipeline* means a pipe system in which a free water surface is almost never found within the conduit itself. Making this definition more precise is difficult because even in a pressurized pipe system, free surfaces are present within reservoirs and tanks and sometimes–for short intervals of time during transient (i.e., unsteady) events–can occur within the pipeline itself. However, in a pressurized pipeline system, in contrast to the open-channel systems, the pressures within the conveyance system are usually well above atmospheric.

Of central importance to a pressurized pipeline system is its *hydraulic capacity:* that is, its ability to pass a design flow. A related issue is the problem of *flow control:* how design flows are established, modified, or adjusted. To deal adequately with these two topics, this chapter considers head-loss calculations in some detail and introduces the topics of pumping, flow in networks, and unsteady flows. Many of these subjects are treated in greater detail in later chapters or in references such as Chaudhry and Yevjevich (1981).

Rather than simply providing the key equations and long tabulations of standard values, this chapter seeks to provide a context and a basis for hydraulic design. In addition to the relations discussed, such issues as why certain relations rather than others are used, what various equations assume, and what can go wrong if a relation is used incorrectly also are considered. Although derivations are not provided, some emphasis is placed on understanding both the strengths and weaknesses of various approaches. Given the virtually infinite combinations and arrangements of pipe systems, such information is essential for the pipeline professional.

2.2 IMPORTANCE OF PIPELINE SYSTEMS

Over the past several decades, pressurized pipeline systems have become remarkably competitive as a means of transporting many materials, including water and wastewater. In fact, pipelines can now be found throughout the world transporting fluids through every conceivable environment and over every possible terrain.

There are numerous reasons for this increased use. Advances in construction techniques and manufacturing processes have reduced the cost of pipelines relative to other alternatives. In addition, increases in both population and population density have tended to favor the economies of scale that are often associated with pipeline systems. The need for greater conservation of resources and, in particular, the need to limit losses caused by evaporation and seepage have often made pipelines attractive relative to open-channel conveyance systems. Moreover, an improved understanding of fluid behavior has increased the reliability and enhanced the performance of pipeline systems. For all these reasons, it is now common for long pipelines of large capacity to be built, many of which carry fluid under high pressure. Some of these systems are relatively simple, composed only of series-connected pipes; in other systems, the pipes are joined to form complex networks having thousands of branched and interconnected lines.

Pipelines often form vital links in the process chain, and high penalties may be associated with both the direct costs of failure (pipeline repair, cost of lost fluid, damages associated with rupture, and so forth) and the interruption of service. This is especially evident in industrial applications, such as paper mills, mines, and power plants. Yet, even in municipal systems, a pipe failure can cause considerable property damage. In addition, the failure may lead indirectly to other kinds of problems. For example, a mainline break could flood a roadway and cause a traffic accident or might make it difficult to fight a major fire.

Although pipelines appear to promise an economical and continuous supply of fluid, they pose critical problems of design, analysis, maintenance, and operation. A successful design requires the cooperation of hydraulic, structural, construction, survey, geotechnical, and mechanical engineers. In addition, designers and planners often must consider the social, environmental, and legal implications of pipeline development. This chapter focuses on the hydraulic considerations, but one should remember that these considerations are not the only, nor necessarily the most critical, issues facing the pipeline engineer. To be successful, a pipeline must be economically and environmentally viable as well as technically sound. Yet, because technical competence is a necessary requirement for any successful pipeline project, this aspect is the primary focus.

2.3 NUMERICAL MODELS: BASIS FOR PIPELINE ANALYSIS

The designer of a hydraulic system faces many questions. How big should each pipe be to carry the required flow? How strong must a segment of pipe be to avoid breaking? Are reservoirs, pumps, or other devices required? If so, how big should they be, and where should they be situated?

There are at least two general ways of resolving this kind of issue. The first way is to build the pipe system on the basis of our "best guess" design and learn about the system's performance as we go along. Then, if the original system "as built" is inadequate, successive adjustments can be made to it until a satisfactory solution is found. Historically, a number of large pipe systems have been built in more or less this way. For example, the Romans built many impressive water supply systems with little formal knowledge of fluid mechanics. Even today, many small pipeline systems are still constructed with little or no analysis. The emphasis in this kind of approach should be to design a system that is both flexible and robust.

However, there is a second approach. Rather than constructing and experimenting with the real system, a replacement or *model* of the system is developed first. This model can take many forms: from a scaled-down version of the original to a set of mathematical equations. In fact, currently the most common approach is to construct an abstract numerical representation of the original that is encoded in a computer. Once this model is "operational," experiments are conducted on it to predict the behavior of the real or proposed system. If the design is inadequate in any predictable way, the parameters of the model are changed and the system is retested until design conditions are satisfied. Only when the modeller is reasonably satisfied will the construction of the complete system be undertaken.

In fact, most modern pipelines systems are modeled extensively before they are built. One reason for this is perhaps surprising–experiments performed on a model are sometimes *better* than those done on the prototype. However, we must be careful here, because better is a relative word. On the plus side, modeling the behavior of a pipeline system has a number of intrinsic advantages:

Cost. Constructing and experimenting on the model is often much less expensive than testing the prototype.

Time. The response rate of the model pipe system may be more rapid and convenient than the prototype. For example, it may take only a fraction of a second for a computer program to predict the response of a pipe system after decades of projected growth in the demand for water.

Safety. Experiments on a real system may be dangerous or risky, whereas testing the model generally involves little or no risk.

Ease of modification. Improvements, adjustments, or modifications in design or operating rules can be incorporated more easily in a model, usually by simply editing an input file.

Aid to communication. Models can facilitate communication between individuals and groups, thereby identifying points of agreement, disagreement, misunderstanding, or issues requiring clarification. Even simple sketches, such as Fig. 2.1, can aid discussion.

These advantages are often seen as so overwhelming that the fact that alternative approaches are available is sometimes forgotten. In particular, we must always remember that the model is not reality. In fact, what makes the model useful is precisely its simplicity–it is not as complex or expensive as the original. Stated more forcibly, the model is

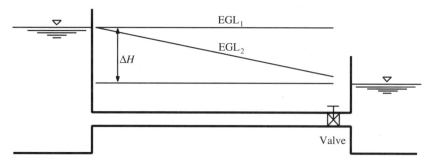

FIGURE 2.1 Energy relations in a simple pipe system.

useful *because* it is wrong. Clearly, the model must be sufficiently accurate for its intended purpose or its predictions will be useless. However, the fact that predictions are imperfect should be no surprise.

As a general rule, systems that are large, expensive, complex, and important justify more complex and expensive models. Similarly, as the sophistication of the pipeline system increases, so do the benefits and advantages of the modeling approach because this strategy allows us to consider the consequences of certain possibilities (decisions, actions, inactions, events, and so on) *before* they occur and to *control* conditions in ways that may be impossible in practice (e.g., weather characteristics, interest rates, future demands, control system failures). Models often help to improve our understanding of cause and effect and to isolate particular features of interest or concern and are our primary tool of prediction.

To be more specific, two kinds of computer models are frequently constructed for a pipeline system, planning models and operational models:

Planning models. These models are used to assess performance, quantity or economic impacts of proposed pipe systems, changes in operating procedures, role of devices, control valves, storage tanks, and so forth. The emphasis is often on selection, sizing, or modification of devices.

Operational models. These models are used to forecast behavior, adjust pressures or flows, modify fluid levels, train operators, and so on over relatively short periods (hours, days, months). The goal is to aid operational decisions.

The basis of both kinds of models is discussed in this chapter. However, before you believe the numbers or graphs produced by a computer program, or before you work through the remainder of this chapter, bear in mind that every model is in some sense a *fake*–it is a replacement, a stand-in, a surrogate, or a deputy for something else. Models are always more or less wrong. Yet it is their simultaneous possession of the characteristics of *both* simplicity and accuracy that makes them powerful.

2.4 MODELING APPROACH

If we accept that we are going to construct computer models to predict the performance of pipeline systems, then how should this be done? What aspects of the prototype can and should be emphasized in the model? What is the basis of the approximations, and what principles constrain the approach? These topics are discussed in this section.

Perhaps surprisingly, if we wish to model the behavior of any physical system, a remarkably small number of fundamental relations are available (or required). In essence, we seek to answer three simple questions: where? what? and how? The following sections provide elaboration.

The first question is resolved most easily. Because flow in a pipe system can almost always be assumed to be one-dimensional, the question of *where* is resolved by *assuming* a direction of flow in each link of pipe. This assumed direction gives a unique orientation to the specification of distance, discharge, and velocity. Positive values of these variables indicate flow in the assumed direction, whereas negative values indicate reverse flows. The issues of *what* and *how* require more careful development.

2.4.1 Properties of Matter (What?)

The question of 'What?' directs our attention to the matter within the control system. In the case of a hydraulic system, this is the material that makes up the pipe walls or fills the interior of a pipe or reservoir or that flows through a pump. Eventually, a modeller must account for all these issues, but we start with the matter that flows, typically consisting mostly of water with various degrees of impurities.

In fact, water is so much a part of our lives that we seldom question its role. Yet water possesses a unique combination of chemical, physical, and thermal properties that makes it ideally suited for many purposes. In addition, although important regional shortages may exist, water is found in large quantities on the surface of the earth. For both these reasons, water plays a central role in both human activity and natural processes.

One surprising feature of the water molecule is its simplicity, formed as it is from two diatomic gases, hydrogen (H_2) and oxygen (O_2). Yet the range and variety of water's properties are remarkable (Table 2.1 provides a partial list). Some property values in the table–especially density and viscosity values–are used regularly by pipeline engineers. Other properties, such as compressibility and thermal values, are used indirectly, primarily to justify modeling assumptions, such as the flow being isothermal and incompressible. Many properties of water depend on intermolecular forces that create powerful attractions (cohesion) between water molecules. That is, although a water molecule is electrically neutral, the two hydrogen atoms are positioned to create a tetrahedral charge distribution on the water molecule, allowing water molecules to be held strongly together with the aid of electrostatic attractions. These strong internal forces–technically called "hydrogen bonds"–arise directly from the nonsymmetrical distribution of charge.

The chemical behavior of water also is unusual. Water molecules are slightly ionized, making water an excellent solvent for both electrolytes and nonelectrolytes. In fact, water is nearly a universal solvent, able to wear away mountains, transport solutes, and support the biochemistry of life. But the same properties that create so many benefits also create problems, many of which must be faced by the pipeline engineer. Toxic chemicals, disinfection by-products, aggressive and corrosive compounds, and many other substances can be carried by water in a pipeline, possibly causing damage to the pipe and placing consumers at risk.

Other challenges also arise. Water's almost unique property of expanding on freezing can easily burst pipes. As a result, the pipeline engineer either may have to bury a line or may need to supply expensive heat-tracing systems on lines exposed to freezing weather, particularly if there is a risk that standing water may sometimes occur. Water's high viscosity is a direct cause of large friction losses and high energy costs, whereas its vapor properties can create cavitation problems in pumps, valves, and pipes. Furthermore, the

TABLE 2.1 Selected Properties of Liquid Water

Physical Properties
1. High density–$\rho_{liq} \approx 1\,000$ kg/m³
2. Density maximum at 4°C–i.e., above freezing!
3. High viscosity (but a Newtonian fluid)–$\mu \approx 10^{-3}$ N · s/m²
4. High surface tension–$\sigma \approx 73$ N/m
5. High bulk modulus (usually assumed incompressible)–$K \approx 2.07$ GPa

Thermal Properties
1. Specific heat–highest except for NH_3–$c \approx 4.187$ kJ/(kg·°C)
2. High heat of vaporization–$c_v \approx 2.45$ MJ/kg
3. High heat of fusion–$c_f \approx 0.36$ MJ/kg
4. Expands on freezing–in almost all other compounds, $\rho_{solid} > \rho_{liq}$
5. High boiling point–c.f., H_2 (20 K), O_2 (90 K) and H_2O (373 K)
6. Good conductor of heat relative to other liquids and nonmetal solids.

Chemical and Other Properties
1. Slightly ionized–water is a good solvent for electrolytes and nonelectrolytes
2. Transparent to visible light; opaque to near infrared
3. High dielectric constant–responds to microwaves and electromagnetic fields

Note: The values are approximate. All the properties listed are functions of temperature, pressure, water purity, and other factors that should be known if more exact values are to be assigned. For example, surface tension is greatly influenced by the presence of soap films, and the boiling point depends on water purity and confining pressure. The values are generally indicative of conditions near 10°C and one atmosphere of pressure.

combination of its high density and small compressibility creates potentially dramatic transient conditions. We return to these important issues after considering how pipeline flows respond to various physical constraints and influences in the next section.

2.4.2 Laws of Conservation (How?)

Although the implications of the characteristics of water are enormous, no mere list of its properties will describe a physical problem completely. Whether we are concerned with water quality in a reservoir or with transient conditions in a pipe, natural phenomena also obey a set of physical laws that contributes to the character and nature of a system's response. If engineers are to make quantitative predictions, they must first understand the physical problem and the mathematical laws that model its behavior.

Basic physical laws must be understood and be applied to a wide variety of applications and in a great many different environments: from flow through a pump to transient conditions in a channel or pipeline. The derivations of these equations are not provided, however, because they are widely available and take considerable time and effort to do properly. Instead, the laws are presented, summarized, and discussed in the pipeline context. More precisely, a quantitative description of fluid behavior requires the application of three essential relations: (1) a kinematic relation obtained from the law of mass conservation in a control volume, (2) equations of motion provided by both Newton's second law and the energy equation, and (3) an equation of state adapted from compressibility considerations, leading to a wavespeed relation in transient flow and justifying the assumption of an incompressible fluid in most steady flow applications.

A few key facts about mass conservation and Newton's second law are reviewed briefly in the next section. Consideration of the energy equation is deferred until steady flow is discussed in more detail, whereas further details about the equation of state are introduced along with considerations of unsteady flow.

2.4.3 Conservation of Mass

One of a pipeline engineer's most basic, but also most powerful, tools is introduced in this section. The central concept is that of conservation of mass, and its key expression is the *continuity* or *mass conservation equation*.

One remarkable fact about changes in a physical system is that not everything changes. In fact, most physical laws are conservation laws: They are generalized statements about regularities that occur in the midst of change. As Ford (1973) said:

> A conservation law is a statement of constancy in nature–in particular, constancy during change. If for an isolated system a quantity can be defined that remains precisely constant, regardless of what changes may take place within the system, the quantity is said to be absolutely conserved.

A number of physical quantities have been found that are conserved in the sense of Ford's quotation. Examples include energy (if mass is accounted for), momentum, charge, and angular momentum. One especially important generalization of the law of mass conservation includes both nuclear and chemical reactions (Hatsopoulos and Keenan, 1965).

2.4.3.1 Law of conservation of chemical species. "Molecular species are conserved in the absence of chemical reactions and atomic species are conserved in the absence of nuclear reactions." In essence, the statement is nothing more than a principle of accounting stating that the number of atoms or molecules that existed before a given change is equal to the number that exists after the change. More powerfully, the principle can be transformed into a statement of revenue and expenditure of some commodity over a definite period of time. Because both hydraulics and hydrology are concerned with tracking the distribution and movement of the Earth's water, which is nothing more than a particular molecular species, it is not surprising that formalized statements of this law are used frequently. These formalized statements are often called water budgets, typically if they apply to an area of land, or continuity relations if they apply in a well-defined region of flow (the region is well-defined; the flow need not be).

The principle of a budget or continuity equation is applied every time we balance a checkbook. The account balance at the end of any period is equal to the initial balance plus all the deposits minus all the withdrawals. In equation form, this can be written as follows:

$$(\text{balance})_f = (\text{balance})_i + \sum \text{deposits} - \sum \text{withdrawals}$$

Before an analogous procedure can be applied to water, the system under consideration must be clearly defined. If we return to the checking-account analogy, this requirement simply says that the deposits and withdrawals included in the equation apply to one account or to a well-defined set of accounts. In hydraulics and hydrology, the equivalent requirement is to define a *control volume*–a region that is fixed in space, completely surrounded by a "control surface," through which matter can pass freely. Only when the region has been precisely defined can the inputs (deposits) and outputs (withdrawals) be identified unambiguously.

2.8 Chapter Two

If changes or adjustments in the water balance (ΔS) are the concern, the budget concept can be expressed as

$$\Delta S = S_f - S_i = (\text{balance})_f - (\text{balance})_i = V_i - V_o \qquad (2.1)$$

where V_i represents the sum of all the water entering an area and V_o indicates the total volume of water leaving the same region. More commonly, however, a budget relation such as Eq. 2.1 is written as a rate equation. Dividing the "balance" equation by Δt and taking the limit as Δt goes to zero produces

$$S' = \frac{dS}{dt} = I - O \qquad (2.2)$$

where the derivative term S' is the time rate of change in storage, S is the water stored in the control volume, I is the rate at which water enters the system (inflow), and O is the rate of outflow. This equation can be applied in any consistent volumetric units (e.g., m³/s, ft³/s, L/s, ML/day.)

When the concept of conservation of mass is applied to a system with flow, such as a pipeline, it requires that the net amount of fluid flowing into the pipe must be accounted for as fluid storage within the pipe. Any mass imbalance (or, in other words, net mass exchange) will result in large pressure changes in the conduit because of compressibility effects.

2.4.3.2 Steady flow. Assuming, in addition, that the flow is steady, (Eq. 2.2) can be reduced further to inflow = outflow or $I = O$. Since the inflow and outflow may occur at several points, this is sometimes rewritten as

$$\underset{\text{inflow}}{\sum V_i A_i} = \underset{\text{outflow}}{\sum V_i A_i} \qquad (2.3)$$

Equation (2.3) states that the rate of flow into a control volume is equal to the rate of outflow. This result is intuitively satisfying since no accumulation of mass or volume should occur in any control volume under steady conditions. If the control volume were taken to be the junction of a number of pipes, this law would take the form of Kirchhoff's current law–the sum of the mass flow in all pipes entering the junction equals the sum of the mass flow of the fluid leaving the junction. For example, in Fig. 2.2, continuity for the control volume of the junction states that

$$Q_1 + Q_2 = Q_3 + Q_4 \qquad (2.4)$$

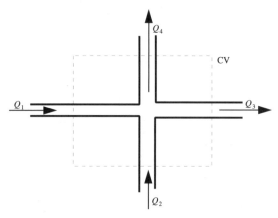

FIGURE 2.2 Continuity at a pipe junction $Q_1 + Q_2 = Q_3 + Q_4$.

2.4.4 Newton's Second Law

When mass rates of flow are concerned, the focus is on a single component of chemical species. However, when we introduce a physical law, such as Newton's law of motion, we obtain something even more profound: a relationship between the apparently unrelated quantities of force and acceleration.

More specifically, Newton's second law relates the changes in motion of a fluid or solid to the forces that cause the change. Thus, the statement that the result of all external forces, including body forces, acting on a system is equal to the *rate of change* of momentum of this system with respect to time. Mathematically, this is expressed as

$$\sum F_{ext} = \frac{d(mv)}{dt} \qquad (2.5)$$

where t is the time and F_{ext} represents the external forces acting on a body of mass m moving with velocity v. If the mass of the body is constant, Eq. (2.5) becomes

$$\sum F_{ext} = m\frac{dv}{dt} = ma \qquad (2.6)$$

where a is the acceleration of the system (the time rate of change of velocity).

In closed conduits, the primary forces of concern are the result of hydrostatic pressure, fluid weight, and friction. These forces act at each section of the pipe to produce the net acceleration. If these forces and the fluid motion are modeled mathematically, the result is a "dynamic relation" describing the transient response of the pipeline.

For a control volume, if flow properties at a given position are unchanging with time, the steady form of the momentum equation can be written as

$$\sum F_{ext} = \int_{cs} \rho v (v \cdot n) \, dA \qquad (2.7)$$

where the force term is the net external force acting on the control volume and the right-hand term gives the net flux of momentum through the control surface. The integral is taken over the entire surface of the control volume, and the integrand is the incremental amount of momentum leaving the control volume.

The control surface usually can be oriented to be perpendicular to the flow, and one can assume that the flow is incompressible and uniform. With this assumption, the momentum equation can be simplified further as follows:

$$\sum F_{ext} = (\rho A v v)_{out} - (\rho A v v)_{in} = \rho Q (v_{out} - v_{in}) \qquad (2.8)$$

where Q is the volumetric rate of flow.

Example: Forces at an elbow. One direct application of the momentum relation is shown in Fig. 2.3, which indicates the flows and forces at an elbow. The elbow is assumed to be mounted in a horizontal plane so that the weight is balanced by vertical forces (not shown).

The reaction forces shown in the diagram are required for equilibrium if the elbow is to remain stationary. Specifically, the force F_x must resist the pressure forces and must account for the momentum-flux term. That is, taking x as positive to the right, direct application of the momentum equation gives

$$(PA)_1 - F_x = -\rho Q V_1 \qquad (2.9)$$

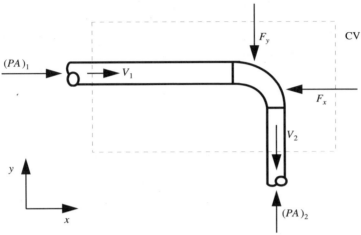

FIGURE 2.3 Force and momentum fluxes at an elbow.

Thus,

$$F_x = (PA)_1 + \rho Q V_1 \tag{2.10}$$

In a similar manner, but taking y as positive upward, direct application of the momentum equation gives

$$(PA)_2 - F_y = \rho Q(-V_2) \tag{2.11}$$

(here the outflow gives a positive sign, but the velocity is in the negative direction). Thus,

$$F_y = (PA)_2 + \rho Q V_2 \tag{2.12}$$

In both cases, the reaction forces are increased above what they would be in the static case because the associated momentum must either be established or be eliminated in the direction shown. Application of this kind of analysis is routine in designing thrust blocks, which are a kind of anchor used at elbows or bends to restrain the movement of pipelines.

2.5 SYSTEM CAPACITY: PROBLEMS IN TIME AND SPACE

A water transmission or supply pipeline is not just an enclosed tube–it is an entire system that transports water, either by using gravity or with the aid of pumping, from its source to the general vicinity of the demand. It typically consists of pipes or channels with their associated control works, pumps, valves, and other components. A transmission system is usually composed of a single-series line, as opposed to a distribution system that often consists of a complex network of interconnected pipes.

As we have mentioned, there are many practical questions facing the designer of such a system. Do the pipes, reservoirs, and pumps have a great enough hydraulic capacity? Can the flow be controlled to achieve the desired hydraulic conditions? Can the system be operated economically? Are the pipes and connections strong enough to with stand both unsteady and steady pressures?

Interestingly, different classes of models are used to answer them, depending on the nature of the flow and the approximations that are justified. More specifically, issues of hydraulic capacity are usually answered by projecting demands (water requirements) and analyzing the system under *steady flow* conditions. Here, one uses the best available estimates of future demands to size and select the primary pipes in the system. It is the hydraulic capacity of the system, largely determined by the effective diameter of the pipeline, that links the supply to the demand.

Questions about the operation and sizing of pumps and reservoirs are answered by considering the gradual variation of demand over relatively short periods, such as over an average day or a maximum day. In such cases, the acceleration of the fluid is often negligible and analysts use a *quasi-steady* approach: That is, they calculate forces and energy balances on the basis of steady flow, but the unsteady form is used for the continuity equation so that flows can be accumulated and stored.

Finally, the issue of required strength, such as the pressure rating of pipes and fittings, is answered by considering *transient conditions*. Thus, the strength of a pipeline is determined at least in part by the pressures generated by a rapid transition between flow states. In this stage, short-term and rapid motions must be taken into account, because large forces and dangerous pressures can sometimes be generated. Here, forces are balanced with accelerations and mass flow rates are balanced with pressure changes. These transient conditions are discussed in more detail in Sec. 2.8 and in chapter 6.

A large number of different flow conditions are encountered in pipeline systems. To facilitate analysis, these conditions are often classified according to several criteria. Flow classification can be based on channel geometry, material properties, dynamic considerations (both kinematic and kinetic), or some other characteristic feature of the flow. For example, on the basis of fluid type and channel geometry, the flow can be classified as open-channel, pressure, or gas flow. Probably the most important distinctions are based on the dynamics of flow (i.e., hydraulics). In this way, flow is classified as steady or unsteady, turbulent or laminar, uniform or nonuniform, compressible or incompressible, or single phase or multiphase. All these distinctions are vitally important to the analyst: Collectively, they determine which physical laws and material properties are dominant in any application.

Steady flow. A flow is said to be *steady* if conditions at a point do not change with time. Otherwise a flow is *unsteady* or *transient*. By this definition, all turbulent flows, and hence most flows of engineering importance, are technically unsteady. For this reason, a more restrictive definition is usually applied: A flow is considered to be steady if the temporal mean velocity does not change over brief periods. Although the assumption is not formally required, pipeline flows are usually considered to be steady; thus, transient conditions represent an "abnormal", or nonequilibrium, transition from one steady-state flow to another. Unless otherwise stated, the initial conditions in transient problems are usually assumed to be steady.

Steady or equilibrium conditions in a pipe system imply a balance between the physical laws. Equilibrium is typified by steady uniform flow in both open channels and closed conduits. In these applications, the rate of fluid inflow to each segment equals the rate of outflow, the external forces acting on the flow are balanced by the changes in momentum, and the external work is compensated for by losses of mechanical energy. As a result, the fluid generally moves down an energy gradient, often visualized as flow in the direction of decreasing hydraulic grade-line elevations (e.g., Fig. 2.1).

Quasi-steady flow. When the flow becomes unsteady, the resulting model that must be used depends on how fast the changes occur. When the rate of change is particularly slow, typically over a period of hours or days, the rate of the fluid's acceleration is negligible.

However, fluid will accumulate or be depleted at reservoirs, and rates of demand for water may slowly adjust. This allows the use of a *quasi-steady* or *extended-duration* simulation model.

Compressible and incompressible. If the density of the fluid ρ is constant–both in time and throughout the flow field–a flow is said to be *incompressible*. Thus, ρ is not a function of position or time in an incompressible flow. If changes in density are permitted or required the flow is compressible.

Surge. When the rate of change in flow is moderate, typically occurring over a period of minutes, a surge model is often used. In North America, the term *surge* indicates an analysis of unsteady flow conditions in pipelines when the following assumptions are made: The fluid is incompressible (thus, its density is constant) and the pipe walls are rigid and do not deform. These two assumptions imply that fluid velocities are not a function of position along a pipe of constant cross-section and the flow is uniform. In other words, no additional fluid is stored in a length of pipe as the pressure changes; because velocities are uniform, the rate at which fluid enters a pipe is always equal to the rate of discharge. However, the acceleration of the fluid and its accumulation and depletion from reservoirs are accounted for in a surge model.

Waterhammer. When rapid unsteady flow occurs in a closed conduit system, the transient condition is sometimes marked by a pinging or hammering noise, appropriately called *waterhammer*. However, it is common to refer to all rapidly changing flow conditions by this term, even if no audible shock waves are produced. In waterhammer models, it is usually assumed that the fluid is slightly compressible and that the pipe walls deform with changes in the internal pressure. Waterhammer waves propagate with a finite speed equal to the velocity of sound in the pipeline.

The speed at which a disturbance is assumed to propagate is the primary distinction between a surge model and a waterhammer model. Because the wavespeed parameter a is related to fluid storage, the wavespeed is *infinite* in surge or quasi-steady models. Thus, in effect, disturbances are assumed to propagate instantly throughout the pipeline system. Of course, they do no such thing because the wavespeed is a finite physical property of a pipe system, much like its diameter, wall thickness, or pipeline material. The implication of using the surge or quasi-steady approximation is that the unsteady behavior of the pipe system is controlled or limited by the rate at which the hydraulic boundary conditions (e.g., pumps, valves, reservoirs) at the ends of the pipe respond to the flow and that the time required for the pipeline itself to react is negligible by comparison.

Although unsteady or transient analysis is invariably more involved than is steady-state modeling, neglecting these effects in a pipeline can be troublesome for one of two reasons: The pipeline may not perform as expected, possibly causing large remedial expenses, or the line may be overdesigned with respect to transient conditions, possibly causing unnecessarily large capital costs. Thus, it is essential for engineers to have a clear physical grasp of transient behavior and an ability to use the computer's power to maximum advantage.

One interesting point is that as long as one is prepared to assume that the flow is compressible, the importance of compressibility does not need to be known a priori. In fact, all the incompressible, quasi-steady, and steady equations are special cases of the full transient equations. Thus, if the importance of compressibility or acceleration effects is unknown, the simulation can correctly assume compressible flow behavior and allow the analysis to verify or contradict this assumption.

Redistribution of water, whatever model or physical devices are used, requires control of the fluid and its forces, and control requires an understanding not only of physical laws but also of material properties and their implications. Thus, an attempt to be more specific and quantitative about these matters will be made as this chapter progresses.

In steady flow, the fluid generally moves in the direction of decreasing hydraulic grade-line elevations. Specific devices, such as valves and transitions, cause local pressure drops and dissipate mechanical energy; operating pumps do work on the fluid and increase downstream pressures, whereas friction creates headlosses more or less uniformly along the pipe length. Be warned, however–in transient applications, this orderly situation rarely exists. Instead, large and sudden variations of both discharge and pressure can occur and propagate in the system, greatly complicating analysis.

2.6 STEADY FLOW

The design of steady flow in pipeline systems has two primary objectives. First, the hydraulic objective is to secure the desired pressure and flow rate at specific locations in the system. Second, the economic objective is to meet the hydraulic requirements with the minimum expense.

When a fluid flows in a closed conduit or open channel, it often experiences a complex interchange of various forms of mechanical energy. In particular, the work that is associated with moving the fluid through pressure differences is related to changes in both gravitational potential energy and kinetic energy. In addition, the flow may lose mechanical energy as a result of friction, a loss that is usually accounted for by extremely small increases in the temperature of the flowing fluid (that is, the mechanical energy is converted to thermal form).

More specifically, these energy exchanges are often accounted for by using an extended version of Bernoulli's famous relationship. If energy losses resulting from friction are negligible, the Bernoulli equation takes the following form:

$$\frac{p_1}{\gamma} + \frac{V_1^2}{2g} + z_1 = \frac{p_2}{\gamma} + \frac{V_2^2}{2g} + z_2 \qquad (2.13)$$

where p_1 and p_2 are the pressures at the end points, γ is the specific weight of the fluid, V_1 and V_2 are the average velocities at the end points, and z_1 and z_2 are the elevations of the end points with respect to an arbitrary vertical datum. Because of their direct graphical representation, various combinations of terms in this relationship are given special labels, historically called *heads* because of their association with vertical distances. Thus:

Head	Definition	Associated with
Pressure head	p/γ	Flow work
Elevation head	z	Gravitational potential energy
Velocity head	$V^2/2g$	Kinetic energy
Piezometric head	$p/\gamma + z$	Pressure + elevation head
Total head	$p/\gamma + z + V^2/2g$	Pressure + elevation + velocity head

A plot of piezometric head along a pipeline forms a line called the *hydraulic grade line* (HGL). Similarly, a plot of the total head with distance along a pipeline is called the *energy grade line* (EGL). In the vast majority of municipally related work, velocity heads are negligible and the EGL and HGL essentially become equivalent.

If losses occur, the situation becomes a little more complex. The headloss h_f is defined to be equal to the difference in total head from the beginning of the pipe to the end over a total distance L. Thus, h_f is equal to the product of the slope of the EGL and the pipe length: $h_f = L \cdot S_f$. When the flow is uniform, the slope of the EGL is parallel to that of the HGL, the difference in piezometric head between the end points of the pipe. Inclusion of a headloss term into the energy equation gives a useful relationship for describing 1-D pipe flow:

$$\frac{p_1}{\gamma} + \frac{V_1^2}{2g} + z_1 = \frac{p_2}{\gamma} + \frac{V_2^2}{2g} + z_2 + h_f \tag{2.14}$$

In this relation, the flow is assumed to be from point 1 to point 2 and h_f is assumed to be positive. Using capital H to represent the total head, the equation can be rewritten as

$$H_1 = H_2 + h_f$$

In essence, a headloss reduces to the total head that would have occurred in the system if the loss were not present (Fig. 2.1). Since the velocity head term is often small, the total head in the above relation is often approximated with the piezometric head.

Understanding head loss is important for designing pipe systems so that they can accommodate the design discharge. Moreover, headlosses have a direct effect on both the pumping capacity and the power consumption of pumps. Consequently, an understanding of headlosses is important for the design of economically viable pipe systems.

The occurrence of headloss is explained by considering what happens at the pipe wall, the domain of *boundary layer theory*. The fundamental assertion of the theory is that when a moving fluid passes over a solid surface, the fluid immediately in contact with the surface attains the velocity of the surface (zero from the perspective of the surface). This *"no slip"* condition gives rise to a velocity gradient in which fluid further from the surface has a larger (nonzero) velocity relative to the velocity at the surface, thus establishing a shear stress on the fluid. Fluid that is further removed from the solid surface, but is adjacent to slower moving fluid closer to the surface, is itself decelerated because of the fluid's own internal cohesion, or viscosity. The shear stress across the pipe section is zero at the center of the pipe, where the average velocity is greatest, and it increases linearly to a maximum at the pipe wall. The distribution of the shear stress gives rise to a parabolic distribution of velocity when the flow is laminar.

More frequently, the flow in a conduit is turbulent. Because turbulence introduces a complex, random component into the flow, a precise quantitative description of turbulent flow is impossible. Irregularities in the pipe wall lead to the formation of eddy currents that transfer momentum between faster and slower moving fluid, thus dissipating mechanical energy. These random motions of fluid increase as the mean velocity increases. Thus, in addition to the shear stress that exists for laminar flow, an apparent shear stress exists because of the exchange of material during turbulent flow.

The flow regime–whether laminar, turbulent, or transitional–is generally classified by referring to the dimensionless *Reynold's number* (Re). In pipelines, Re is given as

$$Re = \frac{VD\rho}{\mu} \tag{2.15}$$

where V is the mean velocity of the fluid, D is the pipe diameter, ρ is the fluid density, and μ is the dynamic viscosity. Although the exact values taken to limit the range of Re vary with author and application, the different flow regimes are often taken as follows: (1) laminar flow: $Re \leq 2000$, (2) transitional flow: $2000 \leq Re \leq 4000$, and (3) turbulent flow: $Re > 4000$. These flow regimes have a direct influence on the headloss experienced in a pipeline system.

2.6.1 Turbulent Flow

Consider an experiment in which a sensitive probe is used to measure flow velocity in a pipeline carrying a flowing fluid. The probe will certainly record the mean or net component of velocity in the axial direction of flow. In addition, if the flow in the pipeline is turbulent, the probe will record many small and abrupt variations in velocity in all three spatial directions. As a result of the turbulent motion, the details of the flow pattern will change randomly and constantly with time. Even in the simplest possible system–a uniform pipe carrying water from a constant-elevation upstream reservoir to a downstream valve–the detailed structure of the velocity field will be unsteady and exceedingly complex. Moreover, the unsteady values of instantaneous velocity will exist even if all external conditions at both the reservoir and the valve are not changing with time. Despite this, the mean values of velocity and pressure will be fixed as long as the external conditions do not change. It is in this sense that turbulent flows can be considered to be steady.

The vast majority of flows in engineering are turbulent. Thus, unavoidably, engineers must cope with both the desirable and the undesirable characteristics of turbulence. On the positive side, turbulent flows produce an efficient transfer of mass, momentum, and energy within the fluid. In fact, the expression to "stir up the pot" is an image of turbulence; it implies a vigorous mixing that breaks up large-scale order and structure in a fluid. But the rapid mixing also may create problems for the pipeline engineer. This "downside" can include detrimental rates of energy loss, high rates of corrosion, rapid scouring and erosion, and excessive noise and vibration as well as other effects.

How does the effective mixing arise within a turbulent fluid? Physically, mixing results from the random and chaotic fluctuations in velocity that exchange fluid between different regions in a flow. The sudden small-scale changes in the instantaneous velocity tend to cause fast moving "packets" of fluid to change places with those of lower velocity and vice versa. In this way, the flow field is constantly bent, folded, and superimposed on itself. As a result, large-scale order and structure within the flow is quickly broken down and torn apart. But the fluid exchange transports not only momentum but other properties associated with the flow as well. In essence, the rapid and continual interchange of fluid within a turbulent flow creates both the blessing and the curse of efficient mixing.

The inherent complexity of turbulent flows introduces many challenges. On one hand, if the velocity variations are ignored by using average or mean values of fluid properties, a degree of uncertainty inevitably arises. Details of the flow process and its variability will be avoided intentionally, thereby requiring empirical predictions of mean flow characteristics (e.g., headloss coefficients and friction factors). Yet, if the details of the velocity field are analyzed, a hopelessly complex set of equations is produced that must be solved using a small time step. Such models can rarely be solved even on the fastest computers. From the engineering view point, the only practical prescription is to accept the empiricism necessitated by flow turbulence while being fully aware of its difficulties–the averaging process conceals much of what might be important. Ignoring the details of the fluid's motion can, at times, introduce significant error even to the mean flow calculations.

When conditions within a flow change instantaneously both at a point and in the mean, the flow becomes unsteady in the full sense of the word. For example, the downstream valve in a simple pipeline connected to a reservoir might be closed rapidly, creating shock waves that travel up and down the conduit. The unsteadiness in the mean values of the flow properties introduces additional difficulties into a problem that was already complex. Various procedures of averaging, collecting, and analyzing data that were well justified for a steady turbulent flow are often questionable in unsteady applications. The entire situation is dynamic: Rapid fluctuations in the average pressure, velocity, and other properties may break or damage the pipe or other equipment. Even in routine applications, special care is required to control, predict, and operate systems in which unsteady flows commonly occur.

The question is one of perspective. The microscopic perspective of turbulence in flows is bewildering in its complexity; thus, only because the macroscopic behavior is relatively predictable can turbulent flows be analyzed. Turbulence both creates the need for approximate empirical laws and determines the uncertainty associated with using them. The great irregularity associated with turbulent flows tends to be smoothed over both by the empirical equations and by a great many texts.

2.6.2 Headloss Caused by Friction

A basic relation used in hydraulic design of a pipeline system is the one describing the dependence of discharge Q (say in m³/s) on headloss h_f (m) caused by friction between the flow of fluid and the pipe wall. This section discusses two of the most commonly used headloss relations: the Darcy-Weisbach and Hazen-Williams equations.

The Darcy-Weisbach equation is used to describe the headloss resulting from flow in pipes in a wide variety of applications. It has the advantage of incorporating a dimensionless friction factor that describes the effects of material roughness on the surface of the inside pipe wall and the flow regime on retarding the flow. The *Darcy-Weisbach equation* can be written as

$$h_{f,DW} = f \frac{L}{D} \frac{V^2}{2g} = 0.0826 \frac{Q^2}{D^5} Lf \qquad (2.16)$$

where $h_{f,DW}$ = headloss caused by friction (m), f = dimensionless friction factor, L = pipe length (m), D = pipe diameter (m), $V = Q/A$ = mean flow velocity (m/s), Q = discharge (m³/s), A = cross-sectional area of the pipe (m²), and g = acceleration caused by gravity (m/s²).

For noncircular pressure conduits, D is replaced by $4R$, where R is the hydraulic radius. The hydraulic radius is defined as the cross-sectional area divided by the wetted perimeter, or $R = A/P$.

Note that the headloss is directly proportional to the length of the conduit and the friction factor. Obviously, the rougher a pipe is and the longer the fluid must travel, the greater the energy loss. The equation also relates the pipe diameter inversely to the headloss. As the pipe diameter increases, the effects of shear stress at the pipe walls are felt by less of the fluid, indicating that wider pipes may be advantageous if excavation and construction costs are not prohibitive. Note in particular that the dependence of the discharge Q on the pipe diameter D is highly nonlinear; this fact has great significance to pipeline designs because headlosses can be reduced dramatically by using a large-diameter pipe, whereas an inappropriately small pipe can restrict flow significantly, rather like a partially closed valve.

For laminar flow, the friction factor is linearly dependent on the Re with the simple relationship $f = 64/Re$. For turbulent flow, the friction factor is a function of both the Re and the pipe's relative roughness. The relative roughness is the ratio of equivalent uniform sand grain size and the pipe diameter (e/D), as based on the work of Nikuradse (1933), who experimentally measured the resistance to flow posed by various pipes with uniform sand grains glued onto the inside walls. Although the commercial pipes have some degree of spatial variance in the characteristics of their roughness, they may have the same resistance characteristics as do pipes with a uniform distribution of sand grains of size e. Thus, if the velocity of the fluid is known, and hence Re, and the relative roughness is known, the friction factor f can be determined by using the Moody diagram or the Colebrook-White equation.

Jeppson (1976) presented a summary of friction loss equations that can be used instead of the Moody diagram to calculate the friction factor for the Darcy-Weisbach equation.

These equations are applicable for *Re* greater than 4000 and are categorized according to the type of turbulent flow: (1) turbulent smooth, (2) transition between turbulent smooth and wholly rough, and (3) turbulent rough.

For turbulent smooth flow, the friction factor is a function of *Re*:

$$\frac{1}{\sqrt{f}} = 2\log(Re\sqrt{f}) \tag{2.17}$$

For the transition between turbulent smooth and wholly rough flow, the friction factor is a function of both *Re* and the relative roughness *e/D*. This friction factor relation is often summarized in the Colebrook-White equation:

$$\frac{1}{\sqrt{f}} = -2\log\left(\frac{e/D}{3.7} + \frac{2.51}{Re\sqrt{f}}\right) \tag{2.18}$$

When the flow is wholly turbulent (large *Re* and *e/D*), the Darcy-Weisbach friction factor becomes independent of *Re* and is a function only of the relative roughness:

$$\frac{1}{\sqrt{f}} = 1.14 - 2\log(e/D) \tag{2.19}$$

In general, Eq. (2.16) is valid for all turbulent flow regimens in a pipe, whereas Eq. (2.19) is merely an approximation that is valid for the hydraulic rough flow. In a smooth-pipe flow, the viscous sublayer completely submerges the effect of *e* on the flow. In this case, the friction factor *f* is a function of *Re* and is independent of the relative roughness *e/D*. In rough-pipe flow, the viscous sublayer is so thin that flow is dominated by the roughness of the pipe wall and *f* is a function only of *e/D* and is independent of *Re*. In the transition, *f* is a function of both *e/D* and *Re*.

The implicit nature of *f* in Eq. (2.18) is inconvenient in design practice. However, this difficulty can be easily overcome with the help of the Moody diagram or with one of many available explicit approximations. The Moody diagram plots *Re* on the abscissa, the resistance coefficient on one ordinate and *f* on the other, with *e/D* acting as a parameter for a family of curves. If *e/D* is known, then one can follow the relative roughness isocurve across the graph until it intercepts the correct *Re*. At the corresponding point on the opposite ordinate, the appropriate friction factor is found; *e/D* for various commercial pipe materials and diameters is provided by several manufacturers and is determined experimentally.

A more popular current alternative to graphical procedures is to use an explicit mathematical form of the friction-factor relation to approximate the implicit Colebrook-White equation. Bhave (1991) included a nice summary of this topic. The popular network-analysis program EPANET and several other codes use the equation of Swanee and Jain (1976), which has the form

$$f = \frac{0.25}{\left[\log\left(\frac{e}{3.7D} + \frac{5.74}{Re^{0.9}}\right)\right]^2} \tag{2.20}$$

To circumvent considerations of roughness estimates and *Re* dependencies, more direct relations are often used. Probably the most widely used of these empirical headloss relations is the Hazen-Williams equation, which can be written as

$$Q = C_u \, CD^{2.63} S^{0.54} \tag{2.21}$$

where C_u = unit coefficient (C_u = 0.314 for English units, 0.278 for metric units), Q = discharge in pipes, gallons/s or m³/s, L = length of pipe, ft or m, d = internal diameter of pipe, inches or mm, C = Hazen-Williams roughness coefficient, and S = the slope of the energy line and equals h_f/L.

The Hazen-Williams coefficient C is assumed to be constant and independent of the discharge (i.e., Re). Its values range from 140 for smooth straight pipe to 90 or 80 for old, unlined, tuberculated pipe. Values near 100 are typical for average conditions. Values of the unit coefficient for various combinations of units are summarized in Table 2.2.

In Standard International (SI) units, the Hazen-Williams relation can be rewritten for headloss as

$$h_{f,HW} = 10.654 \left(\frac{Q}{C}\right)^{\frac{1}{0.54}} \frac{1}{D^{4.87}} L \qquad (2.22)$$

where $h_{f,HW}$ is the Hazen-Williams headloss. In fact, the Hazen-Williams equation is not the only empirical loss relation in common use. Another loss relation, the Manning equation, has found its major application in open-channel flow computations. Like the other expressions, it incorporates a parameter to describe the roughness of the conduit known as Manning's n.

Among the most important and surprisingly difficult hydraulic parameters is the diameter of the pipe. As has been mentioned, the exponent of diameter in headloss equations is large, thus indicating high sensitivity to its numerical value. For this reason, engineers and analysts must be careful to obtain actual pipe diameters, often from manufacturers; the use of nominal diameters is not recommended. Yet another complication may arise, however. The diameter of a pipe often changes with time, typically as a result of chemical depositions on the pipe wall. For old pipes, this reduction in diameter is accounted for indirectly by using an increased value of pipe resistance. Although this approach may be reasonable under some circumstances, it may be a problem under others, especially for unsteady conditions. Whenever possible, accurate diameters are recommended for all hydraulic calculations. However, some combinations of pipes (e.g., pipes in series or parallel; Fig. 2.4) can actually be represented by a single equivalent diameter of pipe.

2.6.3 Comparison of Loss Relations

It is generally claimed that the Darcy-Weisbach equation is superior because it is theoretically based, whereas both the Manning equation and the Hazen-Williams expression use empirically determined resistance coefficients. Although it is true that the functional rela-

TABLE 2.2 Unit Coefficient C_u for the Hazen-Williams Equation

Units of Discharge Q	Units of Diameter D	Unit Coefficient C_u
MGD	ft	0.279
ft³/s	ft	0.432
GPM	in	0.285
GPD	in	405
m³/s	m	0.278

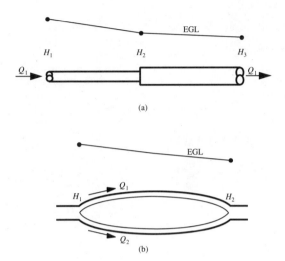

FIGURE 2.4 Flow in series and parallel pipes.

tionship of the Darcy-Weisbach formula reflects logical associations implied by the dimensions of the various terms, determination of the equivalent uniform sand-grain size is essentially experimental. Consequently, the relative roughness parameter used in the Moody diagram or the Colebrook-White equations is not theoretically determined. In this section, the Darcy-Weisbach and Hazen-Williams equations are compared briefly using a simple pipe as an example.

In the hydraulic rough range, the increase in Δh_f can be explained easily when the ratio of Eq. (2.16) to Eq. (2.22) is investigated. For hydraulically rough flow, Eq. (2.18) can be simplified by neglecting the second term 2.51 $(Re\sqrt{f})$ of the logarithmic argument. This ratio then takes the form of

$$\frac{h_{f,HW}}{h_{f,DW}} = 128.94 \left(1.14 - 2 \log \frac{e}{D}\right)^2 \frac{D^{0.13}}{C^{1.852}} \frac{1}{Q^{0.148}} \qquad (2.23)$$

which shows that in most hydraulic rough cases, for the same discharge Q, a larger head loss h_f is predicted using Eq. (2.16) than when using Eq. (2.22). Alternatively, for the same headloss, Eq. (2.22) returns a smaller discharge than does Eq. (2.16).

When comparing headloss relations for the more general case, a great fuss is often made over unimportant issues. For example, it is common to plot various equations on the Moody diagram and comment on their differences. However, such a comparison is of secondary importance. From a hydraulic perspective, the point is this: Different equations should still produce similar head-discharge behavior. That is, the physical relation between headloss and flow for a physical segment of pipe should be predicted well by any practical loss relation. Said even more simply, the issue is how well the h_f versus Q curves compare.

To compare the values of h_f determined from Eq. (2.16) and those from Eq. (2.22), consider a pipe for which the parameters D, L, and C are specified. Using the Hazen-Williams relation, it is then possible to calculate h_f for a given Q. Then, the Darcy-Weisbach f can be obtained, and with the Colebrook formula Eq. (2.18), the equivalent value of roughness e can be found. Finally, the variation of head with discharge can be plotted for a range of flows.

2.20 Chapter Two

This analysis is performed for two galvanized iron pipes with $e = 0.15$ mm. One pipe has a diameter of 0.1 m and a length of 100 m; and the dimensions of the other pipe are $D = 1.0$ m and $L = 1000$ m, respectively. The Hazen-Williams C for galvanized iron pipe is approximately 130. Different C values will be used for these two pipes to demonstrate the shift and change of the range within which Δh_f is small. The results of the calculated $h_f - Q$ relation and the difference Δh_f of the headloss of the two methods for the same discharge are shown in Figs. 2.5 and 2.6.

If $h_{f,DW}$ denotes the headloss determined by using Eq. (2.16) and $h_{f,HW}$, then using Eq. (2.22), Δh_f (m) can be

$$\Delta h_f = h_{f,DW} - h_{f,DW} \tag{2.24}$$

whereby the Darcy-Weisbach headloss $h_{f,DW}$ is used as a reference for comparison.

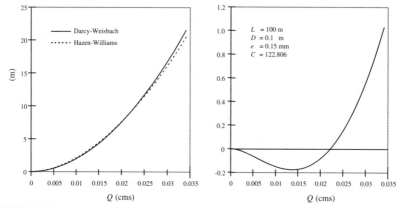

FIGURE 2.5 Comparison of Hazen-Williams and Darcy-Weisbach loss relations (smaller diameter).

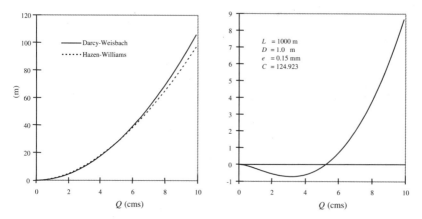

FIGURE 2.6 Comparison of Hazen-Williams and Darcy-Weisbach loss relations (larger diameter).

Figures 2.5 and 2.6 show the existence of three ranges: two ranges, within which $h_{f,DW} > h_{f,DW}$, and the third one for which $h_{f,DW} < h_{f,DW}$. The first range of $h_{f,DW} > h_{f,DW}$ is at a lower headloss and is small. It seems that the difference of Δh_f in this case is the result of the fact that the Hazen-Williams formula is not valid for the hydraulic smooth and the smooth-to-transitional region. Fortunately, this region is seldom important for design purposes. At high headlosses, the Hazen-Williams formula tends to produce a discharge that is smaller than the one produced by the Darcy-Weisbach equation.

For a considerable part of the curve—primarily the range within which $h_{f,DW} > h_{f,DW}$ – Δh_f is small compared with the absolute headloss. It can be shown that the range of small Δh_f changes is shifted when different values of Hazen-Williams's C are used for the calculation. Therefore, selecting the proper value of C, which represents an appropriate point on the head-discharge curve, is essential. If such a C value is used, Δh_f is small, and whether the Hazen-Williams formula or the Darcy-Weisbach equation is used for the design will be of little importance.

This example shows both the strengths and the weaknesses of using Eq. (2.22) as an approximation to Eq. (2.16). Despite its difficulties, the Hazen-Williams formula is often justified because of its conservative results and its simplicity of use. However, choosing a proper value of either the Hazen-Williams C or the relative roughness e/D is often difficult. In the literature, a range of C values is given for new pipes made of various materials. Selecting an appropriate C value for an old pipe is even more difficult. However, if an approximate value of C or e is used, the difference between the headloss equations is likely to be inconsequential.

Headloss also is a function of time. As pipes age, they are subject to corrosion, especially if they are made of ferrous materials and develop rust on the inside walls, which increases their relative roughness. Chemical agents, solid particles, or both in the fluid can gradually degrade the smoothness of the pipe wall. Scaling on the inside of pipes can occur if the water is hard. In some instances, biological factors have led to time-dependent headloss. Clams and zebra mussels may grow in some intake pipes and may, in some cases, drastically reduce discharge capacities.

2.6.4 Local Losses

Headloss also occurs for reasons other than wall friction. In fact, local losses occur whenever changes occur in the velocity of the flow: for example, changes in the direction of the conduit, such as at a bend, or changes in the cross-sectional area, such as an aperture, valve or gauge. The basic arrangement of flow and pressure is illustrated for a venturi contraction in Fig. 2.7.

The mechanism of headloss in the venturi is typical of many applications involving local losses. As the diagram indicates, there is a section of flow contraction into which the flow accelerates, followed by a section of expansion into which the flow decelerates. This aspect of the venturi, or a reduced opening at a valve, is nicely described by the continuity equation. However, what happens to the pressure is more interesting and more important.

As the flow accelerates, the pressure decreases according to the Bernoulli relation. Everything goes smoothly in this case because the pressure drop and the flow are in the same direction. However, in the expansion section, the pressure *increases* in the downstream direction. To see why this is significant, consider the fluid distributed over the cross section. In the center of the pipe, the fluid velocity is high; the fluid simply slows down as it moves into the region of greater pressure. But what about the fluid along the wall? Because it has no velocity to draw on, it tends to respond to the increase in pressure

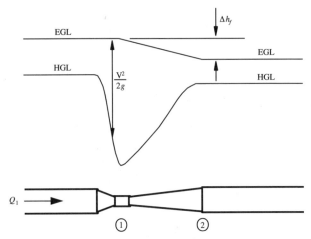

FIGURE 2.7 Pressure relations in a venturí contraction.

in the downstream direction by flowing upstream, counter to the normal direction of flow. That is, the flow tends to separate, which can be prevented only if the faster moving fluid can "pull it along" using viscosity. If the expansion is too abrupt, this process is not sufficient, and the flow will separate, creating a region of recirculating flow within the main channel. Such a region causes high shear stresses, irregular motion, and large energy losses. Thus, from the viewpoint of local losses, nothing about changes in pressure is symmetrical–adverse pressure gradients or regions of recirculating flow are crucially important with regard to local losses.

Local headlosses are often expressed in terms of the velocity head as

$$h_1 = k \frac{V^2}{2g} \tag{2.25}$$

where k is a constant derived empirically from testing the headloss of the valve, gauge, and so on and is generally provided by the manufacturer of the device. Typical forms for this relation are provided in Table 2.3 (Robertson and Crowe, 1993).

2.6.5 Tractive Force

Fluid resistance also implies a flux in momentum and generates a tractive force, which raises a number of issues of special significance to the two-phase (liquid-solid) flows found in applications of transport of slurry and formation of sludge. In these situations, the tractive force has an important influence on design velocities: The velocity cannot be too small or the tractive force will be insufficient to carry suspended sediment and deposition will occur. Similarly, if design velocities are too large, the greater tractive force will increase rates of erosion and corrosion in the channel or pipeline, thus raising maintenance and operational costs. Thus, the general significance of tractive force relates to designing self-cleansing channel and pressure-flow systems and to stable channel design in erodible channels. Moreover, high tractive forces are capable of causing water-quality problems in distribution system piping through the mechanism of biofilm sloughing or suspension of corrosion by-products.

2.6.6 Conveyance System Calculations: Steady Uniform Flow

A key practical concern in the detailed calculation of pressure flow and the estimation of pressure losses. Because the practice of engineering requires competent execution in a huge number of contexts, the engineer will encounter many different applications in practice: compare, for example Fig. 2.4 to 2.8 In fact, the number of applied topics is so large

TABLE 2.3 Local Loss Coefficients at Transitions

Description	Sketch	Additional Data	K		Source*
Pipe entrance $h_L = K_e V^2/2g$		r/d 0.0 0.1 > 0.2	K_e 0.50 0.12 0.03		(1)
Contraction $h_L = K_c V_2^2/2g$		D_2/D_1 0.0 0.20 0.40 0.60 0.80 0.90	K_c $\theta = 60°$ 0.08 0.08 0.07 0.06 0.05 0.04	K_c $\theta = 180°$ 0.50 0.49 0.42 0.32 0.18 0.10	(1)
Expansion $h_L = K_E V_1^2/2g$		D_1/D_2 0.0 0.20 0.40 0.60 0.80	K_E $\theta = 10°$ 0.13 0.11 0.06 0.03	K_E $\theta = 180°$ 1.00 0.92 0.72 0.42 0.16	(1)
90° miter bend	Vanes	Without vanes	$K_b = 1.1$		(26)
		With vanes	$K_b = 0.2$		(26)
90° miter smooth		r/d 1 2 4 6 8 10	$K_b = 0.35$ 0.19 0.16 0.21 0.28 0.32		(3) and (13)
Threaded pipe fittings	Globe valve–wide open Angle valve–wide open Gate valve–wide open Gatevalve–half open Return bend Tee 90° elbow 45° elbow		$K_v = 10.0$ $K_v = 5.0$ $K_v = 0.2$ $K_v = 5.6$ $K_b = 2.2$ $K_t = 1.8$ $K_b = 0.9$ $K_b = 0.4$		(26)

Source: Roberson and Crowe (1993).
*Given in Roberson and Crowe (1993).

2.24 Chapter Two

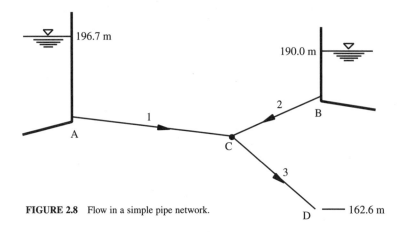

FIGURE 2.8 Flow in a simple pipe network.

that comprehensive treatment is impossible. Therefore, this chapter emphasizes a systematic presentation of the principles and procedures of problem-solving to encourage the engineer's ability to generalize. To illustrate the principles of hydraulic analysis, this section includes an example that demonstrates both the application of the energy equation and the use of the most common headloss equations. A secondary objective is to justify two common assumptions about pipeline flow: namely, that flow is, to a good approximation, incompressible and isothermal.

Problem. A straight pipe 2500 m long is 27 inches in diameter and discharges water at 10°C into the atmosphere at the rate of 1.80 m³/s. The lower end of the pipe is at an elevation of 100 m, where a pressure gauge reads 3.0 MPa. The pipe is on a 4 percent slope.

1. Determine the pressure head, elevation head, total head, and piezometric level at both ends of this pipeline.
2. Determine the associated Darcy-Weisbach friction factor f and Hazen-Williams C for this pipeline and flow.
3. Use the known pressure change to estimate the change in density between the upstream and downstream ends of the conduit. Also estimate the associated change in velocity between the two ends of the pipe, assuming a constant internal diameter of 27 in throughout. What do you conclude from this calculation?
4. Estimate the change in temperature associated with this headloss and flow, assuming that all the friction losses in the pipe are converted to an increase in the temperature of the water. What do you conclude from this calculation?

Solution. The initial assumption in this problem is that both the density of the water and its temperature are constant. We confirm at the end of the problem that these are excellent assumptions (a procedure similar to the predictor-corrector approaches often used for numerical methods). We begin with a few preliminary calculations that are common to several parts of the problem.

Geometry. If flow is visualized as moving from left to right, then the pipeline is at a 100 m elevation at its left end and terminates at an elevation of 100 + 0.04 (2500) = 200 m at its right edge, thus gaining 100 m of elevation head along its length. The hydraulic grade line–representing the distance above the pipe of the pressure head term P/γ–is high above the pipe at the left edge and falls linearly to meet the pipe at its right edge because the pressure here is atmospheric.

Properties. At 10°C, the density of water $\rho = 999.7$ kg/m³, its bulk modulus $K = \rho\Delta\rho/P/\Delta\rho = 2.26$ GPa, and its specific heat $C = 4187$ J/(kg · °C). The weight density is $\gamma = \rho g = 9.81$ kN/m³.

Based on an internal diameter of 27 in, or 0.686 m, the cross-sectional area of the pipe is

$$A = \frac{\pi}{4}D^2 = \frac{\pi}{4}(0.686)^2 = 0.370 \text{ m}^2$$

Based on a discharge $Q = 1.80$ m³/s, the average velocity is

$$V = \frac{Q}{A} = \frac{1.80 \text{ m}^3/\text{s}}{0.370 \text{ m}^2} = 4.87 \text{ m/s}$$

Such a velocity value is higher than is typically allowed in most municipal work.

1. The velocity head is given by

$$h_v = \frac{V^2}{2g} = 1.21 \text{ m}$$

Thus, the following table can be completed:

Variable	Expression	Upstream	Downstream
Pressure (MPa)	P	3.0	0.0
Pressure head (m)	P/γ	305.9	0.0
Elevation head (m)	z	100.0	200.0
Piezometric head (m)	$P/\gamma + z$	405.9	200.0
Total head (m)	$P/\gamma + z + V^2/2g$	407.1	201.2

2. The headloss caused by friction is equal to the net decrease in total head over the length of the line. That is, $h_f = 407.1 - 201.2 = 205.9$ m. Note that because this pipe is of uniform diameter, this value also could have been obtained from the piezometric head terms.

From the Darcy-Weisbach equation, we can obtain the following expression for the dimensionless f:

$$f = \frac{h_f D}{L \frac{V^2}{2g}} = \frac{(205.9)(0.686)}{(2500)(1.21)} = 0.047$$

Alternatively, from the Hazen-Williams equation that $Q = 0.278 \, C \, D^{2.63} \, (h_f/L)^{0.54}$, we obtain the following for the dimensional C:

$$C = \frac{Q}{0.278 \, D^{2.63}(h_f/L)^{0.54}} = \frac{1.8}{0.278(0.686)^{2.63}(205.9/2500)^{0.54}} = 67.2$$

These values would indicate a pipe in poor condition, probably in much need of repair or replacement.

3. In most problems involving steady flow, we assume that the compressibility of the water is negligible. This assumption is easily verified since the density change associated with the pressure change is easily computed.

In the current problem, the pressure change is 3.0 MPa and the bulk modulus is 2200 MPa. Thus, by definition of the bulk modulus K,

$$\frac{\Delta \rho}{\rho} = \frac{\Delta P}{K} = \frac{3}{2200} = 0.0014$$

Clearly, even in this problem, with its unusually extreme pressure changes, the relative change in density is less than 0.7 percent. The density at the higher pressure (upstream) end of the pipe is $\rho_1 = \rho_2 + \Delta \rho = 999.7 (1 + 0.0014) = 1001.1$ kg/m³.

Using the mass continuity equation, we have

$$\rho(AV)_1 = (\rho AV)_2$$

In this case, we assume that the pipe is completely rigid and that the change in pressure results in a change in density only (in most applications, these terms are likely to be almost equally important). In addition, we assume that the velocity we've already calculated applies at the downstream end (i.e., at location 2). Thus, the continuity equation requires

$$V_1 = V_2 \frac{\rho_2}{\rho_1} = 4.87 \frac{999.7}{1001.1} = 4.86 \text{ m/s}$$

Obviously, even in this case, the velocity and density changes are both negligible and the assumption of incompressible flow is an extremely good one.

4. Assuming that the flow is incompressible, the energy dissipated, P_d, can be computed using work done in moving the fluid through a change in piezometric flow (in fact, the headloss is nothing more than the energy dissipating per unit rate of weight of fluid transferred). Thus,

$$P_d = \gamma Q h_f$$

Strictly speaking, this energy is not lost but is transferred to less available forms: typically, heat. Since energy is associated with the increase in temperature of the fluid, we can easily estimate the increase in temperature of the fluid that would be associated with the dissipation of energy, assuming that all the heat is retained in the fluid. That is, $P_f = \rho Q \, c \Delta T = \rho g Q h_f$. Solving for the temperature increase gives

$$\Delta T = \frac{g \, h_f}{c} = \frac{(9.91 \text{m/s}^2)(205.9 \text{ m})}{4187 \text{ J/(kg} \cdot °\text{C)}} = 0.48°\text{C}.$$

We conclude that the assumption of isothermal flow also is an excellent one.

2.6.7 Pumps: Adding Energy to the Flow

Although water is the most abundant substance found on the surface of the earth, its natural distribution seldom satisfies an engineer's partisan requirements. As a result, pumping both water and wastewater is often necessary to achieve the desired distribution of flow. In essence, a pump controls the flow by working on the flowing fluid, primarily by discharging water to a higher head at its discharge flange than is found at the pump inlet. The increased head is subsequently dissipated as frictional losses within the conduit or is delivered further downstream. This section provides a brief introduction to how pumps interact with pipe systems. Further details are found in Chap. 5.

How exactly is the role of a pump quantified? The key definition is the *total dynamic head* (TDH) of the pump. This term describes the difference between the total energy on the discharge side compared with that on the suction side. In effect, the TDH H_p is the difference between the absolute total head at the discharge and suction nozzle of the pump: that is,

$$H_p = \left(h_p + \frac{V^2}{2g}\right)_d - \left(h_p + \frac{V^2}{2g}\right)_s \tag{2.26}$$

where h_p = hydraulic grade line elevation (i.e., pressure-plus-elevation head with respect to a fixed datum), and subscripts d and s refer to delivery and suction flanges, respectively. Typically, the concern is how the TDH head varies with the discharge Q; for a pump, this $H-Q$ relation is called the *characteristic curve*.

What the TDH definition accomplishes can be appreciated better if we consider a typical pump system, such as the one shown in Fig. 2.9. In this relation, the Bernoulli equation relates what happens between points 1 and 2 and between points 3 and 4, but technically it cannot be applied between 2 and 3 because energy is added to the flow. However, the TDH definition spans this gap.

To see this more clearly, the energy relation is written between points 1 and 2 as

$$H_s = H_{PS} + h_{fs} \tag{2.27}$$

where H_s is the head of the suction reservoir, H_{PS} is the total head at the suction flange of the pump, and h_{fs} is the friction loss in the suction line. Similarly, the energy relation is written between points 3 and 4 as

$$H_{PD} = H_D + h_{fd} \tag{2.28}$$

where H_D is the head of the discharge reservoir, H_{PD} is the head at the discharge flange of the pump, and h_{fd} is the friction loss in the discharge line. If Eq. (2.27) is then added to Eq. (2.28), the result can be rearranged as

$$H_{PD} - H_{PS} = H_D - H_S + h_{fd} + h_{fs} \tag{2.29}$$

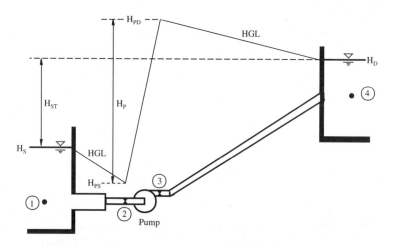

FIGURE 2.9 Definition sketch for pump system relations.

which can be rewritten using Eq. (2.26) as

$$H_p = H_{ST} + h_f \qquad (2.30)$$

where H_{ST} is the total static lift and h_f is the total friction loss. The total work done by the pump is equal to the energy required to lift the water from the lower reservoir to the higher reservoir plus the energy required to overcome friction losses in both the suction and discharge pipes.

2.6.8 Sample Application Including Pumps

Problem. Two identical pumps are connected in parallel and are used to force water into the transmission/distribution pipeline system shown in Fig. 2.10. The elevations of the demand locations and the lengths of $C = 120$ pipe also are indicated. Local losses are negligible in this system and can be ignored. The demands are as follows: $D_1 = 1.2$ m³/s, $D_2 = 1.6$ m³/s, and $D_3 = 2.2$ m³/s. The head-discharge curve for a single pump is approximated by the equation

$$H = 90 - 6Q^{1.70}$$

1. What is the minimum diameter of commercially available pipe required for the 4.2 km length if a pressure head is to be maintained at a minimum of 40 m everywhere in the system? What is the total dynamic head of the pump and the total water horsepower supplied for this flow situation?

2. For the system designed in the previous question, the demand can shift as follows under certain emergency situations: $D_1 = 0.8$ m³/s, $D_2 = 1.2$ m³/s, and $D_3 = 4.2$ m³/s. For this new demand distribution, can the system maintain a residual pressure head of 20 m in the system?

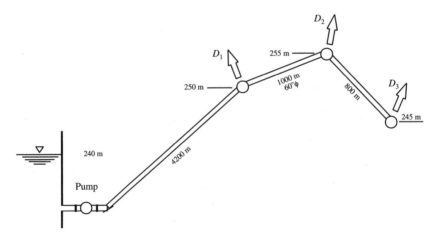

FIGURE 2.10 Example pipe and pump system.

Solution. Total flow is $Q_t = D_1 + D_2 + D_3 = 1.2 + 1.6 + 2.2 = 5.0$ m³/s, and each pump will carry half of this flow: i.e., $Q_{pump} = Q_t/2 = 2.5$ m³/s. The total dynamic head of the pump H_{pump} is

$$H_{pump} = 90 - 6(2.5)^{1.7} = 61.51 \text{ m}$$

which allows the total water power to be computed as

$$\text{Power} = 2\,(Q_{pump} H_{pump} \gamma) = Q_t H_{pump} \gamma$$

Thus, numerically,

$$\text{Power} = \left(5.0 \frac{m^3}{s}\right)(61.51 \text{m})\left(9810 \frac{N}{m^3}\right) = 3017 \text{ kW}$$

which is a huge value. The diameter d_1 of the pipe that is 4.2 km long, the headloss Δh_i caused by friction for each pipe can be determined using the Hazen-Williams formula since the flow can be assumed to be in the hydraulic rough range. Because d_1 is unknown, Δh_2, Δh_3, and Δh_1 are calculated first. The site where the lowest pressure head occurs can be shown to be at node 2 (i.e., the highest node in the system) as follows:

$$\Delta h_3 = L_3 \left(\frac{Q_3}{0.278 C d^{2.63}}\right)^{\frac{1}{0.54}} = 800 \left(\frac{2.2}{0.278(120)(1.067^{2.63})}\right)^{\frac{1}{0.54}} = 3.80 \text{ m}$$

Because the headloss Δh_3 is less than the gain in elevation of 10 m, downstream pressures increase; thus, node 2 (at D_2) will be critical in the sense of having the lowest pressure. Thus, if the pressure head at that node is greater than 40 m, a minimum pressure head of 40 m will certainly be maintained throughout the pipeline.

Continuing with the calculations,

$$\Delta h_2 = L_2 \left(\frac{Q_2}{0.278 C d^{2.63}}\right)^{\frac{1}{0.54}} = 1000 \left(\frac{3.8}{0.278(120)(1.524^{2.63})}\right)^{\frac{1}{0.54}} = 2.30 \text{ m}$$

Now, the pressure head at node 2 is

$$h_{p2} = z_R + H_{pump} - \Delta h_1 - \Delta h_2 - z_2 = 40 \text{ m}$$

which implies that

$$\Delta h_1 = (z_R - z_2) + H_{pump} - \Delta h_2 - 40 = (240 - 255) + 61.51 - 2.30 - 40 = 4.21 \text{ m}$$

where z is the elevation and the subscripts R and 2 denote reservoir and node 2, respectively. Thus, the minimum diameter d_1 is

$$d_1 = \left(\frac{Q}{0.278 C S^{0.54}}\right)^{\frac{1}{2.63}} = \left(\frac{5.0}{0.278(120)\left(\frac{4.21}{4200}\right)^{0.54}}\right)^{\frac{1}{2.63}} = 2.006 \text{ m}$$

Finally, the minimum diameter ($d_1 = 2.134$ m) of the commercially available pipe is therefore 84 in.

Under emergency conditions (e.g., with a fire flow), the total flow is $Q_t = D_1 + D_2 + D_3 = 0.8 + 1.2 + 4.2 = 6.2$ (m³/s). Note that with an increase in flow, the head lost resulting from friction increases while the head supplied by the pump decreases. Both these facts tend to make it difficult to meet pressure requirements while supplying large flows. More specifically,

$$H_{pump} = 90 - 6(3.1)^{1.7} = 48.90 \text{ m}$$

and

$$\Delta h_3 = 800 \left(\frac{4.2}{0.278(120)(1.067)^{2.63}} \right)^{\frac{1}{0.54}} = 12.6 \text{ m}$$

Because this loss now exceeds the elevation change, node 3 (at D_3) now becomes critical in the system; minimum pressure now occurs at the downstream end of the system. Other losses are

$$\Delta h_2 = 1000 \left(\frac{5.4}{0.278(120)(1.524)^{2.63}} \right)^{\frac{1}{0.54}} = 4.4 \text{ m}$$

and

$$\Delta h_1 = 4200 \left(\frac{6.2}{0.278(120)(2.134)^{2.63}} \right)^{\frac{1}{0.54}} = 4.6 \text{ m}$$

Thus, the pressure head at node 3 is

$$h_{p3} = (z_R - z_3) + H_{pump} - \Delta h_1 - \Delta h_2 - \Delta h_3 = -5 + 48.9 - 12.6 - 4.4 - 4.6 = 22.3 \text{ m}$$

Clearly, a residual pressure head of 20 m is still available in the system under emergency situations, and the pressure requirement is still met, though with little to spare!

2.6.9 Networks–Linking Demand and Supply

In water supply and distribution applications, the pipes, pumps, and valves that make up the system are frequently connected into complex arrangements or networks. This topological complexity provides many advantages to the designer (e.g., flexibility, reliability, water quality), but it presents the analyst with a number of challenges. The essential problems associated with "linked" calculations in networks are discussed in Chap. 4.

2.7 QUASI-STEADY FLOW: SYSTEM OPERATION

The hydraulics of pressurized flow is modified and adjusted according to the presence, location, size, and operation of storage reservoirs and pumping stations in the system. This section discusses the criteria for and the approach to these components, introducing the equations and methods that will be developed in later chapters.

A common application of quasi-steady flow arises in reservoir engineering. In this case, the key step is to relate the rate of outflow O to the amount of water in the reservoir (i.e., its total volume or its depth). Although the inflow is usually a known function of time, Eq. (2.2) must be treated as a general first-order differential equation. However, the solution usually can be approximated efficiently by standard numerical techniques, such as the Runge-Kutta or Adams-type methods. This application is especially important when setting operating policy for spillways, dams, turbines, and reservoirs. One simple case is illustrated by the example below.

Hydraulics of Pressurized Flow

Usually, reservoir routing problems are solved numerically, a fact necessitated by the arbitrary form of the input function to the storage system and the sometimes complex nature of the storage-outflow relation. However, there are occasions when the application is sufficiently simple to allow analytical solutions.

Problem. A large water-filled reservoir has a constant free surface elevation of 100 m relative to a common datum. This reservoir is connected by a pipe ($L = 50$ m, $D = 6$ cm, and f = constant = 0.02, $h_f = fLV^2/2gD$) to the bottom of a nearby vertical cylindrical tank that is 3 m in diameter. Both the reservoir and the tank are open to the atmosphere, and gravity-driven flow between them is established by opening a valve in the connecting pipeline.

Neglecting all minor losses, determine the time T (in hours) required to raise the elevation of the water in the cylindrical tank from 75 m to 80 m.

Solution. If we neglect minor losses and the velocity head term, the energy equation can be written between the supply reservoir and the finite area tank. Letting the level of the upstream reservoir be h_r, the variable level of the downstream reservoir above datum being h and the friction losses being h_f, the energy equation takes on the following simple form:

$$h_r = h + h_f$$

This energy relation is called quasi-steady because it does not directly account for any transient terms (i.e., terms that explicitly depend on time).

A more useful expression is obtained if we use the Darcy-Weisbach equation to relate energy losses to the discharge $Q = VA$:

$$h_f = \frac{fL}{D}\frac{V^2}{2g} = \frac{fL}{D}\frac{Q^2}{2gA^2} = \frac{8fL}{g\pi^2 D^5}Q^2$$

What is significant about this expression, however, is that all the terms involved in the last fraction are known and can be treated as a single constant. Thus, we can solve for Q and rewrite it as $Q = C\sqrt{h_r - h}$, where $C^2 = g\pi^2 D^5/8fL$

Thus far, we have a single equation involving *two* unknowns: the head h in the receiving tank and the discharge Q between them. A second relation is required and is given by the continuity equation. Because the flow can be treated as incompressible, the discharge in the tank (i.e., the tank's area A_t times its velocity of dh/dt) must equal the discharge in the pipe Q. in symbols,

$$A_t \frac{dh}{dt} = Q$$

Thus, using the energy equation, we have,

$$\frac{dh}{dt} = \frac{C}{A_t}\sqrt{h_r - h}$$

Separating variables and integrating gives

$$\int_{h_1}^{h_2} \frac{dh}{\sqrt{h_r - h}} = \int_0^t \frac{C}{A_t} dt$$

and performing the integration and using appropriate limits gives

$$2\left(\sqrt{h_r - h_1} - \sqrt{h_r - h_2}\right) = \frac{C}{A_t}t$$

Finally, solving for t gives the final required expression for quasi-steady flow connecting a finite-area tank to a constant head reservoir:

$$t = \frac{2A_t}{C}\left(\sqrt{h_r - h_1} - \sqrt{h_r - h_2}\right)$$

The numerical aspects are now straightforward:

$$C = \left(\frac{g\pi^2\,(0.06)^5}{8 \cdot 1}\right)^{0.5} \text{ or } C = \sqrt{\frac{g\pi^2(0.06)^5}{8 \cdot 1}} \text{ m}^{5/2}/\text{s} = 3.068(10)^{-3}\text{ m}^{5/2}/\text{s}$$

If $h_r = 100$ m, $h_1 = 75$ m, $h_2 = 80$ m, than we have

$$t = \left(\frac{2\frac{\pi}{4}(3\text{ m})^2}{3.068(10)^{-3}\text{m}^{5/2}/\text{s}}\right)(\sqrt{25\text{ m}} - \sqrt{20\text{ m}}) = 2432.6\text{ s}$$

Converting to minutes, this gives a time of about 40.5 min. (0.676 hr).

In problems involving a slow change of the controlling variables, it is often simple to check the calculations. In the current case, a good approximation can be obtained by using the average driving head of 22.5 m (associated with an average tank depth of 77.5 m). This average head, in turn, determines the associated average velocity in the pipeline. Using this "equivalent" steady velocity allows one to estimate how much time is required to fill the tank by the required 5 m. The interested reader is urged to try this and to verify that this approximate time is actually relatively accurate in the current problem, being within 6 s of the "exact" calculation.

2.8 UNSTEADY FLOW: INTRODUCTION OF FLUID TRANSIENTS

Hydraulic conditions in water distribution systems are in an almost continual state of change. Industrial and domestic users often alter their flow requirements while supply conditions undergo adjustment as water levels in reservoirs and storage tanks change or as pumps are turned off and on. Given this dynamic condition, it is perhaps surprising that steady state considerations have so dominated water and wastewater engineering. The following sections provide an introduction to unsteady flow in pipe systems—a topic that is neglected too often in pipeline work. The purpose is not too create a fluid transients expert but to set the stage for Chap. 6, which considers these matters in greater detail.

2.8.1 Importance of Waterhammer

Pressure pipe systems are subjected to a wide range of physical loads and operational requirements. For example, underground piping systems must withstand mechanical forces caused by fluid pressure, differential settlement, and concentrated loads. The pipe must tolerate a certain amount of abuse during construction, such as welding stresses and shock loads. In addition, the pipe must resist corrosion and various kinds of chemical attack. The internal pressure requirement is of special importance, not only because it directly influences the required wall thickness (and hence cost) of large pipes, but also because pipe manufacturers often characterize the mechanical strength of a pipeline by its pressure rating.

The total force acting within a conduit is obtained by summing the steady state and waterhammer (transient) pressures in the line. Transient pressures are most important when the rate of flow is changed rapidly, such as by closing a valve or stopping a pump. Such disturbances, whether caused intentionally or by accident, create traveling pressure and velocity waves that may be of large magnitude. These transient pressures are superimposed on steady-state values to produce the total pressure load on a pipe.

Most people have some experience with waterhammer effects. A common example is the banging or hammering noise sometimes heard when a water faucet is closed quickly. In fact, the mechanism in this simple example typifies all pipeline transients. The kinetic energy carried by the fluid is rapidly converted into strain energy in the pipe walls and fluid. The result is a pulse wave of abnormal pressure that travels along the pipe. The hammering sound indicates that a portion of the original kinetic energy is converted into acoustic form. This and other energy-transformation losses (such as fluid friction) cause the pressure wave to decay gradually until normal (steady) pressures and velocities are once again restored.

It turns out that waterhammer phenomena are the direct means of achieving *all* changes in fluid velocity, gradual or sudden. The difference is that slow adjustments in velocity or pressures produce such small disturbances that the flow appears to change smoothly from one value to another. Yet, even in these cases of near equilibrium, it is traveling pressure waves that satisfy the conservation equations. To illustrate why this must be so, consider the steady continuity equation for the entire pipe. This law requires that the rate at which fluid leaves one end of a conduit must be equal to the rate at which it enters the other end. The coordination between what happens at the two ends of the pipeline is not achieved by chance or conspiracy. It is brought about by the same physical laws and material properties that cause disturbances to propagate in the transient case.

If waterhammer waves were always small, the study of transient conditions would be of little interest to the pipeline engineer. This is not the case. Waterhammer waves are capable of breaking pipes and damaging equipment and have caused some spectacular pipeline failures. Rational design, especially of large pipelines, requires reliable transient analysis. There are several reasons why transient conditions are of special concern for large conduits. Not only is the cost of large pipes greater, but the required wall thickness is more sensitive to the pipe's pressure rating as well. Thus, poor design–whether it results in pipeline failure or the hidden costs of overdesign–can be extremely expensive for large pipes.

Despite their intrinsic importance, transient considerations are frequently relegated to a secondary role when pipeline systems are designed or constructed. That is, only after the pipeline's profile, diameter, and design discharge have been chosen is any thought given to transient conditions. This practice is troublesome. First, the pipeline may not perform as expected, possibly causing large remedial expenses. Second, the line may be overdesigned and thus unnecessarily expensive. This tendency to design for steady-state conditions alone has been particularly common in the water supply industry. In addition, there has been a widely held misconception that complex arrangements of pipelines reflect or dampen waterhammer waves. Although wave reflections in pipe networks do occur, attenuation depends on many factors and cannot be guaranteed. Networks are not intrinsically better behaved than simple pipelines are, and some complex systems may respond even more severely to transient conditions (Karney and McInnis, 1990).

The remainder of this chapter introduces, in a gentle and nonmathematical way, several important concepts relating to transient conditions. Although rigorous derivations and details are avoided, the discussion is physical and accurate. The goal is to answer two key questions: How do transients arise and propagate in a pipeline? and under what circumstances are transient conditions most severe?

Transient conditions in pressure pipelines are modeled using either a "lumped" or "distributed" approach. In distributed systems, the fluid is assumed to be compressible, and the transient phenomena occur in the form of traveling waves propagating with a finite speed a. Such transients often occur in water supply pipes, power plant conduits, and industrial process lines. In a lumped system, by contrast, the flow is considered to be incompressible and the pipe walls are considered to be inelastic. Thus, the fluid behaves as a rigid body in that changes in pressure or velocity at one point in the flow system are assumed to change the flow elsewhere instantaneously. The lumped system approximation can be obtained either directly or in the limit as the wavespeed a becomes unbounded in the distributed model. The slow oscillating water level in a surge tank attached to a short conduit typifies a system in which the effects of compressibility are negligible.

Although the problem of predicting transient conditions in a pipeline system is of considerable practical importance, many challenges face the would-be analyst. The governing partial differential equations describing the flow are nonlinear, the behavior of even commonly found hydraulic devices is complex, and data on the performance of systems are invariably difficult or expensive to obtain. The often-surprising character of pulse wave propagation in a pipeline only makes matters worse. Even the basic question of deciding whether conditions warrant transient analysis is often difficult to answer. For all these reasons, it is essential to have a clear physical grasp of transient behavior.

2.8.2 Cause of Transients

In general, any change in mean flow conditions in a pipeline system initiates a sequence of transient waves. In practice, we are generally concerned with changes or actions that affect hydraulic devices or boundary conditions attached to the conduit. The majority of these devices are used to provide power to the system or to control the flow in some way. The following list illustrates how some transient conditions can originate, although not all of the them are discussed further here:

1. Changes in valve settings (accidental or planned; manual or automatic)
2. Starting or stopping of either supply or booster pumps
3. Changes in the demand conditions, including starting or arresting a fire flow
4. Changes in reservoir level (e.g., waves on a water surface or the slow accumulation of depth with time)
5. Unstable device characteristics, including unstable pump characteristics, valve instabilities, the hunting of a turbine, and so on
6. Changes in transmission conditions, such as when a pipe breaks or buckles
7. Changes in thermal conditions (e.g., if the fluid freezes or if changes in properties are caused by temperature fluctuations)
8. Air release, accumulation, entrainment, or expulsion causing dramatic disturbances (e.g., a sudden release of air from a relief valve at a high point in the profile triggered by a passing vehicle); pressure changes in air chambers; rapid expulsion of air during filling operations
9. Transitions from open channel to pressure flow, such as during filling operations in pressure conduits or during storm events in sewers.
10. Additional transient events initiated by changes in turbine power loads in hydroelectric projects, draft-tube instabilities caused by vortexing, the action of reciprocating pumps, and the vibration of impellers or guide vanes in pumps, fans, or turbines

2.8.3 Physical Nature of Transient Flow

In pipeline work, many approximations and simplifications are required to understand the response of a pipe system following an initialization of a transient event. In essence, this is because the flow is both unsteady in the mean as well as turbulent. Many of these assumptions have been confirmed experimentally to the extent that the resulting models have provided adequate approximations of real flow behavior. Yet, it is wise to be skeptical about any assumption and be cautious about mathematical models. As we have stressed, any model only approximates reality to a greater or lesser extent. Still, even in cases where models perform poorly, they may be the best way of pinpointing sources of uncertainty and quantifying what is not understood.

An air of mystery often surrounds the development, role, and significance of transient phenomena in closed conduits. Indeed, the complexity of the governing differential equations and the dynamic nature of a system's response can be intimidating to the novice. However, a considerable understanding of transient behavior can be obtained with only the barest knowledge about the properties of fluid and a few simple laws of conservation.

When water flows or is contained in a closed conduit so that no free surface is present—for example, in a typical water supply line–the properties of the flowing fluid have some direct implications to the role and significance of transient conditions. For a water pipeline, two properties are especially significant: water's high density and its large bulk modulus (i.e., water is heavy and difficult to compress). Surprisingly, these two facts largely explain why transient conditions in a pipeline can be so dramatic (see also, Karney and McInnis, 1990).

2.8.3.1 Implication 1. Water has a high density. Because water has a high density ($\approx$ 1000 kg/m^3) and because pipelines tend to be long, typical lines carry huge amounts of mass, momentum, and kinetic energy. To illustrate, assume that a pipeline with area $A = 1.0$ m^2 and length $L = 1000$ m is carrying fluid with a velocity $V = 2.0$ m/s. The kinetic energy contained in this pipe is then

$$KE = \tfrac{1}{2} mV^2 = \tfrac{1}{2} \rho LAV^2 \approx 2{,}000{,}000 \ J$$

Now this is a relatively ordinary situation: The discharge is moderate and the pipe is not long. Yet the pipe still contains energy equivalent to, say, 10,000 fast balls or to a pickup truck falling from a 30-story office tower. Clearly, large work interactions are required to change the flow velocity in a pipeline from one value to another.

In addition to kinetic energy, a pipeline for liquid typically transports large amounts of mass and momentum as well. For example, the above pipeline contains $2(10^6)$ kg m/s of momentum. Such large values of momentum imply that correspondingly large forces are required to change flow conditions (Further details can be found in Karney, 1990).

2.8.3.2 Implication 2. Water is only slightly compressible. Because water is only slightly compressible, large head changes occur if even small amounts of fluid are forced into a pipeline. To explain the influence of compressibility in a simple way, consider Fig. 2.11, which depicts a piston at one end of a uniform pipe. If this piston is moved slowly, the volume containing the water will be altered and the confining pressure will change gradually as a result. Just how much the pressure will change depends on how the pipe itself responds to the increasing pressure. For example, the bulk modulus of water is defined as

$$K = \frac{\Delta P}{\Delta \rho / \rho} \approx 2{,}070 \ \text{MPa} \tag{2.31}$$

2.36 Chapter Two

Thus, if the density of the fluid is increased by as little as one-tenth of 1 percent, which is equivalent to moving the imagined piston a meter in a rigid pipe, the pressure will increase by about 200 m of head (i.e., 2 MPa). If the pipe is not rigid, pressure increases are shared between the pipe walls and the fluid, producing a smaller head change for a given motion of the piston. Typical values are shown in the plot in Fig. 2.11. For example, curve 2 indicates typical values for a steel pipe in which the elasticity of the pipe wall and the compressibility of the fluid are nearly equal; in this case, the head change for a given mass imbalance (piston motion) is about half its previous value.

Note that it is important for the conduit to be full of fluid. For this reason, many options for accommodating changes in flow conditions are not available in pipelines that can be used in channels. Specifically, no work can be done to raise the fluid mass against gravity. Also note that any movement of the piston, no matter how slowly it is accomplished, must be accommodated by changes in the density of the fluid, the dimension of the conduit, or both. For a confined fluid, Cauchy and Mach numbers (relating speed of change to speed of disturbance propagation) are poor indexes of the importance of compressibility effects.

2.8.3.3 Implication 3. Local action and control of valves. Suppose a valve or another device is placed at the downstream end of a series-connected pipe system carrying fluid at some steady-state velocity V_0. If the setting of the valve is changed–suddenly say, for simplicity, the valve is instantly closed–the implications discussed above are combined in the pipeline to produce the transient response. We can reason as follows:

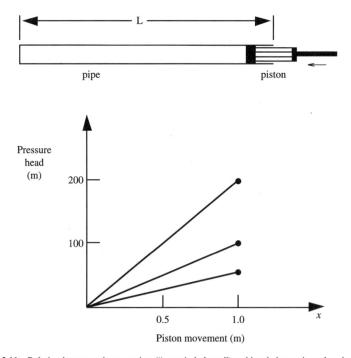

FIGURE 2.11 Relation between piston motion ("mass imbalance") and head change in a closed conduit.

Hydraulics of Pressurized Flow **2.37**

The downstream valve can only act locally, providing a relationship between flow through the valve and the headloss across the valve. In the case of sudden closure, the discharge and velocity at the valve becomes zero the instant the valve is shut. However, for the fluid mass as a whole to be stopped, a decelerating force sufficient to eliminate the substantial momentum of the fluid must be applied. But how is such a force generated? We have already mentioned that gravity cannot help us because the fluid has no place to go. In fact, there is only one way to provide the required decelerating force—the fluid must be compressed sufficiently to generate an increase in pressure large enough to arrest the flow. Because water is heavy, the required force is large; however, since water is only slightly compressible, the wave or disturbance will travel quickly. In a system like the one shown in Fig. 2.11, a pressure wave of nearly 100 m would propagate up the pipeline at approximately 1000 m/s.

In many ways, the response of the system we have described is typical. For closed conduit systems, the only available mechanism for controlling fluid flows is the propagations of shock waves resulting from the elasticity of the fluid and the pipeline. In essence, transient considerations cause us to look at the flow of fluid in a pipeline in a new way: For any flow, we consider not only its present significance but also how this condition was achieved and when it will change because, when change occurs, pressure pulses of high magnitude may be created that can burst or damage pipelines. Although this qualitative development is useful, more complicated systems and devices require sophisticated quantitative analysis. The next section briefly summarizes how more general relations can be obtained. (Greater detail is provided in Chap. 6.)

2.8.4 Equation of State-Wavespeed Relations

In pipeline work, an equation of state is obtained by relating fluid pressure to density through compressibility relations. Specifically, the stresses in the wall of the pipe need to be related to the pressure and density of the fluid. The result is a relationship between the fluid and the properties of the pipe material and the speed at which shock waves are propagated (wavespeed or celerity).

The most basic relation describing the wavespeed in an infinitely elasticly fluid is usually written as follows:

$$a = \gamma \sqrt{\frac{K}{\rho}} \quad (2.32)$$

where a is the wavespeed, γ is the ratio of the specific heats for the fluid, K is the bulk modulus of the fluid, and ρ is the fluid density. If a fluid is contained in a rigid conduit, all changes in density will occur in the fluid and this relation still applies. The following comments relate to Eq. (2.32):

1. As fluid becomes more rigid, K increases and, hence, a increases. If the medium is assumed to be incompressible, the wavespeed becomes infinite and disturbances are transmitted instantaneously from one location to another. This is not, strictly speaking, possible, but at times it is a useful approximation when the speed of propagation is much greater than the speed at which boundary conditions respond.

2. For liquids that undergo little expansion on heating, γ is nearly 1. For example, water at 10°C has a specific heat ratio (γ) of 1.001.

3. Certain changes in fluid conditions can have a drastic effect on celerity (or wavespeed) values. For example, small quantities of air in water (e.g., 1 part in 10,000 by volume) greatly reduce K because gases are so much more compressible than liquids are at normal temperatures. However, density values (ρ) are affected only slightly by the presence of a small quantity of gas. Thus, wavespeed values for gas-liquid mixtures are often much lower than the wavespeed of either component taken alone.

Example: Elastic Pipe. The sonic velocity (a) of a wave traveling through an elastic pipe represents a convenient method of describing a number of physical properties relating to the fluid, the pipe material, and the method of pipe anchoring. A more general expression for the wavespeed is

$$a = \sqrt{\frac{K/\rho_w}{1 + c_1 KD/Ee}} \qquad (2.33)$$

where K is the bulk modulus of the fluid, ρ_w is the density of the fluid, E is the elastic modulus of the pipe material, and D and e are the pipe's diameter and wall thickness, respectively. The constant c_1 accounts for the type of support provided for the pipeline. Typically, three cases are recognized, with c_1 defined for each as follows (μ is the Poison's ratio for the pipe material):

Case a. The pipeline is anchored only at the upstream end:

$$c_1 = 1 - \frac{\mu}{2} \qquad (2.34)$$

Case b. The pipeline is anchored against longitudinal movement.

$$c_1 = 1 - \mu^2 \qquad (2.35)$$

Case c. The pipeline has expansion joints throughout.

$$c_1 = 1 \qquad (2.36)$$

Note that for pipes that are extremely rigid, thick-walled, or both, $c_1 KD/Ee \to 0$ and Eq. (2.33) can be simplified to $a = \sqrt{K/\rho_w}$, which recovers the expression for the acoustic wavespeed in an infinite fluid (assuming that $\gamma = 1$).

For the majority of transient applications, the wavespeed can be regarded as constant. Even in cases where some uncertainty exists regarding the wavespeed, the solutions of the governing equations, with respect to peak pressures, are relatively insensitive to changes in this parameter. It is not unusual to vary the wave celerity deliberately by as much as $\pm$ 15 percent to maintain a constant time step for solution by standard numerical techniques (Wylie and Streeter, 1993). (Again, further details are found in Chap. 6.)

Wavespeeds are sensitive to a wide range of environmental and material conditions. For example, special linings or confinement conditions (e.g., tunnels); variations in material properties with time, temperature, or composition; and the magnitude and sign of the pressure wave can all influence the wavespeed in a pipeline. (For additional details, see Wylie and Streeter, 1993; Chaudhry, 1987; or Hodgson, 1983).

2.8.5 Increment of Head-Change Relation

Three physical relations–Newton's second law, conservation of mass, and the wavespeed relation–can be combined to produce the governing equations for transient flow in a pipeline. The general result is a set of differential equations for which no analytical solution exists. It is these relations that are solved numerically in a numerical waterhammer program.

In some applications, a simplified equation is sometimes used to obtain a first approximation of the transient response of a pipe system. This simple relation is derived with the assumption that headlosses caused by friction are negligible and that no interaction takes place between pressure waves and boundary conditions found at the end of pipe lengths. The resulting head rise equation is called the *Joukowsky relation*:

$$\Delta H = \pm \frac{a}{g} \Delta V \tag{2.37}$$

where ΔH is the head rise, ΔV is the change in velocity in the pipe, a is the wavespeed, and g is the acceleration caused by gravity. The negative sign in this equation is applicable for a disturbance propagating upstream and the positive sign is for one moving downstream. Because typical values of a/g are large, often 100 s or more, this relation predicts large values of head rise. For example, a head rise of 100 m occurs in a pipeline if $a/g = 100$ s and if an initial velocity of 1 m/s is suddenly arrested at the downstream end.

Unfortunately, the Joukowsky relation is misleading in a number of respects. If the equation is studied, it seems to imply that the following relations are true:

1. The greater the initial velocity (hence, the larger the maximum possible ΔV), the greater the transient pressure response.
2. The greater the wavespeed a, the more dramatic the head change.
3. Anything that might lower the static heads in the system (such as low reservoir levels or large headlosses caused by friction) will tend to lower the total head (static plus dynamic) a pipe system is subject to.

Although these implications are true when suitable restrictions on their application are enforced, all of them can be false or misleading in more complicated hydraulic systems. It is important to be skeptical about simple rules for identifying "worst case" scenarios in transient applications. Karney and McInnis (1990) provide further elaboration of this point. However, before considering even a part of this complexity, one must clarify the most basic ideas in simple systems.

2.8.6 Transient Conditions in Valves

Many special devices have been developed to control and manage flows in pipeline systems. This is not surprising because the inability to control the passage of water in a pipeline can have many consequences, ranging from the minor inconvenience of restrictive operating rules to the major economic loss associated with pipeline failure or rupture. The severity of the problem clearly depends on the timing and magnitude of the failure.

In essence, control valves function by introducing a specified and predictable relationship between discharge and pressure drop in a line. When the setting of a valve (or, for that matter, the speed of a pump) is altered, either automatically or by manual action, it is the head-discharge relationship that is controlled to give the desired flow characteristics. The result of the change may be to increase or reduce the pressure or discharge, maintain a preset pressure or flow, or respond to an emergency or unusual condition in the system.

It is a valve control function that creates most difficulties encountered by pipeline designers and system operators. Valves control the rate of flow by restricting the passage of the flow, thereby inducing the fluid to accelerate to a high velocity as it passes through the valve even under steady conditions. The large velocities combine with the no-slip condition at the solid boundaries to create steep velocity gradients and associated high shear stresses in the fluid; in turn, these shear stresses, promote the rapid conversion of mechanical energy into heat through the action of turbulence of the fluid in the valve. The net result is a large pressure drop across the valve for a given discharge through it; it is this Δh-Q relationship for a given opening that makes flow control possible. However, the same high velocities also are responsible for the cavitation, noise, energy loss, wear, and maintenance problems often associated with valves even under steady conditions.

This section presents an overview of control valve hydraulics and considers the basic roles that control valves play in a pipeline. Valves are often classified by both their function and their construction. Valves can be used for on/off control or for pressure or flow control, and the physical detail of a valve's construction varies significantly depending on its application. The kind of valves used can range from traditional gate and globe valves to highly sophisticated slow-closing air valves and surge-anticipating valves. The actuator that generates the valve's motion also varies from valve to valve, depending on whether automatic or manual flow control is desired. Many kinds of valves can be used in a single pipeline, creating challenging interactions for the transient analyst to sort out. The most basic of these interactions is discussed in more detail in the following section.

2.8.6.1 Gate discharge equation. Among the most important causes of transient conditions in many pipelines is the closure of regulating and flow control valves. The details of how these valves are modeled can be influential in determining the maximum pressure experienced on the lines. For this reason, and because some knowledge of valve behavior is required to interpret the output from a simulation program, it is worthwhile to briefly review valve theory.

Consider a simple experiment in which a reservoir, such as the one shown in Fig. 2.12, has a valve directly attached to it. If we initially assume the valve is fully open, the discharge through the valve Q_0 can be predicted with the usual orifice equation:

$$Q_0 = (C_d A_v)_0 \sqrt{2g\Delta H_0} \tag{2.38}$$

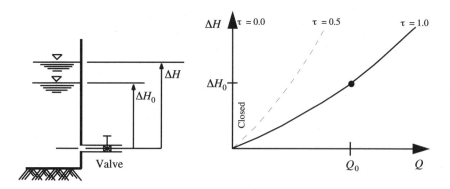

FIGURE 2.12 Relation between head and discharge in a valve.

where C_d is the discharge coefficient, A_v is the orifice area, g is the acceleration caused by gravity, ΔH_0 is the head difference across the valve, and the subscript 0 indicates that the valve is fully open. If the valve could completely convert the head difference across it into velocity, the discharge coefficient C_d would be equal to 1. Since full conversion is not possible, C_d values are inevitably less than 1, with values between 0.7 and 0.9 being common for a fully open valve. The product of the orifice area A_v and the discharge coefficient C_d is often called the "effective area" of the valve. The effective area, as determined by details of a valve's internal construction, controls the discharge through the valve.

Equation (2.38) is valid for a wide range of heads and discharges: For example, the solid curve in the plot above depicts this relation for a fully open valve. Yet, clearly the equation must be altered if the setting (position) of the valve is altered because both the discharge coefficient and the orifice valve area would change. Describing a complete set of a valve's characteristics would appear to require a large set of tabulated $C_d A_v$ values. Fortunately, a more efficient description is possible.

Suppose we take a valve at another position and model its discharge in a way that is analogous to the one shown in Eq. (2.38). That is:

$$Q = (C_d A_v) \sqrt{2g\Delta H} \qquad (2.39)$$

where both C_d and A_v will, in general, have changed from their previous values. If Eq. (2.39) is divided by Eq. (2.38), the result can be written as

$$Q = E_s \tau \sqrt{\Delta H} \qquad (2.40)$$

In Eq. (2.40), E_s is a new valve constant representing the ratio of the fully open discharge to the root of the fully open head difference:

$$E_s = \frac{Q_0}{\sqrt{\Delta H_0}}$$

In essence, E_s "scales" the headlosses across a fully open valve for its size, construction, and geometry. In addition, τ represents the nondimensional effective gate opening:

$$\tau = \frac{C_d A_v}{(C_d A_v)_0}$$

Using τ-values to represent gate openings is convenient because the effective range is from 0.0 (valve fully closed) to 1.0 (valve fully open).

The precise way the τ-value changes as a valve is closed varies from valve to valve. The details of this "closure curve" determine the head-discharge relationship of the valve and thus often have a marked influence on transient conditions in a pipeline.

2.8.6.2 Alternate valve representation. In the literature relating to valves, and as was introduced earlier in this chapter, it is common to model local losses as a multiplier of the velocity head:

$$\Delta H = \psi \frac{V^2}{2g} \qquad (2.41)$$

where V is the average velocity in the pipeline upstream of the valve and ψ is the alternative valve constant. This apparently trivial change has a detrimental effect on numerical calculations: ψ now varies from some minimum value for a fully open valve to infinity for a closed valve. Such a range of values can cause numerical instabilities in a transient program. For this reason, the reciprocal relationship involving τ-values is almost always preferred in transient applications.

2.8.6.3 Pressure regulating valves. In many applications, the valve closure relations are even more complicated than is the case in the situation just described. Pressure-regulating valves are often installed to maintain a preset pressure on their downstream side; they accomplish this function by partially closing, thus inducing a greater pressure drop across the valve. However, if a power failure or other transient condition were now to occur in the line, any "active" pressure-regulating valve would start from an already partially closed position. Depending on its initial setting, a pressure-regulating valve may close in a time that is much less than its design or theoretical value. The influence of the initial valve position is most severe for regulating valves breaking the largest pressures, which are often associated with relatively low headlosses in the remainder of the line. Thus, when a pressure-regulating valve is used, the most severe transient conditions can occur in a system transmitting small flows.

2.8.7 Conclusion

Transient fluid flow, variously called waterhammer, oil hammer, and so on, is the means of achieving a change in steady-state flow and pressure. When conditions in a pipeline are changed, such as by closing a valve or starting a pump, a series of waves are generated. These disturbances propagate with the velocity of sound within the medium until they are dissipated down to the level of the new steady state by the action of some form of damping or friction. In the case of flow in a pipeline, these fluid transients are the direct means of achieving *all* changes in fluid velocity, gradual or sudden. When sudden changes occur, however, the results can be dramatic because pressure waves of considerable magnitude can occur and are capable of destroying the pipe. Only if the flow is regulated extremely slowly is it possible to go smoothly from one steady state to another without large fluctuations in pressure head or pipe velocity.

Clearly, flow control actions can be extremely important, and they have implications not only for the design of the hydraulic system but also for other aspects of system design and operation. Such problems as selecting the pipe layout and profile, locating control elements within the system, and selecting device operating rules as well as handling the ongoing challenges of system management are influenced by the details of the control system. A rational and economic operation requires accurate data, carefully calibrated models, ongoing predictions of future demands and the response of the system to transient loadings, and correct selection of both individual components and remedial strategies. These design decisions cannot be regarded as an afterthought to be appended to a nearly complete design. Transient analysis is a fundamental and challenging part of rational pipeline design.

REFERENCES

Bhave, P. R., *Analysis of Flow in Water Distribution Networks*, Technomic Publishing Inc., Lancaster, PA, 1991.

Chaudhry, H. M., *Applied Hydraulic Transients*, Van Nostrand Reinhold, New York, NY, 1987.

Chaudhry, M. H., and V. Yevjevich, *Closed Conduit Flow,* Water Resources Publications, Littleton, CO, 1981.

Ford, K. W., *Classical and Modern Physics*, Vol. 1, Xerox College Publishing, Lexington, MA, 1973.

Hatsopoulos, N., and J. H. Keenan, *Principles of General Thermodynamics*, John Wiley and Sons, New York, 1965.

Hodgson, J., *Pipeline Celerities,* Master's of Engineering Thesis, University of Alberta, Edmonton, Alberta, Canada. 1983.

Jeppson, R.W. *Analysis of Flow in Pipe Networks*, Ann Arbor Science Publishers, Stoneham, MA, 1976.

Karney, B. W., "Energy Relations in Transient Closed Conduit Flow.," *Journal of Hydraulic Engineering*, 116:1180–1196, 1990.

Karney, B. W., and D. M. McInnis, "Transient Analysis of Water Distribution Systems," *ASCE Journal of the American Water Works Association,* 82(7):62–70, 1990.

Nikuradse, "Strom ungs gesetze in rauhen Rohre." Forsch-Arb, Ing.-Wes. Itett 361, 1933.

Roberson, J. A., and C. T. Crowe, *Engineering Fluid Mechanics,* Houghton Mifflin, Boston, MA, 1993.

Swamee, P. K, and A. K. Jain, "Explicit Equations for Pipe Flow Problems," *ASCE Journal of Hydraulic Engineering,* 102: 657–664, 1997.

Wylie, B. E., and V. L. Streeter, *Fluid Transients in Systems,* Prentice-Hall, Englewood Cliffs, NJ, 1993.

CHAPTER 3
SYSTEM DESIGN: AN OVERVIEW

Mark A. Ysusi
Montgomery Watson
Fresno, CA

3.1 INTRODUCTION

The primary purpose of a water distribution system is to deliver water to the individual consumer in the required quantity and at sufficient pressure. Water distribution systems typically carry potable water to residences, institutions, and commercial and industrial establishments. Though a few municipalities have separate distribution systems, such as a high-pressure system for fire fighting or a recycled wastewater system for nonpotable uses, most municipal water distribution systems must be capable of providing water for potable uses and for nonpotable uses, such as fire suppression and irrigation of landscaping.

The proper function of a water distribution system is critical to providing sufficient drinking water to consumers as well as providing sufficient water for fire protection. Because these systems must function properly, the principals of their planning, design, and construction need to be understood. This chapter focuses on the critical elements of planning and design of a water distribution system. The information presented primarily discusses typical municipal water distribution systems; however, the hydraulic and design principles presented can be easily modified for the planning and design of other types of pressure distribution systems, such as fire protection and recycled wastewater.

3.1.1 Overview

Municipal water systems typically consist of one or more sources of supply, appropriate treatment facilities, and a distribution system. Sources of supply include surface water, such as rivers or lakes; groundwater; and, in some instances, brackish or seawater. The information contained in this chapter is limited to the planning and design of distribution systems and does not address issues related to identifying and securing sources of supply or designing and constructing appropriate water treatment facilities. *Water distribution systems* usually consist of a network of interconnected pipes to transport water to the consumer, storage reservoirs to provide for fluctuations in demand, and pumping facilities.

3.1.2 Definitions

Many of the frequently used terms in water distribution system planning and design are defined here.

Average day demand. The total annual quantity of water production for an agency or municipality divided by 365.

Maximum day demand. The highest water demand of the year during any 24-h period.

Peak hour demand. The highest water demand of the year during any 1-h period.

Peaking factors. The increase above average annual demand, experienced during a specified time period. Peaking factors are customarily used as multipliers of average day demand to express maximum day and peak hour demands.

Distribution pipeline or main. A smaller diameter water distribution pipeline that serves a relatively small area. Water services to individual consumers are normally placed on distribution pipelines. Distribution system pipelines are normally between 150 and 400 mm (6–16 in.).

Transmission pipeline or main. A larger-diameter pipeline, designed to transport larger quantities of water during peak demand-periods. Water services for small individual consumers are normally not placed on transmission pipelines. Transmission mains are normally pipelines larger than 400 mm (16 in.).

3.2 DISTRIBUTION SYSTEM PLANNING

The basic question to be answered by the water distribution system planner/designer is, "How much water will my system be required to deliver and to where?" The answer to this question will require the acquisition of basic information about the community, including historical water usage, population trends, planned growth, topography, and existing system capabilities, to name just a few. This information can then be used to plan for logical extension of the existing system and to determine improvements necessary to provide sufficient water at appropriate pressure.

3.2.1 Water Demands

The first step in the design of a water distribution system is the determination of the quantity of water that will be required, with provision for the estimated requirements for the future.

In terms of the total quantity, the water demand in a community is usually estimated on the basis of per capita demand. According to a study published by the U.S. Geological Survey, the average quantity of water withdrawn for public water supplies in 1990 was estimated to be about 397 L per day per capita (Lpdc) or 105 gal per day per capita (gpdc). The withdrawals by state are summarized in Table 3.1. The reported water usage shown in Table 3.1 illustrates a wide variation. Per capita water use varies from a low use in

TABLE 3.1 Estimated Use of Water in the United States in 1990

State	L/per Capita/per day	gal/per Capita/per day
Alabama	379	100
Alaska	299	79
Arizona	568	150
Arkansas	401	106
California	556	147
Colorado	549	145
Connecticut	265	70
Delaware	295	78
District of Columbia	678	179
Florida	420	111
Georgia	435	115
Hawaii	450	119
Idaho	704	186
Illinois	341	90
Indiana	288	76
Iowa	250	66
Kansas	326	86
Kentucky	265	70
Louisiana	469	124
Maine	220	58
Maryland	397	105
Massachusetts	250	66
Michigan	291	77
Minnesota	560	148
Mississippi	466	123
Missouri	326	86
Montana	488	129
Nebraska	435	115
Nevada	806	213
New Hampshire	269	71
New Jersey	284	75
New Mexico	511	135
New York	450	119
North Carolina	254	67
North Dakota	326	86
Ohio	189	50
Oklahoma	322	85
Oregon	420	111
Pennsylvania	235	62
Rhode Island	254	67
South Carolina	288	76
South Dakota	307	81

TABLE 3.1 *(Continued)*

State	L/per Capita/per day	gal/per Capita/per day
Tennessee	322	85
Texas	541	143
Utah	825	218
Vermont	303	80
Virginia	284	75
Washington	522	138
West Virginia	280	74
Wisconsin	197	52
Wyoming	617	163
Puerto Rico	182	48
Virgin Islands	87	23
United States total	397	105

Source: Solley et al. (1993).

Pennsylvania of just over 60 gpcd to over 200 gpcd in Nevada. These variations depend on geographic location, climate, size of the community, extent of industrialization, and other influencing factors unique to most communities. Because of these variations, the only reliable way to estimate future water demand is to study each community separately, determining existing water use characteristics and extrapolating future water demand using population trends.

In terms of how the total water use is distributed within a community throughout the day, perhaps the best indicator is land use. In a metered community, the best way to determine water demand by land use is to examine actual water usage for the various types of land uses. The goal of examining actual water usage is to develop *water "duties"* for the various types of land uses that can be used for future planning. Water duties are normally developed for the following land uses:

- Single-family residential (some communities have low, medium, and high-density zones)
- Multifamily residential
- Commercial (normally divided into office and retail categories)
- Industrial (normally divided into light and heavy categories and separate categories for very high users
- Public (normally divided into park, or open space, and schools)

Water duties are normally expressed in gallons per acre per day. Table 3.2 shows typical water duties in the western United States. It should be noted that the definitions of land use terms like "low-density residential," "medium-density residential," and so on, will vary by community and should be examined carefully.

Another method of distributing water demand is to examine the water usage for individual users. This is particularly the case when an individual customer constitutes a significant portion of the total system demand. Table 3.3 presents water use for many different establishments. Although the rates vary widely, they are useful in estimating total water use for individual users when no other data are available.

TABLE 3.2 Typical Water Duties

Land Use	Water Duty (gal/day/acre)		
	Low	High	Average
Low–density residential	400	3300	1670
Medium–density residential	900	3800	2610
High–density residential	2300	12000	4160
Single-family residential	1300	2900	2300
Multifamily residential	2600	6600	4160
Office commercial	1100	5100	2030
Retail commercial	1100	5100	2040
Light industrial	200	4700	1620
Heavy industrial	200	4800	2270
Parks	400	3100	2020
Schools	400	2500	1700

Source: Adapted from Montgomery Watson study of data of 28 western U.S. cities.
Note: gal × 3.7854 = L.

TABLE 3.3 Typical Rates of Water Use for Various Establishments

User	Range of Flow	
	L/person or unit/day	gal/person or unit/day
Airport, per passenger	10–20	3–5
Assembly hall, per seat	6–10	2–3
Bowling alley, per alley	60–100	16–26
Camp		
Pioneer type	80–120	21–32
Children's, central toilet and bath	160–200	42–53
Day, no meals	40–70	11–18
Luxury, private bath	300–400	79–106
Labor	140–200	37–53
Trailer with private toilet and bath, per unit (2 1/2 persons)	500–600	132–159
Country clubs		
Resident type	300–600	79–159
Transient type serving meals	60–100	16–26
Dwelling unit, residential		
Apartment house on individual well	300–400	79–106
Apartment house on public water supply, unmetered	300–500	79–132
Boardinghouse	150–220	40–58
Hotel	200–400	53–106

TABLE 3.3 *(Continued)*

User	Range of Flow L/person or unit/day	gal/person or unit/day
Lodging house and tourist home	120–200	32–53
Motel	400–600	106–159
Private dwelling on individual well or metered supply	200–600	53–159
Private dwelling on public water supply, unmetered	400–800	106–211
Factory, sanitary wastes, per shift	40–100	11–26
Fairground (based on daily attendance)	2–6	1–2
Institution		
Average type	400–600	106–159
Hospital	700–1200	185–317
Office	40–60	11–16
Picnic park, with flush toilets	20–40	5–11
Restaurant (including toilet)		
Average	25–40	7–11
Kitchen wastes only	10–20	3–5
Short order	10–20	3–5
Short order, paper service	4–8	1–2
Bar and cocktail lounge	8–12	2–3
Average type, per seat	120–180	32–48
Average type 24 h, per seat	160–220	42–58
Tavern, per seat	60–100	16–26
Service area, per counter seat (toll road)	1000–1600	264–423
Service area, per table seat (toll road)	600–800	159–211
School		
Day, with cafeteria or lunchroom	40–60	11–16
Day, with cafeteria and showers	60–80	16–21
Boarding	200–400	53–106
Self-service laundry, per machine	1000–3000	264–793
Store		
First 7.5 m ($\approx$ 25 ft) of frontage	1600–2000	423–528
Each additional 7.5 m of frontage	1400–1600	370–423
Swimming pool and beach, toilet and shower	40–60	11–16
Theater		
Indoor, per seat, two showings per day	10–20	3–5
Outdoor, including food stand, per car (3 1/3 persons)	10–20	3–5

Source: Adapted from Metcalf and Eddy (1979).

3.2.2 Planning and Design Criteria

To plan and design a water distribution system effectively, criteria must be developed and adopted against which the adequacy of the existing and planned system can be compared. Typical criteria elements include the following:

- Supply
- Storage
- Fire demands
- Distribution system analysis
- Service pressures

3.2.2.1 Supply. In determining the adequacy of water supply facilities, the source of supply must be large enough to meet various water demand conditions and be able to meet at least a portion of normal demand during emergencies, such as power outages and disasters. At a minimum, the source of supply should be capable of meeting the maximum day system demand. It is not advisable to rely on storage to make up any shortfall in supply at maximum day demand. The fact that maximum day demand may occur several days consecutively must be considered by the system planner/designer. It is common for communities to provide a source of supply that meets the maximum day demand, with the additional supply to meet peak hour demand coming from storage. Some communities find it more economical to develop a source of supply that not only meets maximum day but also peak hour demand.

It is also good practice to consider standby capability in the source of supply. If the system has been designed so the entire capacity of the supply is required to meet the maximum demand, any portion of the supply that is placed out of service due to malfunction or maintenance will result in a deficient supply. For example, a community that relies primarily on groundwater for its supply should, at a minimum, be able to meet its maximum day demand with at least one of its largest wells out of service.

3.2.2.2 Storage. The principal function of storage is to provide reserve supply for (1) operational equalization, (2) fire suppression reserves, and (3) emergency needs.

Operational storage is directly related to the amount of water necessary to meet peak demands. The intent of operational storage is to make up the difference between the consumers' peak demands and the system's available supply. It is the amount of desirable stored water to regulate fluctuations in demand so that extreme variations will not be imposed on the source of supply. With operational storage, system pressures are typically improved and stabilized. The volume of operational storage required is a function of the diurnal demand fluctuation in a community and is commonly estimated at 25 percent of the total maximum day demand.

Fire storage is typically the amount of stored water required to provide a specified fire flow for a specified duration. Both the specific fire flow and the specific time duration vary significantly by community. These values are normally established through the local fire marshall and are typically based on guidelines established by the Insurance Service Office, a nonprofit association of insurers that evaluate relative insurance risks in communities.

Emergency storage is the volume of water recommended to meet demand during emergency situations, such as source of supply failures, major transmission main failures, pump failures, electrical power outages, or natural disasters. The amount of emergency storage included with a particular water system is an owner option, typically based on an assessment of risk and the desired degree of system dependability. In

TABLE 3.4 Typical Fire Flow Requirements

Land Use	Fire Flow Requirements, gal/m*
Single–family residential	500–2000
Multifamily residential	1500–3000
Commercial	2500–5000
Industrial	3500–10,000
Central business district	2500–15,000

Note: gal × 3.7854 = L.

considering emergency storage, it is not uncommon to evaluate providing significantly reduced supplies during emergencies. For example, it is not illogical to assume minimal demand during a natural disaster.

3.2.2.3 Fire demands. The rate of flow to be provided for fire flow is typically dependent on the land use and varies by community. The establishment of fire flow criteria should always be coordinated with the local fire marshall. Typical fire flow requirements are shown in Table 3.4.

3.2.2.4 Distribution system analysis. In evaluating an existing system or planning a proposed system, it is important to establish the criteria of operational scenarios against which the system will be compared. Any system can be shown to be inadequate if the established criteria are stringent enough. Most systems are quite capable of meeting the average day conditions. It is only when the system is stressed that deficiencies begin to surface. The degree to which the system will be realistically stressed is the crux of establishing distribution system analysis criteria. In evaluating a system, it is common to see how the system performs under the following scenarios:

- Peak hour demand
- Maximum day demand plus fire flow

Evaluating the system at peak hour demand gives the designer a look at system-wide performance. Placing fire flows at different locations in the system during a "background" demand equivalent to maximum day demand will highlight isolated system deficiencies. Obviously, it is possible for fires to occur during peak hour demand, but since this simultaneous occurrence is more unlikely than for a fire to occur sometime during the maximum day demand, this is not usually considered to be an appropriate criterion for design of the system.

3.2.2.5 Service pressures. There are differences in the pressures customarily maintained in the distribution systems in various communities. It is necessary that the water pressure in a consumer's residence or place of business be neither too low nor too high. Low pressures, below 30 psi, cause annoying flow reductions when more than one water-using device is in service. High pressures may cause faucet's to leak, valve seats to wear out quickly, or hot water heater pressure relief valves to discharge. In addition, abnormally high pressures can result in water being wasted in system leaks. The Uniform Plumbing Code requires that water pressures not exceed 80 psi at service connections, unless the service is provided with a pressure-reducing device. Another pressure criterion, related to

System Design: An Overview   **3.9**

TABLE 3.5 Typical Service Pressure Criteria

Condition	Service Pressure Criteria (psi)
Maximum pressure	65–75
Minimum pressure during maximum day	30–40
Minimum pressure during peak hour	25–35
Minimum pressure during fires	20

Note: psi × 6.895 = kPa.

fire flows, commonly requires a minimum of 20 psi at the connecting fire hydrant used for fighting the fire. Table 3.5 presents typical service pressure criteria.

3.2.3 Peaking Coefficients

Water consumption changes with the seasons, the days of the week, and the hours of the day. Fluctuations are greater in (1) small than in large communities and (2) during short rather than during long periods of time. Variations in water consumption are usually expressed as ratios to the average day demand. These ratios are commonly called *peaking coefficients*. Peaking coefficients should be developed from actual consumption data for an individual community, but to assist the reader, Table 3.6 presents typical peaking coefficients.

3.2.4 Computer Models and System Modeling

Modeling water distribution systems with computers is a proved, effective, and reliable technology for simulating and analyzing system behavior under a wide range of hydraulic conditions. The network model is represented by a collection of pipe lengths interconnected in a specified topological configuration by node points, where water can enter and exit the system. Computer models utilize laws of conservation of mass and energy to determine pressure and flow distribution throughout the network. Conservation of mass dictates that for each node the algebraic sum of flows must equal zero. Conservation of energy requires that along each closed loop, the accumulated energy loss must be zero. These laws can be expressed as nonlinear algebraic equations in terms of either pressures (node formulation) or volumetric flow rates (loop and pipe formulation). The nonlinearity reflects the relationship between pipe flow rate and the pressure drop across its length. Due to the presence of nonlinearity in these equations, numerical solution methods are iterative. Initial estimated values of pressure or flow are repeatedly adjusted until the difference between two successive iterates is within an acceptable tolerance. Several numerical iterative solution techniques have been suggested, from which the Newtonian method is the most widely used. See chapter 9 for more details on modeling.

TABLE 3.6 Typical Peaking Coefficients

Ratio of Rates	U.S. Range	Common Range
Maximum day: average day	1.5–3.5:1	1.8–2.8:1
Peak hour: average day	2.0–7.0:1	2.5–4.0:1

3.2.4.1 History of computer models. Prior to computerization, tedious, and time-consuming manual calculations were required to solve networks for pressure and flow distribution. These calculations were carried out using the Hardy-Cross numerical method of analysis for determinate networks. Only simple pipeline systems consisting of a few loops were modeled and under limited conditions because of the laborious effort required to obtain a solution. The first advent of computers in network modeling was with electric analogues, followed by large mainframe digital computers and smaller microcomputers. The computational power of a laptop computer today is vastly superior to the original computing machines that would fill several floors in an office building and at a fraction of the cost.

Many of the early computer models did not have interactive on-screen graphics, thus limiting the ability of engineers to develop and interpret model runs. The user interface was very rudimentary and often an afterthought. Input was either by punch cards or formatted American Standard Code for Information Interchange (ASCII) files created with a text editor. Errors were commonplace, and just getting a data file that would run could involve days, if not weeks, of effort, depending on the size and complexity of the network being modeled. Model output was usually a voluminous tabular listing of key network results. Interpretation of the results was time consuming and typically involved hand plotting of pressure contours on system maps.

Because of the widespread use of microcomputers during the past two decades, network modeling has taken on new dimensions. Engineers today rely on computer models to solve a variety of hydraulic problems. The use of interactive on-screen graphics to enter and edit network data and to color code and display network maps, attributes, and analysis results has become commonplace in the water industry. This makes it much easier for the engineer to construct, calibrate, and manipulate the model and visualize what is happening in the network under various situations, such as noncompliance with system performance criteria. The engineer is now able to spend more time thinking and evaluating system improvements and less time flipping through voluminous pages of computer printouts, thus leading to improved operation and design recommendations. The new generation of computer models have greatly simplified the formidable task of collecting and organizing network data and comprehending massive results.

3.2.4.2 Software packages. Many of the software packages available offer additional capabilities beyond standard hydraulic modeling, such as water quality assessment (both conservative and reactive species), multiquality source blending, travel time determination, energy and power cost calculation, leakage and pressure management, fire flow modeling, surge (transient) analysis, system head curve generation, automated network calibration, real-time simulation, and network optimization. Some sophisticated models can even accommodate the full library of hydraulic network components including pressure-regulating valves, pressure-sustaining valves, pressure breaker valves, flow control valves, float valves, throttle control valves, fixed-and variable speed pumps, turbines, cylindrical and variable cross-sectional area tanks, variable head reservoirs, and multiple inlet/outlet tanks and reservoirs. Through their predictive capabilities, computer network models provide a powerful tool for making informed decisions to support many organizational programs and policies. Modeling is important for gaining a proper understanding of system dynamic behavior, training operators, optimizing the use of existing facilities, reducing operating costs, determining future facility requirements, and addressing water quality distribution issues.

There is an abundance of network modeling software in the marketplace today. Some are free and others can be purchased at a nominal cost. Costs can vary significantly between models, depending on the range of the features and capabilities provided. The four major sources of computer models include consulting firms, commercial software companies, universities, and government agencies. Many of the programs available from these sources have been on the market for several years and have established track records. Most of the recent computer models, however, provide very sophisticated and intuitive graphical user interfaces and results presentation environment, as well as direct linkages with information management systems, such as relational databases and geographic information systems. Table 3.7 lists the names, addresses, and phone numbers of network modeling software vendors, along with their primary modeling products.

3.2.4.3 Development of a system model. As was just indicated, the computerized tools available to the engineer today are impressive and powerful. Once appropriate software is selected, data must be then input to the software to develop a computer model of the water system under study. Input data include the physical attributes of the system, such as pipe sizes and lengths, topography, and reservoir and pump characteristics, as well as the anticipated nodal demands.

Development of the nodal demands normally involves distributing the average day flow throughout the system in proportion to land use. This is commonly accomplished by determining a demand area for each node, measuring the area of each different land use within the demand area, multiplying the area of each land use within the demand area by its respective average day water duty (converted to gal/min or L/sec), and summing the water duties for each land use within the demand area and applying the sum at the node. In the past this effort required extensive mapping and determining the land use areas by planimeter or hand measurement. Today, with the advent of *graphical information system software* (GIS), the development of nodal demands is normally an activity involving computer-based mapping. The elements of the system, the demand areas, and the land uses are all mapped in separate layers in the GIS software. The GIS software capability of "polygon processing" intersects the different layers and automatically computes the land-use sums with the various demand areas. When the water duties are multiplied by their respective land use, the result is the average day system demand, proportioned to each node by land use. The water system computer model is then used to apply global peaking factors as described above.

3.3 PIPELINE PRELIMINARY DESIGN

The purpose of performing the water system planning tasks as outlined above is to develop a master plan for correcting system deficiencies and providing for future growth. Normally the system improvements are prioritized and a schedule or capital improvement program is developed based upon available (now or future) funding. As projects leave the advanced planning stage, the process of preliminary design begins. During preliminary design, the considerations of pipeline routing (alignment), subsurface conflicts, and rights-of-way are considered.

3.3.1 Alignment

In deciding upon an appropriate alignment for a pipeline, important considerations include right-of-way (discussed further below), constructability, access for future maintenance, and separation from other utilities. Many communities adopt standardized locations for utility pipelines (such as that water lines will generally be located 15 ft north and east of the street centerline). Such standards complement alignment considerations.

TABLE 3.7 Distribution System Modeling Software

Software	Vendor	Address	City	State	Country	Zip Code	Telephone	Fax
AQUA	Computer Modeling, Inc.	2121 Front Street	Cuyahoga Falls	OH	USA	44221	(330) 929-7886	(330) 929-2756
AVWater	CEDRA	65 West Broad Street	Rochester	NY	USA	14614	(716) 232-6998	(716) 262-2042
BOSS EMS	BOSS International	6612 Mineral Point Road	Madison	WI	USA	53705-4200	1-800-488-4775	(608) 258-9943
CYBERNET	Haestad Methods, Inc.	37 Brookside Road	Waterbury	CT	USA	06708	1-800-727-6555	(203) 597-1488
EPANET	US EPA	26 W. Martin Luther King Drive	Cincinnati	OH	USA	45268	(513) 569-7603	(513) 569-7185
FAAST	Faast Software	3062 East Avenue	Livermore	CA	USA	94550-4738	(510) 455-8086	(510) 455-8087
FLOW NETWORK ANALYSIS	Kelix Software Systems	11814 Coursey Blvd., Avenue, Suite 220	Baton Rouge	LA	USA	70816	(504) 769-6785	—
H$_2$ONET	MW Soft, Inc.	300 N. Lake Avenue, Suite 1200	Pasadena	CA	USA	91101	(626) 568-6868	(626) 568-6619
HYDRONET	Tahoe Designs Software	P.O. Box 8128	Truckee	CA	USA	92162	(530) 582-1525	(530) 582-8579
InWater	Intergraph	One Madison Industrial Park	Huntsville	AL	USA	35898	1-800-345-4856	(205) 730-6109
KYPIPE	University of Kentucky	Civil Engineering Software Center, University of Kentucky	Lexington	KY	USA	40506	(606) 257-3436	(606) 257-8005
PICCOLO	SAFEGE Consulting Engrs.	P.O. Box 727 Parc de l'Ile, 15-27 rue du Port	NANTERRE Cedex		France	92007	01133146147181	01133147247202
PIPE-FLO	Engineered Software, Inc.	4531 Intelco Loop	Lacey	WA	USA	98503	(360) 412-0702	(360) 412-0672
PIPES	Watercom Pty Ltd.	105 Queen Victoria Street	Bexley, NSW 2217		Australia		612-9587-5384	612-9587-5384
RINCAD	CEDEGER	1417 rue Michelin	Laval	Quebec	Canada	H7L4S2	(514) 629-8888	(514) 382-3077
RJN CASS WORKS	RJN Computer Services, Inc.	200 W. Front Street	Wheaton	IL	USA	60187	(630) 682-4700	(630) 682-4754
SDP	Charles Howard & Assoc. Ltd.	852 Fort St. 2nd Floor	Victoria	BC	Canada	V8W 2H7	(250) 385-0206	(250) 385-7737
Stoner Workstation	Stoner Associates, Inc.	P.O. Box 86	Carlisle	PA	USA	17013	(717) 243-1900	(717) 243-5564
TDHNET	TDH Engineering	607 Ninth Street	Laurel	MD	USA	20707	(301) 490-4515	(301) 490-4515
USU-NETWK	Utah State University	Utah State University, Dept. of Civil & Environmental Engineering	Logan	UT	USA	84322-4100	(801) 797-2943	(801) 750-1185
WADISO — Water Distribution Systems: Simulation and Sizing	Lewis Publishers - 1990	2000 Corporate Blvd. NW	Boca Raton	FL	USA	33431-9868	1-800-272-7737	1-800-374-3401
Water Works	Syntex Systems Corporation	800-1188 West Georgia Street	Vancouver	BC	Canada	V6E 4A2	(604) 688-8271	(604) 688-1286
WATER/WGRAPH	Municipal Hydraulics Ltd.	2474 Pylades Drive RR3	Ladysmith	BC	Canada	VOR2EO	(250) 722-3810	(250) 722-3088
WaterMax	The Pitometer Associates	20 N. Wacker Drive, Suite 1530	Chicago	IL	USA	60606	(312) 236-5655	(312) 580-2691
WATINET	WRc	8 Neshaminy Interplex, Suite 219	Trevose	PA	USA	19053	(215) 244-9972	(215) 244-9977
WATSYS	Expertware Dev. Corp.	27 Linden Avenue	Victoria	BC	Canada	V8V 4C9	(250) 384-5955	(250) 383-1692

3.3.2 Subsurface Conflicts

A critical element of developing a proposed pipeline alignment is an evaluation of subsurface conflicts. To, evaluate subsurface conflicts properly, it will be necessary for the designer to identify the type, size, and accurate location of all other underground utilities along the proposed pipeline alignment. This information must be considered in the design and accurately placed on the project plans so that the contractor (or whoever is constructing the line) is completely aware of potential conflicts.

It is good practice to thoroughly investigate potential utility conflicts. For example, it is not enough simply to determine that the proposed pipeline route will cross an electrical conduit. The exact location and dimensions of electrical conduits also need to be determined and the proposed water pipeline designed accordingly. What is shown on a utility company plat as a single line representing an electrical conduit may turn out to be a major electrical line with several conduits encased in concrete having a cross section 2 ft wide and 4 ft deep! Or, what is shown as a buried 3/4-inch telephone line may turn out to be a fiber-optic telecommunications cable that, if severed during construction, will result in exorbitant fines being levied by the communications utility.

Another water pipeline alignment consideration is the lateral separation of the line from adjacent sanitary sewer lines. Many state and local health officials require a minimum of 10 ft of separation (out-to-out) between potable water and sanitary sewer lines.

3.3.3 Rights-of-Way

The final location of a pipeline can be selected and construction begun only after appropriate rights-of-way are acquired. Adequate rights-of-way both for construction and for future access are necessary for a successful installation. Water lines are commonly located in streets and roadways dedicated to public use. On occasion, it is necessary to obtain rights-of-way for transmission-type pipelines across private lands. If this is the case, it is very important to properly evaluate the width of temporary easement that will be required during construction and the width of permanent easement that will be required for future access. If a pipeline is to be installed across private property, it is also very important for the entity that will own and maintain the pipeline to gain agreements that no permanent structures will be constructed within the permanent easements and to implement a program of monitoring construction on the private property to ensure that access to the pipeline is maintained. Otherwise, as the property changes hands in the future, the pipeline stands a good chance of becoming inaccessible.

3.4 PIPING MATERIALS

The types of pipe and fittings commonly used for pressurized water distribution systems are discussed in this section. The types of pipeline materials are presented first and then factors effecting the types of materials selected by the designer are presented in Sec. 3.4.7. The emphasis throughout this section is on pipe 100 mm (4 in.) in diameter and larger.

References to a standard or to a specification are given here in abbreviated form—code letters and numbers only, such as American National Standards Institute (ANSI) B36.10. Double designations, such as ANSI/AWWA C115/A21.15, indicate that American Waterworks Association (AWWA) C115 is the same as ANSI A21.15. Most standards are revised periodically, so it is advisable for the designer to obtain the latest edition.

3.14 Chapter Three

3.4.1 Ductile Iron Pipe (DIP)

Available in sizes 100–1350 mm (4–54 in.), DIP is widely used throughout the United States in water distribution systems. On the East Coast and in the Midwest, DIP is commonly used for both smaller distribution mains and larger transmission mains. On the West Coast, DIP is generally used for distribution pipelines 40 mm (16 in) and smaller, with alternative pipeline materials often selected for larger pipelines due to cost. Detailed descriptions of DIP, fittings, joints, installation, thrust restraint, and other factors related to design, as well as several important ANSI/AWWA specifications, are contained in the Ductile Iron Pipe Research Association (DIPRA) handbook (DIPRA, 1984).

3.4.1.1 Materials. DIP is a cast-iron product. Cast-iron pipe is manufactured of an iron alloy centrifugally cast in sand or metal molds. Prior to the early 1970s, most cast-iron pipe and fittings were gray iron, a brittle material that is weak in tension. But now all cast-iron pipe, except soil pipe (which is used for nonpressure plumbing applications) is made of ductile iron. Ductile iron is produced by the addition of magnesium to molten low-sulfur base iron, causing the free graphite to form into spheroids and making it about as strong as steel. Regular DIP (AWWA C151) has a Brinell hardness (BNH) of about 165.

Tolerances, strength, coatings and linings, and resistance to burial loads are given in ANSI/AWWA C151/A21.51.

3.4.1.2 Available sizes and thicknesses. DIP is available in sizes from 100 to 1350 mm (4–54 in). The standard length is 5.5 m (18 ft) in pressure ratings from 1380 to 2400 kPa (200–350 lb/in^2).

Thickness is normally specified by class, which varies from Class 50 to Class 56 (see DIPRA, 1984 or ANSI/AWWA C150/A21.50). Thicker pipe can be obtained by special order.

3.4.1.3 Joints. For DIP, rubber gasket push-on and mechanical are the most commonly used for buried service. These joints allow for some pipe deflection (about 2–5° depending on pipe size) without sacrificing water tightness. Neither of these joints is capable of resisting thrust across the joint and requires thrust blocks or some other sort of thrust restraint at bends and other changes in the flow direction.

Flanged joints (AWWA C115 or ANSI B16.1) are sometimes used at fitting and valve connections. Grooved end joints (AWWA C606) are normally used for exposed service and are seldom used for buried service. Flanged joints are rigid and grooved end joints are flexible. Both are restrained joints and do not typically require thrust restraint. Other types of restrained joints, such as restrained mechanical joints, are also available for buried service.

Various types of ductile iron pipe joints are shown in Fig. 3.1.

3.4.1.4 Gaskets. Gaskets for ductile iron push-on and mechanical joints, described in AWWA C111, are vulcanized natural or vulcanized synthetic rubber. Natural rubber is suitable for water pipelines but deteriorates when exposed to raw or recycled wastewater.

Gaskets for DIP flanges should be rubber, 3.2 mm (1/8 in) thick.

Gaskets for grooved end joints are available in ethylene propylene diene monomer (EPDM), nitrile (Buna-N), halogenated butyl rubber, Neoprene™, silicone, and fluorelastomers. EPDM is commonly used in water service and Buna-N in recycled wastewater.

3.4.1.5 Fittings. Some standard ductile or gray iron fittings are shown in Fig. 3.2. A list of standard and special fittings is also given in Table 3.8. *Ductile iron fittings* are normally

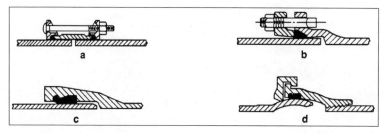

FIGURE 3.1 Couplings and joints for ductile iron pipe: (*a*) flexible coupling; (*b*) mechanical joint; (*c*) push-on joint; (*d*) ball joint. Adapted from Sanks et al. (1989).

available only in standard configurations as described in AWWA C110. Greater cost and longer delivery times can be expected for special fittings. Fittings are designated by the size of the openings, followed (where necessary) by the deflection angle. A 90° elbow for 250 mm (10 in) pipe would be called a 250 mm (10 in) 90° bend (or elbow). Reducers, reducing tees, or reducing crosses are identified by giving the pipe diameter of the largest opening first, followed by the sizes of other openings in sequence. Thus, a reducing tee on a 300 mm (12 in) line for a 150 mm (6 in) fire hydrant run might be designated as a 300 mm × 150 mm × 300 mm (12 in × 6 in × 12 in) tee.

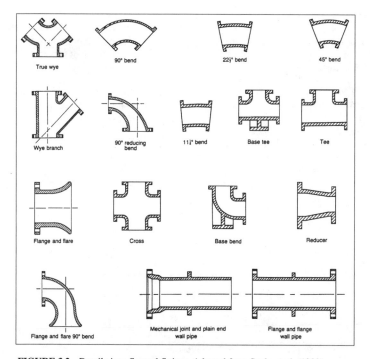

FIGURE 3.2 Ductile iron flanged fittings. Adapted from Sanks et al. (1989).

TABLE 3.8 Ductile Iron and Gray Cast-Iron Fittings, Flanged, Mechanical Joint, or Bell and Spigot*

Standard Fittings	Special Fittings
Bends (90°, 45°, 22.5°, 11.25°)	Reducing bends (90°)
Base bends	Flared bends (90°, 45°)
Caps	Flange and flares
Crosses	Reducing tees
Blind flanges	Side outlet tees
Offsets	Wall pipes
Plugs	True wyes
Reducers	Wye branches
Eccentric reducers	
Tees	
Base tees	
Side outlet tees	
Wyes	

*Size from 100 to 350 mm (4–54 in).

Standard ductile iron fittings are commonly available in flanged, mechanical joint, and push-on ends. It is considered good practice to include sufficient detail in construction plans and specifications to illustrate the type of joints that are expected at connections. The failure to detail a restrained joint when one is required by the design could result in an unstable installation.

3.4.1.6 Linings. Considering its low cost, long life, and sustained smoothness, cementmortar lining for DIP in water distribution systems is the most useful and common. Standard thicknesses for shop linings specified in AWWA C104 are given in Table 3.9.

TABLE 3.9 Thickness of Shop-Applied Cement-Mortar Linings

Nominal Pipe Diameter		Lining Thickness			
		Ductile Iron Pipe*		Steel Pipe†	
mm	in	mm	in	mm	in
100–250	4–10	1.6	1/16	6.4	1/4
300	1–2	1.6	1/16	7.9	5/16
350–550	14–22	2.4	3/32	7.9	5/16
600	24	2.4	3/32	9.5	3/8
750–900	30–36	3.2	1/8	9.5	3/8
1050–1350	42–54	3.2	1/8	12.7	1/2

*Single thickness per AWWA C104. Linings of double thickness are also readily available.
†Per AWWA C205.

TABLE 3.10 Thickness of Cement-Mortar Linings of Pipe in Place per AWWA C602

Nominal Pipe Diameter		DIP or Gray Cast Iron (New or Old Pipe)		Steel Pipe			
				Old Pipe		New Pipe	
mm	in	mm	in	mm	in	mm	in
100–250	4–10	3.2	1/8	6.4	1/4	4.8	3/16
300	12	4.8	3/16	6.4	1/4	4.8	3/16
350–550	14–22	4.8	3/16	7.9	5/16	6.4	1/4
600–900	24–36	4.8	3/16	9.5	3/8	6.4	1/4
1050–1350	42–54	6.4	1/4	9.5	3/8	9.5	3/8
1500	60	—	—	9.5	3/8	9.5	3/8
1650–2250	66–90	—	—	12.7	1/2	11.1	7/16
>2250	>90	—	—	12.7	1/2	12.7	1/2

Pipe can also be lined in place with the thicknesses given in Table 3.10. Because the standard, shop-applied mortar linings are relatively thin, some designers prefer to specify shop linings in double thickness. The designer should also be careful in specifying mortar lining thickness to match the pipe inside diameter (ID) with system valve IDs, particularly with short-body butterfly valves where the valve vane protrudes into the pipe. If the pipe ID is too small, the valve cannot be fully opened.

Although cement-mortar lining is normally very durable, it can be slowly attacked by very soft waters with low total dissolved solids content (less than 40 mg/L), by high-sulfate waters, or by waters undersaturated in calcium carbonate. For such uses, the designer should carefully investigate the probable durability of cement mortar and consider the use of other linings. Other linings and uses are shown in Table 3.11. In general, the cost of cement mortar is about 20 percent of that of other linings, so other linings are not justified except where cement mortar would not provide satisfactory service.

3.4.1.7 Coatings. Although DIP is relatively resistant to corrosion, some soils (and peat, slag, cinders, muck, mine waste, or stray electric current) may attack the pipe. In these

TABLE 3.11 Linings for Ductile Iron and Steel Pipe

Lining Material	Reference Standard	Recommended Service
Cement mortar	AWWA C104, C205	Potable water, raw water and sewage, activated and secondary sludge
Glass	None	Primary sludge, very aggressive fluids
Epoxy	AWWA C210	Raw and potable water
Fusion-bonded epoxy	AWWA C213	Potable water, raw water and sewage
Coal-tar epoxy	AWWA C210	Not recommended for potable water
Coal-tar enamel	AWWA C203	Potable water
Polyurethane	None	Raw sewage, water
Polyethylene	ASTM D 1248	Raw sewage

applications, ductile iron manufacturers recommend that the pipe be encased in loose-fitting, flexible polyethylene tubes 0.2 mm (0.008 in) thick (see ANSI/AWWA C105/A21.5). These are commonly known as "baggies." An asphaltic coating approximately 0.25 mm (0.001 in) thick is a common coating for ductile iron pipe in noncorrosive soils. In some especially corrosive applications, a coating, such as adhesive, hot-applied extruded polyethylene wrap, may be required.

In corrosive soils, the following coatings may be appropriate for protecting the pipe:

- Adhesive, extruded polyethylene wrap
- Plastic wrapping (AWWA C105)
- Hot-applied coal-tar enamel (AWWA C203)
- Hot-applied coal-tar tape (AWWA C203)
- Hot-applied extruded polyethylene [ASTM D 1248 (material only)]
- Coal-tar epoxy (MIL-P-23236)
- Cold-applied tape (AWWA C209)
- Fusion-bonded epoxy (AWWA C213)

Each of the above coatings is discussed in detail in the referenced specifications. Each coating system has certain limited applications and should be used in accordance with the NACE standards or as recommended by a competent corrosion engineer.

3.4.2 Polyvinyl Chloride (PVC) Pipe

In the United States, where it is used in both water and wastewater service, *polyvinyl chloride* (PVC) is the most commonly used plastic pipe for municipal water distribution systems. Because of its resistance to corrosion, its light weight and high strength to weight ratio, its ease of installation, and its smoother interior wall surface, PVC has enjoyed rapid acceptance for use in municipal water distribution systems since the 1960s. There are several other types of plastic pipe, but PVC is the most common plastic pipe selected for use in municipal systems and will be the only type of plastic pipe addressed in this section. There are also several different PVC pipe specifications. Only those having AWWA approval will be addressed in this section, since only those should be used for municipal water distribution systems. Highdensity polyethylene pipe (HDPE) is discussed in Sec. 3.4.5.

3.4.2.1 Materials. PVC is a polymer extruded under heat and pressure into a thermoplastic that is nearly inert when exposed to most acids, alkalis, fuels, and corrosives, but it is attacked by ketones (and other solvents) sometimes found in industrial wastewaters. Basic properties of PVC compounds are detailed in ASTM D 1784. ASTM D 3915 covers performance characteristics of concern, or cell classification, for PVC compounds to be used in pressure pipe applications. Generally, PVC should not be exposed to direct sunlight for long periods. The impact strength of PVC will decrease if exposed to sunlight and should not be used in above-ground service.

In North America, PVC pipe is rated for pressure capacity at 23°C (73.4°F). The pressure capacity of PVC pipe is significantly related to its operating temperature. As the temperature falls below 23°C (73.4°F), such as in normal buried service, the pressure capacity of PVC pipe increases to a level higher than its pressure rating or class. In practice, this increase is treated as an unstated addition to the working safety factor but is not otherwise considered in the design

TABLE 3.12 Thermal Derating Factors for PVC Pressure Pipes and Fittings

Maximum Service Temperature		Multiply the Pressure Rating or Pressure Class at 73.4°F (23°C) by These Factors
°C	°F	
27	80	0.88
32	90	0.75
38	100	0.62
43	110	0.50
49	120	0.40
54	130	0.30
60	140	0.22

Source: Handbook of PVC Pipe (1991).

process. On the other hand, as the operating temperature rises above 23°C (73.4°F), the pressure capacity of PVC pipe decreases to a level below its pressure rating or class. Thermal derating factors, or multipliers, are typically used if the PVC pipe will be used for higher temperature services. Recommended thermal derating factors are shown in Table 3.12. The pressure rating or class for PVC pipe at service temperature of 27°C (80°F) would need to multiplied by a thermal derating factor of 0.88. The pressure rating or class for PVC pipe at service temperature of 60°C (140°F) would need to multiplied by a thermal derating factor of 0.22.

3.4.2.2 Available sizes and thicknesses. AWWA C900 covers PVC pipe in sizes 100 to 300 mm (4−12 in.). AWWA C905 covers PVC pipe in sizes 350–900 mm (14–36 in). There are important differences in these two specifications that should be understood by the designer. AWWA C900 PVC pipe is manufactured in three "pressure classes" (100, 150, and 200). The pressure class selected is typically the highest normal operating system pressure in psi. AWWA C900 PVC pipe design is based on a safety factor of 2.5 plus an allowance for hydraulic transients (surge). AWWA C905 does not provide for "pressure classes" but refers to PVC pressure pipe in terms of "pressure rating." As with pressure class, pressure rating also refers to system pressure in psi. While AWWA C905 covers six pressure rating categories (100, 125, 160, 165, 200, and 235), the most commonly available pressure ratings are 165 and 235. The design of AWWA C905 PVC pipe is based on a safety factor of 2.0 and does not include an allowance for surge. In view of this important difference between the two specifications, designers often specify higher pressure ratings of C905 PVC pipe than system pressure would tend to indicate in order to allow for the reduced factor of design safety.

Both C900 and C905 contain required pipe dimension ratios. Dimension ratios define a constant ratio between the outside diameter and the wall thickness. For a given dimension ratio, pressure capacity and pipe stiffness remain constant, independent of pipe size. Table 3.13 presents dimension ratios (DR) with corresponding pressure classes as defined in AWWA C900. Table 3.14 presents dimension ratios with corresponding pressure ratings as defined in AWWA C905.

3.4.2.3 Joints. For PVC pipe, a rubber gasket bell and spigot type joint is the most commonly joint used for typical, municipal buried service. The bell and spigot joint

TABLE 3.13 Pressure Class versus DR-AWWA C900

DR	Pressure Class at Safety Factor = 2.5 psi (kPa)
14	200 (1380)
18	150 (1030)
25	100 (690)

TABLE 3.14 Pressure Rating versus DR-AWWA C905

DR	Pressure Rating at Safety Factor = 2.0 psi (kPa)
18	235*
21	200
25	165*
26	160
32.5	125
41	100

*Most commonly used ratings for municipal systems.

allows for some pipe deflection *(Handbook of PVC Pipe*, 1991) without sacrificing water tightness. This joint is not capable of resisting thrust across the joint and requires thrust blocks or some other sort of thrust restraint at bends and other changes in the direction of flow. Mechanical restraining devices are commonly used to provide restraint at PVC pipe joints where necessary.

PVC pipe joints are specified in ASTM D 3139. At connections to fittings and other types of piping, it is also common to detail a plain end (field-cut pipe) PVC pipe. Plain-end pipes are used to connect to mechanical joint ductile iron fittings and to flange adapters.

3.4.2.4 Gaskets. Gaskets for PVC joints are specified in ASTM F 477. As with gaskets for DIP, gaskets for PVC pipe are vulcanized natural rubber or vulcanized synthetic rubber. Natural rubber is suitable for water pipelines but deteriorates when exposed to raw or recycled wastewater. EPDM is commonly used in water service and nitrile (Buna N), in recycled wastewater.

3.4.2.5 Fittings. AWWA C900 and C905 PVC pipe for municipal use are manufactured in ductile iron pipe OD sizes, so ductile iron fittings, conforming to AWWA C110, are used in all available sizes. See Sec. 3.4.1.5 for a discussion on ductile iron fittings. Although not widely used, PVC fittings, in configurations similar to ductile iron fittings, are also available for smaller line sizes. AWWA C907 covers PVC pressure pipe fittings for pipe sizes 100–200 mm (4–8 in.) in pressure classes 100 and 150.

3.4.2.6 Linings and Coatings. PVC pipe does not require lining or coating.

3.4.3 Steel Pipe

Steel pipe is available in virtually any size, from 100 m through 3600 mm (4–144 in), for use in water distribution systems. Though rarely used for pipelines smaller than 400 mm (16 in), it is widely used in the western United States for transmission pipelines in sizes larger than 600 mm (24 in). The principal advantages of steel pipe include high strength, the ability to deflect without breaking, the ease of installation, shock resistance, lighter weight than ductile iron pipe, the ease of fabrication of large pipe, the availability of special configurations by welding, the variety of strengths available, and the ease of field modification.

3.4.3.1 Materials. Conventional nomenclature refers to two types of steel pipe: (1) mill pipe and (2) fabricated pipe.

Mill pipe includes steel pipe of any size produced at a steel pipe mill to meet finished pipe specifications. Mill pipe can be seamless, furnace butt welded, electric resistance welded, or fusion welded using either a straight or spiral seam. Mill pipe of a given size is manufactured with a constant outside diameter and a variable internal diameter, depending on the required wall thickness.

Fabricated pipe is steel pipe made from plates or sheets. It can be either straight or spiral-seam, fusion-welded pipe, and it can be specified in either internal or external diameters. Spiral-seam, fusion-welded pipe may be either mill pipe or fabricated pipe.

Steel pipe may be manufactured from a number of steel alloys with various yield and ultimate tensile strengths. Internal working pressure ratings vary from 690 to 17,000 kPa (100–2500 lb/in^2), depending on alloy, diameter, and wall thickness. Steel piping in water distribution systems should conform to AWWA C200, in which there are many ASTM standards for materials (see ANSI B31.1 for the manufacturing processes).

3.4.3.2 Available sizes and thicknesses. Sizes, thicknesses, and working pressures for pipe used in water distribution systems range from 100 m to 3600 mm (4–144 in.) as shown in Table 4.2 of AWWA M11 (American Water Works Association). The standard length of steel water distribution pipe is 12.2 m (40 ft).

Manufacturers should be consulted for the availability of sizes and thicknesses of steel pipe. Table 4.2 of AWWA M11 allows a great variety of sizes and thicknesses.

According to ANSI B36.10,

- Standard weight (STD) and Schedule 40 are identical for pipes up to 250 mm (10 in). Standard weight pipe 300 mm (12 in) and larger have walls 0.5 mm (3/8 in) thick. For standard weight pipe 300 mm (12 in) and smaller, the ID equals approximately the nominal pipe diameter. For pipe larger than 300 mm (12 in), the outside diameter (OD) equals the nominal diameter.

- Extra strong (XS) and Schedule 80 are identical for pipes up to 200 mm (8 in). All larger sizes of extra strong-weight pipe have walls 12.7 mm (1/2 in) thick.

- Double extra strong (XXS) applies only to steel pipe 300 mm (12 in) and smaller. There is no correlation between XXS and schedule numbers. For wall thicknesses of XXS, which (in most cases) is twice that of XS, see ANSI B36.10.

For sizes of 350 mm (14 in) and larger, most pipe manufacturers use spiral welding machines and, in theory, can fabricate pipe to virtually any desired size, In practice, however, most steel pipe manufacturers have selected and built equipment to produce

3.22 Chapter Three

given ID sizes. Any deviation from manufacturers' standard practices is expensive, so it is always good practice for the designer to consult pipe manufacturers during the design process. To avoid confusion, the designer should also show a detail of the specified pipe size on the plans or tabulate the diameters in the specifications. For cement-mortar-lined steel pipelines, AWWA C200, C205, C207, and C208 apply.

Steel pipe must sometimes either be reinforced at nozzles and openings (tees, wye branches) or a greater wall thickness must be specified. A detailed procedure for determining whether additional reinforcing is required is described in Chap. II and Appendix H of ANSI B31.3. If additional reinforcement is necessary, it can be accomplished by a collar or pad around the nozzle or branch, a wrapper plate, or crotch plates. These reinforcements are shown in Fig. 3.3, and the calculations for design are given in AWWA M11 (American Water Works Association, 1989).

3.4.3.3 Joints. For buried service, *bell* and *spigot joints* with rubber gaskets or mechanical couplings (with or without thrust harnesses) are common. *Welded joints* are also common for pipe 600 mm (24 in) and larger. Linings are locally destroyed by the heat of welding, so the ends of the pipe must be bare and the linings field applied at the joints. The reliability of field welds is questionable without careful inspection, but when properly made field welds are stronger than other joints. A steel pipeline project specification involving field welding should always include a carefully prepared section on quality assurance and testing of the welds. Different types of steel pipe joints are shown in Fig. 3.4.

3.4.3.4 Gaskets. Gaskets for steel flanges are usually made of cloth-inserted rubber either 1.6 mm (1/16 in) or 3.2 mm (1/8 in) thick and are of two types:

- ring (extending from the ID of the flange to the inside edge of the bolt holes)
- full face (extending from the ID of the flange to OD)

Gaskets for mechanical and push-on joints for steel pipe are the same as described in Sec. 3.4.1.4 for ductile iron pipe.

3.4.3.5 Fittings. For steel pipe 100 mm (4 in) and larger, specifications for steel fittings can generally be divided into two classes, depending on the joints used and the pipe size:

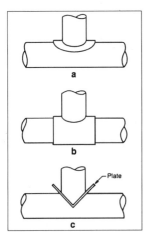

FIGURE 3.3 Reinforcement for steel pipe openings. (*a*) collar plate; (*b*) wrapper plate; (*c*) crotch plates. Adapted from Sanks et al. (1989).

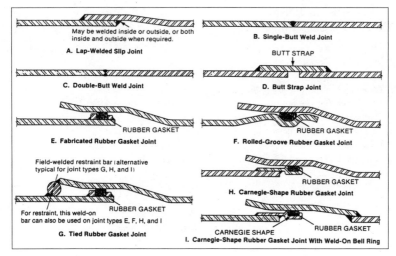

FIGURE 3.4 Welded and rubber-gasketed joints for steel pipe, (AWWA MI, 1989).

- Flanged, welded (ANSI B16.9)
- Fabricated (AWWA C208)

Fittings larger than 100 mm (4 in) should conform to ANSI B16.9 ("smooth" or wrought) or AWWA C208 (mitered). Threaded fittings larger than 100 mm (4 in) should be avoided. The ANSI B16.9 fittings are readily available up to 300–400 mm (12–16 in) in diameter. Mitered fittings are more readily available and cheaper for larger fittings.

The radius of a mitered elbow can range from 1 to 4 pipe diameters. The hoop tension concentration on the inside of elbows with a radius less than 2.5 pipe diameters may exceed the safe working stress. This tension concentration can be reduced to safe levels by increasing the wall thickness, as described in ANSI B31, AWWA C208, and Piping Engineering (Tube Turns Division, 1974). Design procedures for mitered bends are described in ANSI B31.1 and B31.3. Types of steel fittings are shown in Table 3.15 and in Figs.3.5 and 3.6.

3.4.3.6 Linings and coatings. Cement mortar is an excellent lining for steel pipe. Tables 3.9 and 3.10 show required thicknesses for steel pipe.

Steel pipe can also be coated with cement mortar. Recommended mortar coating thicknesses are shown in AWWA C205. These thicknesses, however, are often thinner than those required to provide adequate protection. Many designers specify a minimum cement mortar coating thickness of at least 19 mm (3/4 in).

In corrosive soils, the following coatings may be appropriate for protecting steel pipe:

- Hot-applied coal-tar enamel (AWWA C203)
- Cold-applied tape system (AWWA C214)
- Fusion-bonded epoxy (AWWA C213)
- Coal-tar epoxy (AWWA C210)
- Hot-applied extruded polyethylene [ASTM D 1248 (material only)]

TABLE 3.15 Steel Fittings

Mitered Fittings	Wrought Fittings
Crosses	Caps
Two–piece elbows, 0–30° bend	45° elbows
Three–piece elbows, 31–60° bend	90° elbows, long radius
Four–piece elbows, 61–90° bend	90° elbows, short radius
Four–piece, long radius elbows	90° reducing elbows, long radius
Laterals, equal diameters	
Laterals, unequal diameters	Multiple-outlet fittings
Reducers	Blind flanges
Eccentric reducers	Lap joint flanges
Tees	Slip-on flanges
Reducing tees	Socket-type welding flanges
True wyes	Reducing flanges
	Threaded flanges
	Welding neck flanges
	Reducers
	Eccentric reducers
	180° returns, long radius
	Saddles
	Reducing outlet tees
	Split tees
	Straight tees
	True wyes

As another alternative, epoxy-lined and coated steel pipe can be used. Because this lining is only 0.3–0.6 mm (0.12–0.20 in) thick, the ID of the bare pipe is only slightly reduced by such linings. Epoxy-lined steel pipe is covered by AWWA C203, C210, and C213 standards. Before specifying epoxy lining and coating, pipe suppliers must be consulted to determine the limitations of sizes and lengths of pipe that can be lined with epoxy. Flange faces should not be coated with epoxy if flanges with serrated finish per AWWA C207 are specified.

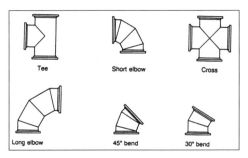

FIGURE 3.5 Typical mitered steel fittings. Adapted from Sanks et al. (1989).

System Design: An Overview **3.25**

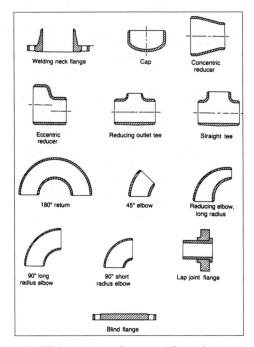

FIGURE 3.6 Wrought (forged) steel fittings for use with welded flanges. Adapted from Sanks et al. (1989).

3.4.4 Reinforced Concrete Pressure Pipe (RCPP)

Several types of RCPP are manufactured and used in North America. These include steel cylinder (AWWA C300), prestressed, steel cylinder (AWWA C301), noncylinder (AWWA C302), and pretensioned, steel cylinder (AWWA 303). Some of these types are made for a specific type of service condition and others are suitable for a broader range of service conditions. A general description of RCPP types is shown in Table 3.16. The designer should be aware that not all RCPP manufacturers make all of the types of pipe listed.

TABLE 3.16 General Description of Reinforced Concrete Pressure Pipe

Type of Pipe	AWWA Standard	Steel Cylinder	Reinforcement	Design Basis*
Steel cylinder	C300	X	Mild reinforcing steel	Rigid
Prestressed, steel cylinder	C301	X	Prestressed wire	Rigid
Noncylinder	C302	None	Mild reinforcing steel	Rigid
Pretensioned, steel cylinder	C303	X	Mild reinforcing steel	Semirigid

* "Rigid" and "semirigid" are terms used in AWWA M9 and are intended to differentiate between two design theories. Rigid pipe does not depend on the passive resistance of the soil adjacent to the pipe for support of vertical loads. Semirigid pipe requires passive soil resistance for vertical load support. The terms "rigid" and "semirigid" as used here should not be confused with the definitions stated by Marston in Iowa State Experiment Station Bulletin No. 96.

3.4.4.1 Steel cylinder pipe, AWWA C300. Prior to the introduction of prestressed steel cylinder pipe (AWWA C301), in the early 1940s, most of the RCPP in the United States was steel cylinder type pipe. New installations of steel cylinder pipe have been declining over the years as AWWA C301 and C303 pipes have gained acceptance. Steel cylinder pipe is manufactured in diameters of 750–3600 mm (30–44 in). Standard lengths are 3.6–7.2 meters (12–24 ft).

AWWA C300 limits the reinforcing steel furnished in the cage(s) to no less than 40 percent of the total reinforcing steel in the pipe. The maximum loads and pressures for this type of pipe depend on the pipe diameter, wall thickness, and strength limitations of the concrete and steel. The designer should be aware that this type of pipe can be designed for high internal pressure, but is limited in external load capacity.

A cross section of AWWA C300 pipe and a typical joint configuration is shown in Fig. 3.7.

3.4.4.2 Prestressed steel cylinder pipe, AWWA C301. Prestressed steel cylinder pipe has been manufactured in the United States since 1942 and is the most widely used type of concrete pressure pipe, except in the western United States. Due to cost considerations, AWWA C301 pipe is often used for high-pressure transmission mains, but it has also been used for distribution mains and for many other low-and high-pressure uses.

A distressing number of failures of this pipe occurred in the United States primarily during the 1980s. The outer shell of the concrete cracked, allowing the reinforcement to corrode and subsequently fail. These failures have resulted in significant revisions in the standards covering this pipe's design. Even so, the designer should not necessarily depend solely on AWWA specifications or on manufacturers' assurances, but should make a careful analysis of internal pressure (including waterhammer) and external loads. Make certain that the tensile strain in the outer concrete is low enough so that cracking will either not occur at all or will not penetrate to the steel under the worst combination of external and internal loading.

Prestressed cylinder pipe has the following two general types of fabrication: (1) a steel cylinder lined with a concrete core or (2) a steel cylinder embedded in concrete

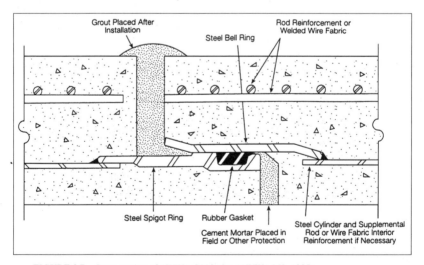

FIGURE 3.7 Cross section of AWWA C300 pipe (AWWA M9, 1995).

core. Lined cylinder pipe is commonly available in IDs from 400 to 1200 mm (16–48 in). Sizes through 1500 mm (60 in) are available through some manufacturers. Embedded cylinder pipe is commonly available in inside diameters 1200 mm (48 in) and larger. Lengths are generally 4.9–7.3 m (16–24 ft), although longer units can be furnished.

AWWA C304, *Standard for Design of Prestressed Concrete Cylinder Pipe* covers the design of this pipe. The maximum working pressure for this pipe is normally 2758 kPa (400 psi). The design method is based on combined loading conditions (the most critical type of loading for rigid pipe) and includes surge pressure and live loads.

Cross sections of AWWA C301 pipe (lined and embedded) and typical joint configurations are shown in Fig. 3.8.

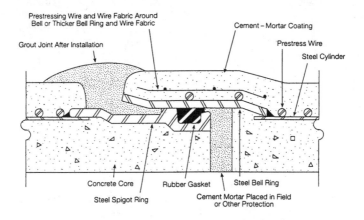

A. Lined cylinder pipe

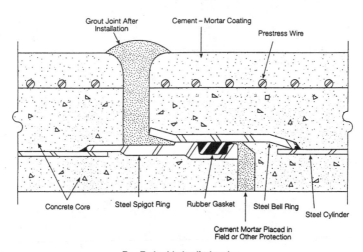

B. Embedded cylinder pipe

FIGURE 3.8 Cross section of AWWA C301 pipe (AWWA M9, 1995).

3.4.4.3 Noncylinder pipe, AWWA C302. The maximum working pressure of noncylinder pipe is 379 kPa (55 psi) and is generally not suitable for typical municipal systems.

Noncylinder pipe is commonly furnished in diameters of 300 to 3600 mm (12–144 in), but larger diameters can be furnished if shipping limitations permit. Standard lengths are 2.4–7.3 m (8–24 ft) with AWWA C302 limiting the maximum length that can be furnished for each pipe size.

Cross sections of AWWA C302 pipe with steel and concrete joint ring configurations are shown in Fig. 3.9.

3.4.4.4 Pretensioned steel cylinder, AWWA C303. Pretensioned steel cylinder, commonly called *concrete cylinder pipe* (CCP), is manufactured in Canada and in the western and southwestern areas of the United States. It is commonly available in diameters of 300–1350 mm (12–54 in). Standard lengths are generally 7.3 to 12.2 m(24–40 ft). With maximum pressure capability up to 2758 kPa (400 psi), the longer laying length, and the overall lighter handling weight, AWWA C303 is a popular choice among many designers for various applications, including municipal transmission and distributions mains.

Manufacture of CCP begins with a fabricated steel cylinder with joint rings that is hydrostatically tested. A cement-mortar lining is then placed by the centrifugal process inside the cylinder. The nominal lining thickness is 13 mm (1/2 in), for sizes up to and including 400 mm (16 in), and 19 mm (3/4 in) for larger sizes. After the lining is cured, the cylinder is wrapped, typically in a helical pattern, with a smooth, hot-rolled steel bar, using a moderate tension in the bar. The size and spacing of the bar, as well as the thickness of the steel cylinder, are proportioned to provide the required pipe strength. The cylinder and bar wrapping are then covered with a cement slurry and a dense mortar coating that is rich in cement.

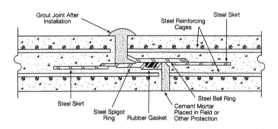

A. AWWA C302-type pipe with steel joint rings

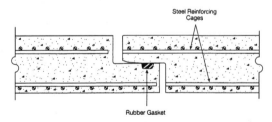

B. AWWA C302-type pipe with concrete joint rings

FIGURE 3.9 Cross section of AWWA C302 pipe, (AWWA M9, 1995).

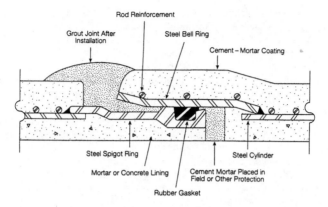

FIGURE 3.10 Cross section of AWWA C303 pipe (AWWA M9, 1995).

The design of CCP is based on a semirigid pipe theory in which internal pressure and external load are designed for separately but not in combination. Since the theory of semirigid pipe design for earth loads above the pipe is based on the passive soil pressure adjacent to the sides of the pipe, the design must be closely coordinated with the installation conditions.

A cross section of AWWA C303 pipe and a typical joint configuration are shown in Fig. 3.10.

3.4.5 High-Density Polyethylene (HDPE) Pipe

Polyethylene pressure pipe has been used in the United States by various utilities in urban environments for several years. Nearly all natural gas-distribution pipe installed in the United States since 1970 is polyethylene. It has only recently, however, become available as an AWWA-approved transmission and distribution system piping material. AWWA Standard C906, *Polyethylene Pressure Pipe and Fittings, 4 in. through 63 in., for Water Distribution,* became effective March 1, 1992. AWWA Standard C901, *Polyethylene Pressure Pipe, Tubing, and Fittings, 1/2 in. through 3 in., for Water Service,* has been in effect since 1978. Prior to 1992, the use of polyethylene pipe in municipal water distribution systems was normally limited to water services. Since AWWA approval in 1992, however, polyethylene pipe is now being used in transmission and distribution system applications. Because of its resistance to corrosion, its light weight and high strength to weight ratio, its resistance to cracking, its smoother interior wall surface, and its demonstrated resistance to damage during seismic events, HDPE pipe is gaining acceptance for use in municipal water systems.

3.4.5.1 Materials. Low-density polyethylene was first introduced in the 1930s and 1940s in England and then in the United States. This first material was commonly used for cable coatings. Pipe grade resins were developed in the 1950s and have evolved to today's high-density, extra-high-molecular weight materials. AWWA C906 specifies several different resins, but today, all HDPE water pipe manufactured in the United States, is made with a material specified in ASTM D 3350 by a cell classification 345434C.

HDPE pipe is rated for pressure capacity at 23°C (73.4°F). Since it is a thermoplastic, its pressure capacity is related to its operating temperature. Through the normal range of municipal water system temperatures, 0°C–24° (32°F–75° F), the pressure rating of HDPE pipe remains relatively constant. As the operating temperature rises above 23°C (73.4°F), however, the pressure capacity of HDPE pipe decreases to a level below its pressure class. The pressure rating for HDPE pipe at a service temperature of 60°C (140°F) would be about half its rating at 23°C (73.4°F).

3.4.5.2 Available sizes and thicknesses. AWWA C906 covers HDPE pipe in sizes 100–600 mm (4–63 in). The design, according to AWWA C906, of HDPE pipe is similar to AWWA C900 PVC pipe in that HDPE pipe is rated according to "pressure classes." The pressure classes detailed in AWWA C906 include allowance for pressure rises above working pressure due to occasional positive pressure transients not exceeding two times the nominal pressure class and recurring pressure surges not exceeding one and one-half times the nominal pressure class. AWWA C906 lists HDPE pipe sizes according to the IPS (steel pipe) and the ISO (metric) sizing systems. Ductile iron pipe sizes are also available.

As with AWWA C900 and C905 (PVC pipe), AWWA C906 contains dimension ratio (outside diameter to wall thickness) specifications. Table 3.17 presents dimension ratios with corresponding pressure classes as defined in AWWA C906 for commonly available HDPE pipe.

3.4.5.3 Joints. HDPE pipe can be joined by thermal butt-fusion, flange assemblies, or mechanical methods as may be recommended by the pipe manufacturer. HDPE is not to be joined by solvent cements, adhesives (such as epoxies), or threaded-type connections. Thermal butt-fusion is the most widely used method for joining HDPE piping. This procedure uses portable field equipment to hold pipe and/or fittings in close alignment while the opposing butt-ends are faced, cleaned, heated and melted, fused together, and then cooled under fusion parameters recommended by the pipe manufacturer and fusion equipment supplier.

For each polyethylene material there exists an optimum range of fusion conditions, such as fusion temperature, interface pressure, and cooling time. Thermal fusion should be conducted only by persons who have received training in the use of the fusion equipment according to the recommendations of the pipe manufacturer and fusion equipment supplier. In situations where different polyethylene piping materials must be joined by the thermal butt-fusion process, both pipe manufacturers should be consulted to determine the appropriate fusion procedures. ASTM D 2657 covers thermal butt-fusion of HDPE pipe.

TABLE 3.17 Pressure Class versus DR–AWWA C906

DR*	Pressure Class, Safety Factor = 2.0 psi (kPa)
11	160 (1100)
13.5	130 (900)
17	100 (690)
21	80 (550)

*These DRs are from the standard dimension ratio series established by ASTM F 412.

HDPE pipe is normally joined above ground and then placed in the pipeline trench. The thermal butt-fusion joint is not subject to movement due to thrust and does not require thrust restraint, such as thrust blocks.

Flanged and mechanical joint adapters are available for joining HDPE pipe to valves and ductile iron fittings. The designer should always consult the pipe manufacturer to ensure a proper fit between pipe and fittings. The designer should also make sure that when connecting to a butterfly valve, the valve disc will freely swing to the open position without hitting the face of the stub end or flange adapter.

3.4.5.4 Gaskets. Gaskets are not necessary for HDPE pipe using thermal butt-fusion joints.

3.4.5.5 Fittings. AWWA C906 HDPE pipe for municipal use is manufactured in ductile iron pipe OD sizes, so ductile iron fittings, conforming to AWWA C110, can be used in all available sizes. See Sec. 3.4.1.5 for a discussion on ductile iron fittings. Although not widely used, HDPE fittings, in configurations similar to ductile iron fittings, are also available. AWWA C906 covers HDPE pressure pipe fittings.

3.4.5.6 Linings and coatings. HDPE pipe does not require lining or coating.

3.4.6 Asbestos-Cement Pipe (ACP)

Asbestos-cement pipe (ACP), available in the United States since 1930, is made by mixing Portland cement and asbestos fiber under pressure and heating it to produce a hard, strong, yet machinable product. It is estimated that over 480,000 km (300,000 mi) of ACP is now in service in the United States.

In the late 1970s, attention was focused on the hazards of asbestos in the environment and, particularly, in drinking water. There was a significant debate on the issue, with one set of experts advising of the potential dangers and a second set of experts claiming that pipes made with asbestos do not result in increases in asbestos concentrations in the water. Studies have shown no association between water delivered by ACP and any general disease; however, the general fear that resulted from the controversy had a tremendous negative impact on ACP use in the United States. The debate on the health concerns of using ACP, along with the introduction of PVC pipe into the municipal water system market, has reduced the use of ACP significantly in the past several years.

3.4.6.1 Available sizes and thicknesses. ACP is available in diameters of 100–1050 mm (4–42 in). Refer to ASTM C 296 and AWWA C401, C402, and C403 for thickness and pressure ratings and AWWA C401 and C403 for detailed design procedures. AWWA C401 for 100–400 mm (4–16 in) pipe is similar to AWWA C403 for 450–1050 mm (18–42 in) pipe. The properties of asbestos-cement for distribution pipe (AWWA C400) and transmission pipe (AWWA C402) are identical. However, under AWWA C403 (transmission pipe), the suggested minimum safety factor is 2.0 for operating pressure and 1.5 for external loads, whereas the safety factors under AWWA C402 (distribution pipe) are 4.0 and 2.5, respectively. So the larger pipe has the smaller safety factors.

Section 4 in AWWA C403 justifies this on the basis that surge pressure in large pipes tends to be less than those in small pipes. However, surge pressures are not necessarily a function of pipe diameter (Chap. 6). The operating conditions, including surge pressures, for any proposed pipeline installation should be closely evaluated before the pipe class is selected. It is the engineer's design prerogative to select which of the safety factors should apply. AWWA C400 specifies that safety factors should be no less than

4.0 and 2.5 if no surge analysis is made. The low safety factors given in AWWA C403 should be used only if all loads (external, internal, and transient) are carefully and accurately evaluated.

3.4.6.2 Joints and fittings. The joints are usually push-on, twin-gasketed couplings, although mechanical and rubber gasket push-on joints can be used to connect ACP to ductile iron fittings.

Ductile iron fittings conforming to ANSI/AWWA C110/A21.10 are used with ACP, and adapters are available to connect ACP to flanged or mechanical ductile iron fittings. Fabricated steel fittings with rubber gasket joints can also be used.

3.4.7 Pipe Material Selection

Buried piping for municipal water transmission and distribution must resist internal pressure, external loads, differential settlement, and corrosive action of both soils and, potentially, the water it carries. General factors to be considered in the selection of pipe include the following:

- Service conditions
    - Pressure (including surges and transients)
    - Soil loads, bearing capacity of soil, potential settlement
    - Corrosion potential of soil
    - Potential corrosive nature of some waters

- Availability
    - Local availability and experienced installation personnel
    - Sizes and thicknesses (pressure ratings and classes)
    - Compatibility with available fittings

- Properties of the pipe
    - Strength (static and fatigue, especially for waterhammer)
    - Ductility
    - Corrosion resistance
    - Fluid friction resistance (more important in transmission pipelines)

- Economics
    - Cost (installed cost, including freight to job site and installation)
    - Required life
    - Cost of maintenance and repairs

The items listed above are general factors related to pipe selection to be considered during the design of any pipeline. Since most municipal water system projects are either let out to competitive bid or are installed as a part of private land development, the

TABLE 3.18 Comparison of Pipe for Municipal System Service

Pipe	Advantages	Disadvantages/Limitations
Ductile iron (DIP)	Yield strength: 290,000 kPa (42,000 lb/in^2); $E = 166 \times 10^6$ kPa (24×10^6 lb/in^2); ductile, elongation ≈ 10%; good corrosion resistance, wide variety of available fittings and joints; available sizes: 100–1350 mm (4–54 in); ID, wide range of available thicknesses, good resistance to waterhammer, high strength for supporting earth loads	Maximum pressure = 2400 kPa (350 lb/in^2); high cost especially for long freight hauls, no diameters above 1350 mm (54 in); difficult to weld, may require wrapping or cathodic protection in corrosive soils
Steel	Yield strengths: 207,000–414,000 kPa (30,000–60,000 lb/in^2); ultimate strengths: 338,000–518,000 kPa (49,000–75,000 lb/in^2); $E = 207 \times 10^6$ kPa (30×10^6 lb/in^2); ductile, elongation varies from 17 to 35%, pressure rating to 17,000 kPa (2500 lb/in^2); diameters to 3.66 m (12 ft); widest variety of available fittings and joints, custom fittings can be mitered and welded, excellent resistance to waterhammer, low cost, high strength for supporting earth loads	Poor corrosion resistance unless both lined and coated or wrapped, may require cathodic protection in corrosive soils, higher unit cost in smaller diameters
Polyvinyl chloride (PVC)	Tensile strength (hydrostatic design basis) = 26,400 kPa (4000 lb/in^2); $E = 2,600,000$ kPa (400,000 lb/in^2); light weight, very durable, very smooth, liners and wrapping not required, can use ductile iron fittings with adapters, diameters from 100 to 375 mm (4–36 in)	Maximum pressure = 2400 kPa (350 lb/in^2); waterhammer not included in AWWA C905; limited resistance to cyclic loading, unsuited for outdoor use above ground
High-density polyethylene (HDPE)	Tensile strength (hydrostatic design basis) = 11,000 kPa (1600 lb/in^2); $E = 896,000$ kPa (130,000 lb/in^2); lightweight, very durable, very smooth, liners and wrapping not required, can use ductile iron fittings, diameters from 100 to 1600 mm (4 to 63 in)	Maximum pressure = 1750 kPa (250 lb/in^2); relatively new product, 750 mm (30 in) is largest size available for municipal system pressures, thermal butt-fusion joints, requires higher laborer skill
Reinforced concrete pressure (RCPP)	Several types available to suit different conditions, high strength for supporting earth loads, wide variety of sizes from 300 to 3600 mm (24–144 in)	Attacked by soft water, acids, sulfides, sulfates, and chlorides, often requires protective coatings; waterhammer can crack outer shell, exposing reinforcement to corrosion and destroying its strength with time; maximum pressure = 1380 kPa (200 lb/in^2)
Asbestos-cement (ACP)	Yield strength: not applicable; design based on crushing strength, see ASTM C 296 and C 500; $E = 23,500,000$ kPa (3,400,000 lb/in^2); rigid, lightweight in long lengths, low cost; diameters from 100 to 1050 mm (4–42 in), compatible with cast-iron fittings, pressure ratings from 1600 to 3100 kPa (225–450 lb/in^2) for large pipe 450 mm (18 in) or more	Attacked by soft water, acids, sulfates; requires thrust blocks at elbows tees, and dead ends; maximum pressure = 1380 kPa (200 lb/in^2) for pipe up to 400 mm (16 in); health hazards of asbestos in potable water service are controversial

designer will find that the installed cost, lacking specific service conditions that require otherwise, will tend to dictate pipe selection. For example, steel pipe and reinforced concrete pressure pipe are both available in 300 mm (12 in) diameter. However, the installed cost of ductile iron pipe or PVC pipe is typically lower (typical municipal use) in the 300 mm (12 in) size. Therefore, if the service conditions do not require the high-pressure capabilities of steel or reinforced concrete pressure pipe, the logical choice for 300 mm pipe (12 in) will optionally be ductile iron or PVC. Conversely, if the proposed pipeline is 900 mm (36 in) in diameter, the installed cost of both steel and reinforced concrete pressure pipe, depending on location, tend to be much more competitive.

A general comparison of the various types of pipe used in municipal water systems is shown in Table 3.18.

3.5 PIPELINE DESIGN

This section will discuss typical issues that are addressed during the design of water distribution and transmission pipelines. Pressure pipelines must primarily be able to resist internal pressures, external loads (earth and impact loads), forces transferred along the pipe when pipe-to-soil friction is used for thrust restraint, and handling during construction. Each of these design issues will be discussed and appropriate formulas presented.

3.5.1 Internal Pressures

The *internal pressure of a pipeline* creates a circumferential tension stress, frequently termed hoop stress, that governs the pipeline thickness. In other words, the pipe must be thick enough to withstand the pressure of the fluid within. The internal pressure used in design should be that to which the pipe may subjected during its lifetime. In a distribution system this pressure may be the maximum working pressure plus an allowance for surge. It may also be the pipeline testing pressure or the shutoff head of an adjacent pump. In a transmission pipeline, the pressure is measured by the vertical distance between the pipe centerline and the hydraulic grade line. Potential hydraulic grade lines on transmission pipelines should be carefully considered. The static hydraulic grade line is potentially much higher than the dynamic grade line if a downstream valve is closed.

Hoop tensile stress is given by the equation

$$s = \frac{pD}{2t} \tag{3.1}$$

where s = allowable circumferential stress in kPa (lb/in^2), p = pressure in kPa (lb/in^2), D = the outside diameter of the pipe in mm (in), and t = thickness of the pipe in mm (in). It should be noted that this equation is the basis for determining the circumferential stress in steel and reinforced pressure pipe and for determining the pressure classes and pressure ratings for virtually all other different types of pressure pipe.

3.5.2 Loads on Buried Pipe

Buried pipes must support external superimposed loads, including the weight of the soil above plus any live loads, such as wheel loads due to vehicles or equipment. The two broad categories for external structural design are rigid and flexible pipe. *Rigid pipe* supports external loads because of the strength of the pipe itself. *Flexible pipe* distributes the external loads to surrounding soil and/or bedding material. For rigid pipe, the soil between the pipe and the trench wall is more compressible than the pipe. This causes the pipe to carry most of the load across the width of the trench. For a flexible pipe, the fact that the pipe deflects causes the soil directly over the pipe to settle more than the adjacent soil. This settlement produces shearing forces that tend to reduce the load on the flexible pipe. DIP, steel pipe, PVC pipe, and HDPE pipe should be considered flexible and be designed accordingly. AWWA C300, C301, and C302 pipe and AC pipe should be considered rigid. AWWA C303 pipe is designed for external loads, according to AWWA M9 (AWWA, 1995), as "semirigid" using rigid pipe formulas to determine the pipe load and controlled pipe deflection as with a flexible pipe.

Supporting strengths for flexible conduits are generally given as loads required to produce a deflection expressed as a percentage of the diameter. Ductile iron pipe may be designed for deflections up to 3 percent of the pipe diameter according to ANSI A21.50. Historically, plastic pipe manufacturers generally agreed that deflections up to 5 percent of the diameter were acceptable, but some manufacturers suggest that deflections up to 7 percent are permissible. Many engineers, however, believe these values are much too liberal and use 2 to 3 percent. Recommended design deflections for flexible pipe are shown in Table 3.19.

The following generally describes the analysis of superimposed loads on buried pipes. As will be seen, the design involves the stiffness of the pipe, the width and depth of the trench, the kind of bedding, the kind of surrounding soil, and the size of the pipe. There are several different types of pipeline installation conditions that should be recognized by the design engineer because different installation conditions will result in different loads on the pipeline. In this text, the only type of installation condition addressed is commonly referred to as a trench condition, where the width of the trench for the pipeline is no larger than two times the width of the pipe. This condition, naturally, requires that the surrounding soil will hold a vertical (or nearly vertical) wall. The subject of how a buried pipe resists earth loads is a subject that should be thoroughly understood by the pipeline design engineer. A complete presentation of this subject is outside the scope of this text; however, further discussions are given in AWWA M11 (American Water Works Association, 1984), AWWA M9 (American Water Works Association, 1995), the *Handbook of PVC Pipe* (Uni-Bell PVC Pipe Association, 1991), the DIPRA handbook (Ductile Iron Pipe Research Association, 1984), and in many other publications.

3.5.2.1 Earth loads. The *Marston theory* is generally used to determine the loads imposed on buried pipe by the soil surrounding it. This theory is applicable to both flexible and rigid pipes installed in a variety of conditions.

TABLE 3.19 Recommended Maximum Deflections for Flexible Pipe

Type of Pipe	Maximum Deflection[*,†]
DIP	2–3%
PVC	3–5%
HDPE	3–5%
Steel, mortar lined and coated	1.5–2%
Steel, mortar lined and flexible coated (tape)	2–3%
Steel, flexible coating and lining	3–5%
AWWA C303	$D^2/4000$

[*]Percentages are of pipe diameter.
[†]D in AWWA C303 is pipe diameter.

Trench conduits are installed in relatively narrow excavations in passive or undisturbed soil and then covered with earth backfill to the original ground surface. The trench load theory is based on the following assumptions:

- Load on the pipe develops as the backfill settles because the backfill is not compacted to the same density as the surrounding earth.
- The resultant load on an underground structure is equal to the weight of the material above the top of the conduit minus the shearing or friction forces on the sides of the trench. These shearing forces are computed in accordance with Rankine's theory.
- Cohesion is assumed to be negligible because (1) considerable time must elapse before effective cohesion between the backfill material and the sides of the trench can develop, and (2) the assumption of no cohesion yields the maximum probable load on the conduit.
- In the case of rigid pipe, the side fills may be relatively compressible and the pipe itself will carry practically all the load developed over the entire width of the trench.

When a pipe is placed in a trench, the prism of backfill placed above it will tend to settle downward. Frictional forces will develop along the sides of the trench walls as the backfill settles and act upward against the direction of the settlement. The fill load on the pipe is equal to the weight of the mass of fill material less the summation of the frictional load transfer.

3.5.2.2 Rigid pipe. The load on buried rigid pipe is expressed by the following formula:

$$W_d = C_d w B_d^2 \tag{3.2}$$

where W_d = trench fill load, pounds per linear foot (lb/Lft), C_d = trench load coefficient, w = unit weight of fill material (lb/ft³), B_d = width of trench at the top of the pipe in ft, and C_d is further defined as:

$$C_d = \frac{1 - e^{-2Ku'(H/B_d)}}{2Ku'} \tag{3.3}$$

where C_d = trench load coefficient, e = base of natural logarithms, K = tan (45° - ϕ'/2) = Rankine's ratio of active lateral unit pressure to vertical unit pressure, with ϕ' = friction angle between backfill and soil, u' = tan ϕ' = friction coefficient of friction between fill material and sides of trench, H = height of fill above top of pipe (ft) B_d = width of trench at the top of the pipe (ft).

Recommended values for the product of Ku' for various soils are:

$Ku' = 0.1924$ for granular materials without cohesion

$Ku' = 0.1650$ maximum for sand and gravel

$Ku' = 0.1500$ maximum for saturated top soil

$Ku' = 0.1300$ maximum for ordinary clay

$Ku' = 0.1100$ maximum for saturated clay

For very deep trenches, the load coefficient C_d approaches a value of $Ku'/2$, so an accurate selection of the appropriate Ku' value becomes more important. The design engineer can benefit greatly from the expert services of an experienced geotechnical engineer who can provide these data to the designer. Generally, though, when the character of the soil is uncertain, it is adequate to assume, for preliminary design, that Ku' = 0.150 and w = 120 lb/ft³ (1922 kg/m³).

Study of the load formula shows that an increase in trench width, B_d (B_d is measured at the top of the pipe), will cause a marked increase in load. Consequently, the value of B should be held to the minimum that is consistent with efficient construction operations and safety requirements. If the trench sides are sloped back or if the width of the trench is large in comparison with the pipe, B_d and the earth load on the pipe can be decreased by constructing a narrow subtrench at the bottom of the wider trench.

As trench width increases, the upward frictional forces become less effective in reducing the load on the pipe until the installation finally assumes the same properties as a positive projecting embankment condition, where a pipe is installed with the top of the pipe projecting above the surface of the natural ground (or compacted fill) and then covered with earth fill. This situation is common when B_d is approximately equal to or greater than H. The positive projection embankment condition represents the severest load to which a pipe can be subjected. Any further increase in trench width would have no effect on the trench load. The maximum effective trench width, where transition to a positive projecting embankment condition occurs, is referred to as the "transition trench width." The trench load formula does not apply when the transition trench width has been exceeded.

3.5.2.3 Flexible pipe.
For a *flexible pipe*, the ability to deflect without cracking produces a situation where the central prism of soil directly over the pipe settles more than the adjacent soil prisms between the pipe and the trench wall. This differential settlement produces shearing forces that reduce the load on a flexible pipe. If the flexible pipe is buried in a trench less than two times the width of the pipe, the load on the pipe may be computed as follows:

$$W_d = C_d w B_d^2 \left(\frac{B_c}{B_d}\right) = C_d w B_d B_c \tag{3.4}$$

where W_d = trench fill load, in pounds per linear foot, C_d = trench load coefficient as defined above in Eq. (3.4) w = unit weight of fill material (lb/ft³), B_d = width of trench at the top of the pipe (ft) and B_c = outside diameter of the pipe (ft).

The deflection of a properly designed flexible pipe installation is limited by the pipe stiffness and the surrounding soil. Under soil loads, the pipe tends to deflect and develop passive soil support at the sides of the pipe. Recommended deflection limits for various types of pipe are shown above. The Iowa deflection formula was first proposed by M. G. Spangler. It was later modified by Watkins and Spangler and has been frequently rearranged by others. In one of its most common forms, deflection is calculated as follows:

$$\Delta x = D \left(\frac{KWr^3}{EI + 0.061E'r^3}\right) \tag{3.5}$$

where Δx = horizontal deflection of pipe (in), D = deflection lag factor (see further definition below) (1.0−1.5), K = bedding constant (0.1), W = load per unit of pipe length, in pounds per linear inch (lb/Lin), r = radius (in), E = modulus of elasticity of pipe (lb/in²) I = transverse moment of inertia per unit length of pipe wall, [(in⁴/(Lin) = and in³), and E' = modulus of soil reaction, see further definition below (lb/in²).

In pipe soil systems, as with all engineering systems involving soil, the soil consolidation at the sides of the pipe continues with time after the maximum load reaches the top of the pipe. Spangler recognized that some pipe deflections increased by as much as 30 percent over a period of 40 years. For this reason, he recommended the addition of a deflection lag factor (D_l) of 1.5 as a conservative design procedure. Others recommend using an ultimate load with a D_l equal to unity.

One attempt to develop information on values of E' was conducted by Amster K. Howard of the U.S. Bureau of Reclamation. Howard reviewed both laboratory and field data from many sources. Using information from over 100 laboratory and field tests, he compiled a table of average E' values for various soil types and densities. Howard's data are reproduced in Table 3.20. These data can be used in designing pipe soil installations.

3.5.3 Thrust Restraint

Thrust forces are unbalanced forces in pressure pipelines that occur at changes in direction (such as in bends, wyes, and tees), at changes in cross-sectional area (such as in reducers), or at pipeline terminations (such as at bulkheads). If not adequately restrained, these forces tend to disengage nonrestrained joints. Two types of thrust forces are (1) hydrostatic thrust due to internal pressure of the pipeline and (2) hydrodynamic thrust due to the changing momentum of flowing water. Since most water lines operate at relatively low velocities, the dynamic force is insignificant and is usually ignored when computing thrust. For example, the dynamic force created by water flowing at 2.4 m/s (8 ft/s) is less than the static force created by 6.9 kPa (1 psi).

Typical examples of hydrostatic thrust are shown in Fig. 3.11. The thrust in dead ends, outlets, laterals, and reducers is a function of the internal pressure P and the cross-sectional area A at the pipe joint. The resultant thrust at a bend is also a function of the deflection angle Δ and is given by the following:

$$T = 2PA \sin(\Delta/2) \tag{3.6}$$

where T = hydrostatic thrust, in pounds, P = internal pressure, in pounds per square inch, A = cross-sectional area of the pipe joint, in square inches, and Δ = deflection angle of bend, in degrees.

For buried pipelines, thrust resulting from small angular deflections at standard and beveled pipe with rubber-gasket joints is resisted by dead weight or frictional drag of the pipe, and additional restraint is not usually needed. Thrust at in-line fittings, such as valves and reducers, is usually restrained by frictional drag on the longitudinally compressed downstream pipe. Other fittings subjected to unbalanced horizontal thrust have the following two inherent sources of resistance: (1) frictional drag from the dead weight of the fitting, earth cover, and contained water and (2) passive resistance of soil against the back of the fitting. If frictional drag and/or passive resistance is not adequate to resist the thrust involved, then it must be supplemented either by increasing the supporting area on the bearing side of the fitting with a thrust block or by increasing frictional drag of the line by "tying" adjacent pipe to the fitting.

Unbalanced uplift thrust at a vertical deflection is resisted by the dead weight of the fitting, earth cover, and contained water. If that is not adequate to resist the thrust involved, then it must be supplemented either by increasing the dead weight with a gravity-type thrust block or by increasing the dead weight of the line by "tying" adjacent pipe to the fitting.

TABLE 3.20 Average Values of Modulus of Soil Reaction, E' (for Initial Flexible Pipe Deflection)*

Soil Type–Pipe Bedding Material (Unified Classification System)† (1)	E' for Degree of Compaction of Bedding, (psi)			
	Dumped (2)	Slight, <85% Proctor, <40% relative density (3)	Moderate, 85%–95% Proctor, 40–70% relative density (4)	High, >95% Proctor, >70% relative density (5)
Fine-grained soils (LL>50) Soils with medium to high plasticity, CH, MH, CH, MH	No data available, consult a competent soils engineer, otherwise use $E' = 0$			
Fine-grained soils (LL<50) Soils with medium to no plasticity, CL. ML, ML–CL, with less than 25% coarse–grained particles	50	200	400	1000
Fine-grained soils (LL<50) Soils with medium to no plasticity, CL. ML, ML–CL, with less than 25% coarse–grained particles	100	400	1000	2000
Coarse-grained soils with fines GM, GC, SM, SC contains more than 12% fines Coarse-grained soils with little or no fines GW, GP, SW, SP contains less than 12% fines	200	1000	2000	3000
Crushed rock	1000	3000	3000	3000
Accuracy in terms of percentage deflection	±2	±2	±1	±.05

†ASTM Designation D 2487, USBR Designation E-3.
*Or any borderline soil beginning with one of these symbols (i.e. GM-GC, GC0SC).
*For ± 1% accuracy and predicted deflection of 3%, actual deflection would be between 2% and 4%.
Note: Values applicable only for fills less than 50 ft (15 m). Table does not include any safety factor. For use in predicting initial deflections only; appropriate deflection lag factor must be applied for long-term deflections. If bedding falls on the borderline between two compaction categories, select lower E' value or average the two values. Percentage Proctor based on laboratory maximum dry density from test standards using about 12,500 ft-lb/ft³ (598,000 J/m³) American Society for Testing of Material (ASTM D698, AASHTO T-99, USBR Designation E-11). 1 psi = 6.9 kPa.
Source: Howard, A. K., *Soil Reaction for Buried Flexible Pipe*, U.S. Bureau of Reclamation, Denver, CO. Reprinted with permission from American Society of Civil Engineers.
Abbreviations: CH, CH-MH, CL, GC, GM, GP, GW, LL, liquid limit; MH, SC, SM, SP, SW.

When a high water table or submerged conditions are encountered, the effects of buoyancy on all materials should be considered.

3.5.3.1 Thrust blocks. Thrust blocks increase the ability of fittings to resist movement by increasing the bearing area. Typical thrust blocking is shown in Fig. 3.12. Thrust block size can be calculated based on the bearing capacity of the soil as follows:

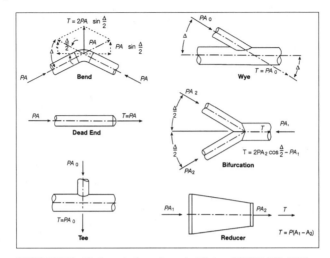

FIGURE 3.11 Hydrostatic thrust for typical fittings (AWWA M9, 1995).

$$\text{Area of block} = L_B \times H_B = (T/\sigma) \qquad (3.7)$$

where

$L_B \times H_B$ = area of bearing surface of thrust block (ft²)

T = thrust force (lb)

σ = safe bearing value for soil (lb/ft²)

If it is impractical to design the block for the thrust force to pass through the geometric center of the soil-bearing area, then the design should be evaluated for stability.

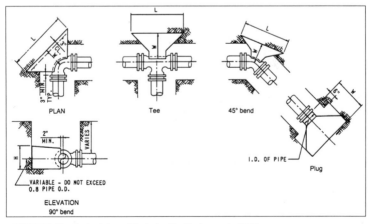

FIGURE 3.12 Typical thrust block details (Sanks, et al., 1989).

Determining the safe bearing value is the key to "sizing" a thrust block. Values can vary from less than 1000 lb/ft² (49.7 kN/m²) for very soft soils to several tons per square foot (kN/m²) for solid rock. Determining the safe bearing value of soil is beyond the scope of this text. It is recommended that a qualified geotechnical expert, knowledgeable of local conditions, be consulted whenever the safe bearing value of a soil is in question.

Most thrust block failures can be attributed to improper construction. Even a correctly sized block can fail if it is not properly constructed. The thrust block must be placed against undisturbed soil, and the face of the block must be perpendicular to the direction of and centered on the line of action of the thrust. Many people involved in construction do not realize the magnitude of the thrusts involved. As an example, a thrust block behind a 36 in (900 mm), 90° bend operating at 100 psi (689 kPa) must resist a thrust force in excess of 150,000 lb (667 kN). Another factor frequently overlooked is that thrust increases in proportion to the square of pipe diameter. A 36-in (900 mm) pipe produces about four times the thrust produced by an 18-in (450-mm) pipe operating at the same internal pressure.

Even a properly designed and constructed thrust block can fail if the soil behind it is disturbed. Thrust blocks of proper size have been poured against undisturbed soil only to fail because another excavation immediately behind the block collapsed when the line was pressurized. The problems of later excavation behind thrust blocks and simply having "chunks" of buried concrete in the pipeline right-of-way have led some engineers to use tied joints only.

3.5.3.2 Restrained joints. Many engineers choose to restrain thrust from fittings by tying adjacent pipe joints. This method fastens a number of pipe joints on each side of the fitting to increase the frictional drag of the connected pipe and resist the fitting thrust.

Much has been written about the length of pipe that is necessary to resist hydrostatic thrust. Different formulas are recommended in various references, particularly with respect to the length of pipe needed at each leg of a horizontal bend. AWWA M9 (AWWA, 1995) provides the following:

$$L = \frac{PA \sin(\Delta/2)}{f(2W_e + W_p + W_w)} \tag{3.8}$$

where L = length of pipe tied to each bend leg (ft), P = internal pressure (lb/in²), A = cross-sectional area of first unrestrained pipe joint (in²), Δ = deflection angle of bend (°), f = coefficient of friction between pipe and soil, W_e = weight of soil prism above pipe (lb/Lft), W_p = weight of pipe (lb/Lft), and W_w = weight of water in pipe (lb/Lft).

This formula assumes that the thrust is resolved by a frictional force acting directly opposite to the hydrostatic thrust. It also assumes that the weight of the earth on top of the pipe can be included in the frictional calculation no matter what the installation depth. The factor $2W_e$ appearing in the denominator indicates that the weight of the earth is acting both on the top and the bottom of the pipe.

AWWA M11 (AWWA, 1989) presents the following alternative formula:

$$L = \frac{PA(1-\cos\Delta)}{f(W_e + W_p + W_w)} \tag{3.9}$$

This formula assumes that the thrust is resolved by frictional forces acting along the length of the pipe and does not assume that the weight of the earth acts both on the top and bottom of the pipe.

The resolution of hydrostatic thrust, acting on a bend of angle Δ, by a length of restrained-joint pipe on each side of the bend is governed by the following:

- Restraint of the pipe is accomplished by a combination of indeterminate forces, including friction between the pipe and soil along the pipe and passive soil pressure perpendicular to the pipe.
- When the pipe is pressurized, the thrust T is not counteracted until the elbow (and length of restrained pipe) moves an amount sufficient (albeit very small) to mobilize friction and passive pressures.
- To develop the frictional resistance at the top of the pipe, it is necessary that the prism of earth above the pipe be restrained from movement.

Given the above statements, the reader is directed to Fig. 3.13. The following should be noted:

- The method proposed in AWWA M9 (AWWA, 1995) does not directly relate the resolution of thrust to a frictional force acting along a length of pipe.
- The method proposed in AWWA M11 (AWWA, 1989) relates the resolution of thrust to a frictional force acting along one leg of the pipe to be restrained. In practice, both legs on each side of the bend are restrained.
- The resolved forces method presents a rational distribution of forces acting to resolve the thrust.

According to Fig. 3.13, the resolution of thrust T by frictional forces acting along the pipe is given by the following equation:

$$R_f = PA \, sin^2 \, \Delta/2 \qquad (3.10)$$

where

$R_f =$ frictional forces acting along each leg of pipe from bend of angle Δ

Inspection of Fig. 3.13 also indicates the following:

- The $PA \, sin \, \Delta/2 \, cos \, \Delta/2$ force is a shearing force acting across the pipe joint. These forces can be significant, even for small Δs and should be considered in the design.
- The length of pipe to be restrained, using AWWA M9 (AWWA, 1995) or AWWA M11 (AWWA, 1989) formulas, is always greater than the lengths using the resolved forces method, $PA \, sin^2 \, \Delta/2$, so either AWWA method can be used as a conservative analysis.

As stated above, to develop the frictional resistance at the top of the pipe, it is necessary that the prism of earth above the pipe be restrained from movement. This can be assumed to be true and the frictional resistance at the top of the pipe included in the calculation if the following can be shown:

$$2P_o \, tan \, \varphi \geq Wf \qquad (3.11)$$

P_o is further defined as:

$$P_o = (\gamma/2) \, H^2 \, k_o$$

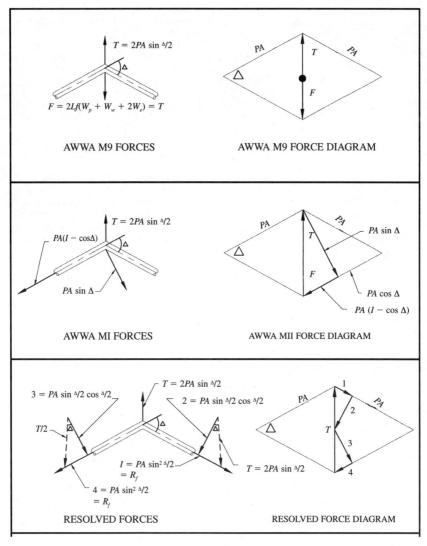

FIGURE 3.13 Frictional thrust restraint.

where P_o = force available from earth prism to provide frictional resistance (lb), W = weight of soil prism above pipe (lb), f = coefficient of friction between pipe and soil, H = height of cover over top of pipe (ft), k_o = coefficient of soil at rest: 0.4 for crushed rock, 0.6 for saturated silty sands, and φ = soil internal angle of friction (varies with soil type) consult geotechnical expert.

Therefore, if the soil prism above the pipe is restrained from movement and can be included in the restraint calculation, the formula for the length of pipe with restrained joints, in accordance with Fig. 3.13, becomes:

$$L = \frac{PA \sin^2 \Delta/2}{f(2W_e + W_p + W_w)} \qquad (3.12)$$

If the soil prism above the pipe is not restrained from movement and cannot be included in the restraint calculation, the formula for the length of pipe with restrained joints, in accordance with Figure 3.13, becomes:

$$L = \frac{PA \sin^2 \Delta/2}{f(W_e + W_p + W_w)} \qquad (3.13)$$

In all the above equations, the value of the coefficient of friction f between the pipe and soil affects the length of pipe that will be required to be restrained. Tests and experience indicate that the value of f is not only a function of the type of soil, but it is also greatly affected by the degree of compaction and moisture content of the backfill, the pipe exterior, and even the pipe joint configuration. Therefore, care should be exercised in the selection of f. Coefficients of friction are generally in the range of 0.2 for PVC or a polyethylene bag to 0.35– 0.4 for a cement mortar coating.

3.6 DISTRIBUTION AND TRANSMISSION SYSTEM VALVES

There are many different types of valves used in municipal water systems, particularly when water treatment plants and pumping station plants are included. There are, however, relatively few different types of valves common to water distribution systems and transmission mains. The discussion in this section will be limited to those valves normally used in municipal distribution and transmission systems.

3.6.1 Isolation Valves

Isolation valves, as their name indicates, are placed into the system to isolate a portion of the system for repair, inspection, or maintenance. They are normally either fully closed or fully opened. Valves that remain in one position for extended periods become difficult (or even impossible) to operate unless they are "exercised" from time to time. Valves should be exercised at least once each year (more often if the water is corrosive or dirty).

In a distribution system, isolation valves are normally installed at junctions. The normal "rule of thumb" for how many valves to install at a junction is one less valve than there are legs at the junctions. In other words, a cross junction (four legs) would require three valves, and a tee junction (three legs) would require two valves, and so on. The designer is encouraged, however, to evaluate seriously the need to isolate a critical line segment and require appropriate valving. For example, isolation of a critical line segment between two cross junctions could require the closing of six isolation valves if the "rule of thumb" is used in design. This obvious inconvenience has caused some designers to adopt a more conservative approach and specify one valve for each leg at junctions.

On large-diameter transmission pipelines, it is common to require the installation of isolation valves at periodic points on the pipeline to minimize the amount of pipeline that must be drained for inspection or maintenance. Depending on the size of the pipeline, isolation valve spacing of up to 5 mi is not uncommon.

In municipal water distribution and transmission systems, the two most common types of isolation valves are gate valves and butterfly valves.

3.6.1.1 Gate valves. For distribution systems, where line sizes are typically between 150 and 400 mm (6–16 in), gate valves are very common. The *gate valve* has a disc sliding in a bonnet at a right angle to the direction of flow. Common gate valve types include the following subtypes:

- Double disc
- Solid wedge resilient seated
- Knife

The *double-disc gate valve* is one of the most popular types for municipal distribution systems. After the discs drop into their seats, further movement of the stem wedges the discs outward to produce a leak proof shutoff even at pressure exceeding 1700 kPa (250 lb/in^2). Opening the valve reverses the procedure. Hence, the discs do not slide until the wedging is relaxed, and sliding and grinding between the disc rings and body rings are minimized.

Solid wedge resilient seated gate valves are a very popular valve choice and are gaining acceptance throughout the United States, particularly when the water contains even small amounts of sand or sediment. The seat of a *gate valve* is a pocket that can entrap solids and prevent the valve from fully closing. The resilient seat type greatly reduces this problem because it has no pocket in the body in which the gate seats (see AWWA C590). Instead, the rubber edge of the disc seats directly on the valve body. The disc is encapsulated with a resilient material (usually vulcanized rubber) that presses against the smooth, prismatic body of the valve. Because there is no pocket for the disc at the bottom of the valve to collect grit, the resilient seated gate valve is suitable for grit-laden waters as well as clean water service.

The *knife gate valve* is lighter than other types of gate valves and is capable of handling more debris than other gate valve types, but does not shut off as effectively and is subject to leakage around the stem packing. The knife gate valve is suitable only when some leakage can be tolerated and when the maximum pressures are around 170–350 kPa (25–50 lb/in^2). This valve is not often seen in municipal water system service.

Gate valves are available in rising-stem and nonrising stem designs. Most gate valves for buried service are nonrising stem, furnished with a 2 in^2 nut that can be accessed from the ground surface to operate the valve.

3.6.1.2 Butterfly valves. A *butterfly valve* is a quarter-turn valve in which a disc is rotated on a shaft so that the disc seats on a ring in the valve body. The seat is usually an elastomer, bonded or fastened either to the disc or to the body. Most, if not all, manufacturers have now standardized on the short body style (see AWWA C504). In the long-body style, the disc is contained entirely within the valve body when the disc is in the fully open position. In the short-body style, the disc protrudes into the adjacent piping when in the open position. In using the short-body-style butterfly valve, the designer must make sure that the pipe, including interior lining, is large enough to accept the disc.

The designer should pay close attention to the seat design if the valve will be subject to throttling. AWWA C504 standards alone do not ensure butterfly valve seats that are adequate for severe throttling (as in pump control valves) where seats must be very rugged for longevity.

Some agencies use butterfly valves for all isolation valve sizes. Because the disc of the butterfly valve prevents passage of a line pig and pigging distribution system lines is considered by many to be a normal maintenance activity, many agencies use gate valves only for 300 mm (12 in) and smaller distribution system lines. Due to economic considerations, butterfly valves are much more commonly used than gate valves on transmission mains where the valve size is larger than 300 mm (12 in).

3.6.2 Control Valves

Special *control valves* are sometimes used to modulate flow or pressure by operating in a partly open position, creating headloss or pressure differential between upstream and downstream locations. Some control valves are manually operated (e.g., needle valves used to control the flow of a fluid in a valued actuator). Some control valves are power-operated by programmed controllers. These are several different varieties of special control valves commonly used in distribution and transmission systems: *pressure reducing*, *pressure sustaining*, *flow control*, *altitude*, and *pressure relief*.

Control valves are selected on the basis of the requirements of the hydraulic system. The designer should use great care in selecting both the type and size of control valve that will be used by carefully evaluating the range of flows that will be handled by the valve. If a valve that is too large is selected, the headloss through the valve may not be enough for the valve to function properly. On the other hand, a large differential pressure across the valve may cause cavitation, which will cause noise, vibration, fluttering of the valve disc, and excessive wearing of the valve seats. While some valves can accommodate sustained velocities of up to 6.1 m/s (20 ft/s), designs for flow velocities between 2.4 and 3.7 m/s (8–12 ft/s) are common. If the designer finds that the expected range of flows is too great to be handled by one size, valve, installations incorporating two of more different sizes of valves are common. With this type of installation, the smaller valves are set to operate at low flows, and the larger valves, become active only during higher flow periods.

Most special control valves have the same body. Only the exterior piping (or more appropriately tubing) to the hydraulic actuator (diaphragm or piston) in the bonnet is changed to effect the type of control wanted, whether it is constant flow, constant pressure, or proportional flow.

Most control valves are either angle or globe pattern. *Angle valves* and *globe valves* are similar in construction and operation except that in an angle valve, the outlet is at 90° to the inlet and the headloss is typically half as great as in the straight-through globe valve. An angle valve is useful if it can serve the dual purpose of a 90° elbow and a valve. Conversely, an angle valve should not be used in a straight piping run where a globe valve should be used. As in the angle valve, a globe valve has a disc or plug that moves vertically in a bulbous body. Flow through a globe valve is directed through two 90° turns (upward then outward) and is controlled or restricted by the disc or plug. The pressure drop across a globe valve is higher than a comparably sized angle valve. Globe valves are either diaphragm or piston operated.

3.6.2.1 Pressure-reducing valve. *Pressure-reducing valves* are often used to establish lower pressure in systems with more than one pressure zone. The pressure-reducing valve will modulate to maintain a preset downstream pressure independent of the upstream pressure. As upstream pressure increases, the valve will close, creating more headloss across the valve, until the target pressure is obtained. Conversely, as upstream pressure decreases, the valve will open. If the upstream pressure decreases to a point lower than the target pressure, the valve would be wide open.

3.6.2.2 Pressure-sustaining valves. Pressure-sustaining valves serve a unique purpose-and their application is limited. In essence they operate in an opposite fashion than a pressure reducing valve. The pressure-sustaining valve functions to maintain a minimum upstream pressure, closing as the upstream pressure drops and opening as the upstream pressure rises. They are sometimes used in multiple pressure zone systems when the downstream zone demand can create pressures that are too low in the upstream zone if not controlled. In these cases, the upstream pressure can be maintained by a pressure-sustaining valve. Naturally, the demand in the downstream zone would need to be met by another source.

3.6.2.3 Flow-control valves. Flow-control valves, like pressure-reducing valves, modulate to maintain a downstream flow characteristic, but, rather than pressure, they will modulate to maintain a preset flow. The flow can be determined by a number of alternatives. Completely hydraulic valves can be operated in response to an orifice plate (factory sized for the design flow) in the piping, or the valve can be operated by an electric operator with some type of flow meter driving the electric operator. Venturi, magnetic, and propeller meter installations are common. As upstream pressure varies, the flow-control valve will open or close to deliver the preset flow.

3.6.2.4 Altitude valves. Altitude valves are used to add water to reservoirs and to one-way tanks used in surge control (Chap. 12). Altitude valves are made in many variations of two functional designs:

- The first, in which the valve closes upon high water level in the tank and does not open again until the water leaves through a separate line and the water level in the tank falls.
- The second, in which the valve closes upon high water level in the tank and opens to allow water to flow out of the tank when pressure on the valve inlet falls below a preset level or below the reservoir pressure on the downstream side of the valve.

3.6.2.5 Pressure-relief valves. As the name implies, *pressure-relief valves* serve to release fluid in a pressure system before high pressure can develop and overstress piping and valves. They are often used in pumping station piping or in other locations where valve operation may induce higher pressures than can be tolerated by the system. Pressure-relief valves are set to open at a preset high pressure. They normally vent to atmosphere. To function properly, they must be positioned so that their discharge is handled in a safe and environmentally sound manner.

3.6.3 Blow-offs

Most water distribution systems, in spite of the best planning efforts, have dead ends where the water can become stagnant. These locations may include cul-de-sacs beyond the last customer service or portions of the distribution system not yet connected to the remainder of the system. At such dead-end locations, it is common to install a *blow-off* that can be periodically opened to allow the stagnant water to be removed from the system. Blow-offs often consist of a small diameter pipe, extended to the surface and terminated in a valve box with a valve than can be operated to allow removal of water from the system. Blow-offs sizes between 50 and 100 mm (2–4 in) are common.

The function of blow-offs on transmission pipelines is often to allow draining of the line for maintenance or inspection and for flushing of the pipeline during construction. To function as a flushing element, the blow-off must be sized to allow cleansing velocity in the main pipeline.

3.6.4 Air-Release and Vacuum-Relief Valves

Air valves are installed with pipelines to admit or vent air. There are basically two types: air-release valves and air-and-vacuum valves. In addition, a combination air valve is available that combines the functions of an air-release and an air-and-vacuum valve.

Air-release valves are used to release air entrained under pressure at high points of a pipeline where the pipe slopes are too steep for the air to be carried through with the flow. The accumulation of air can become so large that it impairs the pipelines flow capacity. Air-release valves are installed at high points to provide for the continuous venting of accumulated air. An air-release valve consists of a chamber in which a float operates through levers to open a small air vent in the chamber top as air accumulates and to close the vent as the water level rises. The float must operate against an air pressure equal to the water pressure and must be able to sustain the maximum pipeline pressure.

Air-and-vacuum valves are used to admit air into a pipe to prevent the creation of a vacuum that may be the result of a valve operation, the rapid draining or failure of a pipe, a column separation, or other causes. Although uncommon, a vacuum in a pipeline can cause the pipe to collapse from atmospheric pressure. Air-and-vacuum valves also serve to vent air from the pipeline while it is filling with water. An air-and-vacuum valve consists of a chamber with a float that is generally center guided. The float opens and closes against a large air vent. As the water level recedes in the chamber, air is permitted to enter; as the water level rises, air is vented. Air-and-vacuum valves are often used as surge control devices and must be carefully sized when used for this purpose (Chap. 12). The air-and-vacuum valve does not vent air under pressure.

Air-release valves and air-and-vacuum valves, if not installed directly over the pipe, may be located adjacent to the pipeline. A horizontal run of pipe connects the air valve and the pipeline. The connecting pipe should rise gradually to the air valve to permit flow of the air to the valve for venting. The performance requirements of the valves are based on the venting capacity and the pressure differential across the valves (system water pressure less atmospheric pressure). The valves must be protected against freezing, and the vents from these valves must be located above ground or positioned so as to prevent contamination when operating. Manufacturers' catalogs should be consulted for accurate sizing information.

REFERENCES

American Water Works Association, AWWA M11, *Steel Pipe—A Guide for Design and Installation*, 3rd ed., American Water Works Association, Denver, CO, 1989.

American Water Works Association, AWWA M9, *Concrete Pressure Pipe*, 2d ed., American Water Works Association, Denver, CO, 1995.

Ductile Iron Pipe Research Association, DIPRA, *Handbook of Ductile Iron Pipe*, 6th ed., Ductile Iron Pipe Research Association, Birmingham, AL, 1984.

Handbook of PVC Pipe, Design and Construction, 3rd ed., Uni-Bell PVC Pipe Association, Dallas, TX, 1991.

Lyne, C., *Updates on Polyethylene Pipe Standard AWWA C906*, CA/NV AWWA Spring Conference, 1997.

Manganaro, C. A., *Harnessed Joints for Water Pipe*, AWWA Annual Conference, San Diego, CA, 1969.

Metcalf & Eddy, Inc., *Water Resources and Environmental Engineering*, 2d ed., McGraw-Hill, New York, 1979.

Piping Engineering, Tube Turns Division, Louisville, KY, 1974.

Sanks, R. L., et al., *Pumping Station Design*, Butterworths, 1989.

Spangler, M. G., *Soil Engineering*, 2d ed., International Textbook Company, Scranton, PA, 1960.

Solley, W. B., R. R. Pierce, and H. A. Perlman, Estimated use of water in the United States in 1990, U. S. Geological Survey Circular 1081, Washington, D. C. 1993.

CHAPTER 4
HYDRAULICS OF WATER DISTRIBUTION SYSTEMS

Kevin Lansey
Department of Civil Engineering and Engineering Mechanics
University of Arizona
Tucson, AZ

Larry W. Mays
Department of Civil and Environmental Engineering
Arizona State University
Tempe, AZ

4.1 INTRODUCTION

In developed countries, water service is generally assumed to be reliable and utility customers expect high-quality service. Design and operation of water systems require an understanding of the flow in complex systems and the associated energy losses. This chapter builds on the fundamental flow relationships described in Chap. 2 by applying them to water distribution systems. Flow in series and parallel pipes is presented first and is followed by the analysis of pipe networks containing multiple loops. Water-quality modeling is also presented. Because solving the flow equations by hand for systems beyond a simple network is not practical, computer models are used. Application of these models is also discussed.

4.1.1 Configuration and Components of Water Distribution Systems

A water distribution system consists of three major components: pumps, distribution storage, and distribution piping network. Most systems require pumps to supply lift to overcome differences in elevation and energy losses caused by friction. Pump selection and analysis is presented in Chap. 5. Storage tanks are included in systems for emergency supply or for balancing storage to reduce energy costs. Pipes may contain flow-control

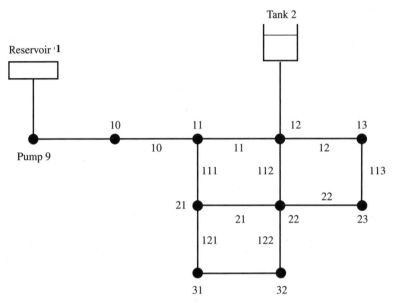

FIGURE 4.1 Network schematic (from EPANET User's Manual, Rossman, 1994).

devices, such as regulating or pressure-reducing valves. A schematic of a distribution system is shown in Fig. 4.1.

The purpose of a distribution system is to supply the system's users with the amount of water demanded under adequate pressure for various loading conditions. A *loading condition* is a spatial pattern of demands that defines the users' flow requirements. The flow rate in individual pipes results from the loading condition and is one variable that describes the network's hydraulic condition. The piezometric and pressure heads are other descriptive variables. The *piezometric* or *hydraulic head* is the surface of the hydraulic grade line or the pressure head (p/γ) plus the elevation head (z):

$$h = \frac{p}{\gamma} + z \qquad (4.1)$$

Because the velocity is relatively small compared to the pressure in these systems, the velocity head typically is neglected. Heads are usually computed at junction nodes. A junction node is a connection of two or more pipes or a withdrawal point from the network. A *fixed-grade node* (FGN) is a node for which the total energy is known, such as a tank.

The loading condition may remain constant or vary over time. A distribution system is in steady state when a constant loading condition is applied and the system state (the flow in all pipes and pressure head at all nodes) does not vary in time. Unsteady conditions, on the other hand, are more common and hold when the system's state varies with time. Extended-period simulation (EPS) considers time variation in tank elevation conditions or demands in discrete time periods. However, within each time period, the flow within the network is assumed to be in steady state. The only variables in the network that are carried between time steps of an EPS are the tank conditions that are updated by a conservation of mass relationship.

Dynamic modeling refers to unsteady flow conditions that may vary at a point and between points from instant to instant. Transient analysis is used to evaluate rapidly varying changes in flow, such as a fast valve closure or switching on a pump. Gradually varied conditions assume that a pipe is rigid and that changes in flow occur instantaneously along a pipe so that the velocity along a pipe is uniform but may change in time. Steady, extended period simulation, and gradually temporally varied conditions are discussed in this chapter. Transient analysis is described in Chap. 6.

4.1.2 Conservation Equations for Pipe Systems

The governing laws for flow in pipe systems under steady conditions are conservation of mass and energy. The *law of conservation of mass* states that the rate of storage in a system is equal to the difference between the inflow and outflow to the system. In pressurized water distribution networks, no storage can occur within the pipe network, although tank storage may change over time. Therefore, in a pipe, another component, or a junction node, the inflow and outflow must balance. For a junction node,

$$\Sigma Q_{in} - \Sigma Q_{out} = q_{ext} \tag{4.2}$$

where Q_{in} and Q_{out} are the pipe flow rates into and out of the node and q_{ext} is the external demand or supply.

Conservation of energy states that the difference in energy between two points is equal to the frictional and minor losses and the energy added to the flow in components between these points. An energy balance can be written for paths between the two end points of a single pipe, between two FGNs through a series of pipes, valves, and pumps, or around a loop that begins and ends at the same point. In a general form for any path,

$$\sum_{i \in I_p} h_{L,i} + \sum_{j \in J_p} h_{p,j} = \Delta E \tag{4.3}$$

where $h_{L,i}$ is the headloss across component i along the path, $h_{p,j}$ is the head added by pump j, and ΔE is the difference in energy between the end points of the path.

Signs are applied to each term in Eq. (4.3) to account for the direction of flow. A common convention is to determine flow directions relative to moving clockwise around the loop. A pipe or another element of energy loss with flow in the clockwise direction would be positive in Eq. (4.3), and flows in the counterclockwise direction are given a negative sign. A pump with flow in the clockwise direction would have a negative sign in Eq. (4.3), whereas counterclockwise flow in a pump would be given a positive sign. See the Hardy Cross method in Sec. 4.2.3.1 for an example.

4.1.3 Network Components

The primary network component is a pipe. Pipe flow (Q) and energy loss caused by friction (h_L) in individual pipes can be represented by a number of equations, including the Darcy-Weisbach and Hazen-Williams equations that are discussed and compared in Sec. 2.6.2. The general relationship is of the form

$$h_L = KQ^n \tag{4.4}$$

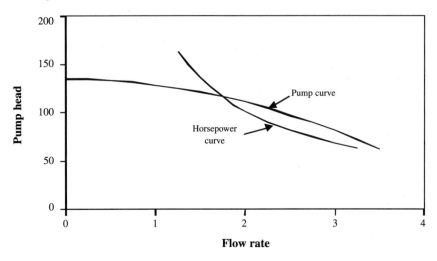

FIGURE 4.2 Typical pump curve.

where K is a pipe coefficient that depends on the pipe's diameter, length, and material and n is an exponent in the range of 2. K is a constant in turbulent flow that is commonly assumed to occur in distribution systems.

In addition to pipes, general distribution systems can contain pumps, control valves, and regulating valves. Pumps add head h_p to flow. As shown in Fig. 4.2, the amount of pump head decreases with increasing discharge. Common equations for approximating a pump curve are

$$h_p = AQ^2 + BQ + h_c \qquad (4.5)$$

or

$$h_p = h_c - CQ^m \qquad (4.6)$$

where A, B, C, and m are coefficients and h_c is the maximum or *cutoff head*. A pump curve can also be approximated by the pump horsepower relationship (Fig. 4.2) of the form

$$H_p = \frac{\gamma Q h_p}{550} \qquad (4.7)$$

where H_p is the pump's water horsepower. Further details about pumps and pump selection are discussed in Chap. 5.

Valves and other fittings also appear within pipe networks. Most often, the headloss in these components is related to the square of the velocity by

$$h_m = K_v \frac{V^2}{2g} = K_v \frac{Q^2}{A^2 2g} \qquad (4.8)$$

where h_m is the headloss, and K_v is an empirical coefficient. Table 2.2 lists K_v values for a number of appurtenances.

Pressure-regulating valves *(PRVs)* are included in many pipe systems to avoid excessive pressure in networks covering varying topography or to isolate pressure zones for reliability and maintaining pressures. Pressure regulators maintain a constant pressure at the downstream side of the valve by throttling flow. Mathematical representation of *PRVs* may be discontinuous, given that no flow can pass under certain conditions.

4.2 STEADY-STATE HYDRAULIC ANALYSIS

4.2.1 Series and Parallel Pipe Systems

The simplest layouts of multiple pipes are series and parallel configurations (Fig. 4.3). To simplify analysis, these pipes can be converted to an equivalent single pipe that will have the same relationship between headloss and flow rate as the original complex configuration.

Series systems, as shown in the Fig. 4.3A, may consist of varying pipe sizes or types. However, because no withdrawals occur along the pipe, the discharge through each pipe is the same. Since the pipes are different, headlosses vary between each segment. The total headloss from a to b is the sum of the headlosses in individual pipes,

$$h_L = \sum_{i \in I_p} h_{L,i} = \sum_{i \in I_p} K_i Q^{n_i} \qquad (4.9)$$

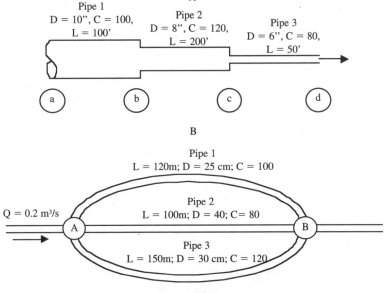

FIGURE 4.3 Pipe systems. A: Series pipe system (not to scale), B: Branched pipe.

where I_p are the set of pipes in the series of pipes. Assuming turbulent flow conditions and a common equation, with the same n_i for all pipes, a single equivalent pipe relationship can be substituted:

$$h_L = K_e Q^n \tag{4.10}$$

where K_e is the pipe coefficient for the equivalent pipe. K_e can be determined by combining Eqs. (4.9) and (4.10):

$$K_e = K_1 + K_2 + K_3 + \ldots = \sum_{i \in I_p} K_i \tag{4.11}$$

Note that no assumption was made regarding Q, so K_e is independent of the flow rate.
Problem. For the three pipes in series in Fig. 4.3, (1) find the equivalent pipe coefficient, (2) calculate the discharge in the pipes is if the total headloss is 10 ft, and (3) determine the piezometric head at points b, c, and d if the total energy at the inlet (pt. a) is 95 ft

Solution. For English units, the K coefficient for the Hazen-Williams equation is

$$K = \frac{\phi L}{C^{1.85} D^{4.87}} \tag{4.12}$$

where ϕ is a unit constant equal to 4.73, L and D are in feet, and C is the Hazen-Williams coefficient. Substituting the appropriate values gives $K_1 = 0.229$, $K_2 = 0.970$, and $K_3 = 2.085$. The equivalent K_e is the sum of the individual pipes (Eq. 4.11), or $K_e = 3.284$.

Using the equivalent loss coefficient, the flow rate can be found by Eq. (4.10), or $h_L = K_e Q^{1.85}$. For h_L equals 10 feet and K_e equals 3.284, the discharge is 1.83 cfs. This relationship and K_e can be used for any flow rate and head loss. Thus, if the flow rate was 2.2 cfs, the headloss by Eq. (4.10) would be $h_L = 3.284\,(2.2)^{1.85} = 14.1$ ft.

The energy at a point in the series pipes can be determined by using a path headloss equation of the form of Eq. (4.3). The total energy at point b is the total energy at the source minus the headloss in the first pipe segment, or

$$H_b = H_a - h_{L,1} = 95 - K_1 Q^{1.85} = 95 - 0.229(1.83)^{1.85} = 94.3 \text{ ft}$$

Similarly, the headlosses in the second and third pipes are 2.97 and 6.38 ft., respectively. Thus, the energy at c and d are 91.33 and 84.95 ft, respectively.

Two or more parallel pipes (Fig. 4.3B) can also be reduced to an equivalent pipe with a similar K_e. If the pipes are not identical in size, material, and length, the flow through each will be different. The energy loss in each pipe, however, must be the same because they have common end points, or

$$h_A - h_B = h_{L,1} = h_{L,2} = h_{L,j} \tag{4.13}$$

Since flow must be conserved, the flow rate in the upstream and downstream pipes must be equal to the sum of the flow in the parallel pipes, or

$$Q = Q_1 + Q_2 + \ldots = \sum_{m \in M_p} Q_m \tag{4.14}$$

where pipe m is in the set of parallel pipes, M_p. Manipulating the flow equation (Eq. 4.4), the flow in an individual pipe can be written in terms of the discharge by $Q = (h_L/K)^{1/n}$. Substituting this in Eq. (4.14) gives

$$Q = \left(\frac{h_{L,1}}{K_1}\right)^{1/n_1} + \left(\frac{h_{L,2}}{K_2}\right)^{1/n_2} + \left(\frac{h_{L,3}}{K_3}\right)^{1/n_3} + \ldots \qquad (4.15)$$

As is noted in Eq. (4.13), the headloss in each parallel pipe is the same. If the same n is assumed for all pipes, Eq. (4.15) can be simplified to

$$Q = h_L^{1/n}\left[\left(\frac{1}{K_1}\right)^{1/n} + \left(\frac{1}{K_2}\right)^{1/n} + \left(\frac{1}{K_3}\right)^{1/n} + \ldots\right] = h_L^{1/n} \sum_{m \in M_p} \left(\frac{1}{K_i}\right)^{1/n} = h_L^{1/n}\left(\frac{1}{K_e}\right)^{1/n} \qquad (4.16)$$

Dividing by $h_L^{1/n}$ isolates the following equivalent coefficient:

$$\sum_{m \in M_p} \left(\frac{1}{K_i}\right)^{1/n} = \left(\frac{1}{K_e}\right)^{1/n} \qquad (4.17)$$

Because the K values are known for each pipe based on their physical properties, K_e can be computed, then substituted in Eq. (4.10) to determine the headloss across the parallel pipes, given the flow in the main pipe.

Problem. Determine the headloss between points A and B for the three parallel pipes. The total system flow is 0.2 m³/s. Also find the flow in each pipe.

Solution. The headloss coefficient K for each pipe is computed by Eq. (4.12), with ϕ equal to 10.66 for SI units and L and D in meters, or $K_1 = 218.3$, $K_2 = 27.9$, and $K_3 = 80.1$. The equivalent K_e is found from Eq. (4.17):

$$\left(\frac{1}{K_1}\right)^{\frac{1}{1.85}} + \left(\frac{1}{K_2}\right)^{\frac{1}{1.85}} + \left(\frac{1}{K_3}\right)^{\frac{1}{1.85}} = \left(\frac{1}{K_e}\right)^{\frac{1}{1.85}} = 0.313$$

or $K_e = 8.58$. By Eq. (4.10), the head loss is $h_L = K_e Q^{1.85} = 8.58*(0.2)^{1.85} = 0.437$ m.

The flow in each pipe can be computed using the individual pipe's flow equation and K. For example, $Q_1 = (h_L/K_1)^{1/1.85} = (0.437/218.3)^{1/1.85} = 0.035$ m³/s. Similarly, Q_2 and Q_3 are 0.105 and 0.060 m³/s, respectively. Note that the sum of the flows is 0.2 m³/s, which satisfies conservation of mass.

4.2.2 Branching Pipe Systems

The third basic pipe configuration consists of branched pipes connected at a single junction node. As shown in Fig. 4.4, a common layout is three branching pipes. Under steady conditions, the governing relationship for this system is conservation of mass applied at the junction. Since no water is stored in the pipes, the flow at the junction must balance

$$Q_1 + Q_2 - Q_3 = 0 \qquad (4.18)$$

where the sign on the terms will come from the direction of flow to or from the node. In addition to satisfying continuity at the junction, the total head at the junction is unique.

4.8 Chapter Four

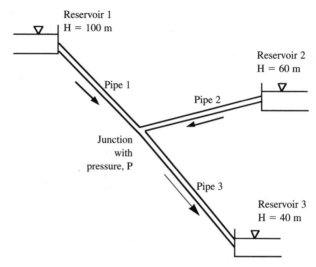

FIGURE 4.4 Branched pipe system.

Given all the pipe characteristics for each system in Fig. 4.4, the seven possible unknowns are the total energy at each source (3), the pipe flows (3), and the junction node's total head P (1). Four equations relating these variables are available: conservation of mass (Eq. 4.18) and the three energy loss equations. Thus, three of the seven variables must be known. Two general problems can be posed.

First, if a source energy, the flow from that source, and one other flow or source energy are is known, all other unknowns can be solved directly. For example, if the flow and source head for reservoir 1 and pipe 1 are known, the pipe flow equation can be used to find P by the following equation (when flow is toward the junction):

$$H_{s1} - P = h_{L,1} = K_1 Q_1^n \tag{4.19}$$

If a flow is the final known (e.g., Q_2), Q_3 can be computed using Eq. (4.18). The source energies can then be computed using the pipe flow equations for Pipes 2 and 3, in the form of Eq. (4.19), with the computed P.

If the final known is a source head, the discharge in the connecting pipe can be computed using the pipe equation in the form of Eq. (4.19). The steps in the previous paragraph are then repeated for the last pipe.

In all other cases when P is unknown, all unknowns can be determined after P is computed. P is found most easily by writing Eq. (4.18) in terms of the source heads. From Eq. (4.19),

$$Q_1 = \text{sign}(H_{s1} - P)\left(\frac{|H_{s1} - P|}{K_1}\right)^{1/n} \tag{4.20}$$

where a positive sign indicates flow to the node. Substituting Eq. (4.20) for each pipe in Eq. (4.18) gives

$$F(P) = \text{sign}(H_{s1} - P)\left(\frac{|H_{s1} - P|}{K_1}\right)^{\frac{1}{n}} + \text{sign}(H_{s2} - P)\left(\frac{|H_{s2} - P|}{K_2}\right)^{\frac{1}{n}} +$$

Hydraulics of Water Distribution Systems **4.9**

$$\text{sign } (H_{s3} - P) \left(\frac{|H_{s3} - P|}{K_3} \right)^{\frac{1}{n}} = 0 \qquad (4.21)$$

If a pipe's flow rate is known, rather than the source head, the flow equation is not substituted; instead the actual flow value is substituted in Eq. (4.21). The only unknown in this equation is P, and it can be solved by trial and error or by a nonlinear equation solution scheme, such as the Newton-Raphson method.

The *Newton-Raphson method* searches for roots of an equation $F(x)$. At an initial estimate of x, a Taylor series expansion is used to approximate F:

$$0 = F(x) + \frac{\partial F}{\partial x}\bigg|_x \Delta x + \frac{\partial^2 F}{\partial x^2}\bigg|_x \Delta x^2 + \ldots \qquad (4.22)$$

where Δx is the change in x required to satisfy $F(x)$. Truncating the expansion at the first-order term gives a linear equation for Δx:

$$\Delta x = -\frac{F(x)}{\partial F/\partial x|_x} \qquad (4.23)$$

The estimated x is updated by $x = x + \Delta x$. Since the higher-order terms were dropped from Eq. (4.22), the correction may not provide an exact solution to $F(x)$. So several iterations may be necessary to move to the solution. Thus, if Δx is less than a defined criterion, the solution has been found and the process ends. If not, the updated x is used in Eq. (4.23) and another Δx is computed. In the three-reservoir case, $x = P$ and the required gradient $\partial F/\partial P$ is:

$$\frac{\partial F}{\partial P} = -\frac{1}{n} \left[\left(\frac{|H_{s1} - P|}{K_1} \right)^{\left(\frac{1}{n}-1\right)} + \left(\frac{|H_{s2} - P|}{K_2} \right)^{\left(\frac{1}{n}-1\right)} + \left(\frac{|H_{s3} - P|}{K_3} \right)^{\left(\frac{1}{n}-1\right)} \right]$$

$$= -\left(\frac{1}{nK_1|Q_1|^{n-1}} + \frac{1}{nK_2|Q_2|^{n-1}} + \frac{1}{nK_3|Q_3|^{n-1}} \right) \qquad (4.24)$$

$F(P)$ is computed from Eq. (4.21) using the present estimate of P. ΔP is then computed using $\Delta P = -F(P)/(\partial F/\partial P)$, and P is updated by adding ΔP to the previous P. The iterations continue until ΔP is less than a defined value. The Newton-Raphson method can also be used for multiple equations, such as the nodal equations (Sec. 4.2.3.3). A matrix is formed of the derivatives of each equation and the update vector is calculated.

Problem. Determine the flow rates in each pipe for the three-pipe system shown in Fig. 4.4. The friction factors in the table below assume turbulent flow conditions through a concrete pipe ($\varepsilon = 0.08$ cm).

Solution. Using the Darcy-Weisbach equation ($n = 2$), the K coefficients are computed using

$$K = \frac{8fL}{\pi^2 D^5 g}$$

Pipe	1	2	3
D (cm)	80	40	40
L (m)	1000	600	700
$f[\ ]$	0.0195	0.0235	0.0235
K	4.9	113.8	132.7
Reservoir elevation H (m)	100	60	40

4.10   Chapter Four

In addition to the three discharges, the energy at the junction P also is unknown. To begin using the Newton-Raphson method, an initial estimate of P is assumed to be 80 m, and Eq. (4.21) is evaluated as follows:

$$F(P = 80m) = \left(sign(100 - 80)\left(\frac{|100 - 80|}{4.9}\right)^{\frac{1}{2}}\right)_{pipe1} + \left(sign(60 - 80)\left(\frac{|60 - 80|}{113.8}\right)^{\frac{1}{2}}\right)_{pipe2}$$

$$+ \left((40 - 80)\left(\frac{|40 - 80|}{132.7}\right)^{\frac{1}{2}}\right)_{pipe3} = 2.020 - 0.419 - 0.549 = 1.052 \text{ m}^3/\text{s}$$

which states that flow enters the node at more than 1.052 m³/s, then leaves through pipes 2 and 3 with $P = 80$ m. Therefore, P must be increased. The correction is computed by Eq. (4.23) after computing $\partial F/\partial P$ using Eq. (4.24):

$$\frac{\partial F}{\partial P} = -\left(\frac{1}{2*4.9*|2.020|^{(2-1)}} + \frac{1}{2*113.8*|-0.419|^{(2-1)}} + \frac{1}{2*132.7*|-0.549|^{(2-1)}}\right) =$$

$$-(0.0505 + 0.0105 + 0.0069) = -0.0679$$

The correction is then

$$\Delta P = -\frac{F(P)}{\frac{\partial F}{\partial P}\bigg|_P} = -\frac{1.052}{-0.0679} = 15.5 \text{ m}$$

The P for the next iteration is then $P = 80 + 15.5 = 95.5$ m. The following iterations give

Iteration 2: $F(P = 95.5 \text{ m}) = -0.247$; $\partial F/\partial P|_{P = 95.5} = -0.120$; $\Delta P = -2.06$ m, $P = 93.44$ m

Iteration 3: $F(P = 93.44 \text{ m}) = -0.020$; $\partial F/\partial P|_{P = 93.44} = -0.102$; $\Delta P = -0.20$ m, $P = 93.24$ m.

Iteration 4: $F(P = 93.24 \text{ m}) = 7.\text{x}10^{-4}$; $\partial F/\partial P|_{P = 93.24} = -0.101$; $\Delta P = -0.006$ m, $P = 93.25$ m.

Stop based on $F(P)$ or ΔP, with $P = 93.25$ m.

Problem. In the same system, the desired flow in pipe 3 is 0.4 m³/s into the tank. What are the flows in the other pipes and the total energy required in Tank 3?

Solution.   First, P is determined with Q_1 and Q_2 using Eq. (4.21). Then H_{s3} can be calculated by the pipe flow equation. Since Q_3 is known, Eq. (4.21) is

$$F(P) = Q_1 + Q_2 + Q_3 = sign(H_{s1} - P)\left(\frac{|H_{s1} - P|}{K_1}\right)^{\frac{1}{n}} + sign(H_{s2} - P)\left(\frac{|H_{s2} - P|}{K_2}\right)^{\frac{1}{n}} - 0.4 = 0$$

Iteration 1
  Using an initial trial of P equal to 90 m, $F(P) = 0.514$ m³/s. When evaluating Eq. (4.24), only the first two terms appear since the flow in pipe 3 is defined, or

$$\frac{\partial F}{\partial P} = -\left(\frac{1}{nK_1|Q_1|} + \frac{1}{nK_2|Q_2|}\right)\bigg|_{P = 90m} = -0.080$$

The correction for the first iteration is then $-(0.514/-0.080) = 6.42$ m, and the new P is 96.42 m. The next two iterations are

Iteration 2: $F(P = 96.42 \text{ m}) = -0.112$; $\partial F/\partial P|_{P=96.42m} = -0.127$; $\Delta P = -0.88$ m, $P = 95.54$ m

Iteration 3: $F(P = 95.54 \text{ m}) = -0.006$; $\partial F/\partial P|_{P=95.54m} = -0.115$; $\Delta P = -0.05$ m, $= 95.49$ m

To determine H_{s3}, the pipe flow equation (Eq. 4.20) is used with the known discharge, or

$$Q_3 = -0.4 = sign(H_{s3} - 95.49) \left(\frac{|H_{s3} - 95.49|}{132.7} \right)^{1/2} \Rightarrow H_{s3} = 74.26 \text{ m}$$

4.2.3 Pipe Networks

A hydraulic model is useful for examining the impact of design and operation decisions. Simple systems, such as those discussed in Secs. 4.2.1 and 4.2.2, can be solved using a hand calculator. However, more complex systems require more effort even for steady state conditions, but, as in simple systems, the flow and pressure-head distribution through a water distribution system must satisfy the laws of conservation of mass and energy (Eqs. 4.2 and 4.3). These relationships have been written in different ways to solve for different sets of unknowns.

Using the energy loss-gain relationships for the different components, the conservation equations can be written in three forms: the node, loop, and pipe equations. All are non-linear and require iterative solution schemes. The form of the equations and their common solution methods are described in the next four sections. Programs that implement these solutions are known as network solvers or simulators.

4.2.3.1 Hardy Cross method. The *Hardy Cross method* was developed in 1936 by Cross before the advent of computers. Therefore, the method is amenable to solution by hand but, as a result, is not computationally efficient for large systems. Essentially, the method is an application of Newton's method to the loop equations.

Loop equations. The *loop equations* express conservation of mass and energy in terms of the pipe flows. Mass must be conserved at a node, as discussed in Sec. 4.2.2 for branched pipes. For all N_j junction nodes in a network, it can be written as

$$\sum_{i \in I_j} Q_i = q_{ext} \qquad (4.25)$$

Conservation of energy (Eq. 4.3) can be written for closed loops that begin and end at the same point ($\Delta E = 0$) and include pipes and pumps as

$$\sum_{i \in I_L} K_i Q_i^n - \sum_{ip \in I_p} (A_{ip} Q_{ip}^2 + B_{ip} Q_{ip} + C_{ip}) = 0 \qquad (4.26)$$

This relationship is written for N_l independent closed loops. Because loops can be nested in the system, the smallest loops, known as primary loops, are identified, and each pipe may appear twice in the set of loops at most. The network in Fig. 4.1 contains three primary loops.

Energy also must be conserved between points of known energy (fixed-grade nodes). If N_f FGNs appear in a network, $N_f - 1$ independent equations can be written in the form of

$$\sum_{i \in I_L} K_i Q_i^n - \sum_{ip \in I_p} (A_{ip} Q_{ip}^2 + B_{ip} Q_{ip} + C_{ip}) = \Delta E_{FGN} \qquad (4.27)$$

where ΔE_{FGN} is the difference in energy between the two FGNs. This set of equations is solved by the Hardy Cross method (Cross, 1936) by successive corrections to the pipe flows in loops and by the linear theory method by solving for the pipe flows directly (Sec. 4.2.3.2).

Solution method. To begin the Hardy Cross method, a set of pipe flows is assumed that satisfies conservation of mass at each node. At each step of the process, a correction ΔQ_L is determined for each loop. The corrections are developed so that they maintain conservation of mass (Eq. 4.25), given the initial set of flows. Since continuity will be preserved, those relationships are not included in the next steps.

The method then focuses on determining pipe flows that satisfy conservation of energy. When the initial flows are substituted in Eqs. (4.26) and (4.27), the equations are not likely to be satisfied. To move toward satisfaction, a correction factor ΔQ_L is determined for each loop by adding this term to the loop equation or for a general loop

$$\sum_{i \in I_L} K_i (Q_i + \Delta Q_L)^n - \sum_{ip \in I_p} (A_{ip} (Q_{ip} + \Delta Q_L)^2 + B_{ip} (Q_{ip} + \Delta Q_L) + C_{ip}) = \Delta E \qquad (4.28)$$

Note that ΔE equals zero for a closed loop and signs on terms are added as described in Sec. 4.1.3. Expanding Eq. (4.28) and assuming that ΔQ_L is small so that higher-order terms can be dropped gives

$$\sum_{i \in I_L} K_i Q_i^n + n \sum_{i \in I_L} \left| K_i Q_L^{n-1} \Delta Q_L \right| - \sum_{ip \in I_p} (A_{ip} Q_{ip}^2 + B_{ip} Q_{ip} + C_{ip}) +$$

$$\sum_{ip \in I_p} \left| (2 A_{ip} Q_{ip} \Delta Q_L + B_{ip} \Delta Q_L) \right| = \Delta E \qquad (4.29)$$

Given $Q_{i,k}$ the flow estimates at iteration k, Eq. (4.29) can be solved for the correction for loop L as

$$\Delta Q_L = - \frac{\left(\sum_{i \in I_L} K_i Q_{i,k}^n - \sum_{ip \in I_p} (A_{ip} Q_{ip,k}^2 + B_{ip} Q_{ip,k} + C_{ip}) - \Delta E \right)}{n \sum_{i \in I_L} \left| K_i Q_{i,k}^{n-1} \right| + \sum_{ip \in I_p} \left| (2 A_{ip} Q_{ip,k} + B_{ip}) \right|} \qquad (4.30)$$

In this form, the numerator of Eq. (4.30) is the excess headloss in the loop and should equal zero by conservation of energy. The terms are summed to account for the flow direction and component. The denominator is summed arithmetically without concern for direction. Most texts present networks with only closed loops and no pumps. Equation (4.30) simplifies this case by dropping the pump terms and setting ΔE to zero, or

$$\Delta Q_L = \frac{- \sum_{i \in I_L} K_i Q_{i,k}^n}{n \sum_{i \in I_L} \left| K_i Q_{i,k}^{n-1} \right|} = \frac{- \sum_{i \in I_L} h_{L,i}}{n \sum_{i \in I_L} \left| h_{L,i} / Q_{i,k} \right|} = \frac{-F(Q_k)}{\partial F / \partial Q \big|_{Q_{i,k}}} \qquad (4.31)$$

Comparing Eq. (4.31) with Eq. (4.23) shows that the Hardy Cross correction is essentially Newton's method.

The ΔQ_L corrections can be computed for each loop in sequence and can be applied before moving to the next loop (Jeppson, 1974) or corrections for all loops can be determined and applied simultaneously. Once the correction has been computed, the estimates for the next iteration are computed by

$$Q_{i,k+1} = Q_{i,k} + \Delta Q_L \qquad (4.32)$$

Q_{k+1} is then used in the next iteration. The process of determining corrections and updating flows continues until the ΔQ_L for each loop is less than some defined value. After the flows are computed, to determine the nodal heads, headlosses or gains are computed along a path from fixed-grade nodes to junction nodes.

The Hardy Cross method provides an understanding of principles and a tool for solving small networks by hand. However, it is not efficient for large networks compared with algorithms presented in the following sections.

Problem. List the loop equations for the network shown in Fig. 4.5 using the direction of flow shown. Then determine the flow in each pipe and the total energy at Nodes 4 and 5.

Solution. The loop equations consist of conservation of mass at the five junction nodes and the loop equations for the two primary loops and one pseudo-loop. In the mass balance equations, inflow to a node is positive and outflow is negative.

Node 1. $Q_1 - Q_2 - Q_5 = 0$

Node 2. $Q_2 + Q_3 - Q_6 = 2$

Node 3. $-Q_3 + Q_4 - Q_7 = 0$

Node 4. $Q_5 - Q_8 = 1$

Node 5. $Q_6 + Q_7 + Q_8 = 2$

Loop I. $h_{L,2} + h_{L,6} - h_{L,8} - h_{L,5} = 0 = K_2 Q_2^2 + K_6 Q_6^2 - K_8 Q_8^2 - K_5 Q_5^2$

Loop II. $h_{L,7} - h_{L,6} - h_{L,3} = 0 = K_7 Q_7^2 - K_6 Q_6^2 - K_3 Q_3^2$

Pseudo-loop. $h_{L,4} - h_p + h_{L,3} - h_{L,2} - h_{L,1} - E_{FGN,2} + E_{FGN,1} = 0$

$\qquad = K_4 Q_4^2 - (A_p Q_4^2 + B_p Q_4 + C_p) + K_3 Q_3^2 - K_2 Q_2^2 - K_1 Q_1^2$

$\qquad - E_{FGN,2} + E_{FGN,1}$

Because the Darcy-Weisbach equation is used, n equals 2. The loop equations assume that flow in the clockwise direction is positive. Flow in pipe 5 is moving counterclockwise and is given a negative sign for loop I. Flow in pipe 6 is moving clockwise relative to loop I (positive sign) and counterclockwise relative to loop II (negative sign). Although flow is moving counterclockwise through the pump in the pseudo-loop, h_p is given a negative sign because it adds energy to flow. To satisfy conservation of mass, the initial set of flows given below is assumed, where the values of K for the Darcy-Weisbach equation are given by

$$K_{DW} = \frac{fL}{A^2 D g} = \frac{8fL}{\pi^2 D^5 g} \qquad (4.33)$$

The concrete pipes are 1 ft in diameter and have a friction factor of 0.032 for turbulent flow.

4.14 Chapter Four

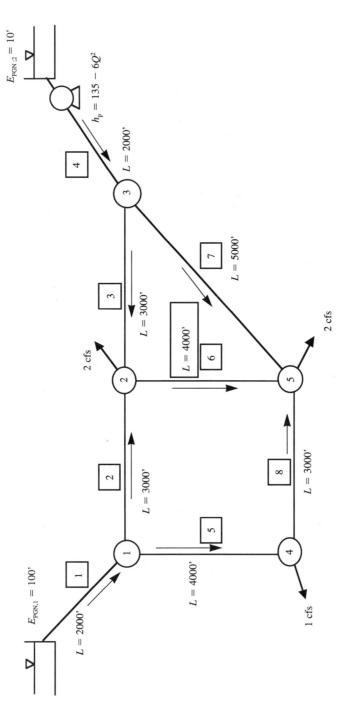

FIGURE 4.5 Example network (Note all pipes have diameters of 1 ft and friction factors equal to 0.032).

Pipe	1	2	3	4	5	6	7	8
K	1.611	2.417	2.417	1.611	3.222	3.222	4.028	2.417
Q	2.5	1.0	1.5	2.5	1.5	0.5	1.0	0.5

Also, $A_p = -6$, $B_p = 0$, and $C_p = 135$

Iteration 1. To compute the correction for the pseudo-loop, the numerator of Eq. (4.30) is

$$K_4 Q_4^2 - (A_p Q_4^2 + C_p) + K_3 Q_3^2 - K_2 Q_2^2 - K_1 Q_1^2 - E_{FGN,2} + E_{FGN,1} =$$
$$1.611(2.5)^2 - (-6(2.5)^2 + 135) + 2.417(1.5)^2 - 2.417(1.0)^2 - 1.611(2.5)^2 - (10 + 100)$$
$$= -4.48$$

The denominator is

$$nK_4 Q_4 + 2A_p Q_4 + nK_3 Q_3 + nK_2 Q_2 + nK_1 Q_1$$
$$= 2(1.611(2.5)) + |2(-6)2.5| + 2(2.417(1.5)) + 2\left|2.417(1.0)\right| + 2\left|1.611(2.5)\right| = 58.20$$

Thus, the correction for the pseudo-loop ΔQ_{PL} is

$$\Delta Q_{PL} = -\frac{(-4.48)}{58.20} = 0.077$$

The correction for loop I is computed next. The numerator of Eq. (4.30) is

$$K_2 Q_2^2 + K_6 Q_6^2 - K_8 Q_8^2 - K_5 Q_5^2 =$$
$$2.417(1.0)^2 + 3.222(0.5)^2 - 2.417(0.5)^2 - 3.222(1.5)^2 = -4.63$$

and the denominator is

$$nK_2 Q_2 + nK_6 Q_6 + nK_8 Q_8 =$$
$$2(2.417(1.0)) + 2(3.222(0.5)) + 2(2.417(0.5)) + 2(3.222(1.5)) = 20.14$$

Thus, the correction for loop 1, ΔQ_I is

$$\Delta Q_I = -\frac{(-4.63)}{20.14} = 0.230$$

Finally to adjust loop II from the numerator of Eq. (4.30) is

$$K_7 Q_7^2 - K_6 Q_6^2 - K_3 Q_3^2 =$$
$$4.028(1.0)^2 - 3.222(0.5)^2 - 2.417(1.5)^2 = -2.22$$

and the denominator is

$$nK_7 Q_7 + nK_6 Q_6 + nK_3 Q_3 = 2(4.028(1.0)) + 2(3.222(0.5)) + 2(2.417(1.5)) = 18.53$$

Thus, the correction for the loop II, ΔQ_{II}, is

$$\Delta Q_{II} = -\frac{(-2.22)}{18.53} = 0.120$$

The pipe flows are updated for iteration 2 as follows:

4.16 Chapter Four

Pipe	1	2	3	4 and pump	5	6	7	8
ΔQ	−0.077	0.230+	0.077−	0.077	−0.230	0.230	0.120	−0.230
		−(0.077)	0.120			−0.120		
Q	2.42	1.15	1.46	2.58	1.27	0.61	1.12	0.27

Because the flow direction for pipe 1 is counterclockwise relative to the pseudo-loop, the correction is given a negative sign. Similarly, pipe 2 receives a negative correction for the pseudo-loop. Pipe 2 is also in Loop I and is adjusted with a positive correction for that loop since flow in the pipe is in the clockwise direction for loop I. Pipes 3 and 6 also appear in two loops and receive two corrections.

Iteration 2. The adjustment for the pseudo-loop is

$$\Delta Q_{pL} = -\frac{K_4 Q_4^2 - (A_p Q_4^2 + C_p) + K_3 Q_3^2 - K_2 Q_2^2 - K_1 Q_1^2 + (E_{FGN,2} - E_{FGN,1})}{nK_4 Q_4 + 2A_p Q_4 + nK_3 Q_3 + nK_2 Q_2 + nK_1 Q_1} =$$

$$\frac{1.611(2.58)^2 - (-6(2.58)^2 + 135) + 2.417(1.46)^2 - 2.417(1.15)^2 - 1.611(2.42)^2 - 10 + 100}{2(1.611(2.58)) + 2|-6(2.58)| + 2(2.417(1.46)) + 2(2.417(1.15)) + 2(1.611(2.42))} =$$

$$= -\left(\frac{-1.82}{59.69}\right) = 0.030$$

In the correction for loop I, the numerator of eq. (4.30) is

$$K_2 Q_2^2 + K_6 Q_6^2 - K_8 Q_8^2 - K_5 Q_5^2 =$$
$$2.417(1.15)^2 + 3.222(0.61)^2 - 2.417(0.27)^2 - 3.222(1.27)^2 = -0.978$$

and the denominator is

$$nK_2 Q_2 + nK_6 Q_6 + nK_8 Q_8 + nK_5 Q_5 =$$

$$2(2.417(1.15)) + 2(3.222(0.61)) + 2(2.417(0.27)) + 2(3.222(1.27)) = 18.98$$

Thus the correction for the pseudo-loop, ΔQ_I is

$$\Delta Q_I = -\frac{(-0.978)}{18.98} = 0.052$$

Finally, to correct loop II, the numerator of Eq. (4.30) is

$$K_7 Q_7^2 - K_6 Q_6^2 - K_3 Q_3^2 = 4.028(1.12)^2 - 3.222(0.61)^2 - 2.417(1.46)^2 = -1.30$$

and the denominator is

$$nK_7 Q_7 + nK_6 Q_6 + nK_3 Q_3 =$$
$$2*[4.028(1.12)] + 2[3.222(0.61)] + 2[2.417(1.46)] = 20.01$$

Thus, the correction for the pseudo-loop, ΔQ_{II} is

$$\Delta Q_{II} - \frac{(-1.30)}{20.01} = 0.065$$

Hydraulics of Water Distribution Systems   **4.17**

The pipe flows are updated for iteration 3 as follows:

Pipe	1	2	3	4 and pump	5	6	7	8
ΔQ	−0.030	0.052+	0.030	0.030	−0.052	0.052	0.065	−0.052
		(−0.030)	−0.065			−0.065		
Q	2.39	1.17	1.43	2.61	1.22	0.60	1.18	0.22

Iteration 3. The corrections for iteration 3 are 0.012, 0.024, and 0.024 for the pseudo-loop, loop I, and loop II, respectively. The resulting flows are as follows:

Pipe	1	2	3	4 and pump	5	6	7	8
ΔQ	−0.012	0.024	+0.012	0.012	−0.024	0.024	0.024	−0.024
		(−0.012)	−0.024			−0.024		
Q	2.38	1.18	1.42	2.62	1.20	0.60	1.20	0.20

After two more iterations, the changes become small, and the resulting pipe flows are as follows. Note that the nodal mass balance equations are satisfied at each iteration.

Pipe	1	2	3	4 and pump	5	6	7	8
Q	2.37	1.19	1.41	2.63	1.18	0.60	1.21	0.18

The total energy at nodes 4 and 5 can be computed by path equations from either FGN to the nodes. For example, paths to node 4 consist of pipes 1 and 5 or of pipes 4 (with the pump), 7, and 8. For the path with pipes 1 and 5, the path equation is

$$100 - K_1 Q_1^2 - K_5 Q_5^2 = 100 - 1.611(2.37)^2 - 3.222(1.19)^2 = 100 - 9.05 - 4.56 = 86.39 \text{m}$$

For the path containing pipes 4, 7, and 8 the result is

$$10 + (135 - 6(2.63)^2) - 1.611(2.63)^2 - 4.028(1.22)^2 + 2.417(0.19)^2 =$$
$$10 + 93.50 - 11.14 - 6.00 + 0.09 = 86.45 \text{m}$$

This difference can be attributed to rounding errors. Note that pipe 8 received a positive sign in the second path equation. Because the flow in pipe 8 is the opposite of the path direction, the energy along the path is increasing from nodes 5 to 4. The total energy at node 5 can be found along pipes 4 and 7 or 86.36 m or along the path of pipes 1-2-6, giving $(100 - 9.05 - 3.42 - 1.16 = 86.37 \text{m})$.

4.2.3.2 Linear theory method. *Linear theory* solves the loop equations or Q equations (Eqs. 4.25 to 4.27). N_p equations ($N_j + N_1 + N_f - 1$) can be written in terms of the N_p unknown pipe flows. Since these equations are nonlinear in terms of Q, an iterative procedure is applied to solve for the flows. Linear theory, as described in Wood and Charles (1972), linearizes the energy equations (Eqs. 4.26 and 4.27) about $Q_{i,k+1}$, where the subscript $k+1$ denotes the current iteration number using the previous iterations $Q_{i,k}$ as known values. Considering only pipes in this derivation, these equations are

$$\sum_{i \in I_j} Q_{i,k+1} = q_{ext} \quad \text{for all } N_j \text{ nodes} \qquad (4.34)$$

$$\sum_{i \in I_L} K_i Q_{i,k}^{n-1} Q_{i,k+1} = 0 \quad \text{for all } N_1 \text{ closed loops} \tag{4.35}$$

and

$$\sum_{i \in I_L} K_i Q_{i,k}^{n-1} Q_{i,k+1} = \Delta E_{\text{FGN}} \quad \text{for all } N_f - 1 \text{ independent pseudo-loops} \tag{4.36}$$

These equations form a set of linear equations that can be solved for the values of $Q_{i,k+1}$. The absolute differences between successive flow estimates are computed and compared to a convergence criterion. If the differences are significant, the counter k is updated and the process is repeated for another iteration. Because of oscillations in the flows around the final solution, Wood and Charles (1972) recommended that the average of the flows from the previous two iterations should be used as the estimate for the next iterations. Once the pipe flows have been determined, the nodal piezometric heads can be determined by following a path from a FGN and accounting for losses or gains to all nodes.

Modified linear theory: Newton method. Wood (1980) and his collaborators at the University of Kentucky developed the KYPIPE program but essentially modified the original linear theory to a Newton's method. However, rather than solve for the change in discharge (ΔQ), Q_{k+1} is determined.

To form the equations, the energy equations (Eq. 4.3) are written in terms of the current estimate of Q_k, including pipes, minor losses and pumps, as

$$f(Q_k) = \sum_{i \in I_L} K_i Q_k^n + \sum_{im \in I_m} K_{im} Q_k^2 + \sum_{ip \in I_p} (A_{ip} Q_k^2 + B_{ip} Q_k + C_{ip}) - \Delta E \tag{4.37}$$

where for simplicity the subscripts i, im, and ip denoting the pipe, minor loss component, and pump, respectively, are dropped from the flow terms and k again denotes the iteration counter. This equation applies to both closed loops ($\Delta E = 0$) and pseudo-loops ($\Delta E = \Delta E_{\text{FGN}}$), but, in either case, $f(Q_k)$ should equal zero at the correct solution.

To move toward the solution, the equations are linearized using a truncated Taylor series expansion:

$$f(Q_{k+1}) = f(Q_k) + \left.\frac{\partial f}{\partial Q}\right|_{Q_k} (Q_{k+1} - Q_k) = f(Q_k) + G_k(Q_{k+1} - Q_k) \tag{4.38}$$

Note that f and Q are now vectors of the energy equations and pipe flow rates, respectively, and G_k is the matrix of gradients that are evaluated at Q_k. Setting Eq. (4.38) to zero and solving for Q_{k+1} gives

$$0 = f(Q_k) + G_k(Q_{k+1} - Q_k)$$

or

$$G_k Q_{k+1} = G_k Q_k + f(Q_k) \tag{4.39}$$

This set of ($N_1 + N_f - 1$) equations can be combined with the N_j junction equations in Eq. (4.34) that also are written in terms of Q_{k+1} to form a set of N_p equations. This set of linear equations is solved for the vector of N_p flow rates using a matrix procedure. The values of Q_{k+1} are compared with those from the previous iteration. If the largest absolute difference is below a defined tolerance, the process stops. If not, Eq. (4.39) is formed using Q_{k+1} and another iteration is completed.

4.2.3.3 Newton-Raphson method and the node equations. The *node equations* are the conservation of mass relationships written in terms of the unknown nodal piezometric heads. This formulation was described in Sec. 4.2.2 for the branching pipe system.

In Fig. 4.4, if P and the pipe flows are unknown, the system is essentially a network with one junction node with three FGNs. In a general network, N_j junction equations can be written in terms of the N_j nodal piezometric heads. Once the heads are known, the pipes flows can be computed from the pipes headloss equations.

Other network components, such as valves and pumps, are included by adding junction nodes at each end of the component. Node equations are then written using the flow relationship for the component.

Solution method. Martin and Peters (1963) were the first to publish an algorithm using the Newton-Raphson method for solving the node equations for a pipe network. Shamir and Howard (1968) showed that pumps and valves could be incorporated and unknowns other than nodal heads could be determined by the method. Other articles have been published that have attempted to exploit the sparse matrix structure of this problem.

At iteration k, the Newton-Raphson method is applied to the set of junction equations $F(h_k)$ for the nodal heads h_k. After expanding the equations and truncating higher-order terms, the result is

$$F(h_k) + \left.\frac{\partial F}{\partial h}\right|_{h_k} \Delta h_k = 0 \tag{4.40}$$

where F is the set of node equations evaluated at h_k, the vector of nodal head estimates at iteration k. $\partial F/\partial h$ is the Jacobian matrix of the gradients of the node equations with respect to the nodal heads. This matrix is square and sparse because each nodal head appears in only two nodal balance equations. The unknown corrections Δh_k can be determined by solving the set of linear equations:

$$F(h_k) = -\left.\frac{\partial F}{\partial h}\right|_{h_k} \Delta h_k \tag{4.41}$$

The nodal heads are then updated by

$$h_{k+1} = h_k + \Delta h_k \tag{4.42}$$

As in previous methods, the magnitude of the change in nodal heads is examined to determine whether the procedure should end. If the heads have not converged, Eq. (4.41) is reformulated with h_{k+1} and another correction vector is computed. If the final solution has been found, the flow rates are then computed using the component relationships with the known heads.

As in all formulations, at least one FGN must be hydraulically connected to all nodes in the system. Some convergence problems have been reported if poor initial guesses are made for the nodal heads. However, the node equations result in the smallest number of unknowns and equations of all formulations.

Problem. Write the node equations for the system in Fig. 4.5.

Node 1:

$$sign(100 - h_1)\left(\frac{1100 - h_1|}{K_1}\right)^{\frac{1}{n}} + sign(h_2 - h_1)\left(\frac{|h_2 - h_1|}{K_2}\right)^{\frac{1}{n}} + sign(h_4 - h_1)\left(\frac{|h_4 - h_1|}{K_5}\right)^{\frac{1}{n}} = 0$$

Node 2 (note that the right-hand side is equal to the external demand):

$$sign(h_1 - h_2)\left(\frac{|h_1 - h_2|}{K_2}\right)^{\frac{1}{n}} + sign(h_3 - h_2)\left(\frac{|h_3 - h_2|}{K_3}\right)^{\frac{1}{n}} + sign(h_5 - h_2)\left(\frac{|h_5 - h_2|}{K_6}\right)^{\frac{1}{n}} = 2$$

Node 3:

$$\text{sign}(h_2 - h_3)\left(\frac{|h_2 - h_3|}{K_3}\right)^{\frac{1}{n}} + \text{sign}(h_5 - h_3)\left(\frac{|h_5 - h_3|}{K_7}\right)^{\frac{1}{n}} + \text{sign}(h_{pd} - h_3)\left(\frac{|h_{pd} - h_3|}{K_4}\right)^{\frac{1}{n}} = 0$$

Node 4:

$$\text{sign}(h_1 - h_4)\left(\frac{|h_1 - h_4|}{K_5}\right)^{\frac{1}{n}} + \text{sign}(h_5 - h_4)\left(\frac{|h_4 - h_5|}{K_8}\right)^{\frac{1}{n}} = 1$$

Node 5:

$$\text{sign}(h_2 - h_5)\left(\frac{|h_2 - h_5|}{K_6}\right)^{\frac{1}{n}} + \text{sign}(h_4 - h_5)\left(\frac{|h_4 - h_5|}{K_8}\right)^{\frac{1}{n}} + \text{sign}(h_3 - h_5)\left(\frac{|h_5 - h_3|}{K_7}\right)^{\frac{1}{n}} = 2$$

New node for the pump:

$$\text{sign}(h_3 - h_{pd})\left(\frac{|h_3 - h_{pd}|}{K_4}\right)^{\frac{1}{n}} + \left(\frac{(h_{pd} - 10) - 135}{-6}\right)^{\frac{1}{2}} = 0$$

The first term in the pump node equation is the outflow from the pump toward node 3 in pipe 4. The second term is the discharge relationship for the pump, written in terms of the total energy at the outlet of the pump h_{pd}.

Because the pump relationship is different from that for pipe 4, this new node with zero demand was added at the outlet of the pump (assuming that the pump inlet is the tank). This type of node must be added for every component (valve, pipe, or pump); therefore, one must know the precise location of the component. For example, if a valve appears within a pipe, to be exact in system representation, new nodes would be added on each side of the valve, and the pipe would be divided into sections upstream and downstream of the valve.

In summary, six equations can be written for the system to determine six unknowns (the total energy for nodes 1 to 5 and for the pump node). Using the solution from the Hardy Cross method gives the following nodal heads, the values of which can be confirmed to satisfy the node equations:

Node	1	2	3	4	5	Pump
Total head (m)	90.95	87.54	92.35	86.45	86.38	103.50

Pipe	1	2	3	4	5	6	7	8	Pump
Pipe flow (m³/s)	2.37	1.19	1.41	2.63	1.18	0.60	1.22	0.18	2.63

4.2.3.4 Gradient algorithm. *Pipe equations.* Unlike the node and loop equations, the pipe equations are solved for Q and h simultaneously. Although this requires a larger set of equations to be solved, the gradient algorithm by Todini and Pilati (1987) has been shown to be robust to the extent that this method is used in EPANET (Rossman, 1994).

To form the pipe equations, conservation of energy is written for each network component in the system in terms of the nodal heads. For example, a pipe equation is

$$h_a - h_b = KQ^n \tag{4.43}$$

and, using a quadratic approximation, a pump equation is

$$h_b - h_a = AQ^2 + BQ + C \tag{4.44}$$

where h_a and h_b are the nodal heads at the upstream and downstream ends of the component. These equations are combined with the nodal balance relationships (Eq. 4.2) to form $N_j + N_p$ equations with an equal number of unknowns (nodal heads and pipe flows).

Solution method. Although conservation of mass at a node is linear, the component flow equations are nonlinear. Therefore, an iterative solution scheme, known as the gradient algorithm, is used. Here the component flow equations are linearized using the previous flow estimates Q_k. For pipes,

$$KQ_k^{n-1} Q_{k+1} + (h_a - h_b) = 0 \tag{4.45}$$

In matrix form, the linearized equations are

$$A_{12}h + A_{11}Q + A_{10}h_0 = 0 \tag{4.46}$$

and

$$A_{21}Q - q_{\text{ext}} = 0 \tag{4.47}$$

where Eq. (4.46) is the linearized flow equations for each network component and Eq. (4.47) is the nodal flow balance equations. A_{12} ($= A_{21}T$) is the incidence matrix of zeros and ones that identify the nodes connected to a particular component and A_{10} identifies the FGNs. A_{11} is a diagonal matrix containing the linearization coefficients (e.g., $|KQ_k^{n-1}|$).

Differentiating eqs. (4.46) and (4.47) gives:

$$\begin{bmatrix} NA_{11} & A_{12} \\ A_{21} & 0 \end{bmatrix} \begin{bmatrix} dQ \\ dh \end{bmatrix} = \begin{bmatrix} dE \\ dq \end{bmatrix} \tag{4.48}$$

where dE and dq are the residuals of Eqs. (4.2) and (4.43-44) evaluated at the present solution, Q_k and h_k. N is a diagonal matrix of the exponents of the pipe equation (n). Eq. 4.48 is a set of linear equations in terms of dQ and dh. Once solved Q and h are updated by

$$Q_{k+1} = Q_k + dQ \tag{4.49}$$

and

$$h_{k+1} = h_k + dh \tag{4.50}$$

Convergence is checked by evaluating dE and dq, and additional iterations are completed as necessary.

Todini and Pilati (1987) applied an alternative efficient recursive scheme for solving for Q_{k+1} and h_{k+1}. The result is

$$h_{k+1} = -(A_{21}N^{-1}A_{11}^{-1} A_{12})^{-1}\{A_{12}N^{-1}(Q_k + A_{11}^{-1} A_{10}H_0) + (q_{\text{ext}} - A_{21}Q_k)\} \tag{4.51}$$

then using h_{k+1}, Q_{k+1} by is determined:

$$Q_{k+1} = (1 - N^{-1})Q_k - N^{-1} A_{11}^{-1} (A_{12}H_{k+1} + A_{10}H_0) \tag{4.52}$$

where A_{11} is computed at Q_k. Note that N and A_{11} are diagonal matrices, so the effort for inversion is negligible. Yet, one full matrix must be inverted in this scheme.

Problem. Write the pipe equations for the network in Fig. 4.5.

Solution. The pipe equations include mass balance equations for each node in the system. The network contains five junction nodes plus an additional node downstream of the pump. The pump is considered to be a link and is assumed to be located directly after the FGN.

Conservation of energy equations are written for each pipe and pump link. Eight pipe equations and one pump equation are written. The total number of equations is then 15, which equals the 15 unknowns, including 8 pipe flows, 1 pump flow, and 6 junction node heads, including the additional nodal head at the pump outlet h_p.

Node 1:	$Q_1 - Q_2 - Q_5 = 0$	Pipe 1:	$100 - h_1 = K_1 Q_1^n$
Node 2:	$Q_2 + Q_3 - Q_6 = 2$	Pipe 2:	$h_1 - h_2 = K_2 Q_2^n$
Node 3:	$-Q_3 + Q_4 - Q_7 = 0$	Pipe 3:	$h_3 - h_2 = K_3 Q_3^n$
Node 4:	$Q_5 - Q_8 = 1$	Pipe 4:	$h_p - h_3 = K_4 Q_4^n$
Node 5:	$Q_6 + Q_7 + Q_8 = 2$	Pipe 5:	$h_1 - h_4 = K_5 Q_5^n$
Pump Node:	$Q_p - Q_4 = 0$	Pipe 6:	$h_2 - h_5 = K_6 Q_6^n$
Pump:	$h_p - 10 = 135 - 6Q_p^2$	Pipe 7:	$h_3 - h_5 = K_7 Q_7^n$
		Pipe 8:	$h_4 - h_5 = K_8 Q_8^n$

4.2.3.5 Comparison of solution methods. All four methods are capable of solving the flow relationships in a system. The loop equations solved by the Hardy Cross method are inefficient compared with the other methods and are dropped from further discussion. The Newton-Raphson method is capable of solving all four formulations, but because the node equations result in the fewest equations, they are likely to take the least amount of time per iteration. In applications to the node equations, however, possible convergence problems may result if poor initial conditions are selected (Jeppson, 1974).

Linear theory is reportedly best for the loop equations and should not be used for the node or loop equations with the ΔQ corrections, as used in Hardy Cross (Jeppson, 1974). Linear theory does not require initialization of flows and, according to Wood and Charles (1972), always converges quickly.

A comparative study of the Newton-Raphson method and the linear theory methods was reported by Holloway (1985). The Newton-Raphson scheme was programmed in two codes and compared with KYPIPE that implemented the linear theory. For a 200-pipe network, the three methods converged in eight or nine iterations, with the Newton-Raphson method requiring the least amount of computation time.

Salgado, Todini, and O'Connell (1987) compared the three methods for simulating a network under different levels of demand and different system configurations. Four conditions were analyzed and are summarized in Table 4.1. Example A contains 66 pipes and 41 nodes but no pumps. Example B is similar to Example A, but 6 pumps are introduced and a branched connection has been added. Example C is the same network as in Example B with higher consumptions, whereas Example D has the same network layout but the valves are closed in two pipes. Closing these pipes breaks the network into two systems. The results demonstrate that all methods can simulate the conditions, but the gradient method for solving the pipe equations worked best for the conditions analyzed.

All comparisons and applications in this chapter are made on the basis of assuming reasonably sized networks. Given the speed and memory available in desktop

TABLE 4.1 Comparison of Solution Methods

Example	Special conditions	Node equations	Solution method: Loop equations	Pipe equations
A	Low velocities	Converged Iterations = 16, T = 70 s	Converged Iterations = 17, T = 789 s	Converged Iterations = 16, T = 30 s
B	Pumps and branched network	Converged Iterations = 12, T = 92 s	Slow convergence Iterations = 13, T = 962 s	Converged Iterations = 10, T = 34 s
C	Example B with high demand	Converged Iterations = 13, T = 100 s	Slow convergence Iterations = 15, T = 1110 s	Converged Iterations = 12, T = 39 s
D	Closed pipes	Converged Iterations = 21, T = 155 s Some heads not available	Converged Iterations = 21, T = 1552 s Some heads not available	Converged Iterations = 19, T = 57 s

Source: Modified from Todini and Pilati (1987).

computers, it is likely that any method is acceptable for these networks. To solve extremely large systems with several thousand pipes, alternative or tailored methods are necessary. Discussion of these approaches is beyond the scope of this chapter. However, numerical simulation of these systems will become possible, as discussed in Chap. 14 on network calibration, but good representation of the system with accurate parameters may be difficult.

4.2.3.6 Extended-period simulation. As noted earlier, time variation can be considered in network modeling. The simplest approach is extended-period simulation, in which a sequence of steady-state simulations are solved using one of the methods described earlier in this section. After each simulation period, the tank levels are updated and demand and operational changes are introduced.

Tank levels or water-surface elevations are used as known energy nodes. The levels change as flow enters or leaves the tank. The change in water height for tanks with constant geometry is the change in volume divided by the area of the tank, or

$$\Delta H_T = \frac{V_T}{A_T} = \frac{Q_T \Delta t}{A_T} \qquad (4.53)$$

where Q_T and V_T are the flow rate and volume of flow that entered the tank during the period, respectively; Δt is the time increment of the simulation; A_T is the tank area; and ΔH_T is the change in elevation of the water surface during period T. More complex relationships are needed for noncylindrical tanks. With the updated tank levels, the extended-period simulation continues with these levels as known energy nodes for the next time step. The process continues until all time steps are evaluated. More complex unsteady analyses are described in the next section.

4.3 UNSTEADY FLOW IN PIPE NETWORK ANALYSIS

In steady-state analysis or within an extended-period simulation, changes in the distributions of pressure and flow are assumed to occur instantaneously after a change in external stimulus is applied. Steady conditions are then reached immediately. In some cases, the time to reach steady state and the changes during this transition may be important. Recently, work has proceeded to model rapid and gradual changes in flow conditions. Rapid changes resulting in transients under elastic column theory are discussed in Chap. 6. Two modeling approaches for gradually varied unsteady flow under a rigid-column assumption are described in this section.

4.3.1 Governing Equations

In addition to conservation of mass, the governing equations for unsteady flow under rigid pipe assumptions are developed from conservation of momentum for an element (Fig. 4.6). *Conservation of momentum* states that the sum of the forces acting on the volume of fluid equals the time rate of change of momentum, or

$$\sum F = F_1 - F_2 - F_f = \frac{d(mv)}{dt} \tag{4.54}$$

where F_1 and F_2 are the forces on the ends of the pipe element, F_f is the force caused by friction between the water and the pipe, and m and v are the mass and velocity of the fluid in the pipe element.

The end forces are equal to the force of the pressure plus the equivalent force caused by gravity or for the left-hand side of the element:

$$F_1 = \gamma A\left(\frac{p_1}{\gamma} + z_1\right) = \gamma A h_1 \tag{4.55}$$

The friction force is the energy loss times the volume of fluid, or

$$F_f = \gamma A h_L \tag{4.56}$$

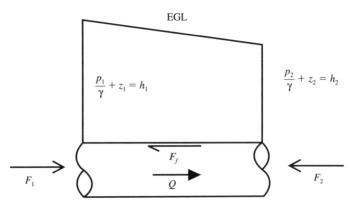

FIGURE 4.6 Force balance on a pipe element.

The change of momentum can be expanded to

$$\frac{d(mv)}{dt} = \frac{d(\rho Vv)}{dt} = \frac{d\left(\frac{\gamma ALv}{g}\right)}{dt} = \frac{\gamma L}{g}\frac{d(Av)}{dt} = \frac{\gamma L}{g}\frac{dQ}{dt} \tag{4.57}$$

where the mass equals $\rho V = \frac{\gamma}{g} AL$, in which all terms are constants with respect to time and can be taken out of the differential. Note that under the rigid-water-column assumption, the density is a constant as opposed to elastic-water-column theory. Substituting these terms in the momentum balance gives

$$\gamma A(h_1 - h_2 - h_L) = \frac{\gamma L}{g}\frac{dQ}{dt} \tag{4.58}$$

Assuming that a steady-state friction loss relationship can be substituted for h_L and dividing each side by γA,

$$h_1 - h_2 - KQ^n = \frac{L}{gA}\frac{dQ}{dt} \tag{4.59}$$

With conservation of mass (Eq. 4.2), this ordinary differential equation and its extensions for loops have been used to solve for time-varying flow conditions.

4.3.2 Solution Methods

4.3.2.1 *Loop formulation.* Holloway (1985) and Islam and Chaudhry (1998) extended the momentum equation (Eq. 4.59) to loops as follows:

$$\sum_{i \in I_p}(h_{1i} - h_{2i}) - \sum_{i \in I_p} K_i Q_i^n = \sum_{i \in I_p}\frac{L_i}{gA_i}\frac{dQ_i}{dt} \tag{4.60}$$

Separating variables and integrating over time gives

$$\int_t^{t+\Delta t}\left[\sum_{i \in I_p}(h_{1i} - h_{2i})\right]dt - \int_t^{t+\Delta t}\left[\sum_{i \in I_p} K_i Q_i^n\right]dt = \int_{Q_t}^{Q_{t+\Delta t}}\sum_{i \in I_p}\frac{L_i}{gA_i}dQ_i \tag{4.61}$$

At any instant in time, the headloss around a closed loop must equal zero, so the first term can be dropped. Dropping this term also eliminates the nodal piezometric heads as unknowns and leaves only the pipe flows.

One of several approximations for the friction loss term can be used:

$$KQ^{t+\Delta t}|Q^t|^{n-1}\Delta t \tag{4.62}$$

$$K[(Q^{t+\Delta t} + Q^t)|Q^{t+\Delta t} + Q^t|^{n-1}/2^n]\Delta t \tag{4.63}$$

$$K[(Q^{t+\Delta t}|Q^{t+\Delta t}|^{n-1} + Q^t|Q^t|^{n-1})/2^n]\Delta t \tag{4.64}$$

Holloway (1985) obtained results using Eq. (4.62), known as the integration approximation, that compared favorably with the other two nonlinear forms. Using this form in Eq. (4.61),

$$\sum_{i \in I_p} \frac{L_i}{gA_i} Q_i^t - \sum_{i \in I_p} K_i Q_i^{t+\Delta t} |Q_i^t|^{n-1} \Delta t = \sum_{i \in I_p} \frac{L_i}{gA_i} Q_i^{t+\Delta t} \qquad (4.65)$$

This equation is written for each loop and is used with the nodal conservation of mass equations to given N_p equations for the N_p unknown pipe flows. Note that these equations are linear in terms of $Q^{t+\Delta t}$ and can be solved at each time step in sequence using the previous time step for the values in the constant terms.

4.3.2.2 Pipe formulation with gradient algorithm. An alternative solution method developed by Ahmed and Lansey (1999) used the momentum equation for a single pipe (Eq. 4.59) and the nodal flow balance equations to form a set of equations similar to those developed in the gradient algorithm. An explicit backward difference is used to solve the equations. The right-hand side of Eq. (4.59) is written in finite difference form as

$$\frac{L_i}{gA_i} \frac{dQ_i}{dt} = \frac{L_i}{gA_i} \frac{(Q_i^{t+\Delta t} - Q_i^t)}{\Delta t} \qquad (4.66)$$

The left-hand side of Eq. (4.9) is written in terms on the unknowns h and Q at time step $t + \Delta t$. After substituting and rearranging a general algebraic equation for pipe between two nodes results in

$$\left[K_i |Q_i^t|^{n-1} \Delta t - \frac{L_i}{gA_i} \right] Q_i^{t+\Delta t} + [h_{1i}^{t+\Delta t} - h_{2i}^{t+\Delta t}] \Delta t = \left[-\frac{L_i}{gA_i} \right] Q_i^t \qquad (4.67)$$

N_p equations of this form can be written for each pipe or other component. With the N_j nodal flow balance equations, a total of $N_j + N_p$ equations can be written in terms of an equal number of unknown pipe flows and nodal heads. Given an initial condition at time t, the pipe flows and nodal heads at time $t + \Delta t$ by solving Eq. (4.67) and Eq. (4.2). The new values are then used for the next time step until all times have been evaluated. Unlike the loop formulation, in the form above, Eq. (4.67) is nonlinear with respect to the unknowns. In addition, like the loop equation, the time step will influence the accuracy of the results.

4.4 COMPUTER MODELING OF WATER DISTRIBUTION SYSTEMS

Because the numerical approaches for analyzing distribution systems cannot be completed by hand except for the smallest systems, computer-simulation models have been developed. These models solve the system of nonlinear equations for the pipe flows and nodal heads. In addition to the equation solver, many modeling packages have sophisticated input preprocessors, which range from spreadsheets to tailored full-page editors, and output postprocessors, including links with computer-aided drafting software and geographic information systems. Although these user interfaces ease the use of the simulation models, a dependable solver and proper modeling are crucial for accurate mathematical models of field systems.

An array of packages is available, and the packages vary in their level of sophistication. The choice of a modeling package depends on the modeling effort. Modeling needs range from designing subdivisions with fewer than 25 pipes to modeling large water utilities that possibly involve several thousand links and nodes. Users should select the package that best meets their objectives.

4.4.1 Applications of Models

Clark et al. (1988) identified a series of seven steps that are necessary to develop and apply a water distribution simulation model:

1. *Model selection:* Definition of modeling requirements including the model's purpose. The desired use of a model is important when selecting one model (hydraulic or water quality) because the necessary accuracy of the model and the level of detail required will vary, depending on its expected use.
2. *Network representation:* Determination of how the components of a system will be represented in the numerical model. Step 2 includes skeletonizing the piping system by not including some pipes in the model or making assumptions regarding the parameter values for pipes, such as assuming that all pipes of a certain type have the same roughness value. The degree of model simplification depends on what problems the model will be used to help address.
3. *Calibration:* Adjustment of nonmeasurable model parameters, with emphasis on the pipe roughness coefficients, so that predicted model results compare favorably with observed field data (see Sec. 4.4.2). This step may also require reexamination of the network representation.
4. *Verification:* Comparison of model results with a second set of field data (beyond that used for calibration) to confirm the adequacy of the network representation and parameter estimates.
5. *Problem definition:* Identification of the design or operation problem and incorporation of the situation in the model (e.g., demands, pipe status or operation decisions).
6. *Model application:* Simulation of the problem condition.
7. *Display/analysis of results:* Presentation of simulation results for modeler and other decision-makers in graphic or tabular form. Results are analyzed to determine whether they are reasonable and the problem has been resolved. If the problem is not resolved, new decisions are made at step 5 and the process continues.

4.4.2 Model Calibration

Calibration, step 3 above, is the process of developing a model that represents field conditions for the range of desired conditions. The time, effort, and money expended for data collection and model calibration depend on the model's purpose. For example, a model for preliminary planning may not be extremely accurate because decisions are at the planning level and an understanding of only the major components is necessary. At the other extreme, a model used for engineering decisions regarding a system that involves pressure and water-quality concerns may require significant calibration efforts to provide precise predictions. All models should be calibrated before they are used in the decision-making process.

The *calibration process* consists of data collection, model calibration, and model assessment. Data collection entails gathering field data, such as tank levels, nodal pressures, nodal elevations, pump head and discharge data, pump status and flows, pipe flows, and, when possible, localized demands. These data are collected during one or more loading conditions or over time through automated data logging. Rossman et al. (1994) discussed using water-quality data for calibration. To ensure that a calibration will be successful, the number of measurements must exceed the number of parameters to be estimated in the model. If this condition is not satisfied, multiple sets of parameters that match the field observations can be found: that is, a unique solution may not be determined. Each set may give dramatically different results when predicting under other conditions.

During model calibration, field data are compared with model estimates and model parameters are adjusted so that the model predictions match the field observations. Two stages of model calibration are desirable. The first stage is a gross study of the data and the model predictions. The intent is to ensure that the data are reasonable and that major modeling assumptions are valid. For example, this level would determine if valves assumed to be open are actually closed or if an unexpectedly high withdrawal, possibly caused by leakage, is occurring. Walski (1990) discussed this level of calibration.

After the model representation is determined to be reasonable, the second stage of model calibration begins with the adjustment of individual model parameters. At this level, the two major sources of error in a model are the demands and the pipe roughness coefficients. The demands are uncertain because water consumption is largely unmonitored in the short term and highly variable and because the water is consumed along a pipe, whereas it is modeled as a point of withdrawal. Because pipe roughnesses vary over time and are not directly measurable, they must be inferred from field measurements.

Adjustment of these terms and others, such as valve settings and pump lifts, can be made by trial and error or through systematic approaches. Several mathematical modeling methods have been suggested for solving the model calibration problem (Lansey and Basnet, 1991).

Once a model is believed to be calibrated, an assessment should be completed. The assessment entails a sensitivity analysis of model parameters to identify which parameters have a strong impact on model predictions and future collection should emphasize improving. The assessment also will identify the predictions (nodal pressure heads or tank levels) that are sensitive to calibrated parameters and forecasted demands. Model assessments can simply be plots of model predictions versus parameter values or demand levels, or they can be more sophisticated analyses of uncertainty, as discussed in Araujo (1992) and Xu and Goulter (1998).

REFERENCES

Ahmed, I., Application of the Gradient Method for Analysis of Unsteady Flow in Water Networks, Master's thesis (Civil Engineering), University of Arizona, Tucson, 1997.

Araujo, J.V., A Statistically Based Procedure for Calibration of Water Systems, Doctoral dissertation, Oklahoma State University, Stillwater, 1992.

Ahmed, I., and K. Lansey, "Analysis of Unsteady Flow in Networks Using a Gradient Algorithm Based Method," ASCE Specialty Conference on Water Resources, Tempe, AZ, June 1999.

Cross, H. "Analysis of Flow in Networks of Conduits or Conductors," Bulletin No. 286, University of Illinois Engineering Experimental Station, Urbana, IL, 1936.

Holloway, M. B., *Dynamic Pipe Network Computer Model*, Doctoral dissertation, Washington State University, Pullman, WA, 1985.

Islam R., and M. H. Chaudhry, "Modeling of Constituent Transport in Unsteady Flows in Pipe Networks," *Journal of Hydraulics Division*, 124 (11): 1115–1124, 1998.

Jeppson, R. W., *Analysis of Flow in Pipe Networks*, Ann Arbor Science, Ann Arbor, MI, 1974.

Lansey, K., and C. Basnet, "Parameter Estimation for Water Distribution Systems," *Journal of Water Resources Planning and Management*, 117(1): 126-144, 1991.

Martin, D. W., and G. Peters, "The Application of Newton's Method to Network Analysis by Digital Computer," *Journal of the Institute of Water Engineers*, 17: 115-129, 1963.

Rossman, L. A., "EPANET—Users Manual," EPA-600/R-94/057, U.S. Environmental Protection Agency, Risk Reduction Engineering Laboratory, Cincinnati, OH, 1994.

Salgado, R., E. Todini, and P. E. O'Connell, "Comparison of the Gradient Method with Some Traditional Methods for the Analysis of Water Supply Distribution Networks," Proceedings, *International Conference on Computer Applications for Water Supply and Distribution 1987*, Leicester Polytechnic, UK, September 1987.

Shamir, U., and C. D. Howard, "Water Distribution System Analysis," *Journal of Hydraulics. Division*, 94(1): 219-234, 1965.

Todini, E., and S. Pilati, "A Gradient Method for the Analysis of Pipe Networks," *International Conference on Computer Applications for Water Supply and Distribution 1987*, Leicester Polytechnic, UK, September 1987.

Walski, T., "Hardy–Cross Meets Sherlock Holmes or Model Calibration in Austin, Texas," *Journal of the American Water Works Association*, 82:34–38, March, 1990.

Wood, D. J., *User's Manual—Computer Analysis of Flow in Pipe Networks Including Extended Period Simulations*, Department of Civil Engineering, University of Kentucky, Lexington, KY, 1980.

Wood, D., and C. Charles, "Hydraulic Network Analysis Using Linear Theory," *Journal of Hydraulic Division*, 98, (HY7): 1157-1170, 1972.

Xu, C., and I. Goulter, "Probabilistic Model for Water Distribution Reliability," *Journal of Water Resources Planning and Management*, 124(4): 218-228, 1998.

CHAPTER 5
PUMP SYSTEM HYDRAULIC DESIGN

B. E. Bosserman
Boyle Engineering Corporation
Newport Beach, CA

5.1 PUMP TYPES AND DEFINITIONS

5.1.1 Pump Standards

Pump types are described or defined by various organizations and their respective publications:

- Hydraulics Institute (HI), *American National Standard for Centrifugal Pumps for Nomenclature, Definitions, Application and Operation* [American National Standards Institute (ANSI)/HI 1.1-1.5-1994].
- American Petroleum Institute (API), *Centrifugal Pumps for Petroleum, Heavy Duty Chemical, and Gas Industry Services,* Standard 610, 8th ed., August 1995.
- American Society of Mechanical Engineers (ASME), *Centrifugal Pumps,* Performance Test Code PTC 8.2–1990.

In addition, there are several American National Standards Institute (ANSI) and American Water Works Associations (AWWA) standards and specifications pertaining to centrifugal pumps:

- ANSI/ASME B73.1M-1991, *Specification for Horizontal End Suction Centrifugal Pumps for Chemical Process.*
- ANSI/ASME B73.2M-1991, *Specification for Vertical In-Line Centrifugal Pumps for Chemical Process.*
- ANSI/ASME B73.5M-1995, *Specification for Thermoplastic and Thermoset Polymer Material Horizontal End Suction Centrifugal Pumps for Chemical Process.*

- ANSI/AWWA E 101-88, *Standard for Vertical Turbine Pumps—Lineshaft and Submersible Types.*

5.1.2 Pump Definitions and Terminology

Pump definitions and terminology, as given in Hydraulics Institute (HI) 1.1-1.5-1994 (Hydraulics Institute, 1994), are as follows:

Definition of a centrifugal pump. A centrifugal pump is a kinetic machine converting mechanical energy into hydraulic energy through centrifugal activity.

Allowable operating range. This is the flow range at the specified speeds with the impeller supplied as limited by cavitation, heating, vibration, noise, shaft deflection, fatigue, and other similar criteria. This range to be defined by the manufacturer.

Atmospheric head (h_{atm}). Local atmospheric pressure expressed in ft (m) of liquid.

Capacity. The capacity of a pump is the total volume throughout per unit of time at suction conditions. It assumes no entrained gases at the stated operating conditions.

Condition points

- *Best efficiency point (BEP).* The BEP is the capacity and head at which the pump, efficiency is a maximum.
- *Normal condition point.* The normal condition point applies to the point on the rating curve at which the pump will normally operate. It may be the same as the rated condition point.
- *Rated condition point.* The *rated condition point* applies to the capacity, head, net positive suction head, and speed of the pump, as specified by the order.
- *Specified condition point.* The specified condition point is synonymous with the rated condition point.

Datum. The pump's datum is a horizontal plane that serves as the reference for head measurements taken during a test. Vertical pumps are usually tested in an open pit with the suction flooded. The datum is then the eye of the first-stage impeller (Fig. 5.1).

Optional tests can be performed with the pump mounted in a suction can. Regardless of the pump's mounting, its datum is maintained at the eye of the first-stage impeller.

Elevation head (Z). The potential energy of the liquid caused by its elevation relative to a datum level measuring to the center of the pressure gauge or liquid surface.

Friction head. Friction head is the hydraulic energy required to overcome frictional resistance of a piping system to liquid flow expressed in ft (m) of liquid.

Gauge head (h_g). The energy of the liquid due to its pressure, as determined by a pressure gauge or another pressure-measuring device.

Head. Head is the expression of the energy content of the liquid referred to any arbitrary datum. It is expressed in units of energy per unit weight of liquid. The measuring unit for head is ft (m) of liquid.

High-energy pump. High-energy pump refers to a pump with a head greater than 650 ft (200 m) per stage and requiring more than 300 hp (225 KW) per stage.

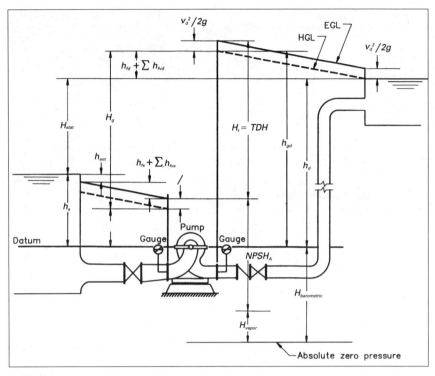

FIGURE 5.1 Terminology for a pump with a positive suction head.

Impeller balancing

- *Single-plane balancing (*also called *static balancing).* Single-plane balancing refers to correction of residual imbalance to a specified maximum limit by removing or adding weight in one correction plane only. This can be accomplished statically using balance rails or by spinning.
- *Two-plane balancing (*also called *dynamic balancing).* Two plane-balancing refers to correction of residual imbalance to a specified limit by removing or adding weight in two correction planes. This is accomplished by spinning on appropriate balancing machines.

Overall efficiency (η_{OA}). This is the ratio of the energy imparted to the liquid (P_w) by the pump to the energy supplied to the (P_{mot}): that is, the ratio of the water horsepower to the power input to the primary driver expressed as a percentage.

Power

- *Electric motor input power* (P_{mot}). This is the electrical input power to the motor.
- *Pump input power* (P_p). This is the power delivered to the pump shaft at the driver to pump coupling. It is also called *brake horsepower.*

- *Pump output power* (P_w). This is the power imparted to the liquid by the pump. It is also called *water horsepower*.

$$P_w = \frac{Q \times H \times s}{3960} \qquad \text{(U.S. units)} \qquad (5.1)$$

$$P_w = \frac{Q \times H \times s}{366} \qquad \text{(S.I. units)} \qquad (5.2)$$

where Q = flow in gal/min (U.S.) or m³/hr (SI), H = head in feet (U.S.) or meters (SI), S = specific gravity and P_w = power in a horsepower (U.S.) or kilowatt (SI)

- *Pump efficiency* (η_p). This is the ratio of the energy imported to the liquid (P_w) to the energy delivered to the pump shaft (P_p) expressed in percent.

Pump pressures

- *Field test pressure.* The maximum static test pressure to be used for leak testing a closed pumping system in the field if the pumps are not isolated. Generally this is taken as 125 percent if the maximum allowable casing working pressure. In cases where mechanical seals are used, this pressure may be limited by the pressure-containing capabilities of the seal.

  Note: See "Maximum allowable casing working pressure" below, consideration of which may limit the field test pressure of the pump to 125 percent of the maximum allowable casing working pressure on the suction split-case pumps and certain other pump types.

- *Maximum allowable casing working pressure.* This is the highest pressure at the specified pumping temperature for which the pump casing is designed. This pressure shall be equal to or greater than the maximum discharge pressure. In the case of some pumps (double suction, vertical turbine, axial split case can, or multistage, for example), the maximum allowable casing working pressure on the suction side may be different from that on the discharge side.

- *Maximum suction pressure.* This is the highest suction pressure to which the pump will be subjected during operation.

- *Working pressure* (p_d). This is the maximum discharge pressure that could occur in the pump, when it is operated at rated speed and suction pressure for the given application.

  Shut off. This is the condition of zero flow where no liquid is flowing through the pump, but the pump is primed and running.

  Speed. This is the number of revolutions of the shaft in a given unit of time. Speed is expressed as revolutions per minute.

Suction conditions

- *Maximum suction pressure.* This is the highest suction pressure to which the pump will be subjected during operation.

- *Net positive suction head available* ($NPSH_A$). Net positive suction head available is the total suction head of liquid absolute, determined at the first-stage impeller datum, less the absolute vapor pressure of the liquid at a specific capacity:

$$NPSH_A = h_{sa} - h_{vp} \tag{5.3}$$

where h_{sa} = total suction head absolute = $h_{atm} + h_s$ (5.4)

or
$$NPSH_A = h_{atm} + h_s - h_{vp} \tag{5.5}$$

- *Net positive suction head required* ($NPSH_R$). This is the amount of suction head, over vapor pressure, required to prevent more than a 3 percent loss in total head from the first stage of the pump at a specific capacity.
- *Static suction lift* (I_s). Static suction lift is a hydraulic pressure below atmospheric pressure at the intake port of the pump.
- *Submerged suction.* A submerged suction exists when the center line of the pump inlet is below the level of the liquid in the supply tank.
- *Total discharge head* (h_d). The total discharge head (h_d) is the sum of the discharge gauge head (h_{gd}) plus the velocity head (h_{vd}) at the point of gauge attachment plus the elevation head (Z_d) from the discharge gauge center line to the pump datum:

$$h_d = h_{gd} + h_{vd} + Z_d \tag{5.6}$$

- *Total head* (*H*). This is the measure of energy increase per unit weight of the liquid imparted to the liquid by the pump and is the difference between the total discharge head and the total suction head. This is the head normally specified for pumping applications since the complete characteristics of a system determine the total head required.
- *Total suction head* (h_s), *closed suction test.* For closed suction installations, the pump suction nozzle may be located either above or below grade level.
- *Total suction head* (h_s), *open suction.* For open suction (wet pit) installations, the first stage impeller of the bowl assembly is submerged in a pit. The total suction head (h_s) at datum is the submergence (Z_w). If the average velocity head of the flow in the pit is small enough to be neglected, then:

$$h_s = Z_w \tag{5.7}$$

where Z_w = vertical distance in feet from free water surface to datum.

The total suction head (h_s) referred to the eye of the first-stage impeller is the algebraic sum of the suction gauge head (h_{vs}) plus the velocity head (h_{vs}) at the point of gauge attachment plus the elevation head (Z_s) from the suction gauge center line (or manometer zero) to the pump datum:

$$h_s = h_{gs} + h_{vs} + Z_s \tag{5.8}$$

The suction head (h_s) is positive when the suction gauge reading is above atmospheric pressure and negative when the reading is below atmospheric pressure by an amount exceeding the sum of the elevation head and the velocity head.

Velocity head (h_v). This is the kinetic energy of the liquid at a given cross section. Velocity head is expressed by the following equation:

$$h_v = \frac{v^2}{2g} \tag{5.9}$$

5.6 Chapter Five

where v is obtained by dividing the flow by the cross-sectional area at the point of gauge connection.

5.1.3 Types of Centrifugal Pumps

The HI and API standards do not agree on these definitions of types of centrifugal pumps (Figs. 5.2 and 5.3). Essentially, the HI standard divides centrifugal pumps into two types (overhung impeller and impeller between bearings), whereas the API standard divides them into three types (overhung impeller, impeller between bearings, and vertically suspended). In the HI standard, the "vertically suspended" type is a subclass of the "overhung impeller" type.

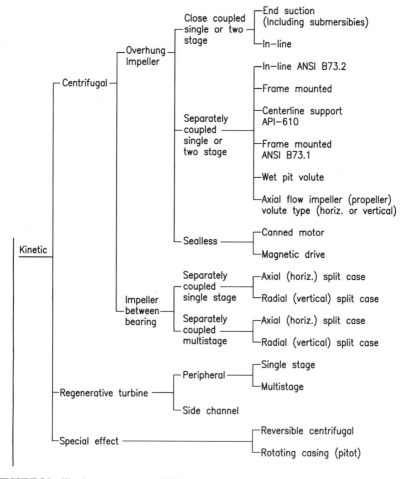

FIGURE 5.2 Kinetic type pumps per ANSI/HI-1.1-1.5-1994.

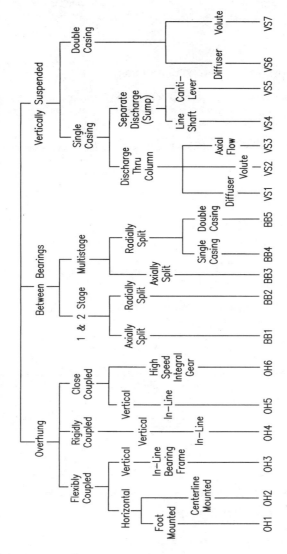

FIGURE 5.3 Pump class type identification per API 610.

5.2 PUMP HYDRAULICS

5.2.1 Pump Performance Curves

The head that a centrifugal pump produces over its range of flows follows the shape of a downward facing or concave curve (Fig. 5.4). Some types of impellers produce curves that are not smooth or continuously decreasing as the flow increases: that is, there may be dips and valleys in the pump curve.

5.2.2 Pipeline Hydraulics and System Curves

A *system curve* describes the relationship between the flow in a pipeline and the head loss produced: see Fig. 5.5 for an example. The essential elements of a system curve include the following:

- The static head of the system, as established by the difference in water surface elevations between the reservoir the pump is pumping from and the reservoir the pump is pumping to.
- The friction or headloss in the piping system. Different friction factors representing the range in age of the pipes from new to old should always be considered.

The system curve is developed by adding the static head to the headlosses that occur as flow increases. Thus, the system curve is a hyperbola with its origin at the value of the static head.

The three most commonly used procedures for determining friction in pipelines are the following:

5.2.2.1 Hazen-Williams equation. The Hazen-Williams procedure is represented by the equation

$$V = 1.318C\ R^{0.63}S^{0.54} \text{ (U.S. units)} \tag{5.10a}$$

where V = velocity (ft/s), C = roughness coefficient, R = hydraulic radius (ft), and S = friction head loss per unit length or the slope of the energy grade line (ft/ft). In SI units, Eq. (5.10a) is

$$V = 0.849CR^{0.63}S^{0.54} \tag{5.10b}$$

where V = velocity (m/s), C = roughness coefficient, R = hydraulic radius (m), and S = friction headloss per unit length or the slope of the energy grade line in meters per meter.

A more convenient form of the Hazen-Williams equation for computing headloss or friction in a piping system is

$$H_L = \frac{4.72}{D^{4.86}}\left(\frac{Q}{C}\right)^{1.85} \tag{5.11a}$$

where H_L = headloss (ft), L = length of pipe (ft), D = pipe internal diameter (ft), Q = flow (ft³/s), and C = roughness coefficient or friction factor. In SI units, the Hazen-Williams equation is

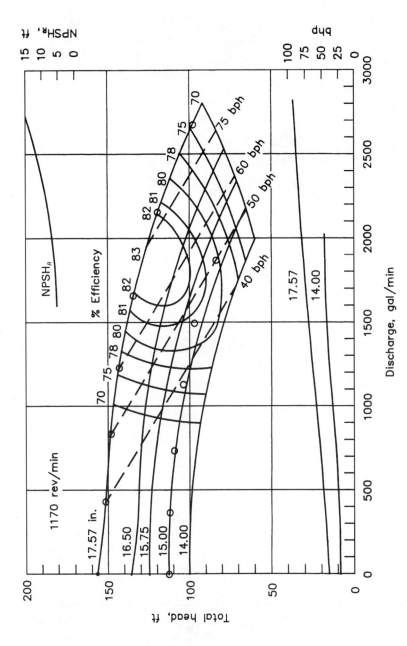

FIGURE 5.4 Typical discharge curves.

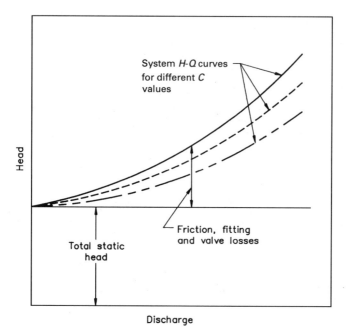

FIGURE 5.5 Typical system head-capacity curves.

$$H_L = \left(\frac{L}{1000}\right)\left(\frac{151Q}{CD^{2.63}}\right)^{1.85} = \frac{10.74L}{D^{4.86}}\left(\frac{Q}{C}\right)^{1.85} \quad (5.11b)$$

where H_L = headloss (m), Q = flow (m³/s), D = pipe diameter (m), and L = pipe length (m). The C coefficient typically has a value of 80 to 150; the higher the value, the smoother the pipe. C values depend on the type of pipe material, the fluid being conveyed (water or sewage), the lining material, the age of the pipe or lining material, and the pipe diameter. Some ranges of values for C are presented in Table 5.1 for differing pipe materials.

TABLE 5.1 Hazen-Williams Coefficients

Pipe Material	C Value for Water	C Value for Sewage
PVC	135–150	130–145
Steel (with mortar lining)	120–145	120–140
Steel (unlined) 120 to 140	110–130	110–130
Ductile iron (with mortar lining)	100–140	100–130
Asbestos cement	120–140	110–135
Concrete pressure pipe	130–140	120–130
Ductile iron (unlined)	80–120	80–110

AWWA Manual M11, *Steel Pipe—A Guide for Design and Installation* (AWWA, 1989), offers the following relationships between C factors and pipe diameters for water service:

$$C = 140 + 0.17d \text{ for new mortar-lined steel pipe (U.S. units)} \quad (5.12)$$

$$= 140 + 0.0066929d \text{ (SI units, } d \text{ in (mm))}$$

$$C = 130 + 0.16d \text{ (U.S. units) for long-term considerations of lining} \quad (5.13)$$
deterioration, slime buildup, and so on.

$$= 130 + 0.0062992d \text{ (SI units, } d \text{ in mm),}$$

where C = roughness coefficient or friction factor (See Table 5.1) and d = pipe diameter, inches or millimeters, as indicated above.

5.2.2.2 Manning's equation. Manning's procedure is represented by the equation

$$V = \frac{1.486}{n} R^{2/3} S^{1/2} \text{ (U.S. units)} \quad (5.14)$$

$$V = \frac{1}{n} R^{2/3} S^{1/2} \text{ (SI units)}$$

where V = velocity (f/s or m/s), n = roughness coefficient, R = hydraulic radius (ft or m), and S = friction headloss per unit length or the slope of the energy grade line in feet per foot or meters per meter.

A more convenient form of Manning's equation for computing headloss or friction in a pressurized piping system is

$$H_L = \frac{4.66 \, L \, (nQ)^2}{D^{16/3}} \quad \text{(U.S. units)} \quad (5.15)$$

$$= \frac{5.29 L (nQ)^2}{D^{16/3}} \quad \text{(SI units)}$$

where n = roughness coefficient, H_L = headloss (ft or m), L = length of pipe (ft or m), D = pipe internal diameter (ft or m), and Q = flow (cu³/s or m³/s).

Values of n are typically in the range of 0.010 – 0.016, with n decreasing with smoother pipes.

5.2.2.3 Darcy-Weisbach equation. The Darcy-Weisbach procedure is represented by the equation

$$H_L = f \frac{L}{D} \frac{V^2}{2g} \quad (5.16)$$

where f = friction factor from Moody diagram, g = acceleration due to gravity = 32.2 (ft/s) (U.S. units) = 9.81 m/s² (SI units), H_L = headloss (ft or m), L = length of pipe (ft or m), D = pipe internal diameter (ft or m), and V = velocity (ft/s or m/s).

Sanks et al. (1998) discuss empirical equations for determining f values. A disadvantage of using the Darcy-Weisbach equation is that the values for f depend on both roughness (E/D) and also on the Reynolds number (Re):

$$Re = \frac{VD}{v} \quad (5.17)$$

where Re = Reynolds number (dimensionless), V = fluid velocity in the pipe (ft/s or m/s), D = pipe inside diameter (ft or m), and v = kinematic viscosity (ft²/s or m²/s).

Values for f as a function of Re can be determined by the following equations:

Re less than 2000:
$$f = \frac{64}{Re} \tag{5.18}$$

Re = 2000–4000:
$$\frac{1}{\sqrt{f}} = 2 \log_{10}\left(\frac{E/D}{3.7} + \frac{2.51}{Re\sqrt{f}}\right) \tag{5.19}$$

Re greater than 4000:
$$f = \frac{0.25}{\left[\log_{10}\left(\frac{E/D}{3.7} + \frac{5.74}{Re^{0.9}}\right)\right]^2} \tag{5.20}$$

where E/D = roughness, with E = absolute roughness (ft or m), and D = pipe diameter, (ft or m). Equation (5.19) is the Colebrook-White equation, and Eq. (5.20) is an empirical equation developed by Swamee and Jain, in Sanks et al., (1998). For practical purposes, f values for water works pipelines typically fall in the range of 0.016 to about 0.020.

5.2.2.4 Comparisons of f, C, and n. The Darcy-Weisbach friction factor can be compared to the Hazen-Williams C factor by solving both equations for the slope of the hydraulic grade line and equating the two slopes. Rearranging the terms gives, in SI units,

$$f = \left(\frac{1}{C^{1.85}}\right)\left(\frac{134}{V^{0.15}D^{0.167}}\right) \tag{5.21a}$$

where V is in meters per second and D is in meters. In U.S. customary units, the relationship is

$$f = \left(\frac{1}{C^{1.85}}\right)\left(\frac{194}{V^{0.15}D^{0.167}}\right) \tag{5.21b}$$

where V is in fps and D is in feet (Sanks et al., 1998).

For pipes flowing full and under pressure, the relationship between C and n is

$$n = 1.12 \frac{D^{0.037}}{CS^{0.04}} \tag{5.22a}$$

in SI units, where D is the inside diameter in meters. In U.S. customary units, the equation is

$$n = 1.07 \frac{D^{0.037}}{CS^{0.04}} \tag{5.22b}$$

where D is the (inside diameter) in feet.

5.2.3 Hydraulics of Valves

The effect of headlosses caused by valves can be determined by the equation for minor losses:

$$h_L = K \frac{V^2}{zg} \tag{5.23}$$

where h_L = minor loss (ft or m), K = minor loss coefficient (dimensionless), V = fluid velocity (ft/s or m/s), and g = acceleration due to gravity (= 32.2 ft·s/s or 9.81 m·s/s).

Headloss or pressure loss through a valve also is determined by the equation

$$Q = C_v \sqrt{\Delta P} \text{ (U.S. units)} = 0.3807 C_v \sqrt{\Delta P} \quad \text{(S.I. units)} \tag{5.24}$$

where Q = flow through valve (gal/m or m³/s), C_V = valve capacity coefficient, and ΔP = pressure loss through the valve (psi or kPa). The coefficient C_V varies with the position of the valve plug, disc, gate, and so forth. C_V indicates the flow that will pass through the valve at a pressure drop of 1 psi. Curves of C_V versus plug or disc position (0–90, with 0 being in the closed position) must be obtained from the valve manufacturer's catalogs or literature.

C_V and K are related by the equation

$$C_V = 29.85 \frac{d^2}{\sqrt{K}} \quad \text{(U.S. units)} \tag{5.25}$$

where d = valve size (in). Thus, by determining the value for C_V from the valve manufacturer's data, a value for K can then be calculated from Eq. (5.25). This K value can then be used in Eq. (5.23) to calculate the valve headloss.

5.2.4 Determination of Pump Operating Points-Single Pump

The system curve is superimposed over the pump curve (Fig. 5.6). The pump operating points occur at the intersections of the system curves with the pump curves. It should be observed that the operating point will change with time. As the piping ages and becomes rougher, the system curve will become steeper, and the intersecting point with the pump curve will move to the left. Also, as the impeller wears, the pump curve moves downward.

Thus, over a period of time, the output capacity of a pump can decrease significantly. See Fig. 5.7 for a visual depiction of these combined effects.

5.2.5 Pumps Operating in Parallel

To develop a composite pump curve for pumps operating in parallel, add the flows together that the pumps provide at common heads (Fig. 5.8). This can be done with identical pumps (those having the same curve individually) as well as with pumps having different curves.

5.2.6 Variable-Speed Pumps

The pump curve at maximum speed is the same as the one described above. The point on a system-head curve at which a variable-speed pump will operate is similarly determined by the intersection of the pump curve with the system curve. What are known as the *pump affinity laws* or *homologous laws* must be used to determine the pump curve at reduced speeds. These affinity laws are described in detail in Chap. 6. For the discussion here, the relevant mathematical relationships are in Sanks et al. (1998).

$$\frac{Q_1}{Q_2} = \frac{n_1}{n_2} \tag{5.26}$$

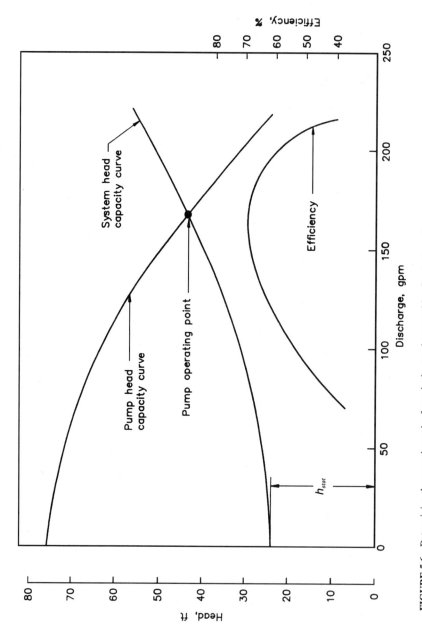

FIGURE 5.6 Determining the operating point for a single-speed pump with a fixed value of h_{stat}.

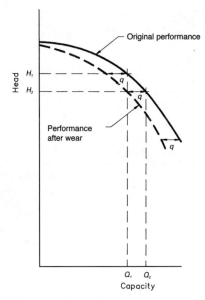

FIGURE 5.7 Effect of impeller wear.

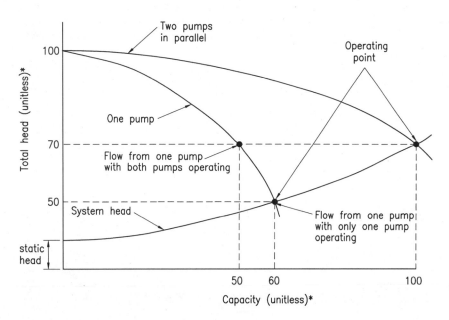

FIGURE 5.8 Pumps operating in parallel.

$$\frac{H_1}{H_2} = \left(\frac{n_1}{n_2}\right)^2 \tag{5.27}$$

$$\frac{P_1}{P_2} = \left(\frac{n_1}{n_2}\right)^3 \tag{5.28}$$

where Q = flow rate, H = head, P = power, n = rotational speed, and subscripts 1 and 2 are only for corresponding points. Equations (5.26) and (5.27) must be applied simultaneously to ensure that point 1 "corresponds" to point 2. Corresponding points fall on parabolas through the original. They do not fall on system H-Q curves. These relationships, known collectively as the affinity laws, are used to determine the effect of changes in speed on the capacity, head, and power of a pump.

The affinity laws for discharge and head are accurate because they are based on actual tests for all types of centrifugal pumps, including axial-flow pumps. The affinity law for power is not as accurate because efficiency increases with an increase in the size of the pump.

When applying these relationships, remember that they are based on the assumption that the efficiency remains the same when transferring from a given point on one pump curve to a homologous point on another curve. Because the hydraulic and pressure characteristics at the inlet, at the outlet, and through the pump vary with the flow rate, the errors produced by Eq. (5.28) may be excessive, although errors produced by Eqs. (5.26) and (5.27) are extremely small. See Fig. 5.9 for an illustration of the pump curves at different speeds.

Example. Consider a pump operating at a normal maximum speed of 1800 rpm, having a head-capacity curve as described in Table 5.2. Derive the pump curve for operating speeds of 1000–1600 rpm at 200-rpm increments.

The resulting new values for capacity (Q) and head (H) are shown in Table 5.2. The values are derived by taking the Q values for the 1800 rpm speed and multiplying them by the ratio (n_1/n_2) and by taking the H values for the 1800 rpm speed and multiplying them by the ratio (n_1/n_2)^2.

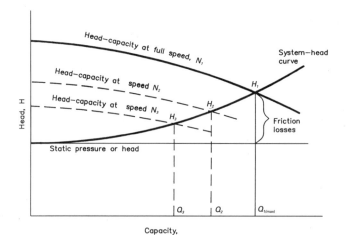

FIGURE 5.9 Typical discharge curves for a variable speed pump.

TABLE 5.2 Example calculations

Speed (rpm)	Ratio n_1/n_2	Ratio $(n_1/n_2)^2$	Head—Capacity at Various Points							
			Point 1		Point 2		Point 3		Point 4	
			Q (gpm)	H(feet)	Q (gpm)	H(feet)	Q (gpm)	H(feet)	Q (gpm)	H(feet)
1,800	1,00	1,00	0	200	1,000	180	2,000	160	3,000	130
1,600	0,889	0,790	0	158	889	142	1,778	126	2,667	103
1,400	0,777	0,605	0	121	778	109	1,556	97	2,333	79
1,200	0,667	0,444	0	89	667	80	1,333	71	2,000	58
1,000	0,555	0,309	0	62	556	56	1,111	49	1,667	40

Speed (rpm)	Ratio n_1/n_2	Ratio $(n_1/n_2)^2$	Head—Capacity at Various Points							
			Point 1		Point 2		Point 3		Point 4	
			Q (gpm)	H(feet)	Q (gpm)	H(feet)	Q (gpm)	H(feet)	Q (gpm)	H(feet)
1,800	1,00	1,00	0	60	63	55	126	49	189	40
1,600	0,889	0,790	0	47	56	44	112	39	168	32
1,400	0,777	0,605	0	36	49	33	98	30	147	24
1,200	0,667	0,444	0	27	42	24	84	22	126	18
1,000	0,555	0,309	0	19	33	17	70	15	105	12

5.3 CONCEPT OF SPECIFIC SPEED

5.3.1 Introduction: Discharge-Specific Speed

The *specific speed* of a pump is defined by the equation

$$N_s = \frac{nQ^{0.50}}{H^{0.75}} \quad (5.29)$$

where N_s = specific speed (unitless), n = pump rotating speed (rpm), Q = pump discharge flow (gal/mm, m³/s, L/s, m³h) (for double suction pumps, Q is one-half the total pump flow, For example, if the specific speed is expressed in metric units (e.g., N = rev/min, Q = m³/s, and H = m), the corresponding value expressed in U.S. customary units (e.g., N = rev/min, Q = gal/min, and H = feet) is obtained by multiplying the metric value by 51.64 and H = total dynamic head (ft or m) (for multistage pumps, H is the head per stage).

The relation between specific speeds for various units of discharge and head is given in Table 5.3, wherein the numbers in bold type are those customarily used (Sank et al., 1998).

Pumps having the same specific speed are said to be geometrically similar. The specific speed is indicative of the shape and dimensional or design characteristics of the pump impeller (HI, 1994). Sanks et al. (1998) also gives a detailed description and discussion of impeller types as a function of specific speed. Generally speaking, the various types of impeller designs are as follows:

Type of Impeller	Specific Speed Range (U.S. Units)
Radial-vane	500–4200
Mixed-flow	4200–9000
Axial-flow	9000–15,000

TABLE 5.3 Equivalent Factors for Converting Values of Specific Speed Expressed in One Set of Units to the Corresponding Values in Another Set of Units

Quantity	Expressed in Units of				
N Q H	(rev/min, L/s, m)	(rev/min, m³/s, m)	(rev/min, m³/h, m)	(rev/min, gal/mn, ft)	(rev/min, ft³/s, ft)
	1.0	0.0316	1.898	1.633	0.0771
	31.62	1.0	60.0	51.64	2.437
	0.527	0.0167	1.0	0.861	0.0406
	0.612	0.0194	1.162	1.0	0.0472
	12.98	0.410	24.63	21.19	1.0

Source: Sanks, et al 1998.

5.3.2 Suction-Specific Speed

Suction-specific speed is a number similar to the discharge specific speed and is determined by the equation

$$S = \frac{nQ^{0.50}}{NPSH_R^{0.75}} \tag{5.30}$$

where S = suction-specific speed (unitless), n = pump rotating speed (rpm), Q = pump discharge flow as defined for Eq. (5.29), and $NPSH_R$ = net positive suction head required, as described in Sec. 5.4 The significance of suction-specific speed is that increased pump speed without proper suction head conditions can result in excessive wear on the pump's components (impeller, shaft, bearings) as a result of excessive cavitation and vibration (Hydraulics Institute, 1994). That is, for a given type of pump design (with a given specific speed), there is an equivalent maximum speed (n) at which the pump should operate.

Rearranging Eq. (5.30) results in

$$n = \frac{S \times NPSH_A^{0.75}}{Q^{0.50}} \tag{5.31}$$

Equation (5.31) can be used to determine the approximate *maximum allowable pump speed* as a function of net positive suction head available and flow for a given type of pump (i.e., a given suction-specific speed). Inspection of Eq. (5.31) reveals that, for a given specific speed, the following pump characteristics will occur:

- The higher the desired capacity (Q), the lower the allowable maximum speed. Thus, a properly selected high-capacity pump will be physically larger beyond what would be expected due solely to a desired increased capacity.

- The higher the $NPSH_A$, the higher the allowable pump speed.

5.4 NET POSITIVE SUCTION HEAD

Net positive suction head, or NPSH, actually consists of two concepts: the net positive suction available ($NPSH_A$), and the net positive suction head required ($NPSH_R$). The definition of $NPSH_A$ and $NPSH_R$, as given by the Hydraulics Institute (1994), were presented in Sec. 5.1.

5.4.1 Net Positive Suction Head Available

Figure 5.1 visually depicts the concept of $NPSH_A$. Since the $NPSH_A$ is the head available at the impeller, friction losses in any suction piping must be subtracted when making the calculation. Thus, the equation for determining $NPSH_A$ becomes

$$NPSH_A = h_{atm} + h_s - h_{vp} - h_L \tag{5.32}$$

where

h_{atm} = atmospheric pressure (ft or m).

h_s = static head of water on the suction side of the pump (ft or m) (h_s is negative if the water surface elevation is below the eye of the impeller).

h_{vp} = vapor pressure of water, which varies with both altitude and temperature (ft or m), and

h_L = friction losses in suction piping (ft or m), typically expressed as the summation of velocity heads ($KV^2/2g$) for the various fittings and pipe lengths in the suction piping.

Key points in determining $NPSH_A$ are as follows (Sanks et al., 1998): (1) the barometric pressure must be corrected for altitude, (2) storms can reduce barometric pressure by about 2 percent, and (3) the water temperature profoundly affects the vapor pressure. Because of uncertainties involved in computing $NPSH_A$, it is recommended that the $NPSH_A$ be at least 5 ft (1.5 m) greater than the $NPSH_R$ or 1.35 times the $NPSH_R$ as a factor of safety (Sanks et al., 1998). An example of calculating $NPSH_A$ is presented in Section 5.5.

5.4.2 Net Positive Suction Head Required by a Pump

The Hydraulics Institute (1994) and Sanks et al. (1998) have discussed the concept and implications of $NPSH_R$ in detail. Their discussions are presented or summarized as follows. The $NPSH_R$ is determined by tests of geometrically similar pumps operated at constant speed and discharge but with varying suction heads. The development of cavitation is assumed to be indicated by a 3 percent drop in the head developed as the suction inlet is throttled, as shown in Fig. 5.10. It is known that the onset of cavitation occurs well before the 3 percent drop in head (Cavi, 1985). Cavitation can develop substantially before any drop in the head can be detected, and erosion indeed occurs more rapidly at a 1 percent change in head (with few bubbles) than it does at a 3 percent change in head (with many bubbles). In fact, erosion can be inhibited in a cavitating

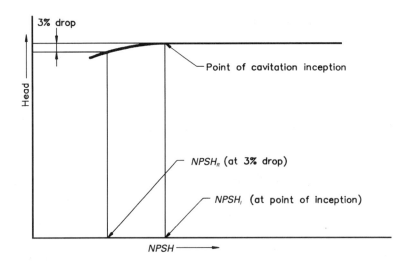

FIGURE 5.10 Net positive suction head criteria as determined from pump test results.

pump by introducing air into the suction pipe to make many bubbles. So, because the 3 percent change is the current standard used by most pump manufacturers to define the $NPSH_R$, serious erosion can occur as a result of blindly accepting data from catalogs. In critical installations where continuous duty is important, the manufacturer should be required to furnish the $NPSH_R$ test results. Typically, $NPSH_R$ is plotted as a continuous curve for a pump (Fig. 5.11). When impeller trim has a significant effect on the $NPSH_R$, several curves are plotted.

The NPSH required to suppress all cavitation is always higher than the $NPSH_R$ shown in a pump manufacturer's curve. The NPSH required to suppress all cavitation at 40 to 60 percent of a pump's flow rate at BEP can be two to five times as is necessary to meet guaranteed head and flow capacities at rated flow (Fig. 5.10; Taylor, 1987). The HI standard (Hydraulics Institute, 1994) states that even higher ratios of $NPSH_A$ to $NPSH_R$ may be required to suppress cavitation: It can take from 2 to 20 times the $NPSH_R$ to suppress incipient cavitation completely, depending on the impeller's design and operating capacity.

If the pump operates at low head at a flow rate considerably greater than the capacity at the BEP, Eq. (5.33) is approximately correct:

$$\frac{NPSH_R \text{ at operating point}}{NPSH_R \text{ at BEP}} = \left(\frac{Q \text{ at operating point}}{Q \text{ at BEP}}\right)^n \quad (5.33)$$

where the exponent n varies from 1.25 to 3.0, depending on the design of the impeller. In most water and wastewater pumps, n lies between 1.8 and 2.8. The $NPSH_R$ at the BEP increases with the specific speed of the pumps. For high-head pumps, it may be necessary either to limit the speed to obtain the adequate NPSH at the operating point or to lower the elevation of the pump with respect to the free water surface on the suction side i to increase the $NPSH_A$.

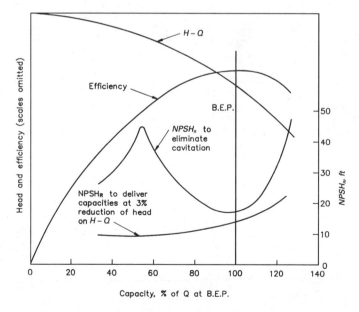

FIGURE 5.11 NPSH required to suppress visible cavitation.

5.4.3 NPSH Margin or Safety Factor Considerations

Any pump and piping system must be designed so that the net positive suction head available ($NPSH_A$) is equal to, or exceeds, the net positive suction head required ($NPSH_R$) by the pump throughout the range of operation. The *margin* is the amount by which $NPSH_A$ exceeds $NPSH_R$ (Hydraulics Institute, 1994). The amount of margin required varies, depending on the pump design, the application, and the materials of construction.

Practical experience over many years has shown that, for the majority of pump applications and designs, $NPSH_R$ can be used as the lower limit for the NPSH available. However, for high-energy pumps, the $NPSH_R$ may not be sufficient. Therefore, the designer should consider an appropriate NPSH margin over $NPSH_R$ for high-energy pumps that is sufficient at all flows to protect the pump from damage caused by cavitation.

5.4.4 Cavitation

Cavitation begins to develop in a pump as small harmless vapor bubbles, substantially before any degradation in the developed head can be detected (Hydraulics Institute, 1994). This is called the *point of incipient cavitation* (Cavi, 1985; Hydraulics Institute, 1994).

Studies on high-energy applications show that cavitation damage with the $NPSH_A$ greater than the $NPSH_R$ can be substantial. In fact, there are studies on pumps that show the maximum damage to occur at $NPSH_A$ values somewhere between 0 and 1 percent head drop (or two to three times the $NPSH_R$), especially for high suction pressures, as required by pumps with high impeller-eye peripheral speeds. There is no universally accepted relationship between the percentage of head drop and the damage caused by cavitation. There are too many variables in the specific pump design and materials, properties of the liquid and system. The pump manufacturer should be consulted about NPSH margins for the specific pump type and its intended service on high-energy, low-$NPSH_A$ applications. According to a study of data contributed by pump manufacturers, no correlation exists, between the specific speed, the suction specific speed, or any other simple variable and the shape of the NPSH curve break-off. The design variables and manufacturing variables are too great. This means that no standard relationship exists between a 3, 2, 1, or 0 percent head drop. The ratio between the NPSH required for a 0 percent head drop and the $NPSH_R$ is not a constant, but it generally varies over a range from 1.05 to 2.5. NPSH for a 0, 1, or 2 percent head drop cannot be predicted by calculation, given $NPSH_R$.

A pump cannot be constructed to resist cavitation. Although a wealth of literature is available on the resistance of materials to cavitation erosion, no unique material property or combination of properties has been found that yields a consistent correlation with cavitation damage rate (Sanks et al., 1998).

5.5 CORRECTED PUMP CURVES

Figures 5.6 and 5.9 depict "uncorrected" pump curves. That is, these curves depict a pump H-Q curve, as offered by a pump manufacturer. In an actual pumping station design, a manufacturer's pump must be "corrected" by subtracting the headlosses that occur in the suction and discharge piping that connect the pump to the supply tank and the pipeline system. See Table 5.4 associated with Fig. 5.12 in the following sample problem in performing these calculations. The example in Table 5.4 uses a horizontal pump. If a vertical turbine pump is used, minor losses in the pump column and discharge elbow also must be included in the analysis. This same example is worked in U.S. units in Appendix 5.A to this chapter.

TABLE 5.4 Calculate Minor Losses

Item in Fig. 5.12	Description	Pipe Size mm	m	Friction Factor K*	C†
1	Entrance	300	0.30	1.0	
2	90° elbow	300	0.30	0.30	
3	4.5 m of straight pipe	300	0.30		140
4	30° elbow	300	0.30	0.20	
5	2 m of straight pipe	300	0.30		140
6	Butterfly valve	300	0.30	0.46	
7	1.2 m of straight pipe	300	0.30		140
8	300 mm × 200 mm reducer	200	0.20	0.25	
9	150 mm × 250 mm increaser	250	0.25	0.25	
10	1 m of straight pipe	250	0.25		140
11	Pump control valve	250	0.25	0.80	
12	1 m of straight pipe	250	0.25		140
13	Butterfly valve	250	0.25	0.46	
14	0.60 m of straight pipe	250	0.25		140
15	90° elbow	250	0.25	0.30	
16	1.5 m of straight pipe	250	0.25		140
17	Tee connection	250	0.25	0.50	

*Typical K values. Different publications present other values.
†Reasonable value for mortar-lined steel pipe. Value can range from 130 to 145.

Problem

1. *Calculation of minor losses.* The principal headloss equation for straight sections of pipe is:

$$H_L = \left(\frac{L}{1000}\right)\left(\frac{151Q}{CD^{2.63}}\right)^{1.85} \quad (5.11a)$$

where L = length (m), D = pipe diameter (m), Q = flow (m3/s), C = Hazen-Williams friction factor.

The principal headloss equation for fittings is

$$H_L = \sum K \frac{V^2}{2g}$$

where K = fitting friction coefficient, V = velocity (m/s), and g = acceleration due to gravity (m·s/s)

Sum of K values for various pipe sizes:

$K_{300} = 1.96$

$K_{200} = 0.25$

$K_{250} = 2.31$

Sum of C values for various pipe sizes:

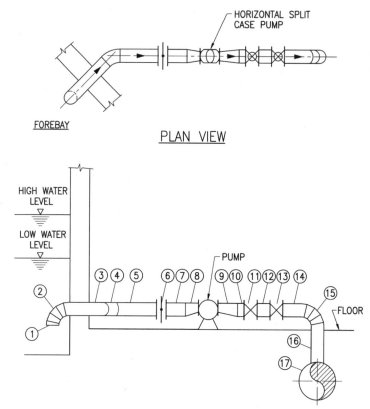

FIGURE 5.12 Piping system used in the example in Table 5.4.

Pipe lengths for 300-mm pipe: $L = 7.7$ m

Pipe lengths for 250-mm pipe: $L = 4.1$ m

Determine the total headloss:

$$H_L = H_{L\,300\,mm} + H_{L\,250\,mm} + K_{300}\frac{V^2_{300\,mm}}{2g} + K_{250\,mm}\frac{V^2_{250\,mm}}{2g} + K_{200\,mm}\frac{V^2_{200\,mm}}{2g}$$

$$H_{L\,300\,mm} = \left(\frac{7.7}{1000}\right)\left(\frac{151Q}{140 \times 0.30^{2.63}}\right)^{1.85} = 3.10\,Q^{1.85}$$

$$H_{L\,250\,mm} = \left(\frac{4.1}{1000}\right)\left(\frac{151Q}{140 \times 0.25^{2.63}}\right)^{1.85} = 4.00\,Q^{1.85}$$

Convert $V^2/2g$ terms to Q^2 terms:

$$\frac{V^2}{2g} = \frac{1}{2g}\left(\frac{Q}{A}\right)^2 = \frac{1}{2g}\left(\frac{1}{A}\right)^2 Q^2 = \frac{1}{2g}\left(\frac{1}{\pi D^2/4}\right)^2 Q^2 = \frac{1}{2g}\left(\frac{16}{\pi^2 D^4}\right)^2 Q^2 = \frac{0.0826}{D^4} Q^2$$

Therefore,

$$K_{300\,mm} \frac{V^2_{300\,mm}}{2g} = 1.96 \left[\frac{0.0826}{(0.30)^4}\right] Q^2 = 19.99\, Q^2$$

$$K_{200\,mm} \frac{V^2_{200\,mm}}{2g} = 0.25 \left[\frac{0.0826}{(0.20)^4}\right] Q^2 = 12.90\, Q^2$$

$$K_{250} \frac{V^2_{250}}{2g} = 2.31 \left[\frac{0.0826}{(0.25)^4}\right] Q^2 = 21.15\, Q^2$$

$$\text{Total } H_L = H_{L\,300\,mm} + H_{L\,250\,mm} + K_{300\,mm}\frac{V^2_{300\,mm}}{2g} + K_{250\,mm}\frac{V^2_{250\,mm}}{2g} + K_{200\,mm}\frac{V^2_{200\,mm}}{2g}$$

$$= 3.10\, Q^{1.85} + 4.00\, Q^{1.85} + 19.99\, Q^2 + 12.90\, Q^2 + 21.15\, Q^2$$

$$= 7.10\, Q^{1.85} + 54.04\, Q^2$$

2. *Modification of pump curve.* Using the above equation for H_L, a "modified" pump curve can then be developed (see Table 5.5) The H values as corrected must then be plotted. The operating point of the pump is the intersection of the corrected H-Q curve with the system curve.

3. *Calculation of $NPSH_A$.* Using the data developed above for calculating the minor losses in the piping, it is now possible to calculate the $NPSH_A$ for the pump. Only the minor losses pertaining to the suction piping are considered: items 1-8 in Fig. 5.12. For this suction piping, we have

$$K_{300\,mm} = 1.96$$
$$K_{200\,mm} = 0.25$$

Sum of the C values: pipe length for 300-mm pipe is $L = 7.7$ m.
Determine the headloss in the suction piping:

TABLE 5.5 Convert Pump Curve Head Values to Include Minor Piping Losses

Q		H (m)	
L/s	m³/s	Uncorrected	Corrected
0	0.	60	60.00
63	0.063	55	54.74
126	0.126	49	47.99
189	0.189	40	37.74
252	0.252	27	23.01

$$H_L = H_{L\,300\,mm} + K_{300}\frac{V_{300}^2}{2g} + K_{200}\frac{V_{200}^2}{2g}$$

$$= H_{L300\,mm} + 1.96\frac{V_{300}^2}{2g} + 0.25\frac{V_{200}^2}{2g}$$

$$= 3.10\,Q^{1.85} + 19.99\,Q^2 + 12.90\,Q^2$$

$$= 3.10\,Q^{1.85} + 32.89\,Q^2$$

For Fig. 5.12, assume that the following data apply:

High-water level = elevation 683 m
Low water level = elevation 675 m
Pump center line elevation = 674 m

Therefore,

Maximum static head = 683 − 674 = 9 m
Minimum static head = 675 − 674 = 1 m

Per Eq. (5.32), with computation of $NPSH_A$ shown in Table 5.6

$$NPSH_A = h_{atm} + h_s - h_{vp} - h_L$$

For this example, use

h_{atm} = 10.35 m
h_{vp} = 0.24 m at 15°C
h_s = 9 m maximum
h_s = 1 m minimum

TABLE 5.6 Computation of $NPSH_A$

Condition	Flow (m³/s)	h_s (m)	h_{atm} (m)	h_{vp} (m)	H_L at Flow (m)	NPSH at Flow (m)
High-static suction head	0.00	9.0	10.35	0.24	0.00	19.11
	0.06	9.0	10.35	0.24	0.12	18.99
	0.12	9.0	10.35	0.24	0.53	18.58
	0.18	9.0	10.35	0.24	1.20	17.91
	0.24	9.0	10.35	0.24	2.12	16.99
	0.30	9.0	10.35	0.24	3.29	15.81
Low-static suction head	0.00	1.0	10.35	0.24	0.00	11.11
	0.06	1.0	10.35	0.24	0.12	10.99
	0.12	1.0	10.35	0.24	0.53	10.58
	0.18	1.0	10.35	0.24	1.20	9.91
	0.24	1.0	10.35	0.24	2.12	8.99
	0.30	1.0	10.35	0.24	3.29	7.82

5.6 HYDRAULIC CONSIDERATIONS IN PUMP SELECTION

5.6.1 Flow Range of Centrifugal Pumps

The flow range over which a centrifugal pump can perform is limited, among other things, by the vibration levels to which it will be subjected. As discussed in API Standard 610 (American Petroleum Institute, n.d.), centrifugal pump vibration varies with flow, usually being a minimum in the vicinity of the flow at the BEP and increasing as flow is increased or decreased. The change in vibration as flow is varied from the BEP depends on the pump's specific speed and other factors. A centrifugal pump's operation flow range can be divided into two regions. One region is termed the *best efficient* or *preferred operating region*, over which the pump exhibits low vibration. The other region is termed the *allowable operating range*, with its limits defined as those capacities at which the pump's vibration reaches a higher but still "acceptable" level.

ANSI/HI Standard 1.1–1.5 points out that vibration can be caused by the following typical sources (Hydraulics Institute, 1994):

1. Hydraulic forces produced between the impeller vanes and volute cutwater or diffuser at vane-passing frequency.

2. Recirculation and radial forces at low flows. This is one reason why there is a definite *minimum* capacity of a centrifugal pump. The pump components typically are not designed for continuous operation at flows below 60 or 70 percent of the flow that occurs at the BEP.

3. Fluid separation at high flows. This is one reason why there is also a definite *maximum* capacity of a centrifugal pump. The pump components typically are not designed for continuous operation of flows above about 120 to 130 percent of the flow that occurs at the BEP.

4. Cavitation due to net positive suction head (NPSH) problems. There is a common misconception that if the net positive suction head available ($NPSH_A$) is equal to or greater than the net positive suction head required ($NPSH_R$) as shown on a pump manufacturer's pump curve, then there will be no cavitation. This is *wrong*! As discussed in ANSI/HI 1.1–1.5-1994 and also by Taylor (1987), it takes a suction head of 2 to 20 times the $NPSH_R$ value to eliminate cavitation completely.

5. Flow disturbances in the pump intake due to improper intake design.

6. Air entrainment or aeration of the liquid.

7. Hydraulic resonance in the piping.

8. Solids contained in the liquids, such as sewage impacting in the pump and causing momentary unbalance, or wedged in the impeller and causing continuous unbalance.

The HI standard then states: The pump manufacturer should provide for the first item in the pump design and establish limits for low flow. "*The system designer is responsible for giving due consideration to the remaining items*" (italics added).

The practical applications of the above discussion can be accomplished by observing what can happen in a plot of a pump curve-system head curve, as discussed in Fig. 5.6. If the intersection of the system curve with the pump H-Q curve occurs too far to the left of the BEP (i.e., at less than about 60 percent of flow at the BEP) or too far to the right of the BEP (i.e., at more than about 130 percent of the flow at the BEP), then the pump will eventually fail as a result of hydraulically induced mechanical damage.

5.6.2 Causes and Effects of Centrifugal Pumps Operating Outside Allowable Flow Ranges

As can be seen in Fig. 5.6, a pump always operates at the point of intersection of the system curve with the pump H-Q curve. Consequently, if too conservative a friction factor is used in determining the system curve, the pump may actually operate much further to the right of the assumed intersection point so that the pump will operate beyond its allowable operating range. Similarly, overly conservative assumptions concerning the static head in the system curve can lead to the pump operating beyond its allowable range. See Fig. 5.13 for an illustration of these effects. The following commentary discusses the significance of the indicated operating points 1 through 6 and the associated flows Q_1 through Q_6:

- Q_1 is the theoretical flow that would occur, ignoring the effects of the minor headlosses in the pump suction and discharge piping. See Fig. 5.12 for an example. Q_1 is slightly to the *right* of the most efficient flow, indicated as 100 units.

- Q_2 is the actual flow that would occur in this system, with the effects of the pump suction and discharge piping minor losses included in the analysis. Q_2 is less than Q_1, and Q_2 is also to the *left* of the point of most efficient flow. As shown in Fig. 5.7, as the impeller wears, this operating point will move even further to the left and the pump will become steadily less efficient.

- Q_1 and Q_2 are the flows that would occur assuming that the system head curve that is depicted is "reasonable": that is, not unrealistically conservative. If, in fact, the system head curve is flatter (less friction in the system than was assumed), then the operating point will be Q_3 (ignoring the effects of minor losses in the pump suction and discharge piping). If these minor losses are included in the analysis, then the true operating point is Q_4. At Q_3, the pump discharge flow in this example is 130 percent of the flow that occurs at the BEP. A flow of 130 percent of flow at the BEP is just at edge of, and may even exceed, the maximum acceptable flow range for pumps (see discussion in Sec. 5.6.1). With most mortar-lined steel or ductile-iron piping systems, concrete pipe, or with plastic piping, reasonable C values should almost always be in the range of 120–145 for water and wastewater pumping systems. Lower C usually would be used only when the pumping facility is connected to existing, old unlined piping that may be rougher.

- If the static head assumed was too conservative, then the actual operating points would be Q_5 or Q_6. Q_5 is 150 percent of the flow at the BEP;. Q_6 is 135 percent of the flow at the BEP. In both cases, it is most likely that these flows are outside the allowable range of the pump. Cavitation, inadequate $NPSH_A$, and excessive hydraulic loads on the impeller and shaft bearings are likely to occur, with resulting poor pump performance and high maintenance costs.

5.6.3 Summary of Pump Selection

In selecting a pump, the following steps should be taken:

1. Plot the system head curves, using reasonable criteria for both the static head range and the friction factors in the piping. Consider all feasible hydraulic conditions that will occur: (a) Variations in static head (b) Variations in pipeline friction factor (C value) Variations in static head result from variations in the water surface elevations (WSE) in the supply reservoir to the pump and in the reservoir to which the pump is pumping. Both minimum and maximum static head conditions should be investigated:

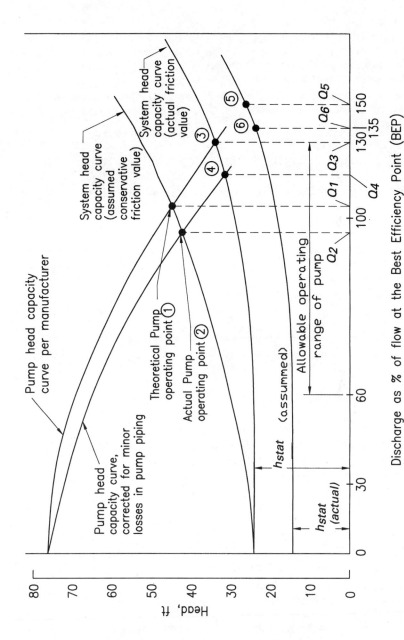

FIGURE 5.13 Determining the operating point for a single-speed pump.

- *Maximum static head.* Minimum WSE in supply reservoir and maximum WSE in discharge reservoir.
- *Minimum static head.* Maximum WSE in supply reservoir and minimum WSE in discharge reservoir.

2. Be sure to develop a corrected pump curve or modified pump curve by subtracting the minor losses in the pump suction and discharge piping from the manufacturer's pump curve (Table 5.5 and Fig. 5.13). The true operating points will be at the intersections of the corrected pump curve with the system curves.
3. Select a pump so that the initial operating point (intersection of the system head curve with the pump curve) occurs to the right of the BEP. As the impeller wears, the pump's output flow will decrease (Fig. 5.7), but the pump efficiency will actually increase until the impeller has worn to the level that the operating point is to the left of the BEP.

    For a system having a significant variation in static head, it may be necessary to select a pump curve so that at high static head conditions, the operating point is to the left of the BEP. However, the operating point for the flows that occur a majority of the time should be at or to the right of the BEP. Bear in mind that high static head conditions normally only occur a minority of the time: The supply reservoir must be at its low water level and the discharge reservoir must simultaneously be at its maximum water level-conditions that usually do not occur very often. Consequently, select a pump that can operate properly at this condition–but also select the pump that has a BEP which occurs at the flow that will occur most often. See Fig. 5.14 for an example.
4. In multiple-pump operations, check the operating point with each combination of pumps that may operate. For example, in a two-pump system, one pump operating alone will produce a flow that is greater than 50 percent of the flow that is produced with both pumps operating. This situation occurs because of the rising shape of the system head curve; see Fig. 5.8. Verify that the pump output flows are within the pump manufacturer's recommended operating range; see Fig. 5.13.
5. Check that $NPSH_A$ exceeds the $NPSH_R$ for all the hydraulic considerations and operating points determined in steps 1 and 3.

5.7 APPLICATION OF PUMP HYDRAULIC ANALYSIS TO DESIGN OF PUMPING STATION COMPONENTS

5.7.1 Pump Hydraulic Selections and Specifications

5.7.1.1 Pump operating ranges. Identify the minimum, maximum, and design flows for the pump based on the hydraulic analyses described above. See Fig. 5.14 as an example.

- The flow at 100 units would be defined as the design point.
- There is a minimum flow of 90 units.
- There is a maximum flow of 115 units.

In multiple-pump operation, the combination of varying static head conditions and the different number of pumps operating in parallel could very likely result in operating points as follows (100 units = flow at BEP; see Table 5.7).

Pump System Hydraulic Design **5.31**

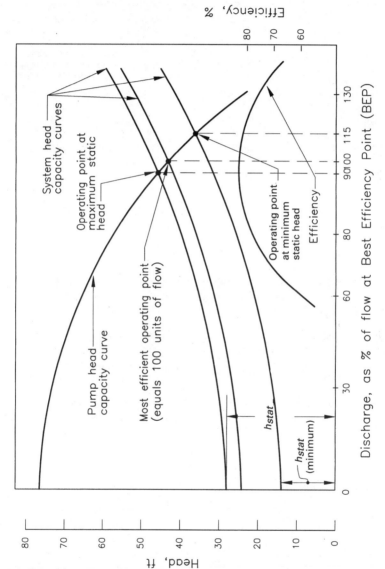

FIGURE 5.14 Determining the operating points for a single-speed pump with variation in values of h_{stat}.

TABLE 5.7 Pump Operating Ranges

Operating Flow Condition	Flow (per Pump)	Comments
Minimum	70	Maximum static head condition, all pumps operating
Normal 1	100	Average or most frequent operating condition: fewer than all pumps operating, average static head condition. Might also be the case of all pumps operating, minimum static head condition
Normal 2	110	Fewer than all pumps operating, minimum static head condition
Maximum 1	115	Maximum static head condition, one pump operating
Maximum 2	125	Minimum static head condition, one pump operating

Some observations of the above example are the following:

- The flow range of an individual pump is about 1.8:1 (125 ÷ 70).
- The pump was deliberately selected to have its most efficient operating point ($Q = 100$) at the most frequent operating condition, not the most extreme condition. This will result in the minimum power consumption and minimum power cost for the system.
- The pump was selected or specified to operate over all possible conditions, not just one or two conditions.

In variable-speed pumping applications, the minimum flow can be much lower than what is shown in these examples. It is extremely important that the minimum flow be identified in the pump specification so that the pump manufacturer can design the proper combination of impeller type and shaft diameter to avoid cavitation and vibration problems.

5.7.1.2 Specific pump hydraulic operating problems. Specific problems that can occur when operating a centrifugal pump beyond its minimum and maximum capacities include the following (Hydraulics Institute, 1994):

- *Minimum flow problems.* Temperature buildup, excessive radial thrust, suction recirculation, discharge recirculation, and insufficient $NPSH_A$.
- *Maximum flow problems.* Combined torsional and bending stresses or shaft deflection may exceed permissible limits; erosion drainage, noise, and cavitation may occur because of high fluid velocities.

5.7.2 Piping

Having selected a pump and determined its operating flows and discharge heads or pressures, it is then desirable to apply this data in the design of the piping. See Fig. 5.12 for typical piping associated with a horizontal centrifugal pump.

5.7.2.1 Pump suction and discharge piping installation guidelines. Section 1.4 in the Hydraulic Institute (HI) publication ANSI/HI 1.1–1.5 (1994) and Chap. 6 in API Recommended Practice 686 (1996) provide considerable discussion and many recommendations on the layout of piping for centrifugal pumps to help avoid the hydraulic problems discussed above.

5.7.2.2 Fluid velocity. The allowable velocities of the fluid in the pump suction and discharge piping are usually in the following ranges:

Suction: 3–9 ft/s (4–6 ft/s most common)

1.0–2.7 m/s (1.2–1.8 m/s most common)

Discharge: 5–15 ft/s (7–10 ft/s most common)

1.5–4.5 m/s (2–3 m/s most common)

Bear in mind that the velocities will vary for a given pump system because the operating point on a pump curve (i.e., intersection of the pump curve with the system curve) varies for the following reasons:

1. Variation in static heads, as the water surface elevations in both the suction and discharge reservoirs vary

2. Long-term variations in pipeline friction factors (Fig. 5.5)

3. Long-term deterioration in impeller (Fig. 5.7)

4. Variation in the number of pumps operating in a multipump system (Fig. 5.8).

A suggested procedure for sizing the suction and discharge piping is as follows:

1. Select an allowable suction pipe fluid velocity of 3–5 ft/s (1.0–1.5 m/s) with all pumps operating at the minimum static head condition. As fewer pumps are used, the flow output of each individual pump will increase (typically by about 20 to 40 percent with one pump operating compared to all pumps operating), with the resulting fluid velocities in the suction piping also increasing to values above the 3–5 ft/s (1.0–1.5 m/s) nominal criteria.

2. Select an allowable discharge pipe fluid velocity of 5–8 ft/s (1.5–2.4 m/s), also with all pumps operating at the minimum static head condition. As discussed above, as fewer pumps are used, the flow output of each individual pump will increase, with the resulting fluid velocities in the discharge piping also increasing in values above the 5–8 ft/s (1.5–2.4 m/s) nominal criteria.

5.7.2.3 Design of pipe wall thickness (pressure design). Metal pipes are designed for pressure conditions by the equation for *hoop tensile strength:*

$$t = \frac{PD}{2SE} \quad (5.34)$$

where

t = wall thickness, in or mm

D = inside diameter, in or mm (although in practice, the outside diameter is often conservatively used, partly because the ID is not known initially and because it is the outside diameter (OD) that is the fixed dimension: ID then varies with the wall thickness)

P = design pressure (psi or kPa)

S = allowable design circumferential stress (psi or kPa)

E = longitudinal joint efficiency

The design value for S is typically 50 percent of the material yield strength for "normal" pressures. For surge or transient pressures in steel piping systems, S is typically allowed to rise to 70 percent of the material yield strength (American Water Works Association 1989).

The factor E for the longitudinal joint efficiency is associated with the effective strength of the welded joint. The ANSI B31.1 and B31.3 codes for pressure piping recommend the values for E given in Table 5.8 (American Society for Mechanical Engineers, 1995, 1996)

The wall thickness for plastic pipes (polyvinyl chloride, PVC; high-density polyethylene, HDPE; and FRP) is usually designed in the United States on what is known as the hydrostatic design basis, or HDB:

$$P_t = \frac{2t}{D-t} \times \frac{HDB}{F} \qquad (5.35)$$

where P_t = total system pressure (operating + surge), t = minimum wall thickness (in), D = average outside diameter (in), HDB = hydrostatic design basis (psi) anh F = factor of safety (2.50–4.00).

5.7.2.4 Design of pipe wall thickness (vacuum conditions). If the hydraulic transient or surge analysis (see Chap. 6) indicates that full or partial vacuum conditions may occur, then the piping must also be designed accordingly. The negative pressure required to collapse a circular metal pipe is described by the equation

$$\Delta P = \frac{2E}{(1 - \mu^2)SF} \left(\frac{e}{D}\right)^3 \qquad (5.36)$$

TABLE 5.8 Weld Joint Efficiencies

Type of Longitudinal Joint	Weld Joint Efficiency Factor (E)
Arc or gas weld (steel pipe)	
Single-butt weld	0.80
Double-butt weld	0.90
Single-or double-butt weld with 100% radiography	1.00
Electric resistance weld (steel pipe)	0.85
Furnace butt weld (steel pipe)	0.60
Most steel water pipelines	0.85
Ductile iron pipe	1.0

where ΔP = difference between internal and external pipeline pressures (psi or kPa), E = modulus of elasticity of the pipe material (psi or kPa), μ = Poisson's ratio, SF = safety factor (typically 4.0), e = wall thickness (in or mm), and D = outside diameter (in or mm).

Because of factors such as end effects, wall thickness variations, lack of roundness, and other manufacturing tolerances, Eq. (5.3b) for steel pipe is frequently adjusted in practice to

$$\Delta P = \frac{50{,}000{,}000}{SF}\left(\frac{e}{D}\right)^3 \quad (5.37)$$

5.7.2.5 Summary of pipe design criteria. The wall thickness of the pump piping system is determined by consideration of three criteria:

1. Normal operating pressure (Eq. 5.34), with S = 50 percent of yield strength
2. Maximum pressure due to surge (static + dynamic + transient rise), using Eq. (5.34) with S = 70 percent of yield strength (in the case of steel pipe)
3. Collapsing pressure, if negative pressures occur due to surge conditions (Eq. 5.36).

5.8 IMPLICATIONS OF HYDRAULIC TRANSIENTS IN PUMPING STATION DESIGN

Hydraulic transient, or surge, analysis is covered in detail in Chap. 6. Surge or hydraulic transient effects must be considered in pump and piping systems because they can cause or result in the following (Sanks et al., 1998):

- rupture or deformation of pipe and pump casings,
- pipe collapse,
- vibration,
- excessive pipe or joint displacements, or
- pipe fitting and support deformation or even failure.

The pressures generated due to hydraulics, thus, must be considered in the pipe design, as was discussed in Sec. 5.7, above.

5.8.1 Effect of Surge on Valve Selection

At its worst, surges in a piping can cause swing check valves to slam closed violently when the water column in the pipeline reverses direction and flows backward through the check valve at a significant velocity before the valve closes completely. Consequently, in pump and piping systems in which significant surge problems are predicted to occur, check valves or pump control valves are typical means to control the rate of closure of the valve. Means of controlling this rate of closure include

- using a valve that closes quickly, before the flow in the piping can reverse and attain a high reverse velocity,
- providing a dashpot or buffer on the valve to allow the valve clapper or disc to close gently, and

- closing the valve with an external hydraulic actuator so that the reverse flowing water column is gradually brought to a halt. This is frequently done with ball or cone valves used as pump control valves.

The pressure rating of the valve (both the check valve or pump control valve and the adjacent isolation) should be selected with a pressure rating to accommodate the predicted surge pressures in the piping system.

5.8.2 Effect of Surge on Pipe Material Selection

Metal piping systems, such as steel and ductile iron, have much better resistance to surge than do most plastic pipes (PVC, HDPE, ABS, and FRP). The weakness of plastic pipes with respect to surge pressures is sometimes not adequately appreciated because the wave velocity (a), and hence the resulting surge pressures, are significantly lower than is the case with metal piping systems. Since the surge pressures in plastic piping are lower than those in metal piping systems, there is sometimes a mistaken belief that the entire surge problem can then be neglected. However, plastic piping systems inherently offer less resistance to hydraulic transients than do metal piping systems, even with the lower pressures. This is particularly the case with solvent- or adhesive-welded plastic fittings.

HDPE has better resistance to surge pressures than do other plastic piping systems. In addition, the joints are fusion-butt welded, not solvent, welded, which results in a stronger joint. However, HDPE is still not as resistant to surge effects as is a properly designed steel or ductile iron piping system.

REFERENCES

American Petroleum Institute, *Centrifugal Pumps for Petroleum, Heavy Duty Chemical, and Gas Industry Services,* API Standard 610, 8th ed American Petroleum Institute, Washington, DC, 1995.

American Petroleum Institute, *Recommended Practices for Machinery Installation and Installation Design,* Practice 686, Washington, DC, 1996.

American Society of Mechanical Engineers (ASME), B31.1, *Power Piping,* ASME, NewYork, 1995.

American Society of Mechanical Engineers (ASME), B31.3, *Process Piping,* ASME, NewYork, 1996.

American Water Works Association, *Steel Pipe—A Guide for Design and Installation, AWWA M11,* 3rd ed., American Water Works Association, Denver, CO, 1989.

Cavi, D., "$NPSH_R$ Data and Tests Need Clarification," *Power Engineering,* 89:47–50, 1985.

Hydraulics Institute, *American National Standard for Centrifugal Pumps for Nomenclature, Definitions, Applications, and Operation,* ANSI/HI 1.1–1.5-1994, Hydraulics Institute, Parsippany, NJ, 1994.

Sanks, R. L., et al., *Pumping Station Design,* 2nd ed., Butterworth London, UK, 1998.

Taylor., "Pump Bypasses Now More Important," *Chemical Engineering,* May 11, 1987.

APPENDIX 5. A.
PUMP SYSTEM HYDRAULIC DESIGN

Calculation of Minor Losses and $NPSH_A$ in Piping and Modification of a Pump Curve (U.S. units)

Part 1. Calculation of Minor Losses

Principal headloss equations

- For straight sections of pipe: $H_L = \dfrac{4.72L}{D^{4.86}} \left(\dfrac{Q}{C}\right)^{1.85}$ [See Eq. (5.11)]

where L = length (ft), D = pipe diameter (ft), Q = flow (ft³/s), and C = Hazen-Williams friction factor

- For fittings: $H_L = \Sigma K \dfrac{V^2}{2g}$ [See Eq. (5.23)]

where K = fitting friction coefficient, V = velocity (ft/s), and g = acceleration due to gravity [(ft·s)/s].

Sum of K values for various pipe sizes: $K_{12} = 1.96$, $K_8 = 0.25$, and $K_{10} = 2.31$.
Sum of C values for various pipe sizes: pipe lengths for 12-in pipe: $L = 26$ ft; pipe lengths for 10-in pipe: $L = 13$ ft.
Determine the total headloss:

$$H_L = H_{L\ 12in} + H_{L\ 10in} + K_{12}\dfrac{V^2_{12}}{2g} + K_{10}\dfrac{V^2_{10}}{2g} + K_8\dfrac{V^2_8}{2g}$$

$$H_{L\ 12in} = \dfrac{4.72(26)}{(12/12)^{4.86}} \left(\dfrac{Q}{140}\right)^{1.85} = 0.013139 Q^{1.85}$$

$$H_{L\ 10in} = \dfrac{4.72(13)}{(10/12)^{4.86}} \left(\dfrac{Q}{140}\right)^{1.85} = 0.015936 Q^{1.85}$$

Convert $\dfrac{V^2}{2g}$ terms to Q^2 terms:

5.38 Chapter Five

Item in Fig. 5.12	Description	Pipe Size (in)	Friction Factor K*	Friction Factor C†
1	Entrance	12	1.0	
2	90° elbow	12	0.30	
3	15 ft of straight pipe	12		140
4	30° elbow	12	0.20	
5	7 ft of straight pipe	12		140
6	Butterfly valve	12	0.46	
7	4 ft of straight pipe	12		140
8	12 in × 8 in reducer	8	0.25	
9	6 in × 10 in increaser	10	0.25	
10	3 ft of straight pipe	10		140
11	Pump control valve	10	0.80	
12	3 ft of straight pipe	10		140
13	Butterfly valve	10	0.46	
14	2 ft of straight pipe	10		140
15	90° elbow	10	0.30	
16	5 ft of straight pipe	10		140
17	Tee connection	10	0.50	

*Typical K values. Different publications present other values.
†Reasonable value for mortar-lined steel pipe. Value can range from 130 to 145.

$$\frac{V^2}{2g} = \frac{1}{2g}\left(\frac{Q}{A}\right)^2 = \frac{1}{2g}\left(\frac{1}{A}\right)^2 Q^2$$

$$= \frac{1}{2g}\left(\frac{1}{\pi D^2/4}\right)^2 Q^2$$

$$= \frac{1}{2g}\left(\frac{16}{\pi^2 D^4}\right)^2 Q^2$$

$$= \frac{0.025173}{D^4} Q^2$$

Therefore,

$$K_{12}\frac{V_{12}^2}{2g} = 1.96\left[\frac{0.025173}{(12/12)^4}\right] Q^2$$

$$= 0.04933\, Q^2$$

$$K_8 \frac{V_8^2}{2g} = 0.25 \left[\frac{0.025173}{(8/12)^4}\right] Q^2$$

$$= 0.031859 \, Q^2$$

$$K_{10} \frac{V_{10}^2}{2g} = 2.31 \left[\frac{0.025173}{(10/12)^4}\right] Q^2$$

$$= 0.12058 \, Q^2$$

$$\text{Total } H_L = H_{L\,12\text{in}} + H_{L\,10\text{in}} + K_{12} \frac{V_{12}^2}{2g} + K_{10} \frac{V_{10}^2}{2g} + K_8 \frac{V_8^2}{2g}$$

$$= 0.013139 \, Q^{1.85} + 0.015936 \, Q^{1.85} + 0.04933 \, Q^2 + 0.12058 \, Q^2 + 0.031859 \, Q^2$$

$$= 0.0291 \, Q^{1.85} + 0.202 \, Q^2$$

Part 2. Modification of Pump Curve

Using the above equation for H_l, a "modified" pump curve can then be developed by converting pump curve head values to include minor piping losses:

Q		H (ft)	
GPM	CFS	Uncorrected	Corrected
0	0.000	200	200.00
1000	2.228	180	178.87
2000	4.456	160	151.52
3000	6.684	130	120.0
4000	8.912	90	72.29

The H values, as corrected, must then be plotted. The operating point of the pump is the intersection of the corrected H-Q curve with the system curve.

Part 3. Calculation of NPSH$_A$

Using the data developed above for calculating the minor losses in the piping, it is now possible to calculate the NPSH$_A$ for the pump. Only the minor losses pertaining to the suction piping are considered: Items 1–8 in Fig. 5.12. For this suction piping, we have: $K_{12} = 1.96$, $K_8 = 0.25$, sum of C values. Pipe length for 12–in pipe: $L = 26$ ft.

Determine the headloss in the suction piping:

$$H_L = H_{L\,12\text{in}} + K_{12} \frac{V_{12}^2}{2g} + K_8 \frac{V_8^2}{2g}$$

$$= H_{L\,12\text{in}} + 1.96 \frac{V_{12}^2}{2g} + 0.25 \frac{V_8^2}{2g}$$

5.40 Chapter Five

$$= 0.013139\, Q^{1.85} + 0.04933 Q^2 + 0.031859 Q^2$$

$$= 0.013139\, Q^{1.85} + 0.081189 Q^2$$

For Fig. 5.12, assume that the following data apply:

High-water level = elevation 2241 ft

Low-water level = elevation 2217 ft

Pump center line elevation = 2212 ft

Therefore,

Maximum static head = 2241 − 2212 = 29 ft.
Minimum static head = 2217 − 2212 = 5 ft.

Per Eq. (5.31), $NPSH_A = h_{atm} + h_s - h_{vp} - h_L$ For this example, use the following:

$h_{atm} = 33.96$ ft
$h_{vp} = 0.78$ ft at 60°F
$h_s = 29$ ft maximum
$h_s = 5$ ft minimum

Compute $NPSH_A$:

Condition	Flow (ft³/s)	h_s (ft)	h_{atm} (ft)	h_{vp} (ft)	H_L at Flow (ft)	$NPSH_A$ at Flow (ft)
High-static suction head	0	29	33.96	0.78	0.00	62.18
	2	29	33.96	0.78	0.37	61.81
	4	29	33.96	0.78	1.47	60.71
	6	29	33.96	0.78	3.28	58.90
	8	29	33.96	0.78	5.81	56.37
	10	29	33.96	0.78	9.05	53.13
Low-static suction head	0	5	33.96	0.78	0.00	38.18
	2	5	33.96	0.78	0.37	37.81
	4	5	33.96	0.78	1.47	36.71
	6	5	33.96	0.78	3.28	34.90
	8	5	33.96	0.78	5.81	32.37
	10	5	33.96	0.78	9.05	29.13

CHAPTER 6
HYDRAULIC TRANSIENT DESIGN FOR PIPELINE SYSTEMS

C. Samuel Martin
School of Civil and Environmental Engineering
Georgia Institute of Technology
Atlanta, GA

6.1 INTRODUCTION TO WATERHAMMER AND SURGING

By definition, *waterhammer* is a pressure (acoustic) wave phenomenon created by relatively sudden changes in the liquid velocity. In pipelines, sudden changes in the flow (velocity) can occur as a result of (1) pump and valve operation in pipelines, (2) vapor pocket collapse, or (3) even the impact of water following the rapid expulsion of air out of a vent or a partially open valve. Although the name waterhammer may appear to be a misnomer in that it implies only water and the connotation of a "hammering" noise, it has become a generic term for pressure wave effects in liquids. Strictly speaking, waterhammer can be directly related to the compressibility of the liquid-primarily water in this handbook. For slow changes in pipeline flow for which pressure waves have little to no effect, the unsteady flow phenomenon is called *surging*.

Potentially, waterhammer can create serious consequences for pipeline designers if not properly recognized and addressed by analysis and design modifications. There have been numerous pipeline failures of varying degrees and resulting repercussions of loss of property and life. Three principal design tactics for mitigation of waterhammer are (1) alteration of pipeline properties such as profile and diameter, (2) implementation of improved valve and pump control procedures, and (3) design and installation of surge control devices.

In this chapter, waterhammer and surging are defined and discussed in detail with reference to the two dominant sources of waterhammer-pump and/or valve operation. Detailed discussion of the hydraulic aspects of both valves and pumps and their effect on hydraulic transients will be presented. The undesirable and unwanted, but often potentially possible, events of liquid column separation and rejoining are a common justification for surge protection devices. Both the beneficial and detrimental effects of free (entrained or entrapped) air in water pipelines will be discussed with reference to waterhammer and surging. Finally, the efficacy of various surge protection devices for mitigation of waterhammer is included.

6.2 FUNDAMENTALS OF WATERHAMMER AND SURGE

The fundamentals of waterhammer, an elastic process, and surging, an incompressible phenomenon, are both developed on the basis of the basic conservational relationships of physics or fluid mechanics. The acoustic velocity stems from mass balance (continuity), while the fundamental waterhammer equation of Joukowsky originates from the application of linear momentum [see Eq. (6.2)].

6.2.1 Definitions

Some of the terms frequently used in waterhammer are defined as follows.

- *Waterhammer.* A pressure wave phenomenon for which liquid compressibility plays a role.
- *Surging.* An unsteady phenomenon governed solely by inertia. Often termed *mass oscillation* or referred to as either *rigid column* or *inelastic effect*.
- *Liquid column separation.* The formation of vapor cavities and their subsequent collapse and associated waterhammer on rejoining.
- *Entrapped air.* Free air located in a pipeline as a result of incomplete filling, inadequate venting, leaks under vacuum, air entrained from pump intake vortexing, and other sources.
- *Acoustic velocity.* The speed of a waterhammer or pressure wave in a pipeline.
- *Joukowsky equation.* Fundamental relationship relating waterhammer pressure change with velocity change and acoustic velocity. Strictly speaking, this equation is only valid for sudden flow changes.

6.2.2 Acoustic Velocity

For wave propagation in liquid-filled pipes the *acoustic (sonic) velocity* is modified by the pipe wall elasticity by varying degrees, depending upon the elastic properties of the wall material and the relative wall thickness. The expression for the wave speed is

$$a = \frac{\sqrt{K/\rho}}{\sqrt{1 + \frac{D}{e}\frac{K}{E}}} = \frac{a_o}{\sqrt{1 + \frac{D}{e}\frac{K}{E}}} \qquad (6.1)$$

where E is the elastic modulus of the pipe wall, D is the inside diameter of the pipe, e is the wall thickness, and a_o is the acoustic velocity in the liquid medium. In a very rigid pipe or in a tank, or in large water bodies, the acoustic velocity a reduces to the well-known relationship $a = a_o = \sqrt{(K/\rho)}$. For water $K = 2.19$ GPa (318,000 psi) and $\rho = 998$ kg/m³ (1.936 slug/ft³), yielding a value of $a_o = 1483$ m/sec (4865 ft/sec), a value many times that of any liquid velocity V.

6.2.3 Joukowsky (Waterhammer) Equation

There is always a pressure change Δp associated with the rapid velocity change ΔV across a waterhammer (pressure) wave. The relationship between Δp and ΔV from the basic physics of linear momentum yields the well-known *Joukowsky equation*

$$\Delta p = -\rho a \Delta V \qquad (6.2)$$

where ρ is the liquid mass density, and a is the sonic velocity of the pressure wave in the fluid medium in the conduit. Conveniently using the concept of head, the *Joukowsky head rise* for instantaneous valve closure is

$$\Delta H = \frac{\Delta p}{\rho g} = -\frac{\rho a \Delta V}{\rho g} = \frac{aV_o}{g} \qquad (6.3)$$

The compliance of a conduit or pipe wall can have a significant effect on modification of (1) the acoustic velocity, and (2) any resultant waterhammer, as can be shown from Eq. (6.1) and Eq. (6.2), respectively. For simple waterhammer waves for which only radial pipe motion (*hoop stress*) effects are considered, the germane physical pipe properties are Young's elastic modulus (E) and Poisson ratio (μ). Table 6.1 summarizes appropriate values of these two physical properties for some common pipe materials.

The effect of the elastic modulus (E) on the acoustic velocity in water-filled circular pipes for a range of the ratio of internal pipe diameter to wall thickness (D/e) is shown in Fig. 6.1 for various pipe materials.

TABLE 6.1 Physical Properties of Common Pipe Materials

Material	Young's Modulus E (GPa)	Poisson's Ratio μ
Asbestos cement	23–24	–
Cast iron	80–170	0.25–0.27
Concrete	14–30	0.10–0.15
Concrete (reinforced)	30–60	–
Ductile iron	172	0.30
Polyethylene	0.7–0.8	0.46
PVC (polyvinyl chloride)	2.4–3.5	0.46
Steel	200–207	0.30

6.4 Chapter Six

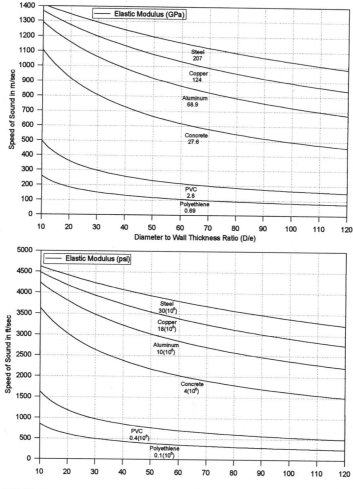

FIGURE 6.1 Effect of wall thickness of various pipe materials on acoustic velocity in water pipes.

6.3 HYDRAULIC CHARACTERISTICS OF VALVES

Valves are integral elements of any piping system used for the handling and transport of liquids. Their primary purposes are flow control, energy dissipation, and isolation of portions of the piping system for maintenance. It is important for the purposes of design and final operation to understand the hydraulic characteristics of valves under both steady and unsteady flow conditions. Examples of dynamic conditions are direct opening or closing of valves by a motor, the response of a swing check valve under unsteady conditions, and the action of hydraulic servovalves. The hydraulic characteristics of valves under either noncavitating or cavitating conditions vary considerably from one type of valve design to another. Moreover, valve characteristics also depend upon particular valve design for a special function, upon absolute size, on manufacturer as well as the type of pipe fitting employed. In this section the fundamentals of valve hydraulics are presented in terms of pressure drop (headloss) characteristics. Typical flow characteristics of selected valve types of control-gate, ball, and butterfly, are presented.

6.3.1 Descriptions of Various Types of Valves

Valves used for the control of liquid flow vary widely in size, shape, and overall design due to vast differences in application. They can vary in size from a few millimeters in small tubing to many meters in hydroelectric installations, for which spherical and butterfly valves of very special design are built. The hydraulic characteristics of all types of valves, albeit different in design and size, can always be reduced to the same basic coefficients, notwithstanding fluid effects such as viscosity and cavitation. Figure 6.2

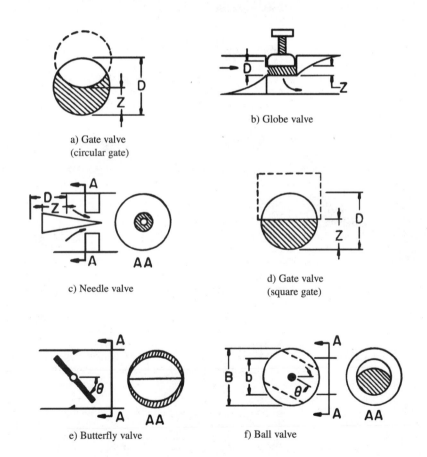

FIGURE 6.2 Cross sections of selected control valves: (From Wood and Jones, 1973).

shows cross sections of some valve types to be discussed with relation to hydraulic performance.

6.3.2 Definition of Geometric Characteristics of Valves

The valve geometry, expressed in terms of cross-sectional area at any opening, sharpness of edges, type of passage, and valve shape, has a considerable influence on the eventual hydraulic characteristics. To understand the hydraulic characteristics of valves it is useful, however, to express the projected area of the valve in terms of geometric quantities. With reference to Fig. 6.2 the ratio of the projected open area of the valve A_v to the full open valve A_{vo} can be related to the valve opening, either a linear measure for a gate valve, or an angular one for rotary valves such as ball, cone, plug, and butterfly types. It should be noted that this geometric feature of the valve clearly has a bearing on the valve hydraulic performance, but should not be used directly for prediction of hydraulic performance –either steady state or transient. The actual hydraulic performance to be used in transient calculations should originate from experiment.

6.3.3 Definition of Hydraulic Performance of Valves

The hydraulic performance of a valve depends upon the flow passage through the valve opening and the subsequent recovery of pressure. The hydraulic characteristics of a valve under partial to fully opened conditions typically relate the volumetric flow rate to a characteristic valve area and the headloss ΔH across the valve. The principal fluid properties that can affect the flow characteristics are fluid density ρ, fluid viscosity μ, and liquid vapor pressure p_v if cavitation occurs. Except for small valves and/or viscous liquids or both, Reynolds number effects are usually not important, and will be neglected with reference to water. A valve in a pipeline acts as an obstruction, disturbs the flow, and in general causes a loss in energy as well as affecting the pressure distribution both upstream and downstream. The characteristics are expressed either in terms of (1) flow capacity as a function of a defined pressure drop or (2) energy dissipation (headloss) as a function of pipe velocity. In both instances the pressure or head drop is usually the difference in total head caused by the presence of the valve itself, minus any loss caused by regular pipe friction between measuring stations.

The proper manner in determining ΔH experimentally is to measure the hydraulic grade line (*HGL*) far enough both upstream and downstream of the valve so that uniform flow sections to the left of and to the right of the valve can be established, allowing for the extrapolation of the energy grade lines (*EGL*) to the plane of the valve. Otherwise, the valve headloss is not properly defined. It is common to express the hydraulic characteristics either in terms of a headloss coefficient K_L or as a discharge coefficient C_f where A_v is the area of the valve at any opening, and ΔH is the headloss defined for the valve. Frequently a discharge coefficient is defined in terms of the fully open valve area. The hydraulic coefficients embody not only the geometric features of the valve through A_v but also the flow characteristics.

Unless uniform flow is established far upstream and downstream of a valve in a pipeline the value of any of the coefficients can be affected by effects of nonuniform flow. It is not unusual for investigators to use only two pressure taps-one upstream and one downstream, frequently 1 and 10 diameters, respectively. The flow characteristics of valves in terms of pressure drop or headloss have been determined for numerous valves by many investigators and countless manufacturers. Only a few sets of data and typical curves will be presented here for ball, butterfly, and gate, valves C_D. For a valve located in the interior of a long continuous pipe, as shown in Fig. 6.3, the presence of the valve disturbs the flow both upstream and downstream of the obstruction as reflected by the velocity distribution, and the pressure variation, which will be non- hydrostatic in the regions of nonuniform flow. Accounting for the pipe friction between upstream and downstream uniform flow sections, the headloss across the valve is expressed in terms of the pipe velocity and a headloss coefficient K_L

$$\Delta H = K_L \frac{V^2}{2g} \qquad (6.4)$$

Often manufacturers represent the hydraulic characteristics in terms of discharge coefficients

$$Q = C_f A_{vo} \sqrt{2g\Delta H} = C_F A_{vo} \sqrt{2gH} \qquad (6.5)$$

where

$$H = \Delta H + \frac{V^2}{2g} \qquad (6.6)$$

Both discharge coefficients are defined in terms of the nominal full-open valve area A_{vo} and a representative head, ΔH for C_f and H for C_Q, the latter definition generally reserved for large valves employed in the hydroelectric industry. The interrelationship between C_f, C_F, and K_L is

$$K_L = \frac{1}{C_f^2} = \frac{1 - C_F^2}{C_F^2} \qquad (6.7)$$

Frequently valve characteristics are expressed in terms of a dimensional flow coefficient C_v from the valve industry

$$Q = C_v \sqrt{\Delta p} \qquad (6.8)$$

where Q is in American flow units of gallons per minute (gpm) and Δp is the pressure loss in pounds per square inch (psi). In transient analysis it is convenient to relate either the loss coefficient or the discharge coefficient to the corresponding value at the fully open valve position, for which $C_f = C_{fo}$. Hence,

$$\frac{Q}{Q_o} = \frac{C_f}{C_{fo}} \sqrt{\frac{\Delta H}{\Delta H_o}} = \tau \sqrt{\frac{\Delta H}{\Delta H_o}} \qquad (6.9)$$

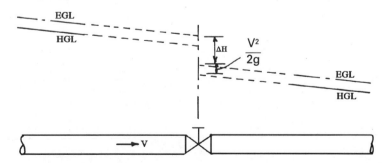

FIGURE 6.3 Definition of headloss characteristics of a valve.

6.8 Chapter Six

Traditionally the dimensionless valve discharge coefficient is termed τ and defined by

$$\tau = \frac{C_f}{C_{fo}} = \frac{C_v}{C_{vo}} = \frac{C_f}{C_{fo}} = \sqrt{\frac{K_{Lo}}{K_L}} \tag{6.10}$$

6.3.4 Typical Geometric and Hydraulic Valve Characteristics

The geometric projected area of valves shown in Fig. 6.2 can be calculated for ball, butterfly, and gate valves using simple expressions. The dimensionless hydraulic flow coefficient τ is plotted in Fig. 6.4 for various valve openings for the three selected valves along with the area ratio for comparison. The lower diagram, which is based on hydraulic mea-

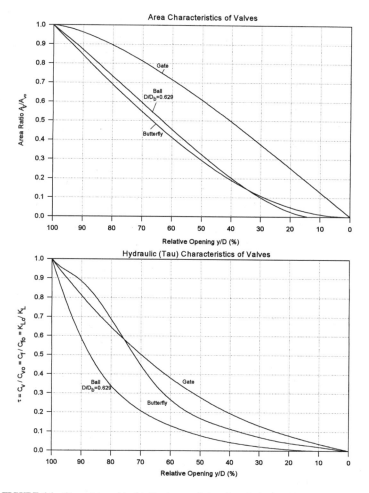

FIGURE 6.4 Geometric and hydraulic characteristics of typical control valves.

TABLE 6.2 Classification of Valve Closure

Time of Closure t_c	Type of Closure	Maximum Head ΔH_{max}	Phenomenon
0	Instantaneous	aV_o/g	Waterhammer
$\leq 2L/a$	Rapid	aV_o/g	Waterhammer
$> 2L/a$	Gradual	$< aV_o/g$	Waterhammer
$>> 2L/a$	Slow	$<< aV_o/g$	Surging

surements, should be used for transient calculations rather than the upper one, which is strictly geometric.

6.3.5 Valve Operation

The instantaneous closure of a valve at the end of a pipe will yield a pressure rise satisfying Joukowsky's equation–Eq. (6.2) or Eq. (6.3). In this case the velocity difference $\Delta V = 0 - V_o$, where V_o is the initial velocity of liquid in the pipe. Although Eq. (6.2) applies across every wavelet, the effect of complete valve closure over a period of time greater than $2L/a$, where L is the distance along the pipe from the point of wave creation to the location of the first pipe area change, can be beneficial. Actually, for a simple pipeline the maximum head rise remains that from Eq. (6.3) for times of valve closure $t_c \leq 2L/a$, where L is the length of pipe. If the value of $t_c > 2L/a$, then there can be a considerable reduction of the peak pressure resulting from beneficial effects of negative wave reflections from the open end or reservoir considered in the analysis. The phenomenon can still be classified as waterhammer until the time of closure $t_c > 2L/a$, beyond which time there are only inertial or incompressible deceleration effects, referred to as *surging*, also known as *rigid column analysis*. Table 6.2 classifies four types of valve closure, independent of type of valve.

Using standard waterhammer programs, parametric analyses can be conducted for the preparation of charts to demonstrate the effect of time of closure, type of valve, and an indication of the physical process-waterhammer or simply inertia effects of deceleration. The charts are based on analysis of valve closure for a simple reservoir-pipe-valve arrangement. For simplicity fluid friction is often neglected, a reasonable assumption for pipes on the order of hundreds of feet in length.

6.4 HYDRAULIC CHARACTERISTICS OF PUMPS

Transient analyses of piping systems involving centrifugal, mixed-flow, and axial-flow pumps require detailed information describing the characteristics of the respective turbomachine, which may pass through unusual, indeed abnormal, flow regimes. Since little if any information is available regarding the dynamic behavior of the pump in question, invariably the decision must be made to use the steady-flow characteristics of the machine gathered from laboratory tests. Moreover, complete steady-flow characteristics of the machine may not be available for all possible modes of operation that may be encountered in practice.

In this section steady-flow characteristics of pumps in all possible zones of operation are defined. The importance of geometric and dynamic similitude is first discussed with respect to both (1) homologous relationships for steady flow and (2) the importance of the assumption of similarity for transient analysis. The significance of the eight zones of operation within each of the four quadrants is presented in detail with reference to three possible modes of data representation. The steady-flow characteristics of pumps are discussed in detail with regard to the complete range of possible operation. The loss of driving power to a pump is usually the most critical transient case to consider for pumps, because of the possibility of low pipeline pressures which may lead to (1) pipe collapse due to buckling, or (2) the formation of a vapor cavity and its subsequent collapse. Other waterhammer problems may occur due to slam of a swing check valve, or from a discharge valve closing either too quickly (column separation), or too slowly (surging from reverse flow). For radial-flow pumps for which the reverse flow reaches a maximum just subsequent to passing through zero speed (locked rotor point), and then is decelerated as the shaft runs faster in the turbine zone, the head will usually rise above the nominal operating value. As reported by Donsky (1961) mixed-flow and axial-flow pumps may not even experience an upsurge in the turbine zone because the maximum flow tends to occur closer to runaway conditions.

6.4.1 Definition of Pump Characteristics

The essential parameters for definition of hydraulic performance of pumps are defined as

- *Impeller diameter.* Exit diameter of pump rotor D_1.
- *Rotational speed.* The angular velocity (rad/s) is ω, while $N = 2\pi\omega/60$ is in rpm.
- *Flow rate.* Capacity Q at operating point in chosen units.
- *Total dynamic head* (*TDH*). The total energy gain (or loss) H across pump, defined as

$$H = \left(\frac{P_d}{\gamma} + z_d\right) - \left(\frac{P_s}{\gamma} + z_s\right) + \frac{V_d^2}{2g} - \frac{V_s^2}{2g} \qquad (6.11)$$

where subscripts s and d refer to suction and discharge sides of the pump, respectively,

6.4.2 Homologous (Affinity) Laws

Dynamic similitude, or dimensionless representation of test results, has been applied with perhaps more success in the area of hydraulic machinery than in any other field involving fluid mechanics. Due to the sheer magnitude of the problem of data handling it is imperative that dimensionless parameters be employed for transient analysis of hydraulic machines that are continually experiencing changes in speed as well as passing through several zones of normal and abnormal operation. For liquids for which thermal effects may be neglected, the remaining fluid-related forces are pressure (head), fluid inertia, resistance, phase change (cavitation), surface tension, compressibility, and gravity. If the discussion is limited to single-phase liquid flow, three of the above fluid effects–cavitation, surface tension, and gravity (no interfaces within machine)–can be eliminated, leaving the forces of pressure, inertia, viscous resistance, and compressibility. For the steady or even transient behavior of hydraulic machinery conducting liquids the effect of compressibility may be neglected.

In terms of dimensionless ratios the three forces yield an *Euler number* (ratio of inertia force to pressure force), which is dependent upon geometry, and a Reynolds number.

For all flowing situations, the viscous force, as represented by the Reynolds number, is definitely present. If water is the fluid medium, the effect of the Reynolds number on the characteristics of hydraulic machinery can usually be neglected, the major exception being the prediction of the performance of a large hydraulic turbine on the basis of model data. For the transient behavior of a given machine the actual change in the value of the Reynolds number is usually inconsequential anyway. The elimination of the viscous force from the original list reduces the number of fluid-type forces from seven to two-pressure (head) and inertia, as exemplified by the Euler number. The appellation geometry in the functional relationship in the above equation embodies primarily, first, the shape of the rotating impeller, the entrance and exit flow passages, including effects of vanes, diffusers, and so on; second, the effect of surface roughness; and lastly the geometry of the streamline pattern, better known as kinematic similitude in contrast to the first two, which are related to geometric similarity. *Kinematic similarity* is invoked on the assumption that similar flow patterns can be specified by congruent velocity triangles composed of peripheral speed U and absolute fluid velocity V at inlet or exit to the vanes. This allows for the definition of a *flow coefficient*, expressed in terms of impeller diameter D_I and angular velocity ω:

$$C_Q = \frac{Q}{\omega D_I^3} \tag{6.12}$$

The reciprocal of the Euler number (ratio of pressure force to inertia force) is the *head coefficient*, defined as

$$C_H = \frac{gH}{\omega^2 D_I^2} \tag{6.13}$$

A *power coefficient* can be defined

$$C_P = \frac{P}{\rho \omega^3 D_I^5} \tag{6.14}$$

For transient analysis, the desired parameter for the continuous prediction of pump speed is the unbalanced torque T. Since $T = P/\omega$, the *torque coefficient* becomes

$$C_T = \frac{T}{\rho \omega^2 D_I^5} \tag{6.15}$$

Traditionally in hydraulic transient analysis to refer pump characteristics to so-called *rated conditions*–which preferably should be the optimum *or best efficiency point* (BEP), but sometimes defined as the *duty, nameplate*, or *design point*. Nevertheless, in terms of *rated conditions*, for which the subscript R is employed, the following ratios are defined;

Flow: $v = \dfrac{Q}{Q_R}$ speed: $\alpha = \dfrac{\omega}{\omega_R} = \dfrac{N}{N_R}$ head: $h = \dfrac{H}{H_R}$ torque: $\beta = \dfrac{T}{T_R}$

Next, for a given pump undergoing a transient, for which D_I is a constant, Eqs. (6.12–6.15) can be written in terms of the above ratios

$$\frac{v}{\alpha} = \frac{C_Q}{C_{QR}} = \frac{Q_R}{Q}\frac{\omega}{\omega_R} \qquad \frac{h}{\alpha^2} = \frac{C_H}{C_{HR}} = \frac{H}{H_R}\frac{\omega_R^2}{\omega^2} \qquad \frac{\beta}{\alpha^2} = \frac{C_T}{C_{TR}} = \frac{T}{T_R}\frac{\omega_R^2}{\omega^2}$$

6.4.3 Abnormal Pump (Four-Quadrant) Characteristics

The performance characteristics discussed up to this point correspond to pumps operating normally. During a transient, however, the machine may experience either a reversal in flow, or rotational speed, or both, depending on the situation. It is also possible that the torque and head may reverse in sign during passage of the machine through abnormal zones of performance. The need for characteristics of a pump in abnormal zones of operation can best be described with reference to Fig. 6.5, which is a simulated pump power failure transient. A centrifugal pump is delivering water at a constant rate when there is a sudden loss of power from the prime mover-in this case an electric motor. For the postulated case of no discharge valves, or other means of controlling the flow, the loss of driving torque leads to an immediate deceleration of the shaft speed, and in turn the flow. The three curves are dimensionless head (h), flow (v), and speed (α). With no additional means of controlling the flow, the higher head at the final delivery point (another reservoir) will eventually cause the flow to reverse ($v < 0$) while the inertia of the rotating parts has maintained positive rotation ($\alpha > 0$). Up until the time of flow reversal the pump has been operating in the normal zone, albeit at a number of off-peak flows.

To predict system performance in regions of negative rotation and/or negative flow the analyst requires characteristics in these regions for the machine in question. Indeed, any peculiar characteristic of the pump in these regions could be expected to have an influence on the hydraulic transients. It is important to stress that the results of such analyses are critically governed by the following three factors: (1) availability of complete pump characteristics in zones the pump will operate, (2) complete reliance on dynamic similitude (homologous) laws during transients, and (3) assumption that steady-flow derived pump characteristics are valid for transient analysis.

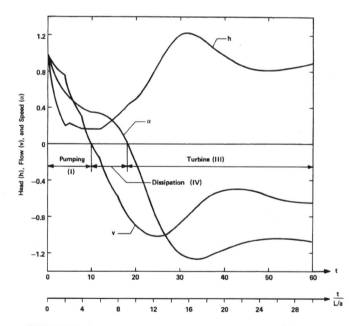

FIGURE 6.5 Simulated pump trip without valves in a single-pipeline system.

Investigations by Kittredge (1956) and Knapp (1937) facilitated the understanding of abnormal operation, as well as served to reinforce the need for test data. Following the work by Knapp (1941) and Swanson (1953), and a summary of their results by Donsky (1961), eight possible zones of operation, four normal and four abnormal, will be discussed here with reference to Fig. 6.6, developed by Martin (1983). In Fig. 6.6 the head H is shown as the difference in the two reservoir elevations to simplify the illustration. The effect of pipe friction may be ignored for this discussion by assuming that the pipe is short and of relatively large diameter. The regions referred to on Fig. 6.6 are termed zones and quadrants, the latter definition originating from plots of lines of constant head

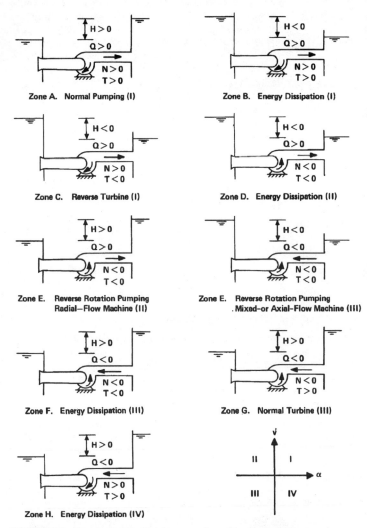

FIGURE 6.6 Four quadrants and eight zones of possible pump operation. (From Martin, 1983).

and constant torque on a flow-speed plane ($v - \alpha$ axes). Quadrants I ($v > 0$, $\alpha > 0$) and III ($v < 0$, $\alpha < 0$) are defined in general as regions of pump or turbine operation, respectively. It will be seen, however, that abnormal operation (neither pump nor turbine mode) may occur in either of these two quadrants. A very detailed description of each of the eight zones of operation is in order. It should be noted that all of the conditions shown schematically in Fig. 6.6 can be contrived in a laboratory test loop using an additional pump (or two) as the master and the test pump as a slave. Most, if not all, of the zones shown can also be experienced by a pump during a transient under the appropriate set of circumstances.

Quadrant I. Zone A (normal pumping) in Fig. 6.6 depicts a pump under normal operation for which all four quantities– Q, N, H, and T are regarded as positive. In this case $Q > 0$, indicating useful application of energy. Zone B (energy dissipation) is a condition of positive flow, positive rotation, and positive torque, but negative head—quite an abnormal condition. A machine could operate in Zone B by (1) being overpowered by another pump or by a reservoir during steady operation, or (2) by a sudden drop in head during a transient caused by power failure. It is possible, but not desirable, for a pump to generate power with both the flow and rotation in the normal positive direction for a pump, Zone C (reverse turbine), which is caused by a negative head, resulting in a positive efficiency because of the negative torque. The maximum efficiency would be quite low due to the bad entrance flow condition and unusual exit velocity triangle.

Quadrant IV. Zone H, labeled energy dissipation, is often encountered shortly after a tripout or power failure of a pump, as illustrated in Fig. 6.5. In this instance the combined inertia of all the rotating elements–motor, pump and its entrained liquid, and shaft—has maintained pump rotation positive but at a reduced value at the time of flow reversal caused by the positive head on the machine. This purely dissipative mode results in a negative or zero efficiency. It is important to note that both the head and fluid torque are positive in Zone H, the only zone in Quadrant IV.

Quadrant III. A machine that passes through Zone H during a pump power failure will then enter Zone G (normal turbining) provided that reverse shaft rotation is not precluded by a mechanical ratchet. Although a runaway machine rotating freely is not generating power, Zone G is the precise mode of operation for a hydraulic turbine. Note that the head and torque are positive, as for a pump but that the flow and speed are negative, opposite to that for a pump under normal operation (Zone A).

Subsequent to the tripout or load rejection of a hydraulic turbine or the continual operation of a machine that failed earlier as a pump, Zone F (energy dissipation) can be encountered. The difference between Zones F and G is that the torque has changed sign for Zone F, resulting in a braking effect, which tends to slow the free-wheeling machine down. In fact the real runaway condition is attained at the boundary of the two zones, for which torque $T = 0$.

Quadrant II. The two remaining zones–D and E–are very unusual and infrequently encountered in operation, with the exception of pump/turbines entering Zone E during transient operation. Again it should be emphasized that both zones can be experienced by a pump in a test loop, or in practice in the event a machine is inadvertently rotated in the wrong direction by improper wiring of an electric motor. Zone D is a purely dissipative mode that normally would not occur in practice unless a pump, which was designed to increase the flow from a higher to lower reservoir, was rotated in reverse, but did not have the capacity to reverse the flow (Zone E, mixed or axial flow), resulting in $Q > 0$, $N < 0$, $T < 0$, for $H < 0$. Zone E, for which the pump efficiency > 0, could occur in practice under steady flow if the preferred rotation as a pump was reversed. There is always the question regarding the eventual direction of the flow. A radial-flow machine will produce positive flow at a much reduced capacity and efficiency compared to $N > 0$ (normal pumping), yielding of course $H > 0$. On the other hand, mixed and axial-flow machines create flow in the opposite direction (Quadrant III), and $H < 0$, which corresponds still to an increase in head across the machine in the direction of flow.

6.4.4 Representation of Pump Data for Numerical Analysis

It is conventional in transient analyses to represent h/α^2 and β/α^2 as functions of v/α, as shown in Fig. 6.7 and 6.8 for a radial-flow pump. The curves on Fig. 6.7 are only for positive rotation ($\alpha > 0$), and constitute pump Zones A, B, and C for $v > 0$ and the region of energy dissipation subsequent to pump power failure (Zone H), for which $v < 0$. The remainder of the pump characteristics are plotted in Fig. 6.8 for $\alpha < 0$. The complete characteristics of the pump plotted in Figs. 6.7 and 6.8 can also be correlated on what is known as a *Karman-Knapp circle diagram*, a plot of lines of constant head (h) and torque (β) on the coordinates of dimensionless flow (v) and speed (α). Fig. 6.9 is such a correlation for the same pump. The complete characteristics of the pump require six curves, three each for head and torque. For example, the h/α^2 curves from Figs. 6.7 and 6.8 can be represented by continuous lines for $h = 1$ and $h = -1$, and two straight lines through the origin for $h = 0$. A similar pattern exists for the torque (β) lines. In addition to the eight zones A–H illustrated in Fig. 6.6, the four Karman-Knapp quadrants in terms of v and, are well defined. Radial lines in Fig. 6.9 correspond to constant values for v/α in Figs. 6.7 and 6.8, allowing for relatively easy transformation from one form of presentation to the other.

In computer analysis of pump transients, Figs. 6.7 and 6.8, while meaningful from the standpoint of physical understanding, are fraught with the difficulty of $|v/\alpha|$ becoming

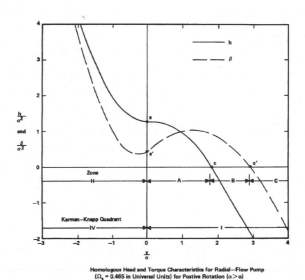

FIGURE 6.7 Complete head and torque characteristics of a radial-flow pump for positive rotation. (From Martin, 1983).

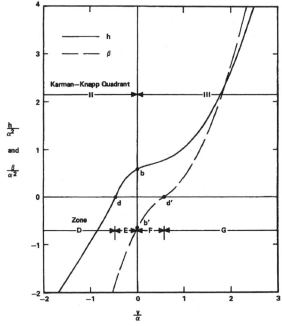

FIGURE 6.8 Complete head and torque characteristics of a radial-flow pump for negative rotation. (From Martin, 1983).

infinite as the unit passes through, or remains at, zero speed ($\alpha = 0$). Some have solved that problem by switching from h/α^2 versus v/α to h/v^2 versus α/v, and likewise for β, for $|v/\alpha| > 1$. This technique doubles the number of curves on Figs. 6.7 and 6.8, and thereby creates discontinuities in the slopes of the lines at $|v/\alpha| = 1$, in addition to complicating the storing and interpolation of data. Marchal et al. (1965) devised a useful transformation which allowed the complete pump characteristics to be represented by two single curves, as shown for the same pump in Fig. 6.10. The difficulty of v/α becoming infinite was eliminated by utilizing the function $\tan^{-1}(v/\alpha)$ as the abscissa. The eight zones, or four quadrants can then be connected by the continuous functions. Although some of the physical interpretation of pump data has been lost in the transformation, Fig. 6.10 is now a preferred correlation for transient analysis using a digital computer because of function continuity and ease of numerical interpolation. The singularities in Figs. 6.7 and 6.8 and the asymptotes in Fig. 6.9 have now been avoided.

6.4.5 Critical Data Required for Hydraulic Analysis of Systems with Pumps

Regarding data from manufacturers such as pump curves (normal and abnormal), pump and motor inertia, motor torque-speed curves, and valve curves, probably the most critical for

Hydraulic Transient Design for Pipeline Systems **6.17**

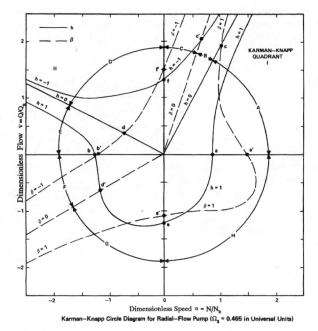

FIGURE 6.9 Complete four-quadrant head and torque characteristics of radial-flow pump. (From Martin, 1983).

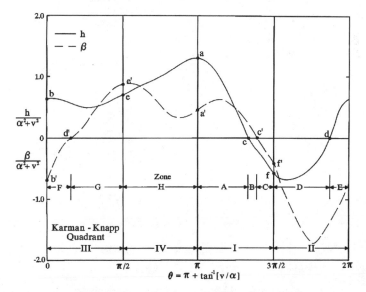

FIGURE 6.10 Complete head and torque characteristics of a radial-flow pump in Suter diagram. (From Martin, 1983).

pumping stations are pump-motor inertia and valve closure time. Normal pump curves are usually available and adequate. Motor torque-speed curves are only needed when evaluating pump startup. For pump trip the inertia of the combined pump and motor is important.

6.5 SURGE PROTECTION AND SURGE CONTROL DEVICES

There are numerous techniques for controlling transients and waterhammer, some involving design considerations and others the consideration of surge protection devices. There must be a complete design and operational strategy devised to combat potential waterhammer in a system. The transient event may either initiate a low-pressure event (*downsurge*) as in the case of a pump power failure, or a high pressure event (*upsurge*) caused by the closure of a downstream valve. It is well known that a downsurge can lead to the undesirable occurrence of water-column separation, which itself can result in severe pressure rises following the collapse of a vapor cavity. In some systems negative pressures are not even allowed because of (1) possible pipe collapse or (2) ingress of outside water or air.

The means of controlling the transient will in general vary, depending upon whether the initiating event results in an upsurge or downsurge. For pumping plants the major cause of unwanted transients is typically the complete outage of pumps due to loss of electricity to the motor. For full pipelines, pump startup, usually against a closed pump discharge valve for centrifugal pumps, does not normally result in significant pressure transients. The majority of transient problems in pumping installations are associated with the potential (or realized) occurrence of *water-column separation* and *vapor-pocket collapse*, resulting from the tripout of one or more pumps, with or without valve action. The pump-discharge valve, if actuated too suddenly, can even aggravate the downsurge problem. To combat the downsurge problem there are a number of options, mostly involving the design and installation of one or more surge protection devices. In this section various surge protection techniques will be discussed, followed by an assessment of the virtue of each with respect to pumping systems in general. The lift systems shown in Fig. 6.11 depict various surge protection schemes.

6.5.1 Critical Parameters for Transients

Before discussing surge protection devices, some comments will be made regarding the various pipeline, pump and motor, control valve, flow rate, and other parameters that affect the magnitude of the transient. For a pumping system the four main parameters are (1) pump flow rate, (2) pump and motor WR^2, (3) any valve motion, and (4) pipeline characteristics. The pipeline characteristics include piping layout-both plan and profile-pipe size and material, and the acoustic velocity. So-called short systems respond differently than long systems. Likewise, valve motion and its effect, whether controlled valves or check valves, will have different effects on the two types of systems.

The pipeline characteristics-item number (4)-relate to the response of the system to a transient such as pump power failure. Clearly, the response will be altered by the addition of one or more surge protection device or the change of (1) the flow rate, or (2) the WR^2, or (3) the valve motion. Obviously, for a given pipe network and flow distribution there are limited means of controlling transients by (2) WR^2 and (3) valve actuation. If these two parameters can not alleviate the problem than the pipeline response needs to be altered by means of surge protection devices.

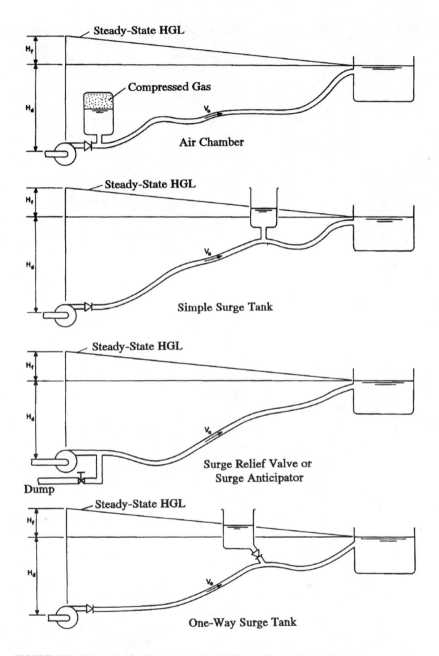

FIGURE 6.11 Schematic of various surge protection devices for pumping installations.

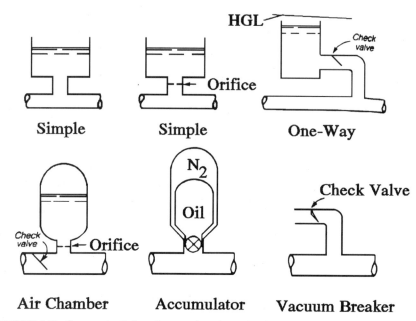

FIGURE 6.12 Cross-sectional view of surge tanks and gas related surge protection devices.

6.5.2 Critique of Surge Protection

For pumping systems, downsurge problems have been solved by various combinations of the procedures and devices mentioned above. Details of typical surge protection devices are illustrated in Figs. 6.12 and 6.13. In many instances local conditions and preferences of engineers have dictated the choice of methods and/or devices. Online devices such as accumulators and simple surge tanks are quite effective, albeit expensive, solutions. One-way surge tanks can also be effective when judiciously sized and sited. Surge anticipation valves should not be used when there is already a negative pressure problem. Indeed, there are installations where surge anticipation functions of such valves have been deactivated, leaving only the surge relief feature. Moreover, there have been occasions for which the surge anticipation feature aggravated the low pressure situation by an additional downsurge caused by premature opening of the valve.

Regarding the consideration and ultimate choice of surge protection devices, subsequent to calibration of analysis with test results, evaluation should be given to simple surge tanks or standpipes, one-way surge tanks, and hydropneumatic tanks or air chambers. A combination of devices may prove to be the most desirable and most economical.

The admittance of air into a piping system can be effective, but the design of air vacuum-valve location and size is critical. If air may be permitted into pipelines careful analysis would have to be done to ensure effective results. The consideration of air-vacuum breakers is a moot point if specifications such as the Ten State Standards limit the pressures to positive values.

Hydraulic Transient Design for Pipeline Systems **6.21**

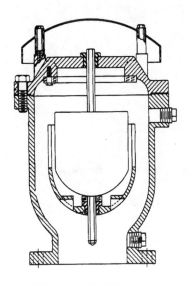

a. Vacuum Breaker Valve

a. Air Release Valve

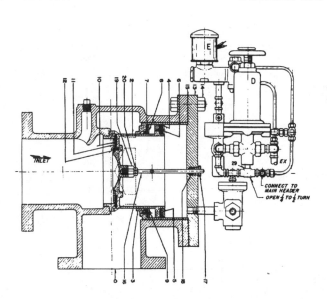

c. Surge Relief or Surge Anticipator Valve

FIGURE 6.13 Cross sections of vacuum breaker, air release and surge relief valves.

6.5.3 Surge Protection Control and Devices

Pump discharge valve operation. In gravity systems the upsurge transient can be controlled by an optimum valve closure-perhaps two stage, as mentioned by Wylie and Streeter (1993). As shown by Fleming (1990), an optimized closing can solve a waterhammer problem caused by pump power failure if coupled with the selection of a surge protection device. For pump power failure a control valve on the pump discharge can often be of only limited value in controlling the downsurge, as mentioned by Sanks (1989). Indeed, the valve closure can be too sudden, aggravating the downsurge and potentially causing column separation, or too slow, allowing a substantial reverse flow through the pump. It should also be emphasized that an optimum controlled motion for single-pump power failure is most likely not optimum for multiple-pump failure. The use of microprocessors and servomechanisms with feedback systems can be a general solution to optimum control of valves in conjunction with the pump and pipe system. For pump discharge valves the closure should not be too quick to exacerbate downsurge, nor too slow to create a substantial flow back through the valve and pump before closure.

Check valves. Swing check valves or other designs are frequently employed in pump discharge lines, often in conjunction with slow acting control valves. As indicated by Tullis (1989), a check valve should open easily, have a low head loss for normal positive flow, and create no undesirable transients by its own action. For short systems, a slow-responding check valve can lead to waterhammer because of the high reverse flow generated before closure. A spring-or counterweight-loaded valve with a dashpot can (1) give the initial fast response followed by (2) slow closure to alleviate the unwanted transient. The proper selection of the load and the degree of damping is important, however, for proper performance.

Check valve slam is also a possibility from stoppage or failure of one pump of several in a parallel system, or resulting from the action of an air chamber close to a pump undergoing power failure. Check valve slam can be reduced by the proper selection of a dashpot.

Surge anticipator valves and surge relief valves. A surge anticipation valve, Fig. 6.13c frequently installed at the manifold of the pump station, is designed to open initially under (1) pump power failure, or (2) the sensing of underpressure, or (3) the sensing of overpressure, as described by Lescovitch (1967). On the other hand, the usual type of surge relief valve opens quickly on sensing an overpressure, then closes slowly, as controlled by pilot valves. The surge anticipation valve is more complicated than a surge relief valve in that it not only embodies the relief function at the end of the cycle, but also has the element of anticipation. For systems for which water-column separation will not occur, the surge anticipation valve can solve the problem of upsurge at the pump due to reverse flow or wave reflection, as reported in an example by White (1942). An example of a surge relief valve only is provided by Weaver (1972). For systems for which water-column separation will not occur, Lundgren (1961) provides charts for simple pipeline systems.

As reported by Parmakian (1968,1982a-b) surge anticipation valves can exacerbate the downsurge problem inasmuch as the opening of the relief valve aggravates the negative pressure problem. Incidents have occurred involving the malfunctioning of a surge anticipation valve, leading to extreme pressures because the relief valve did not open.

Pump bypass. In shorter low-head systems a pump bypass line (Fig. 6.11) can be installed in order to allow water to be drawn into the pump discharge line following power failure and a downsurge. As explained by Wylie and Streeter (1993), there are two possible bypass configurations. The first involves a control valve on the discharge line and a check valve on the bypass line between the pump suction or wet well and the main line. The check valve is designed to open subsequent to the downsurge, possibly alleviating column separation down the main line. The second geometry would reverse the valve locations, having a control valve in the bypass and a check valve in the main line downstream of the pump. The control valve would open on power failure, again allowing water to bypass the pump into the main line.

Open (simple) surge tank. A simple on-line surge tank or standpipe (Fig. 6.11) can be an excellent solution to both upsurge and downsurge problems, These devices are quite common in hydroelectric systems where suitable topography usually exists. They are practically maintenance free, available for immediate response as they are on line. For pumping installations open simple surge tanks are rare because of height considerations and the absence of high points near most pumping stations. As mentioned by Parmakian (1968) simple surge tanks are the most dependable of all surge protection devices. One disadvantage is the additional height to allow for pump shutoff head. Overflowing and spilling must be considered, as well as the inclusion of some damping to reduce oscillations. As stated by Kroon et al. (1984) the major drawback to simple surge tanks is their capital expense.

One-way surge tank. The purpose of a one-way surge tank is to prevent initial low pressures and potential water-column separation by admitting water into the pipeline subsequent to a downsurge. The tank is normally isolated from the pipeline by one or more lateral pipes in which there are one or more check valves to allow flow into the pipe if the HGL is lower in the pipe than the elevation of the water in the open tank. Under normal operating conditions the higher pressure in the pipeline keeps the check valve closed. The major advantage of a one-way surge tank over a simple surge tank is that it does not have to be at the HGL elevation as required by the latter. It has the disadvantage, however, on only combatting initial downsurges, and not initial upsurges. One-way surge tanks have been employed extensively by the U.S. Bureau of Reclamation in pump discharge lines, principally by the instigation of Parmakian (1968), the originator of the concept. Another example of the effective application of one-way surge tanks in a pumping system was reported by Martin (1992), to be discussed in Sec. 6.9.1.

Considerations for design are: (1) location of high points or knees of the piping, (2) check valve and lateral piping redundancy, (3) float control refilling valves and water supply, and other appurtenances. Maintenance is critical to ensure the operation of the check valve(s) and tank when needed.

Air chamber (hydropneumatic surge tank). If properly designed and maintained, an air chamber can alleviate both negative and positive pressure problems in pumping systems. They are normally located within or near the pumping station where they would have the greatest effect. As stated by Fox (1977) and others, an air chamber solution may be extremely effective in solving the transient problem, but highly expensive. Air chambers have the advantage that the tank-sometimes multiple-can be mounted either vertically or horizontally. The principal criteria are available water volume and air volume for the task at hand.

For design, consideration must be given to compressed air supply, water level sensing, sight glass, drains, pressure regulators, and possible freezing. Frequently, a check valve is installed between the pump and the air chamber. Since the line length between the pump and air chamber is usually quite short, check valve slamming may occur, necessitating the consideration of a dashpot on the check valve to cushion closure.

The assurance of the maintenance of air in the tank is essential–usually 50 percent of tank volume, otherwise the air chamber can be ineffective. An incident occurred at a raw water pumping plant where an air chamber became waterlogged due to the malfunctioning of the compressed air system. Unfortunately, pump power failure occurred at the same time, causing water column separation and waterhammer, leading to pipe rupture.

Air vacuum and air release valves. Another method for preventing subatmospheric pressures and vapor cavity formation is the admittance of air from air-vacuum valves (vacuum breakers) at selected points along the piping system. Proper location and size of air-vacuum valves can prevent water-column separation and reduce waterhammer effects, as calculated and measured by Martin (1980). The sizing and location of the valves are critical, as stated by Kroon et al. (1984). In fact, as reported by Parmakian (1982a,-b) the inclusion of air-vacuum valves in a pipeline did not eliminate failures. Unless the air-vacuum system is properly chosen, substantial pressures can still occur due to the compression of the air during resurge, especially if the air is at extremely low pressures within the pipeline when admitted. Moreover, the air must be admitted quickly enough to be effective. Typical designs are shown in Fig. 6.13

As shown by Fleming (1990) vacuum breakers can be a viable solution. The advantage of an air-vacuum breaker system, which is typically less expensive than other measures such as air chambers, must be weighed against the disadvantages of air accumulation along the pipeline and its subsequent removal. Maintenance and operation of valves is critical in order for assurance of valve opening when needed. Air removal is often accomplished with a combined air-release air-vacuum valve. For finished water systems the admittance of air is not a normal solution and must be evaluated carefully. Moreover, air must be carefully released so that no additional transient is created.

Flywheel. Theoretically, a substantial increase in the rotating inertia (WR^2) of a pump-motor unit can greatly reduce the downsurge inasmuch as the machine will not decelerate as rapidly. Typically, the motor may constitute from 75 to 90 percent of the total WR^2. Additional WR^2 by the attachment of a flywheel will reduce the downsurge. As stated by Parmakian (1968), a 100 percent increase in WR^2 by the addition of a flywheel may add up to 20 percent to the motor cost. He further states that a flywheel solution is only economical in some marginal cases. Flywheels are usually an expensive solution, mainly useful only for short systems. A flywheel has the advantage of practically no maintenance, but the increased torque requirements for starting must be considered.

Uninterrupted power supply (UPS). The availability of large uninterrupted power supply systems are of potential value in preventing the primary source of waterhammer in pumping; that is, the generation of low pressures due to pump power failure. For pumping stations with multiple parallel pumps, a UPS system could be devised to maintain one or more motors while allowing the rest to fail, inasmuch as there is a possibility of maintaining sufficient pressure with the remaining operating pump(s). The solution usually is expensive, however, with few systems installed.

6.6 DESIGN CONSIDERATIONS

Any surge or hydraulic transient analysis is subject to inaccuracies due to incomplete information regarding the systems and its components. This is particularly true for a water distribution system with its complexity, presence of pumps, valves, tanks, and so forth, and some uncertainty with respect to initial flow distribution. The ultimate question is how all of the uncertainties combine in the analysis to yield the final solution. There will be offsetting effects and a variation in accuracy in terms of percentage error throughout the system. Some of the uncertainties are as follows.

The simplification of a pipe system, in particular a complex network, by the exclusion of pipes below a certain size and the generation of equivalent pipes surely introduces some error, as well as the accuracy of the steady-state solution. However, if the major flow rates are reasonably well known, then deviation for the smaller pipes is probably not too critical. As mentioned above incomplete pump characteristics, especially during reverse flow and reverse rotation, introduce calculation errors. Valve characteristics that must be assumed rather than actual are sources of errors, in particular the response of swing check valves and pressure reducing valves. The analysis is enhanced if the response of valves and pumps from recordings can be put in the computer model.

For complex pipe network systems it is difficult to assess uncertainties until much of the available information is known. Under more ideal conditions that occur with simpler systems and laboratory experiments, one can expect accuracies when compared to measurement on the order of 5 to 10 percent, sometimes even better. The element of judgment does enter into accuracy. Indeed, two analyses could even differ by this range because of different assumptions with respect to wave speeds, pump characteristics, valve motions, system schematization, and so forth. It is possible to have good analysis and poorer analysis, depending upon experience and expertise of the user of the computer code. This element is quite critical in hydraulic transients. Indeed, there can be quite different results using the same code.

Computer codes, which are normally based on the *method of characteristics* (MOC), are invaluable tools for assessing the response based of systems to changes in surge protection devices and their characteristics. Obviously, the efficacy of such an approach is enhanced if the input data and network schematization is improved via calibration. Computer codes have the advantage of investigating a number of options as well as optimizing the sizing of surge protection devices. The ability to calibrate a numerical analysis code to a system certainly improves the determination of the proper surge protection. Otherwise, if the code does not reasonably well represent a system, surge protection devices can either be inappropriate or under- or oversized.

Computer codes that do not properly model the formation of vapor pockets and subsequent collapse can cause considerable errors. Moreover, there is also uncertainty regarding any free or evolved gas coming out of solution. The effect on wave speed is known, but this influence can not be easily addressed in an analysis of the system. It is simply another possible uncertainty.

Even for complicated systems such as water distribution networks, hydraulic transient calculations can yield reasonable results when compared to actual measurements provided that the entire system can be properly characterized. In addition to the pump, motor, and valve characteristics there has to be sufficient knowledge regarding the piping and flow demands. An especially critical factor for a network is the schematization of the network; that is, how is a network of thousands of pipes simplified to one suitable for computer analysis, say hundreds of pipes, some actual and some equivalent. According to Thorley (1991), a network with loops tends to be more forgiving regarding waterhammer because of the dispersive effect of many pipes and the associated reflections. On the other hand, Karney and McInnis (1990) show by a simple example that wave superposition can cause amplification of transients. Since water distribution networks themselves have not been known to be prone to waterhammer as a rule, there is meager information as to simplification and means of establishing equivalent pipes for analysis purposes. Large municipal pipe networks are good examples wherein the schematization and the selection of pipes characterizing the networks need to be improved to represent the system better.

6.7 NEGATIVE PRESSURES AND WATER COLUMN SEPARATION IN NETWORKS

For finished water transmission and distribution systems the application of 138 kPa (20 psig) as a minimum pressure to be maintained under all conditions should prevent column separation from occurring provided analytical models have sufficient accuracy. Although water column separation and collapse is not common in large networks, it does not mean that the event is not possible. The modeling of water column separation is clearly difficult for a complicated network system. Water column separation has been analytically modeled with moderate success for numerous operating pipelines. Clearly, not only negative pressures, but also water column separation, are unwanted in pipeline systems, and should be eliminated by installation of properly designed surge protection devices.

If the criterion of a minimum pressure of 138 kPa (20 psig) is imposed then the issue of column separation and air-vacuum breakers are irrelevant, except for prediction by computer codes. Aside from research considerations, column separation is simulated for engineering situations mainly to assess the potential consequences. If the consequences are serious, as they often are in general, either operational changes or more likely surge protection devices are designed to alleviate column separation. For marginal cases of column separation the accuracy of pressure prediction becomes difficult. If column separation is not to be allowed and the occurrence of vapor pressure can be adequately predicted, then the simulation of column separation itself is not necessary.

Some codes do not simulate water column separation, but instead only maintain the pressure at cavity location at vapor pressure. The results of such an analysis are invalid, if indeed an actual cavity occurred, at some time subsequent to cavity formation. This technique is only useful to know if a cavity could have occurred, as there can be no assessment of the consequences of column separation. The inability of any code to model water column separation has the following implications: (1) the seriousness of any column separation event, if any, can not be determined, and (2) once vapor pressure is attained, the computation model loses its ability to predict adequately system transients. If negative pressures below 138 kPa (20 psig) are not to be allowed the inability of a code to assess the consequences of column separation and its attendant collapse is admittedly not so serious. The code need only flag pressures below 138 kPa (20 psig) and negative pressures, indicating if there is a need for surge protection devices.

The ability of any model to properly simulate water column separation depends upon a number of factors. The principal ones are

- Accurate knowledge of initial flow rates
- Proper representation of pumps, valves, and piping system
- A vapor pocket allowed to form, grow, and collapse
- Maintenance of vapor pressure within cavity while it exists
- Determination of volume of cavity at each time step
- Collapse of cavity at the instant the cavity volume is reduced to zero

6.8 TIME CONSTANTS FOR HYDRAULIC SYSTEMS

- Elastic time constant

$$t_e = \frac{2L}{a} \tag{6.16}$$

- Flow time constant

$$t_f = \frac{LV_o}{gH_o} \tag{6.17}$$

- Pump and motor inertia time constant

$$t_m = \frac{I\omega_R}{T_R} = \frac{I\omega_R^2}{\rho g Q_R H_R \eta_{RT}} \tag{6.18}$$

- Surge tank oscillation inelastic time constant

$$t_s = 2p \sqrt{\frac{L_T}{g} \frac{A_t}{A_T}} \tag{6.19}$$

6.9 CASE STUDIES

For three large water pumping systems with various surge protection devices waterhammer analyses and site measurements have been conducted. The surge protection systems in question are (1) one-way and simple surge tanks, (2) an air chamber, and (3) air-vacuum breakers.

6.9.1 Case Study with One-way and Simple Surge Tanks

A very large pumping station has been installed and commissioned to deliver water over a distance of over 30 kilometers. Three three-stage centrifugal pumps run at a synchronous speed of 720 rpm, with individual rated capacities of 1.14 m³/sec, rated heads of 165 m, and rated power of 2090 kw. Initial surge analysis indicated potential water-column separation. The surge protection system was then designed with one-way and simple surge tanks as well as air-vacuum valves strategically located.

The efficacy of these various surge protection devices was assessed from site measurements. Measurements of pump speed, discharge valve position, pump flow rate, and pressure at seven locations were conducted under various transient test conditions. The site measurements under three-pump operation allowed for improvement of hydraulic transient calculations for future expansion to four and five pumps. Figure 6.14 illustrates the profile of the ground and the location of the three pairs of surge tanks. The first and second pair of surge tanks are of the one-way (feed tank) variety, while the third pair are simple open on-line tanks.

Pump trip tests were conducted for three-pump operation with cone valves actuated by the loss of motor power. For numerical analysis a standard computer program apply-

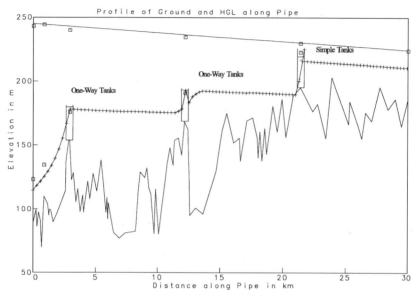

Comparison of Predicted and Measured Hydraulic Grade Line Along Pipe System

FIGURE 6.14 Case study of pump power failure at pumping station with three pair of surge tanks —two pair one way and one pair simple surge tanks Martin(1992).

ing the method of characteristics was employed to simulate the transient events. Figure 6.15 shows the transient pressures for three pump power failure. The transient pressures agree reasonably well for the first 80 seconds. The minimum HGLs in Fig. 6.14 also show good agreement, as well as the comparison of measured and calculated pump speeds in.

6.9.2 Case Study with Air chamber

Hydraulic transients caused by simultaneous tripping of pumps at the pumping station depicted on Fig. 6.16 were evaluated to assess the necessity of surge protection. Without the presence of any protective devices such as accumulators, vacuum breakers, or surge suppressors, water hammer with serious consequences was shown to occur due to depressurization caused by the loss of pumping pressure following sudden electrical outage. In the case of no protection a large vapor cavity would occur at the first high point above the pumping station, subsequently collapsing after the water column between it and the reservoir stops and reverses. This phenomenon, called water-column separation, can be mitigated by maintaining the pressures above vapor pressure.

The efficacy of the 11.6 m (38 ft) diameter air chamber shown in Fig. 6.16 was investigated analytically and validated by site measurements for three-pump operation. The envelope of the minimum HGL drawn on Fig. 6.16 shows that all pressures remained positive. The lower graph compares the site measurement with the calculated pressures obtained by a standard waterhammer program utilizing MOC.

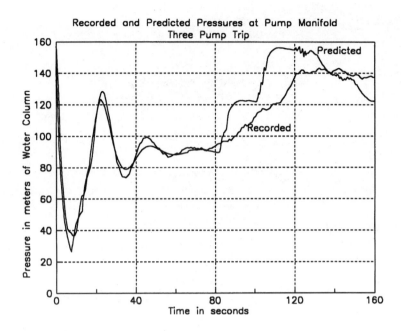

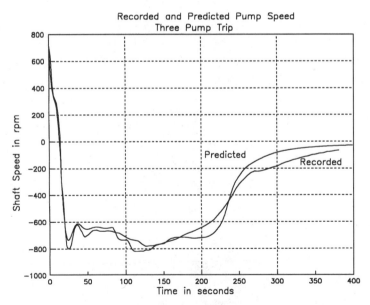

FIGURE 6.15 Case study of pump power failure at pumping station with three pair of surge tanks—two pair one way and one pair simple surge tanks. (From Martin, 1992).

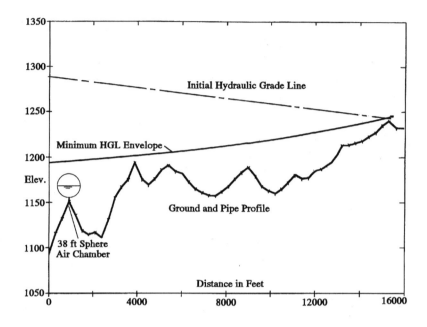
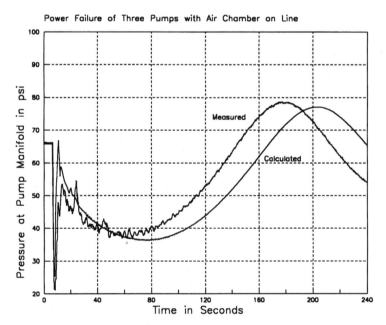

FIGURE 6.16 Case study of air chamber performance for raw water supply.

6.9.3 Case Study with Air-vacuum Breaker

Air-inlet valves or air-vacuum breakers are frequently installed on liquid piping systems and cooling water circuits for the purpose of (1) eliminating the potential of water-column separation and any associated waterhammer subsequent to vapor pocket collapse; (2) protecting the piping from an external pressure of nearly a complete vacuum; and (3) providing an elastic cushion to absorb the transient pressures.

A schematic of the pumping and piping system subject to the field test program is shown in Fig. 6.17. This system provides the cooling water to a power plant by pumping water from the lower level to the upper reservoir level. There are five identical vertical pumps in parallel connected to a steel discharge pipe 1524 mm (60 in) in diameter. On the discharge piping of each pump there are 460 mm (18 in) diameter swing check valves. Mounted on top of the 1524 mm (60 in) diameter discharge manifold is a 200 mm (8 in) diameter pipe, in which is installed a swing check valve with a counter weight. Air enters the vacuum breaker through the tall riser, which extends to the outside of the pump house.

Transient pressures were measured in the discharge header for simultaneous tripout of three, four, and five pumps. The initial prediction of the downsurge caused by pump power failure was based on the method of characteristics with a left end boundary condition at the pumps, junction boundary condition at the change in diameter of the piping, and a constant pressure boundary condition at the right end of the system.

The predicted pressure head variation in the pump discharge line is shown in Fig. 6.17 for a simulated five pump tripout. The predicted peak pressure for the five pump tripout compares favorably with the corresponding measured peak, but the time of occurrence of the peaks and the subsequent phasing vary considerably. Analysis without a vacuum breaker or other protective device in the system predicted waterhammer pressure caused by collapse of a vapor pocket to exceed 2450 kPa (355 psi). The vacuum breaker effectively reduced the peak pressure by 60 per cent. Peak pressures can be adequately predicted by a simplified liquid column, orifice, and air spring system. Water-column separation can be eliminated by air-vacuum breakers of adequate size.

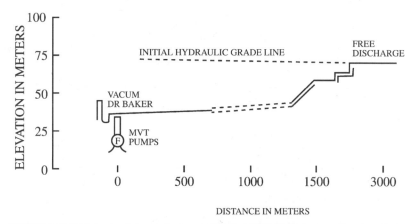

FIGURE 6.17 Case study of vacuum breaker performance for river water system of nuclear plant, Martin (1980).

REFERENCES

Chaudhry, M. H., *Applied Hydraulic Transients*, 2d ed., Van Nostrand Reinhold, New York 1987.

Donsky, B., "Complete Pump Characteristics and the Effects of Specific Speeds on Hydraulic Transients," *Journal of Basic Engineering, Transactions, American Society of Mechanical Engineers*, 83: 685–699, 1961.

Fleming, A. J., "Cost-Effective Solution to a Waterhammer Problem," *Public Works*, 42–44, 1990.

Fox, J. A., *Hydraulic Analysis of Unsteady Flow in Pipe Networks*, John Wiley & Sons, New York 1977.

Karney, B. W., and McInnis, D., "Transient Analysis of Water Distribution Systems," *Journal American Water Works Association*, 82: 62–70, 1990.

Kittredge, C. P., "Hydraulic Transients in Centrifugal Pump Systems," *Transactions, American Society of Mechanical Engineers*, 78: 1307–1322, 1956.

Knapp, R. T., "Complete Characteristics of Centrifugal Pumps and Their Use in Prediction of Transient Behavior," *Transactions, American Society of Mechanical Engineers*, 59:683–689, 1937.

Knapp, R. T., "Centrifugal-Pump Performance Affected by Design Features," *Transactions, American Society of Mechanical Engineers*, 63:251–260, 1941.

Kroon, J. R., Stoner, M. A., and Hunt, W. A., "Water Hammer: Causes and Effects," *Journal American Water Works Association*, 76:39–45, 1984.

Lescovitch, J. E., "Surge Control of Waterhammer by Automatic Valves," *Journal American Water Works Association*, 59:632-644, 1967.

Lundgren, C. W., "Charts for Determining Size of Surge Suppressor for Pump-Discharge Lines," *Journal of Engineering for Power, Transactions, American Society of Mechanical Engineers*, 93:43–47, 1961.

Marchal, M., Flesh, G., and Suter, P., "The Calculation of Waterhammer Problems by Means of the Digital Computer," *Proceedings, International Symposium on Waterhammer in Pumped Storage Projects*, American Society of Mechanical Engineers (ASME), Chicago, 1965.

Martin, C. S., "Entrapped Air in Pipelines," Paper F2, *Second BHRA International Conference on Pressure Surges*, The City University, London, September 22–24, 1976.

Martin, C. S., "Transient Performance of Air Vacuum Breakers," Fourth International Conference on Water Column Separation, Cagliari, November 11–13, 1979. "Transient Performance Air Vacuum Breakers," *L'Energia Elettrica, Proceedings No. 382*, 1980, pp. 174–184.

Martin, C. S., "Representation of Pump Characteristics for Transient Analysis," *ASME Symposium on Performance Characteristics of Hydraulic Turbines and Pumps*, Winter Annual Meeting, Boston, November 13–18, pp. 1–13, 1983.

Martin, C. S., "Experience with Surge Protection Devices," *BHr Group International Conference on Pipelines*, Manchester, England, March pp. 24–26, 171–178, 1992.

Martin, C. S., "Hydraulics of Valves," in J. A. Schetz and A. E. Fuhs, eds. *Handbook of Fluid Dynamics and Fluid Machinery*, Vol. III, McGraw-Hill, New York, pp. 2043–2064, 1996.

Parmakian, J., *Water Hammer Analysis*, Prentice-Hall, New York, 1955.

Parmakian, J., "Unusual Aspects of Hydraulic Transients in Pumping Plants," *Journal of the Boston Society of Civil Engineers*, 55:30–47, 1968.

Parmakian, J., "Surge Control," in M. H. Chaudhry, ed., *Proceedings, Unsteady Flow in Conduits*, Colorado State University, pp. 193–207, 1982.

Parmakian, J., "Incidents, Accidents and Failures Due to Pressure Surges," in M. H. Chaudhry ed., *Proceedings, Unsteady Flow in Conduits*, Colorado State University, pp. 301–311, 1982.

Sanks, R. L., *Pumping Station Design*, Butterworths, Bestar, 1989.

Stepanoff, A. I., *Centrifugal and Axial Flow Pumps*, John Wiley & Sons, New York, 1957.

Swanson, W.M., "Complete Characteristic Circle Diagrams for Turbomachinery," *Transactions, American Society of Mechanical Engineers*, 75:819–826, 1953.

Thorley, A. R. D., *Fluid Transients in Pipeline Systems*, D. & L. George Ltd., 1991.

Tullis, J. P., *Hydraulics of Pipelines*, John Wiley & Sons, New York 1989.

Watters, G. Z., *Modern Analysis and Control of Unsteady Flow in Pipelines*, Ann Arbor Science, Ann Arbor, MI, 1980.

Weaver, D. L., "Surge Control," *Journal American Water Works Association*, 64: 462–466, 1972.

White, I. M., "Application of the Surge Suppressor in Water Systems," *Water Works Engineering*, 45, 304–306, 1942.

Wood, D.J., and Jones, S.E., "Waterhammer Charts for Various Types of Valves," *ASCE, Journal of Hydraulics Division*, HY1, 99:167–178, 1973.

Wylie, E. B., and Streeter, V. L., *Fluid Transients in Systems*, Prentice-Hall, 1993.

CHAPTER 7
OPTIMAL DESIGN OF WATER DISTRIBUTION SYSTEMS

Kevin E. Lansey
Department of Civil Engineering and Engineering Mechanics
University of Arizona
Tucson, AZ

7.1 OVERVIEW

A common design process for a water distribution system is trial and error. An engineer selects alternative designs and simulates these designs using a network solver (see Chap. 12). Because of the complex interactions between components, identifying changes to improve a design can be difficult even for mid-sized systems. In addition, this approach does not provide assurance that a good, let alone optimal, solution has been determined. Thus, significant research effort has been placed in developing approaches to solve for optimal designs of distribution systems with some success. However, given the complexity of the problem and the limitations of mathematical programming tools, the complete problem has not been resolved. This chapter is a brief overview of the general directions of this research, and it highlights representative work in the area. Other reviews can be found in Goulter (1987, 1992), Walski (1985), and Walters (1988).

7.2 PROBLEM DEFINITION

When designing or rehabilitating a water distribution system using trial-and-error methods or with formal optimization tools, a broad range of concerns can be considered. Cost is likely to be the primary emphasis and includes the costs for construction, operation, and maintenance. The initial capital investment is for system components: pipes, pumps, tanks, and valves. Energy consumption occurs over time to operate the system. The main constraints are that the desired demands are supplied with adequate pressure head being maintained at withdrawal locations. Also, the flow of water in a distribution network and the nodal pressure heads must satisfy the governing laws of conservation of energy and mass, as described in Chap. 4.

In summary, the problem can be verbally stated as:
Minimize capital investment plus energy costs,
subject to: Meeting hydraulic constraints,
  Fulfilling water demands, and
  Satisfying pressure requirements

In practice, additional complexities beyond the concise formulation presented above can have a significant impact on both optimal design methods and trial-and-error approaches. For example, the number of demand conditions and the demand distributions to be satisfied must be defined prior to any analysis. It has been shown that an optimal design for a single demand pattern will be a branched pipe network; thus, to introduce reliability and redundancy, multiple demands must be considered. No general guidance is available to select the set of loading conditions to examine during the design process.

In addition to demands, other simple constraints can be added, such as type, size, or material or allowing different rehabilitation alternatives (cleaning, relining, or both).

Further complexity may be added if the system layout is not defined. In most distribution systems, pipes are restricted to be placed beneath roadways or in right-of-ways. In some cases, most notably major supply systems, the layout may be more flexible and must be determined. Another significant factor is budgetary constraints that may require staging of construction over time.

System operations have been included in the formulation above. This alone is a significant problem in both defining the demands and in addressing the large number of constraints and decisions associated with operations. To operate the system with tanks, a time series of demands is considered that multiplies the number of constraints by the number of demand conditions. Decisions are made during each period to define pump operations. From an optimization perspective, these decisions are discrete (integer) and are extremely difficult to determine efficiently. Determining operations for existing networks also has been widely studied (Chap. 15; Ormsbee and Lansey, 1994).

Additional constraints may be added to represent simple limits or complex constraints that are related to both the component sizes and the flow and pressure-head distributions. For example, a common engineering rule of thumb is to limit the range of flow velocities, which are functions of the flow rate and pipe diameter, between 2 and 5 ft/s for normal operating conditions. Water-quality requirements may also be added to the problem. Complex relationships exist to describe changes in water quality in a system. Since water quality decays with time in the network after disinfection, smaller pipes and tanks that would be introduced to improve water quality would conflict with meeting pressure requirements and reducing operation costs.

As noted above, an engineer typically defines one or more demand patterns that are considered during design. The intent in this selection is to produce a design that will operate effectively for the full range of conditions applied to the system. Indirectly, this implies some degree of reliability. Chapter 18 describes other, more rigorous measures for estimating the reliability of a distribution system. The reliability measures are related to the number of components and their sizes. Improving reliability, however, increases system costs and may adversely affect water quality under normal operating conditions. The addition of reliability in optimal design is beyond the scope of this discussion. The interested reader is referred to Mays (1989).

Location of valves for the purposes of reliability and their operation to reduce pressures and leakage have also been considered in the literature (Jowitt and Xu, 1990; Reis et al., 1997). This specialized problem is not discussed further here.

7.3 MATHEMATICAL FORMULATION

Including general relationships to account for the constraints noted above, the overall optimization problem for the design of water distribution systems can be stated mathematically in terms of the nodal pressure heads H and the various design/operational parameters D as follows

Objective: *Minimize Cost* = $f(H, D)$ (7.1)
subject to:

$G(H, D) = 0$	Conservation of mass and energy equations,	(7.2)
$\underline{H} \leq H \leq \overline{H}$	Nodal pressure head bounds,	(7.3)
$\underline{u} \leq u(D) \leq \overline{u}$	Constraints related to design/operational parameters, and	(7.4)
$\underline{w} \leq w(H, D) \leq \overline{w}$	Constraints related to design parameters and pressure heads	(7.5)

where the decision variables D define the dimensions for each component in the system, such as the pipe diameter, pump size, valve setting, and tank volume or elevation. The objective function (Eq. 7.1) can be linear or nonlinear. Since each component may have a term associated with it in the objective function, variability in costs due to installation location or period can be considered. In this formulation, system expansions or new systems can be designed.

Upper and lower limits on nodal pressure heads are given as strict bounds in this formulation (Eq. 7.3). Other reliability-based optimization extensions recognize that slight deviations from absolute ranges may be acceptable, depending on the change in system cost (Cullinane et al. 1992; Halhal et al., 1997; Wagner et al., 1988).

Bounds on the design variables (u) are written in general form (Eq. 7.4). However, most often they are simple bounds defining the size of the component. Decisions, as presented, can be continuous or discrete. Many decisions, such as commercially available pipe diameters, are discrete and make the problem especially difficult to solve. Templeman (1982) showed that solving for discrete pipe sizes is NP-hard. NP-hardness implies that the computation time for an N-pipe system is an exponential function of N; therefore, solution time increases rapidly with the number of pipes. Many researchers consider pipe diameters to be continuous and assume that rounded solutions are nearly optimal, or they split pipes into two sections with varying diameter in the final design. Neither approach is appealing to practitioners.

The general constraint set w includes limits on terms that are functions of both the nodal pressure heads and the design variables (Eq. 7.5). As noted above, pipe velocity, water-quality constraints, budgetary limits, and reliability measures are examples of these terms.

The conservation equations (Eq. 7.2) are the set of nonlinear equations relating the flow and pressure-head distribution to the nodal demands and selected design variables (Chap. 4). These equality constraints are written for each loading condition examined. The external nodal demands are included in these equations. These equations can be written for branched or looped networks for one or more demands, including an extended-period simulation. The conservation equations pose a significant difficulty in solving this optimization problem. Most approaches attempt to simplify these equations or to solve them outside the optimization model to avoid embedding them as constraints. Many methods can theoretically consider multiple demand conditions, but the computational expense may be high.

At this point, a general problem has been presented that can incorporate most concerns of water distribution engineers: that is, cost with budgetary limits and construction staging, water quality, reliability, operations, and discrete solutions. Because of the complexity of the general problem, however, it has not been solved. The majority of effort has been on cost minimization subject to simple bounds (Eqs. 7.3 and 7.4) without operations. This review considers this problem, with minor discussion of the other problems. Clearly, research should be expanded to incorporate realistic and practical concerns that make the results more applicable to engineering practice (Walski, 1996).

7.4 OPTIMIZATION METHODS

Virtually every optimization method has been applied to the problem of water distribution optimization. A progression can be seen over time. Branched networks were considered first, followed by looped networks that considered only pipe sizing. Both were solved using linear programming (LP). Nonlinear programming (NLP) was used later to solve a more general problem. Finally, to overcome continuous NLP decisions, more recent work has focused on stochastic search techniques, such as genetic algorithms and simulated annealing. The following sections describe representative work in each area. Dynamic, integer, and geometric programming methods also have been applied to this problem with limited general applicability and are not discussed further (Kim and Mays, 1991; Schaake and Lai, 1969; Yang et al., 1975).

7.4.1 Branched Systems

To size pipes in branched networks, LP-was applied using split-pipe formulations. Here, a pipe link is broken into segments of different diameters and the pipe lengths of each diameter are optimized (Altinbilek, 1981; Calhoun, 1971; Gupta, 1969; Karmeli et al., 1968). Branched systems can thus be posed without further simplification to linear optimization models. Later, to overcome the limitation of split-pipe lengths, Walters and Lohbeck (1993) applied a genetic algorithm (GA) for the layout and design of a tree network.

While GAs are discussed in more detail later, the LP formulation is presented for introduction. Branched or tree systems, such as irrigation systems, have a single path of flow to each node (Fig. 7.1). Thus, by conservation of mass, the flow rate in each pipe can be computed for a defined set of nodal demands by summing the total demand downstream of the pipe. The energy loss in each pipe is a function of the pipe diameter and the selected pipe material. In addition, the energy loss and pipe cost increase linearly with the length of pipe. Thus, the LP model can be stated as follows:

$$\text{Minimize } Z = \sum_{(i,j) \in I} \sum_{m \in M_{i,j}} c_{i,j,m} X_{i,j,m} \tag{7.6}$$

subject to

$$\sum_{m \in M_{i,j}} X_{i,j,m} = L_{i,j} \quad (i,j) \in I \tag{7.7}$$

$$H_{\min,n} \leq H_s + E_p - \sum_{(i,j) \in I_n} \sum_{m \in M_{i,j}} J_{i,j,m} X_{i,j,m} \leq H_{\max,n} \quad n = 1,\ldots, N \tag{7.8}$$

$$X_{i,j,m} \geq 0 \tag{7.9}$$

where $c_{i,j,m}$, $X_{i,j,m}$, and $J_{i,j,m}$, are the cost per unit length, length, and hydraulic gradient of pipe diameter m connecting nodes i and j, respectively. The objective function sums the

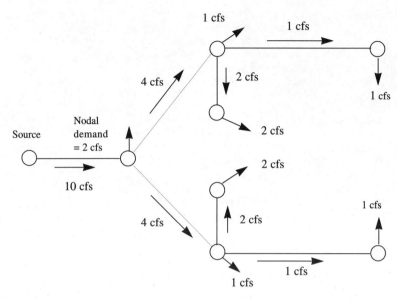

FIGURE 7.1 Branched pipe network.

cost of each pipe segment making up the total pipe length $L_{i,j}$ between nodes i and j, and sums over all pipes I. Constraint Eq. (7.7) requires that the total length of the segments of different diameters in the set of admissible M equals the distance between nodes i and j for all pipes. Equation (7.9) ensures that each pipe segment has a non-negative length.

As noted, the nodal conservation of mass equation defines the flow rate in each pipe. The hydraulic gradients can then be computed using the desired flow equation: for example, for the Darcy-Weisbach equation

$$J = \frac{8fQ^2}{\pi^2 g D^5} \quad (7.10)$$

where f is the pipe friction factor, Q is the flow rate (ft³/s), and D is the pipe diameter (ft). The hydraulic equation is the headloss along the path from the source with total head H_s to each node N. This energy loss is bounded to ensure the nodal pressure head does not exceed minimum and maximum pressures H_{min} and H_{max}, respectively, in Eq. (7.8). To account for pumps, a pump-head added term E_p is included in the path equation.

In this form, the model is linear with respect to the pipe segment lengths and can be solved by standard LP methods. This single source, single load model can be extended to multiple sources and to more than one loading condition. Finally, if the cost of added pump head is linear, it can be included in the objective function (Altinbilek, 1981).

7.4.2 Looped Pipe Systems via Linearization

Some of the most successful looped-system optimization methods linearize the looped network hydraulics so the design problem can be solved quickly by LP (Alperovits and

Shamir 1977; Featherstone and El-Jumaily, 1983; Fujiwara et al., 1987; Morgan and Goulter, 1985a,b; Quindry et al., 1981). Alperovits and Shamir first developed this linearization approach described as the linear programming gradient method. All these methods apply iterative processes by fixing pipe flow rates or pressure heads, optimizing the pipe sizes for the given flow or pressure distribution, updating the flow/pressure head distribution, and reoptimizing. The search for an updated flow/pressure head distribution varies. This process continues until an overall convergence criterion is met. Both discrete diameter and continuous pipe length models have been formulated. Some can handle decisions beyond pipe sizing. Most of these models are limited to considering only a single loading condition during the optimization process.

Morgan and Goulter (1985b) considered multiple demand patterns for pipe sizing. Their work is described below for a single loading condition, and the extension to multiple demands is straightforward. Unlike other linearization schemes and branched network models that determine pipe lengths directly, the decisions here are to determine the length of the pipe segment that will change the present solution's diameter m to an adjacent commercially available pipe diameter $m + 1$ or $m - 1$. The result, however, is still a split pipe solution.

Minimize Z =

$$\sum_{i,j} [(c_{i,j,m+1} - c_{i,j,m}) X_{i,j,m+1} + (c_{i,j,m-1} - c_{i,j,m}) X_{i,j,m-1}] \tag{7.11}$$

subject to

$$\sum_{(i,j) \in I_n} [(J_{i,j,m+1} - J_{i,j,m}) X_{i,j,m+1} + (J_{i,j,m-1} - J_{i,j,m}) X_{i,j,m-1}] \leq H_{\min,n} - H_n \tag{7.12}$$

$$L_{i,j} \geq X_{i,j,m+1} \geq 0 \tag{7.13}$$

$$L_{i,j} \geq X_{i,j,m-1} \geq 0 \tag{7.14}$$

where $X_{i,j,m+1}$ and $X_{i,j,m-1}$ are the pipe segment replacement lengths of the adjacent commercially available pipe diameters to diameter m that makes up the majority of the pipe length $L_{i,j}$ between nodes i and j. The lengths of these segments are restricted to be nonnegative and less than the entire distance (Eqs. 7.13 and 7.14).

For the present set of pipe segment lengths, the flow and nodal pressure-head distribution can be computed by the network solver and the hydraulic gradients can be determined (e.g., Eq. 7.10). Equation (7.12) represents the change in nodal pressure head from H_n toward its limiting value $H_{\min,n}$ along path I_n. If slack exists between H_n and $H_{\min,n}$, the next smaller-diameter pipe may replace some segments of larger pipe, reducing H_n and resulting in a lower objective function value. Since changing the pipe lengths alters the flow and pressure-head distributions, the new flow distribution is computed using the network solver after the LP is solved. New J's are inserted in Eq. (7.12) and another problem is solved. This process is repeated until the final solution has been determined, when no pipe segments have been changed in a linear program. In this method, the choice of paths is a concern since only pipes along the selected path can be modified (Goulter et al., 1986).

All linearization approaches noted above follow similar procedures. Although a linear program is solved at each iteration, the problem is nonlinear and a global optimal solution cannot be guaranteed. Fujiwara and Khang (1990), Eiger et al. (1994), and Loganathan et al. (1995) have developed approaches to move toward global optimal solutions by developing two-phase searches. In their first phase, Fujiwara and Khang solved a linear program for changing the headlosses (and pipe-segment lengths) for a fixed flow pattern. The flow pattern is then changed to find a local optimum using a nonlinear search until a local optimum is determined (Fujiwara et al., 1987). In the second phase for the optimal solution with fixed headlosses and pipe lengths, a concave linear optimization problem is solved to determine a new flow distribution. This new flow pattern will have a lower cost than the previous local optimum and is used as the starting point for another iteration of the first-phase problem. The process continues until no better local optimal solution is found. Loganathan et al. (1995) also used the modified LP gradient method to solve for pipe size in an inner problem while using simulated annealing to solve the inner problem of modifying the pipe flows.

Eiger et al. (1994) considered the problem to be a nonsmooth optimization and used duality theory to converge to a global optimum using a similar two-phase approach. The so-called inner or primal problem is solved given a flow distribution using LP for the optimum pipe segment lengths and an upper bound on the global optimal solution. An outer nonconvex, nonsmooth dual problem is solved that provides a lower bound on the global optimal solution. The difference between the upper and lower bound is known as a *duality gap*. A branch and bound algorithm, in which the inner and dual problems are solved numerous times, is applied to reduce the duality gap to an acceptable level by changing the flow rates in each loop. Ostfeld and Shamir (1996) later modified and applied this approach to consider reliability and water quality in the optimization process. Both of these methods have been shown to provide good results.

7.4.3 General System Design via Nonlinear Programming

Typically, NLP has also been applied to more general problems (Lansey and Mays, 1989; Ormsbee and Contractor, 1981; Shamir, 1974). In these models, a network simulation model was linked with the optimization model so that the hydraulic constraints (Eq. 7.2) could be removed from the optimization problem. The NLP models generally can consider multiple demand conditions and a broad range of design variables. Similar methods also were applied to the optimal operation problem. In addition, the methods can be extended to consider reliability measures—at a heavy computational expense, however.

The general method for solving the design problem efficiently uses an optimal control framework that links a network simulation model with a nonlinear optimizer. In this case, a set of decision variables D, known as the control variables in this formulation, is passed from the optimizer to the network solver. The simulation model solves the hydraulic equations and determines the values of the nodal pressure heads H known as the state variables. The set of hydraulic equations is therefore implicitly satisfied and the other constraints containing the state variables (Eqs. 7.2 and 7.4) can be evaluated. This information is then passed back to the optimizer. D is then modified to move to a lower-cost solution and the process continues until a stopping criterion is met.

In terms of the optimization problem, when the simulator is linked with an optimizer (Fig. 7.2) and assuming that H can be determined for all D, the optimization problem (Eq. 7.15) transformed to a reduced problem:

Objective: *Minimize Cost* $= f(H(D), D) = F(D)$ (7.15)

subject to

$\underline{H} \leq H(D) < \overline{H}$ Nodal pressure head bounds,   (7.16)

$\underline{u} \leq u(D) \leq \overline{u}$ Constraints related to design/operational parameters, and   (7.17)

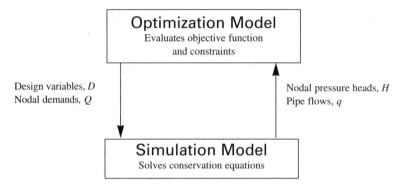

FIGURE 7.2 Optimization-simulation model link for nonlinear programming methodology.

$\underline{w} \leq w(H(D), D) \leq \overline{w}$ Constraints related to design parameters and pressure heads. (7.18)

The problem is reduced since Eqs. (7.2) are no longer included because the network simulator satisfies them.

The means to modify D depend on the optimizer. Direct search methods use information from a few points to determine the next D. More efficiently, gradients of the objective function with respect to changes in the design variables can be computed using the reduced gradient equations to provide information and define a search direction. The movement in the search direction is then guided by one of several possible optimization techniques (e.g., Newton type methods, conjugate gradients).

Difficulties remain in handling constraints containing the state variables (Eq. 7.2 and 7.4). Most typically, they are included in the objective function in the form of a penalty function, such as an augmented Lagrangian term. Violations of those constraints are multiplied by some factor and increase the cost of the solution. Gradients of the objective with the penalty term are then used to change D, and the constraints are directly considered when determining changes in the design parameters.

Since a network simulator is used in this formulation, any decision that can be formulated as a continuous variable in terms of cost can be considered in this model. Also, the number of demand conditions to be evaluated is not limited, with the exception of increased computational effort. Thus, multiple demands for fire conditions or a sequence of demands describing daily operations can be considered. In addition, more complex constraints, such as reliability constraints, can be addressed (Duan et al., 1990; Su et al., 1987).

7.4.4 Stochastic Search Techniques

More recently, evolutionary optimization, genetic algorithms (Dandy et al., 1996; Savic and Walters, 1997; Simpson et al., 1994), and simulated annealing (Loganathan et al., 1995) have been used to solve the pipe-sizing problem. A logical extension is to move to other decision variables (Savic and Walters, 1997). Details on the optimization algorithms are not presented here. For reference, genetic algorithms were proposed by Holland (1975) and further developed by Goldberg (1989) and others. A good description of simulated annealing is found in Loganathan et al. (1995).

In general, when searching for an optimal solution with a stochastic search algorithm, the objective function (Eq. 7.1) is evaluated for a set of solutions (D). Based only on the objective function of the previous solutions, new decision vectors are generated. The generation process varies by approach but always contains a random element. The new solution vectors are evaluated and another set of solutions is generated using that information. The process of evaluating and generating solutions continues until a defined stopping criterion is met.

Stochastic search techniques can directly solve unconstrained optimization problems: that is, without constraints containing the control variables H. Pipe diameters are only selected within their acceptable ranges. In the evaluation step of the network design problem, the hydraulic relationships (Eq. 7.2) and the remaining constraints (Eq. 7.4) are typically also computed by executing a network solver. If a solution does not satisfy Eq. (7.2) or Eq. (7.4), the objective function can be assigned a large value so that the search will not continue near this point or the violations of the constraints can be included in a penalty term in the objective function (e.g., Savic and Walters 1997).

Because many solutions are examined during the search process, the optimal solution and a number of nearly optimal solutions are provided by these techniques. A penalty term has the advantage of allowing slightly unfeasible solutions to be maintained during the search. The decision-maker can then judge the cost of not satisfying the constraints by comparing the cost of the unfeasible solutions with feasible, nearly optimal solutions.

As noted above, stochastic search methods only use objective function values to move to better solutions compared to LP and NLP, which use gradient information. Since stochastic search methods only depend on objective function evaluations, they can easily handle discrete decisions inherent with pipe sizing. Also, in comparison to NLP, although not proved, experience has shown that those methods tend to produce global optimal solutions that cannot be guaranteed by NLP methods. Promising results have been seen in finding low-cost pipe network designs.

The advantage of not requiring gradients, however, is also the methods' shortcoming. Since only objective function information is required, a large number of simulations are necessary to move to the optimal solution. This is time consuming and limits the size of the problem that can be solved. However, this complaint applies to all the optimization techniques discussed. Faster computers and parallel processing provide hope for these general techniques.

7.5 APPLICATIONS

The previous section provided a general view of the present use of and the need for moving the theoretical field of optimal design to practical use. It also is interesting to compare the abilities of the various approaches. Table 7.1, from Savic and Walters (1997), presents solutions for a well-posed problem of adding parallel tunnels to the New York City water supply system. This system, shown in Fig. 7.3, was first examined in 1969 by Schaake and Lai and subsequently was studied by many researchers to test their approaches. The purpose is to identify pipes to install in parallel with the existing 21 pipes to ensure that adequate pressure head is supplied as the demands are increased to a defined level. The network has two loops, two dead-ends and a known constant source of energy. As seen in the table, most methods converged to solutions that had similar costs and introduced many of the same parallel pipes. Thus, most methods work reasonably well for small networks but may be affected by local optima.

A few notes of clarification regarding the table are warranted. First, two GA solutions are provided from Savic and Walters (1997). These solutions account for the range of solutions that can be developed through rounding of constant terms in the Hazen-Williams headloss equation. Differences in solutions can be partially attributed to the coefficients used

7.10 Chapter Seven

TABLE 7.1 Comparison of Results from Various Methods for New York City Water Supply Problem

	Schaake and Lai (1969)	Quindry et al. (1981)	Gessler (1982)	Bhave (1985)	Morgan and Goulter (1985)	Fujiwara and Khang (1990)	Murphy et al. (1993)	Loganathan et al. (1995)	Savic and Walters (Solution 1) (1997)	Savic and Walters (Solution 2) (1997)
Decisions	C	C	D	C	S	C	D	S	D	D
Pipe	D (in)	D (in)	D (in)	D (in)	D (in)	D (in)	D (in)	D (in)	D (in)	D (in)
1	52.02	0	0	0	0	0	0	0	0	0
2	49.90	0	0	0	0	0	0	0	0	0
3	63.41	0	0	0	0	0	0	0	0	0
4	55.59	0	0	0	0	0	0	0	0	0
5	57.25	0	0	0	0	0	0	0	0	0
6	59.19	0	0	0	0	0	0	0	0	0
7	59.06	0	100	0	144	73.62	0	120–132	108	144
8	54.95	0	100	0	0	0	0	0	0	0
9	0	0	0	0	0	0	0	0	0	0
10	0	0	0	0	0	0	0	0	0	0
11	116.21	119.02	0	0	0	0	0	0	0	0
12	125.25	134.39	0	0	0	0	0	0	0	0
13	126.87	132.49	0	0	0	0	0	0	0	0
14	133.07	132.87	0	136.43	0	0	120	0	0	0
15	126.52	131.37	0	87.37	96	99.01	84	96–108	96	84
16	19.52	19.26	100	99.23	96	98.75	96	96–108	96	96
17	91.83	91.71	100	78.17	84	78.97	84	84–96	84	84
18	72.76	72.76	80	54.40	60	83.82	72	72–84	72	72
19	72.61	72.64	60	0	0	0	0	0	0	0
20	0	0	0	0	0	0	0	0	0	0
21	54.82	54.97	80	81.50	84	66.59	72	72–84	72	72
Cost ($ million)	78.09	63.58	41.80	40.18	39.20	36.10	38.80	38.0	37.13	40.42

Note: Decisions are defined as D – Discrete pipe diameters, C – Continuous diameters, S – Split pipe lengths.

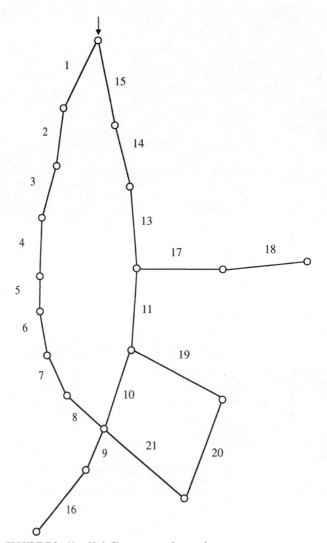

FIGURE 7.3 New York City water supply tunnels.

in the models. Second, Gessler's solution (1982) is not truly optimization; it is an enumeration scheme. Third, the solution by Fujiwara and Khang, the only NLP method and an extension of Alperovits and Shamir (1977), is slightly infeasible with violations of less than 0.2 ft of pressure head. This result shows the sensitivity of optimal cost to the defined nodal pressure requirements. It also points to the importance of setting those values and the potential impact of the uncertainty in forecasting desired service levels and pipe roughness coefficients. An unpublished NLP methodology from the author's work at the University of from the author's work at the University of Texas also arrived at several solutions with similar optimal costs ($37.9 million and $39.3 million). The approach embedded the hydraulic equations as constraints in the NLP model.

The robustness of the solutions in terms of cost and new pipe locations is reassuring and provides confidence in the usefulness of the methods. Savic and Walters (1997) presented similar results for two other pipe networks in the literature containing 8 and 34 pipes under a single loading condition.

A more difficult problem was posed to researchers in the Battle of the Network Models (Walski et al., 1987). Expansion and rehabilitation decisions were necessary for a realistic network in Anytown, USA. Pipe costs were varied by location. Additional tanks and pumps could be added over time to meet demands at three time horizons. Energy costs were computed for daily demands, and a large set of peak and fire conditions were to be satisfied.

Solution approaches varied between modelers. Most often, a single demand condition was used to optimize the network, and other decisions were made via judgment. Energy consumption and pumping decisions also were made using judgment. This approach is a valid use of optimization tools. The tools are intended to provide guidance with regard to reducing costs but are not used without detailed evaluation of the solutions. As noted by Goulter (1992), the end results for Anytown, however, were surprisingly similar in terms of cost (within 12 percent).

A significant difficulty at the time of the Battle was that all the computer programs but one were research oriented and required detailed knowledge of their workings. The gap between development and application resulted from a perceived lack of a market by researchers and other software developers and a lack of interest among practicing engineers. Recently, several commercially available optimization programs based on genetic algorithms have been produced. The models are more intuitive and hopefully begin a broader application of technology that has been developing over the last 30 years.

7.6 SUMMARY

Although much effort has been placed in developing methods, optimization models have seen limited usage in practice. Criticism has followed the use of analytical methods to determine pipe sizes since Camp (1939) presented an approach for economic pipeline design. The criticism of those methods parallels that of optimization methods, as expressed by Lischer (1979), Walski (1985), and Walski (1996). These criticisms focus on the uncertainty in defining design parameters, including cost coefficients, future roughness coefficients, and, most significantly, water demands. Solutions providing continuous diameters or split pipe also cause concern. Optimization models also have been correctly criticized for removing redundancy in design because they focus solely on minimizing costs.

In light of these complaints, Lischer (1979) stated that optimization and mathematical modeling should not be used without experienced engineering judgment. In other words, optimization models are another tool for engineers to determine efficient low-cost solutions for a set of design conditions. The choice of these conditions will have an impact on trial-and-error procedures and optimization models. A sensitivity analysis of the resulting solution can be completed in either case, and modifications can be made to improve it. If viewed from that perspective, continuous pipe sizes are also not a critical problem.

Reliability of least-cost optimal water networks is a critical issue. When designing by judgment, engineers explicitly or implicitly introduce reliability and redundancy. However, this reliability is typically based on their judgment rather than on a systematic and quantifiable method. Guidelines for demands and criteria for acceptable design, such as pressure heads, are lacking and, as Templeman suggested in 1982, a first step is to establish a firm set of design criteria for water distribution systems. Designing for multiple demand conditions will introduce redundancy to a network. Quantifying reliability and including it as a constraint within optimization models is a second way to force redundancy. However, no universal reliability measure or acceptable reliability levels have been developed, and inclusion of such measures comes at a high computation expense. Overcoming these weaknesses of present optimization methods will improve practicing engineers' view of the models. In turn, recognition by these engineers of how optimization models that are presently available can be used as part of a design process should lead to wider application of the tools.

REFERENCES

Alperovits, E., and U. Shamir, "Design of optimal water distribution systems," *Water Resources Research*, 13(6):885–900, 1977.

Altinbilek, H., "Optimum design of branched water distribution networks by linear programming," *Proceedings of the International Symposium on Urban Hydrology, Hydraulics and Sediment Control*, Lexington, KY, 249–254, 1981.

Bhave, P., "Optimal expansion of water distribution systems," *Journal of Environmental Engineering, ASCE*, 111(2):177–197, 1985.

Calhoun, C., "Optimization of pipe systems by linear programming," in J. P. Tullis (ed.), *Control of Flow in Closed Conduits*, Colorado State University, Ft. Collins, 175–192, 1971.

Camp, T., "Economic pipe sizes for water distribution systems," *Transactions of ASCE*, 104: 190–213, 1939.

Cullinane, M., K. Lansey, and L. Mays, "Optimization of availability-based design of water distribution networks," *ASCE, Journal of Hydraulic Engineering*, 118(3): 420–441, 1992.

Dandy, G., A. Simpson, and L. Murphy, "An improved genetic algorithm for pipe network optimization," *Water Resources Research*, 32(2):449–458, 1996.

Duan, N., L. Mays, and K. Lansey, "Optimal reliability-based design of pumping and distribution systems," *ASCE Journal of Hydraulic Engineering*, 116(2):249–268, 1990.

Eiger, G., U. Shamir, and A. Bent Tal, "Optimal design of water distribution networks," *Water Resources Research*, 30(9):2637–2646, 1994.

Featherstone, R., and K. El-Jumaily, "Optimal diameter selection for pipe networks," *ASCE Journal of Hydraulic Engineering*, 109(2):221–233, 1983.

Fujiwara, O., and D. Khang, "A two-phase decomposition method for optimal design of looped water distribution networks," *Water Resources Research*, 26(4):539–549, 1990.

Fujiwara, O., B. Jenchaimahakoon, and N. Edirisinghe, "A modified linear programming gradient method for optimal design of looped water distribution networks," *Water Resources Research*, 23(6):977–982, 1987.

Gessler, J., "Optimization of pipe networks," *International Symposium on Urban Hydrology, Hydraulics and Sediment Control, University of Kentucky*, Lexington, KY, 1982.

Gessler, J., "Pipe network optimization by enumeration," *Proceedings of Specialty Conference on Computer Applications in Water Resources, American Society of Civil Engineers (ASCE)*, New York, 572–581, 1985.

Goldberg, D., *Genetic Algorithms in Search, Optimization, and Machine Learning*, Addison-Wesley, Reading, MA, 1989.

Goulter, I., "Current and future use of systems analysis in water distribution network design," *Civil Engineering Systems*, 4(4):175–184, 1987.

Goulter, I., "Systems analysis in water-distribution network design: From theory to practice," *ASCE Journal of Water Resources Planning and Management*, 118(3):238–248, 1992.

Goulter, I., B. Lussier, and D. Morgan, "Implications of head loss path choice in the optimization of water distribution networks," *Water Resources Research*, 22(5):819–822, 1986.

Gupta, I., "Linear programming analysis of water supply system," *AIIE Transactions*, 1(1): 56–61, 1969.

Halhal, D., G. Walters, D. Ouazar, and D. Savic, "Water network rehabilitation with structured messy genetic algorithm," *ASCE Journal of Water Resources Planning and Management*, 123(3):137–146, 1997.

Holland, J., *Adaptation in Natural and Artificial Systems*, University of Michigan Press, Ann Arbor, MI, 1975.

Jowitt, P., and C. Xu, "Optimal valve control in water-distribution networks," *ASCE Journal of Water Resources Planning and Management*, 116(4):455–472, 1990.

Karmeli, D., Y. Gadish, and S. Meyers, "Design of optimal water distribution networks," *Journal of Pipeline Division, ASCE*, 94(1): 1–9, 1968.

Kim, H., and L. Mays, "Optimal rehabilitation model for water distribution systems," *ASCE Journal of Water Resources Planning and Management*, 120(5):674–692, 1994.

Lansey, K. and L. Mays, "Optimal design of water distribution system design," *ASCE Journal of Hydraulic Engineering*, 115(10):1401–1418, 1989.

Lischer, V., "Discussion of 'Optimal design of water distribution networks' by Cedenese and Mele," *ASCE Journal of Hydraulic Engineering*, 105(1): 113–114, 1979.

Loganathan, G., J. Greene, and T. Ahn, "Design heuristic for globally minimum cost water distribution systems," *ASCE Journal of Water Resources Planning and Management*, 121(2):182–192, 1995.

Mays, L. ed., *Reliability Analysis of Water Distribution Systems*, ASCE, New York, 1989.

Morgan, D., and I. Goulter, "Optimal urban water distribution design," *Water Resources Research*, 21(5):642–652, 1985a.

Morgan, D., and I. Goulter, "Water distribution design with multiple demands," *Proceedings of Specialty Conference on Computer Applications in Water Resources, ASCE*, New York, 582–590, 1985b.

Murphy, L., A. Simpson, and G. Dandy, *Pipe Network Optimization Using an Improved Genetic Algorithm*, Research Report No. R109, Department of Civil and Environmental Engineering, University of Adelaide, Australia, 1993.

Ormsbee, L. and D. Contractor, "Optimization of hydraulic networks," *International Symposium on Urban Hydrology, Hydraulics, and Sediment Control*, Lexington, KY, 255–261, 1981.

Ormsbee, L., and K. Lansey, "Optimal control of water supply pumping systems," *ASCE Journal of Water Resources Planning and Management*, 120(2):237–252, 1991.

Ostfeld, A., and U. Shamir, "Design of optimal reliable multi-quality water-supply systems," *ASCE Journal of Water Resources Planning and Management*, 122(5):322–333, 1996.

Quindry G., E. Brill, and J. Liebman, "Optimization of looped water distribution systems," *ASCE Journal of Environmental Engineering*, 107(4): 665–679, 1981.

Reis, L., R. Porto, and F. Chaudhry, "Optimal location of control valves in pipe networks by genetic algorithm," *ASCE Journal of Water Resources Planning and Management*, 123(6):317–326, 1997.

Savic, D., and G. Walters, "Genetic algorithms for least-cost design of water distribution networks," *ASCE Journal of Water Resources Planning and Management*, 123(2):67–77, 1997.

Schaake, J., and D. Lai, Linear Programming and Dynamic Programming Applications to Water Distribution Network Design, Report 116, Department of Civil Engineering, Massachusetts Institute of Technology, Cambridge, MA, 1969.

Shamir, U., "Optimal design and operation of water distribution systems," *Water Resources Research*, 10(1):27–35, 1974.

Simpson, A., G. Dandy, and L. Murphy, "Genetic algorithms compared to other techniques for pipe optimization," *ASCE Journal of Water Resources Planning and Management*, 120(4):423–443, 1994.

Su, Y., L. Mays, N. Duan, and K. Lansey, "Reliability-based optimization for water distribution systems," *ASCE Journal of Hydraulic Engineering*, 113(12):589–596, 1987.

Templeman, A., "Discussion of 'Optimization of looped water distribution systems,' by Quindry et al," *ASCE Journal of Environmental Engineering,* 108(3):599–596, 1982.

Wagner, J., U. Shamir, and D. Marks, "Water distribution reliability: Simulation methods," *ASCE Journal of Water Resources Planning and Management,* 114(3):276–294, 1988.

Walski, T., "State of the art pipe network optimization," *Proceedings of Specialty Conference on Computer Applications in Water Resources, ASCE,* New York, 559–568, 1985.

Walski, T., "Discussion of 'Design heuristic for globally minimum cost water-distribution systems' by Loganathan et al." *ASCE Journal of Water Resources Planning and Management,* 122(4):313–314, 1996.

Walski, T., E. Brill, J. Gessler, I. Goulter, R. Jeppson, K. Lansey, H-L. Lee, J. Liebman, L. Mays, D. Morgan, and L. Ormsbee, (1987), "Battle of the network models: Epilogue," *ASCE Journal of Water Resources Planning and Management,* 113(2):191–203, 1987.

Walters, G., "Optimal design of pipe networks: A review," Proceedings of the, 1st *International Conference on Computer Methods and Water Resources in Africa, Vol. 2; Computational Hydraulics,"* Computational Mechanics Publications and Springer, Verlag, Southhampton, England, 21–32, 1998.

Walters, G., and T. Lohbeck, "Optimal layout of tree networks using genetic algorithms," *Engineering Optimization,* 22(1):27–48, 1993.

Yang, K., T. Liang, and I. Wu, "Design of conduit systems with diverging branches," *ASCE Journal of Hydraulic Engineering,* 101(1):167–188, 1975.

CHAPTER 8
WATER-QUALITY ASPECTS OF CONSTRUCTION AND OPERATION

Thomas M. Walski
Pennsylvania American Water Co.
Wilkes-Barre, PA

8.1 INTRODUCTION

During construction and operation, numerous factors can have an impact on water quality in the distribution system, including handling and disinfection of new mains; prevention and elimination of cross-connections; elimination of leaks and breaks; disinfection of storage tanks after construction; inspection, or maintenance; installation and operation of blowoffs; air releases, and flushing hydrants; implementation of a flushing program; proper break repair practices; covering and properly venting storage tanks; maintenance of adequate separation from sewers; enforcement of applicable building plumbing codes; and, of course, maintenance of positive pressure at all times. These considerations are described in the appropriate state water systems standards and in numerous references (AWWA, 1986a; Departments of Air Force, Army and Navy, 1984; Great Lakes and Upper Mississippi River Board of State Public Health and Environmental Managers, 1992). AWWA also produces training videos on some of these subjects.

Three of the more important aspects of distribution system water quality, new main disinfection, tank disinfection, cross-connection control, and flushing are covered in more detail in the following sections.

8.2 DISINFECTION OF NEW WATER MAINS

Disinfection of water mains is addressed in AWWA Standard C651–92 (1992), although some utilities may have their own variations on the AWWA Standard.

8.2.1 Need for Disinfection

Ideally, water mains would be delivered to construction sites in sterile condition and be kept that way during pipe installation. Unfortunately, there is no reasonable way to keep pipes sterile during shipping, storage, and installation. Pipes may be left outdoors for months to years before they are installed, with contamination potentially caused by a variety of animal, plant, and microbial life entering the pipe. Pipes may become flooded before and during construction with water of varying quality. Jointing, packing, and sealing material may become contaminated. To the extent possible, pipes and jointing materials should be kept as clean as possible before placing the pipe in service, but only disinfection of the pipe once it is installed can reasonably ensure clean pipes.

8.2.2 Disinfection Chemicals

Numerous disinfection chemicals are available. The following three are used most commonly:

1. Liquid chlorine (Cl_2), which is usually available in 100-lb (45.4-kg) or 150-lb (68-kg) pressurized containers. Liquid chlorine is inexpensive but highly toxic and should be used only by appropriately trained individuals with the proper chlorinators and ejectors.

2. Sodium hypochlorite (NaOCl), which is a liquid stored in glass, rubber-lined, or plastic containers of varying sizes. It is more expensive and bulky than liquid chlorine, but is much safer to handle. It is usually 5–15 percent chlorine, but it has a finite shelf life.

3. Calcium hypochlorite ($Ca(OCl)_2$), which is approximately 65 percent chlorine by weight. It is easy to handle in either tablets or granular form, but is relatively expensive and must be kept dry to prevent degradation. An exothermic reaction yielding oxygen and chlorine can occur if the ($Ca(OCl)_2$) is heated to 350°F.

8.2.3 Disinfection Procedures

Disinfection involves contacting the pipe with a sufficiently high dosage of chlorine for a sufficiently long period of time. The three methods presented in C651-92 are summarized below.

8.2.3.1 The tablet method. This method can be used if mains have been kept clean and dry during construction and involves placing hypochlorite granules or tablets in the pipes during installation at intervals no greater than every 500 ft (150 m). The number of 5-g tablets per length of pipe can be estimated from

$$N = 0.0012 \; L \; D^2$$

where N = number of 5–g tablets, L = length of pipe (ft), and D = diameter of pipe (in). Using this method, the average concentration of chlorine during the test should be approximately 25 mg/L. The main must be filled with potable water at a velocity slower than 1 ft/s (0.3 m/s), making sure to eliminate all air pockets. If the temperature is higher than 5°C, the water must be kept in the pipe for at least 24 h; if the temperature is lower than 5°C, it must be kept in the pipe for 48 h. The tablet method is acceptable for smaller mains, mains without solvent welded, or threaded steel joints.

8.2.3.2 The continuous feed method. This method consists of, first, flushing the main, at a velocity of at least 0.76 m/s (2.5 ft/s) to remove any sediment and air pockets. For larger mains, where flushing may not be effective, brooming or swabbing can be used. Chlorine must then be fed at a rate that maintains a concentration of 25 mg/L for 24 h. At the end of that time, the free chlorine residual must be greater than 10 mg/L at all points in the pipe.

8.2.3.3 The slug method. This method consists of placing hypochlorite granules, as in the tablet method, and flushing the main, as in the continuous feed method. Then a slug of highly chlorinated water with a concentration of at least 100 mg/L is passed through the main so that the slug is in contact with the pipe for at least 3 h, and the concentration of the slug does not drop below 50 mg/L. Pigs or swabs can be used to separate the slug from the potable water in the pipe. The slug method is used most commonly for larger mains, where the volume of water required for the continuous feed method is impractical.

Highly chlorinated water must be disposed of in an environmentally safe manner in compliance with all applicable water–quality regulations. This may necessitate the use of a reducing chemical (e.g., sodium sulfite) at the downstream end of the pipe being disinfected to react with the excess chlorine in the water.

8.2.4 Testing New Mains

After disinfection and flushing, the new main is filled with potable water and is sampled for coliform bacteria at least every 1200 ft (366 m) in accordance with Standard Methods (APHA, 1995). Other tests, such as heterotrophic plate count, may be required. If a sample fails the test, the main should be flushed and the sampling repeated. If flushing does not result in an acceptable test, the main may need to be disinfected again.

8.2.5 Main Repairs

After water main breaks, the mains should be flushed to remove any water in the vicinity of the break and also may need to be disinfected if pressure was lost in the mains. However, because mains are needed to maintain fire protection and sanitation, they are usually placed back in service before the results of testing are available. If there is concern that contamination has occurred during repair, a precautionary "boil water" notice should be posted.

8.2.6 Disposal of Highly Chlorinated Water

Highly chlorinated water can be toxic to living things. Before such water is discharged to the environment, an assessment should be made to determine if the water can be discharged without causing any harm. Ideally, the water should be discharged to a nearby sanitary or combined sewer–collection system, provided the system has the capacity to handle the discharge and the water utility has received permission for the discharge from the sewer utility. Depending on the circumstances, direct discharge into a water body or storm sewer system may be a violation of the Clean Water Act or may require a permit.

If the water cannot be discharged without treatment, then the chlorine must be neutralized before the water is discharged. This is accomplished by adding a reducing agent that reacts with the chlorine. Table 8.1 gives the amount of several disinfection

TABLE 8.1 Dose of Reducing Agent Needed to Neutralize 1 MG of Water with 1 mg/L Chlorine

Chemical	Formula	Dose (lb/MG)	Dose (kg/MG)
Sulfur dioxide	SO_2	8	3.6
Sodium bisulfite	$NaHSO_3$	12	5.4
Sodium sulfite	Na_2SO_3	14	6.4
Sodium thiosulfate	$Na_2S_2O_3\ 5H_2O$	12	5.4

chemicals required to neutralized 1 MG (3.78 ML) of water with a residual chlorine concentration of 1 mg/L based on 100 % purity and sufficient mixing and detention time to complete the reaction. To determine the quantity needed for any volume of water, adjust the quantity from Table 8.1 using

$$\text{Dose} = (V)(C)(100/P)$$

where V = volume of water (Mg), C = chlorine concentration (mg/L), P = purity of reducing agent (percent).

8.3 DISINFECTION OF STORAGE TANKS

Disinfection of tanks is covered in AWWA Standard C652 (1992b). The procedures are summarized below. In the case of water mains, once a main is disinfected, it may never need to be disinfected again. Tanks, however, are occasionally taken out of service for inspection, cleaning, repairs, and painting.

The chemicals used for disinfection of tanks are the same as those used for disinfection of pipelines and are described in Sec. 8.2.2. All tools and equipment are removed from the tank, and the tank is washed, swept, or scrubbed to remove any debris or dirt.

8.3.1 Disinfection Procedures for Filling Tanks

AWWA Standard C652 recognizes three methods for disinfecting storage tanks when they have been taken out of service and drained. It also describes procedures for disinfection during underwater inspections.

8.3.1.1 Method 1. Liquid chlorine or sodium hypochlorite is added to the influent pipe while filling, or calcium hypochlorite is placed on the bottom of the tank before filling so that the chlorine concentration shall be at least 10 mg/L after the retention period. A 6 h retention period is used when chlorine has been fed uniformly with the influent water, whereas a 24-h period is used when the chemicals are mixed into the tank.

8.3.1.2 Method 2. The tank is sprayed or brushed entirely with a solution of 200 mg/L of available chlorine. Drain pipes are filled with a solution of 10 mg/L chlorine, as in method 1. The highly chlorinated solution is left on the surface for at least 30 min. Only surfaces that will be in contact with potable water need to be sprayed or brushed. Upon filling and bacteriological testing, this method may produce water that can be delivered to the distribution system rather than be discharged.

8.3.1.3 Method 3. Chlorine is added using the procedures in method 1, except that the target concentration when the tank is approximately 5 percent full is 50 mg/L. This water is held for 6 h, then the tank is slowly brought to full level with potable water and this water is kept in the tank for an additional 24 h. Following bacteriological testing, this tank may be placed in service, provided any drain lines have been purged of highly chlorinated water and the chlorine residual is at least 2 mg/L.

8.3.2 Underwater Inspection

Underwater inspection by divers or robotic cameras is becoming increasingly popular. All equipment, clothing, and personnel entering the tank must be cleaned thoroughly, and it must be certified that the equipment and clothing has been used only for potable water inspections. Equipment can be cleaned by spraying, submersion, or sponging with a chlorine solution of at least 200 mg/L.

If divers are used, two certified divers should be on site, and an additional diver should be available outside the tank in case of an emergency. Before anyone or any equipment is allowed to enter the tank, the residual chlorine should be checked to determine that it is adequate. Air should be supplied to the divers from external air-supplying equipment.

When feasible, tanks should be valved offline during the inspection and should not be brought back online until a satisfactory bacteriological test result has been received. Tanks should be full during inspections.

8.4 CROSS-CONNECTION CONTROL

8.4.1 Definitions

A *cross-connection* is any connection or potential connection between a potable water system and any source of contamination that can affect the quality of water in such a way that the contamination could enter the potable system under certain circumstances. *Backflow* is the actual reversal of flow in such a way that contamination enters a water system through a cross-connection. Backflow will occur when the head on the potable side of the cross-connection drops below that on the contaminated side. It can occur through *backsiphonage*, a drop in the pressure on the potable side, or *backpressure*, an increase in the pressure on the potentially contaminated side.

Backflow is undesirable because it can result in the introduction of contamination into the potable system. Examples might include water from a neighboring utility, antifreeze in fire-protection piping, toxic chemicals at factories, or etiologic agents from a hospital. Numerous cases of contamination of water systems from cross-connections have been documented (Angele, 1974; AWWA, 1990; Foundation for Cross-Connection, Control and Hydraulic Research; 1988; United States Environmental Protection Agency (USEPA), 1989).

8.4.2 Cross-Connection Control Programs

Cross-connection control consists of the implementation and enforcement of appropriate ordinances and regulations to eliminate backflows. States and provinces require that utilities (usually called water purveyors in cross-connection regulations) must implement a cross-connection control program. These programs usually require the installation of assemblies that prevent backflows consistent with the level of hazard associated with the user, at the user's expense. They also authorize appropriate officials to inspect the user's plumbing at any reasonable time and to terminate service if a hazard has not been corrected.

8.4.3 Backflow Prevention

Backflow preventers are the assemblies or other means used to "prohibit" backflow. Several types of backflow preventers are used, depending on the situation.

8.4.3.1 Air Gap. An *air gap* is a physical separation between the potable water system and the source of contamination. The air gap between the outlet of the potable system and the maximum level of any source of contamination must be at least twice as large as the diameter of the potable water outlet and never bigger than 1in (25 mm). An air gap is considered to be the safest and simplest means of backflow prevention. However, an air gap results in a loss of any head and therefore is not used when the downstream piping must be pressurized from the water source.

8.4.3.2 Reduced-pressure backflow preventers and double-check valve assemblies. These are two devices used to prevent both backflow and backsiphonage. Both consist of two independently acting, tightly closing, resilient seated check valves in series with test ports. The check valves are usually spring loaded (Figs. 8.1 and 8.2). The difference between the two is that the reduced pressure assembly also contains an independently acting pressure-relief valve between the two check valves and lower than the first check valve (Fig. 8.3). The reduced-pressure backflow preventer can be used in highly hazardous situations.

8.4.3.3 Atmospheric and pressure vacuum breakers, and barometric loops. These devices that can prevent backsiphonage, but not backpressure. In the vacuum breakers, water pressure keeps the valve open, as shown in Fig. 8.4. When water pressure stops (i.e., backsiphonage can occur), the internal float seats and siphonage cannot occur.

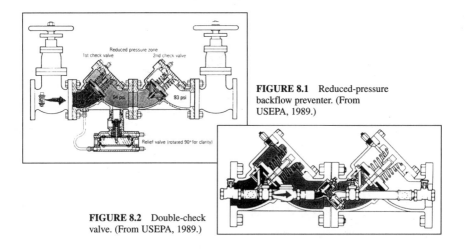

FIGURE 8.1 Reduced-pressure backflow preventer. (From USEPA, 1989.)

FIGURE 8.2 Double-check valve. (From USEPA, 1989.)

FIGURE 8.3 Reduced-pressure backflow assembly application. (Photograph by T. M. Walski.)

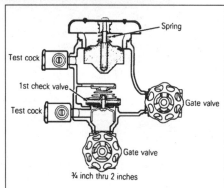

FIGURE 8.4 Vacuum breaker valve. (From USEPA, 1989.)

A pressure vacuum breaker can be tested in line. A *barometric loop* is a continuous piece of pipe that rises at least 35 ft (10.5 m), then returns back to the original level. Water cannot be siphoned over this loop.

8.4.3.4 Single and dual check valves. These types of valves are not recognized as backflow prevention assemblies because they can allow a small amount of leakage and cannot be tested in place. However, they can provide an inexpensive level of protection when the hazard is minimal and an approved backflow preventer is not required.

8.4.4 Application of Backflow Preventers

Although backflow preventers help reduce the risk of contamination, they introduce significant headloss in the piping and additional cost, of installation and required annual testing. These factors are especially troublesome when retrofitting existing services, particularly in cold climates, because the assembly cannot be located outdoors. This is troubling in situations where installation of backflow prevention can compromise fire protection. Duranceau et al. (1998) showed that water quality can seriously deteriorate in fire sprinkler piping.

Hart et al. (1996) performed a risk analysis and concluded that "the risk of death and injury associated with burning dwellings that lack a sprinkler system is higher than the risk of illness associated with unprotected sprinkler systems" and recommended the use of single check valves in residential sprinklers. Wood (1993) highlighted the need for a hydraulic analysis before retrofitting prevention of backflow in an existing fire protection system. Residential sprinklers also can be interconnected with plumbing fixtures to prevent stagnant water. In such a system, whenever someone flushes a toilet or uses a shower, clean water is drawn into the combined sprinkler-domestic system to prevent deterioration of water quality.

8.5 FLUSHING OF DISTRIBUTION SYSTEMS

8.5.1 Background

Distribution system flushing consists of opening up appropriate hydrants or blowoffs to improve water quality and can be done to remove sediment, eliminate low chlorine, solve taste, odor, and turbidity problems, and to remove biofilms from water mains. Flushing can be accomplished as a systematic program or in response to customer complaints (Chaterton et al., 1992; Pattison, 1980). Flushing is usually done by opening fire hydrants.

Flushing has the obvious benefit of bringing water a high concentration of with desinfectants and generally "fresh" water into a portion of the distribution system. Yohe and Gittleman (1986) and Wajon et al. (1988) reported a reduction of taste and odor problems as a result of flushing. Wakeman et al. (1980) and Larson et al. (1983) reported reduced levels of tetrachloroethylene (PCE) after flushing. Shindala and Chisolm (1970) and Lakshman (1981) listed numerous benefits of flushing.

8.5.2 Flushing Procedures

AWWA (1986b), Chadderton et al. (1992, 1993) and California-Nevada Section AWWA (1981) presented a summary of the mechanics of a flushing program. Good planning and public notification, if required, represent the start of successful flushing. In particular, hospitals and laundries should be warned about impending flushing. Crews should have proper safety equipment, including lights and reflective gear if work is done at night. Hydrants should be flushed from the clean water source outward. A rule of thumb used by crews is "always flush with clean water behind you." In general, a large main should not be flushed from a smaller main, and valves and hydrants should be opened and closed slowly to prevent waterhammer. Care must be taken not to reduce pressure below 130 kPa (20 psi) at nearby customers and high local elevations. Flow should be directed in a way that minimizes damage and disruption of traffic. Diffuser outlets (Fig. 8.5) should be used where needed. Flushing should not be conducted when the temperature is likely to drop below freezing before the water can run off roads and sidewalks. Flushing should be conducted at times when water is plentiful and reservoirs are full rather than during drought periods. Flushing is needed most when water consumption is relatively low.

Hydrants should be opened to generate a velocity suitable for scouring solids from pipes. The velocity should be at least 2 ft/s (0.61 m/s) to suspend sediment and should be no more than 10 ft/s (3.1 m/s) to minimize the potential of waterhammer in startup and shutdown. Walski and Draus (1996) presented data showing that most of the scouring is

FIGURE 8.5 Hydrant diffuser. (Photograph by T. M. Walski)

accomplished within the first seconds after maximum velocity is reached and that velocity can then be decreased. Flushing should continue until the water clears up or disinfectant residuals increase.

Flushing also provides maintenance personnel the opportunity to measure static pressure and discharge from the hydrant and to assess the general condition of the hydrants. The amount of water used during flushing should be recorded so that an overall estimate of this nonmetered use can be made. Crews should maintain a record of flushing.

Walski (1991) and Walski and Draus (1996) discussed the importance of modeling the effects of flushing and showed that models can be used to estimate the time it would take for chlorine residual to increase during flushing.

8.5.3 Directional Flushing

Flushing can be enhanced by closing valves to maximize velocity in pipes being flushed. Oberoi (1994) showed that this type of directional flushing is desirable despite the additional labor because it uses less water, creates higher velocity, and provides the opportunity to test distribution valves. By restricting the direction in which water can flow to the open hydrant, it is possible to maximize the velocity in the mains and control the individual main being flushed. Figure 8.6 shows how the velocity to a hydrant can be increased by valving off the direction the water can take in reaching the flowed hydrant.

8.5.4 Alternating of Disinfectants

Some utilities find that switching disinfectants (e.g., from chloramines to free chlorine) during flushing results in more effective cleaning of mains. Such a change in disinfectants, coupled with the high velocities of flow during flushing, can be especially effective in attacking biofilms that have become adapted to a given disinfectant. Public notification is especially important when switching disinfectants because this can have an impact on such items as tropical fish and dialysis machines.

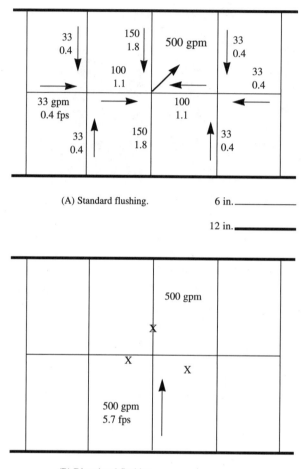

FIGURE 8.6 Effect of valving on directional flushing.

REFERENCES

Angele, G.J., Cross *Connection and Backflow Prevention,* American Water Works Association, (AWWA), Denver, CO, 1974.

APHA, 1995, *Standard Methods for Examination of Water and Wastewater,* American Water Works Association, (AWWA), Denver, CO.

AWWA, 1986a, *Principles and Practices of Water Supply Operation, Vol. 3, Water Distribution,* Denver, CO.

AWWA, 1986b, *Maintenance of Water Distribution System Water Quality,* American Water Works Association, Denver, CO.

AWWA, *Disinfecting Water Mains,* AWWA C651-92, Denver, CO, 1992a.

AWWA, *Disinfection of Water-Storage Facilities,* AWWA C652-92, Denver, CO. 1992b.

AWWA, 1990, *Recommended Practice for Backflow Prevention and Cross-Connection Control,* AWWA Manual M-14, Denver, CO.

California-Nevada AWWA, *Distribution Main Flushing and Cleaning,* California-Nevada Section AWWA, 1981.

Chadderton, R. A., G. L. Christensen, and P. Henry-Unrath, *Implementation and Optimization of Distribution Flushing Programs,* AWWA RF, Denver, CO, 1992.

Chadderton, R. A., G. L. Christensen, and P. Henry-Unrath, *Planning a Distribution System Flushing Program, Journal of the American Water Works Association,* 85:(7)89., 1993.

Departments of Air Force, Army and Navy, 1984, *Maintenance and Operation of Water Supply, Treatment and Distribution Systems,* TM 5-660, Washington, DC, 1994.

Duranceau, S. J., J. V. Foster and J. Poole, *Impact of Wet-Pipe Fire Sprinkler Systems on Drinking Water Quality,* AWWA RF, Denver, CO, 1998.

Foundation for Cross-Connection Control and Hydraulic Research, *Manual of Cross Connection Control,* University of Southern California, Los Angeles, 1988.

Great Lakes and Upper Mississippi River Board of State Public Health and Environmental Managers, "Recommended standards for water works" Albany, NY. 1992.

Hart, F. L., C. Nardini, R. Till and D. Bisson, Backflow Protection and Residential Fire Sprinklers, *Journal of the American Water Works Association,* 88:(10)60, 1996.

Lakshman, B. T., 1981, "Water Main Flushing," *Water Engineering and Management,* 128:(1)28, 1981.

Larson, C. D., O. T. Love, and G. Reynolds, "Terachloroethylene Leached from Lined Asbestos-Cement Pipe Into Drinking Water," *Journal of the American Water Works Association,* 75:(4)184, 1983.

Oberoi, K., "Distribution Flushing Programs: The Benefits and Results," *AWWA Annual Conference,* New York, 1994.

Patison, P. L., "Conducting a Regular Main Flushing Program," *Journal of the American Water Works Association,* 72:(2)88, 1980.

Shindala, A., and C. H. Chisolm, "Water Quality Changes in Distribution Systems," *Water and Wastes Engineering,* 62:(1)35, 1970.

USEPA, *Cross Connection Control Manual,* EPA 570/9-89-007., 1989.

Wajon, J. E., B. B. Kavanach, R. I. Kagi, R. S. Rosich and R. Alexander, "Controlling Swampy Odors in Drinking Water," *Journal of American Water Works Association* 80:(6)77, 1988.

Wakeman, S. G., et al., "Tetrachloroethylene Contamination of Drinking Water by Vinyl-Coated Asbestos Cement Pipe," *Bulletin of Environmental Contamination and Toxicology,* 25:(4)639, 1980.

Walski, T.M., "Understanding Solids Transport in Water Distribution Systems," *In Water Quality Modeling in Distribution Systems,* AWWA RF, 1991.

Walski, T. M. and S. J. Draus, "Predicting Water Quality Changes during Flushing," AWWA Annual Convention, Toronto, Ont., 1996.

Wood, T. R., *A Study of Backflow Prevention and Fire Sprinkler Systems,* National Fire Academy., 1993.

Yohe, T. L. and T. S. Gittleman, "Tastes and Odors in Distribution Systems," in *Water Quality Concerns in Distribution Systems,* AWWA, Denver, CO, 1986.

CHAPTER 9
WATER QUALITY

Walter M. Grayman
Consulting Engineer
Cincinnati, OH

Lewis A. Rossman
U.S. Environmental Protection Agency
National Risk Management Research Laboratory
Cincinnati, OH

Edwin E. Geldreich
Consulting Microbiologist
Cincinnati, OH

9.1 INTRODUCTION

The goal of a drinking water distribution system is to deliver sufficient quantities of water where and when it is needed at an acceptable level of quality. Although water quality may be acceptable when water leaves a treatment plant, transformations can occur as water travels through a distribution system. In the past, distribution systems were designed and operated mainly on the basis of hydraulic reliability and economics, with little attention paid to water-quality concerns except when serious problems arose. This attitude is changing as more water suppliers realize the important influence that time spent in a distribution system can have on water quality. This chapter reviews the most common processes affecting water quality impairment in distribution systems, methods for monitoring water-quality conditions and techniques for modeling water-quality transport and transformations.

9.1.1 Overview

The pipes and storage facilities of a distribution system constitute a complex network of uncontrolled chemical and biological reactors that can produce significant variations in water quality in both space and time. Factors leading to water quality deterioration in distribution systems include the following:

- supply sources going on- and off-line,
- contamination via cross-connections or from leaky pipe joints,
- corrosion of iron pipes and dissolution of lead and copper from pipe walls,
- loss of disinfectant residual in storage facilities with long residence times,
- reactions of disinfectants with organic and inorganic compounds resulting in taste and odor problems,
- bacterial regrowth and harboring of opportunistic pathogens,
- increased turbidity caused by particulate resuspension, and
- formation of disinfection by-products, some of which are suspected carcinogens.

The principal factors affecting water quality in a distribution system are the quality of the treated water fed to the system; the material and condition of the pipes, valves, and storage facilities that make up the system; and the amount of time that water is kept in the system. Regarding this last item, it is important to understand that in a looped pipe network, the water reaching any particular consumer is actually a blend of water parcels that may originate from different sources at different points in time and follow different flow paths. This fact can have enormous influence when trying to understand the relation between residence time and water quality. Actions that can be taken to improve water quality or prevent its deterioration in the distribution system include changes in treatment practices, pipe repair, relining or replacement, and modifications to system operation, such as keeping less water in storage. Finding an optimal combination of these actions can involve trade-offs between cost, hydraulic reliability, and risk of impaired water quality.

9.1.2 Definitions

Many of the frequently used terms pertaining to water quality issues in distribution systems are defined below.

Advective transport. The movement of water quality constituents at the same mean velocity and direction as the bulk carrier fluid.

Aerobic zone. An area where dissolved oxygen is present.

Alkalinity. Ability of water to neutralize changes in pH.

Anaerobic zone. An area devoid of dissolved oxygen.

Biofilm. A consortium of microorganisms attached to a solid surface along with a surrounding slimy matrix of extracellular organic polymers.

Bulk reactions. Reactions that take place within the volume of water not in contact with the wall of a pipe.

Coliform bacteria. A group of bacteria associated with the intestinal tract of warm-blooded animals as well as vegetable matter and soil, the presence of which is considered to be an indirect indication of possible fecal contamination.

Disinfection by-products. Products of the reaction between a water disinfectant (such as chlorine or ozone) and naturally occurring organic matter in water, some of which are suspected human carcinogens.

Dispersive transport. Movement of a water-quality constituent caused by concentration gradients.

Eulerian approach. A modeling framework that assumes a fixed frame of reference when developing conservation of mass, momentum, and energy flow for a fluid control volume.

First-order reaction. A chemical reaction where the rate of growth or decay of a constituent is proportional to its concentration.

Heterotrophic bacteria. The class of single-cell microorganisms that require organic carbon for both respiration and cell synthesis.

Lagrangian approach. A modeling framework that assumes a moving frame of reference when developing conservation of mass, momentum, and energy flow for a fluid control volume.

Opportunistic pathogens. Organisms that may exist as part of the normal body microflora but under certain conditions cause disease in compromised hosts, such as in the elderly, newborns, victims of acquired immunodeficiency syndrome (AIDS), and cancer patients receiving chemotherapy.

Pipe wall reactions. Reactions occurring between water-quality constituents and materials originating from the wall of a pipe, such as released iron or biofilm slime.

Standpipe. A ground-level storage tank whose height is greater than its width.

Tracer chemical. A nonreactive chemical whose presence is used to track the flow path and travel time of water originating from a particular point of addition.

Topological sort. A renumbering of the nodes in a directed node-link network so that all links directed into a given node are connected to nodes of lower number.

Tubercle. An encrustation growing inward from the wall of a pipe caused by the buildup over time of oxidized corrosion products.

Water age. The average amount of time that a parcel of water has been in the distribution system.

9.2 WATER-QUALITY PROCESSES

As water moves through a distribution system, transformations and deterioration in water quality can occur in the bulk water phase and through interaction with the pipe wall. These transformations may be physical, chemical, or microbiological in nature. Figure 9.1 is a conceptual representation of the transformation processes in the distribution system. A schematic of the physical, chemical, and microbiological transformations at the pipe wall is presented in Fig. 9.2. These processes are described in greater detail below.

9.2.1 Loss of Disinfectant Residual

Disinfection is the process of using chemical or physical means to inactivate harmful microorganisms that might be present in water and to protect distributed water from pathogen regrowth or recontamination. The majority of surface water supplies maintain some level of residual chemical disinfectant throughout the distribution system. In the United States, the Surface Water Treatment Rule requires that detectable residuals be maintained for all surface water systems as well as those groundwater systems deemed to be under the influence of surface water. Loss of disinfectant residual can weaken the barrier against microbial contamination resulting from line breaks, cross-connections, or

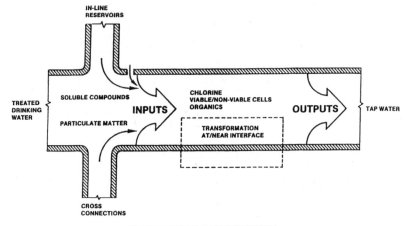

FIGURE 9.1 Water-quality transformations in the distribution system.

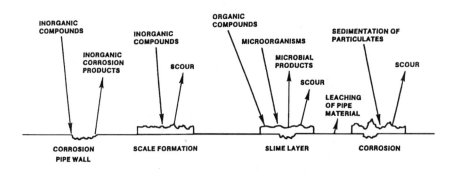

FIGURE 9.2 Water-quality transformations at the pipe wall.

other unforeseen occurrences and can encourage the growth of nuisance and pathogen harboring biofilms.

9.2.1.1 Disinfection methods. Primary disinfection is the treatment step used to actually destroy or inactivate pathogens at the treatment works. It can be accomplished using a variety of chemicals, the most popular being chlorine, chloramines (chlorine combined with ammonia), ozone, and chlorine dioxide. *Secondary or postdisinfection* refers to the practice of adding additional disinfectant to treated water before it is released to the distribution system to maintain a residual disinfecting capability. By their very nature, disinfectant chemicals are extremely reactive and do not persist for long periods. If chlorine or chloramines is used as a primary disinfectant, secondary disinfection may not be necessary. Secondary disinfection will always be needed when using ozone. Chlorine and chloramines are the most commonly used chemicals for maintaining a disinfectant residual in the distribution system. *Booster disinfection* is the practice of adding disinfectant at points in the distribution system beyond the point of raw water treatment. It is typically used at the entrance point to a service area being supplied by a wholesale provider or at distant points in a system subject to large disinfectant demands. Chlorine is the most common chemical used for booster disinfection.

In the United States, CT (residual concentration multiplied by contact time) requirements stipulate what product of disinfectant concentration and contact time is needed to assure a specific percentage kill of target organisms. It governs the doses used for primary disinfection. For secondary disinfection, the Surface Water Treatment Rule requires that at least a 0.2 mg/L disinfectant residual must be present in water entering the distribution system and that a "detectable" residual must be maintained throughout the system. In reality, most utilities find that they have to provide much larger residuals leaving the plant, particularly during the warm weather months, to maintain detectable residuals at the farthest reaches of their systems. On the other hand there are practical limits on the maximum levels of disinfectant residual that can be released into a distribution system. Excessive levels of chlorine will produce taste and odor problems, might accelerate pipe corrosion, might enhance formation of harmful by-products, and has health concerns of its own. The United States Environmental Protection Agency (USEPA) has recently established a maximum limit on free chlorine residual of 4 mg/L leaving the plant.

9.2.1.2 Rates of disinfectant loss. Most waters exhibit a rapid immediate consumption of chlorine when the chemical is first added during primary disinfection. Losses of 50 percent or more over a contact time of several hours are not uncommon. After the contact time used for primary disinfection is completed, rates of loss of chlorine are significantly lower as it reacts with the more recalcitrant organic components. Typical half-lives range from several days to several weeks, depending on water temperature and the amount and reactivity of the organic carbon in the water. These rates refer to losses occurring in the bulk solution. They can be quantified by performing a laboratory bottle test (see Sec. 9.4.4.3).

An additional loss can occur as chlorine reacts with materials on or near the pipe wall, such as iron released because of corrosion or the organic slime associated with a biofilm. The rate of this reaction can be much higher than that caused by bulk reactions, particularly in older, unlined cast iron or steel pipes. Chlorine half-lives might be reduced to only several hours in pipes with large wall demand. Unfortunately, there is no simple way to measure this wall demand, and it might vary considerably throughout a system with a mixture of pipes of different ages and materials. The only in situ measurements made of wall demand have been performed by measuring chlorine at the entrance and exit points of a long length of pipe with no inflow points in between.

Chloramines exhibit a much slower rate of decay than does free chlorine. They also have the advantage of producing much smaller amounts of chlorinated by-products. One disadvantage is that under certain pH conditions and chlorine-to-ammonia ratios, the decomposition of chloramines can produce nitrogen, which in turn can stimulate the development of nitrification in the distribution system. To date there have been no reported measurements of the contribution that wall reactions might make to rates of chloramine loss.

9.2.1.3 Mitigation of disinfectant loss. Utilities have several strategies available to correct problems with excessive loss of chlorine in their distribution systems. First, they can switch to a more stable secondary disinfecting chemical, such as chloramines.

Second, they can undertake a program of pipe replacement, flushing, and relining. Third, they can consider making operational changes to reduce the time that water spends in the system, such as keeping less water in storage, exercising tanks more often, valving off pipes, or looping dead-end sections. Fourth, they can use booster disinfection. And fifth, they might consider changes in treatment to reduce total levels of organic carbon in their water. The optimal choice of mitigation strategy will obviously be highly site-specific and might involve a mixture of these alternatives.

9.2.2 Growth of Disinfection By-products

Chlorine can react with naturally occurring organic matter in treated water, usually of humic origin, to produce literally hundreds of halogenated by-products, most of which are present in only trace quantities. Some of these by-products, such as chloroform and dichloroacetic acid, have been shown to be potential human carcinogens. The by-products produced in greatest quantities as a result of chlorination are the trihalomethanes (THMs). These are methane (CH_4) molecules in which three of the hydrogen atoms have been replaced by some combination of chlorine and bromine atoms. In the United States, community water supplies serving more than 10,000 people have been required since 1979 to maintain THM levels below 100 µg/L in the distribution system. Just recently, this number was lowered to 80 µg/L and was made applicable to all community water systems.

The amount of THMs produced as a result of chlorination has been found to depend on pH, temperature, chlorine dose, amount of organic precursor, bromide concentration, and reaction time. Although production rates are highest after chlorine is first applied at the treatment works, THMs will continue to grow out in the distribution system as long as a chlorine residual and reactable precursor material remain. It is not unusual to see 50 percent or more of the total THMs produced to occur in the distribution system.

THMs constitute only 20 to 40 percent of the total organic halides produced by chlorination. Other classes of by-products of concern include haloacetic acids, haloacetonitriles, cyanogen halides, halopicrins, and chloral hydrate. The use of chloramines as a disinfectant results in greatly reduced levels of by-products. Chlorine dioxide is another possible primary and secondary disinfectant. Its principal by-products of concern are the inorganic species chlorite and chlorate. Ozone, which would be used only as a primary disinfectant, can produce aldehydes, a variety of organic acids, bromoform, and bromate. The aldehydes and acids are readily biodegradable and can serve to enhance biofilm growth in the distribution system. Ozonation also can enhance the production of some chlorinated by-products, such as chloral hydrate and chloropicrin, when free chlorine is used as a secondary disinfectant.

9.2.3 Internal Corrosion

Internal corrosion is the deterioration of the inside wall or wall lining of a pipe caused by reactions with water. Deterioration can be the result of physical actions that erode the lining or surface coating of a pipe, chemical dissolution that leaches a pipe's lining or wall material, or electrochemical reactions that remove metal from the wall of the pipe. Corrosion can result in the leaching of the toxic metals lead and copper, impart a metallic taste to water, cause staining of plumbing fixtures, help harbor nuisance and pathogenic microorganisms, reduce a pipe's hydraulic carrying capacity, and ultimately result in leaks and clogs. Table 9.1 summarizes the corrosive properties of materials frequently used in water distribution systems.

TABLE 9.1 Corrosion Properties of Different Materials used in Distribution Systems

Distribution Material	Corrosion Resistance	Potential Contaminants
Copper	Good overall corrosion resistance; subject to corrosive attack from high velocities, soft water, chlorine, dissolved oxygen, and low pH	Copper
Lead	Corrodes in soft water with low pH	Lead, arsenic, and cadmium
Mild steel	Subject to uniform corrosion; affected primarily by high dissolved oxygen levels	Iron, resulting in turbidity and red-water complaints
Cast or ductile iron (unlined)	Can be subject to surface erosion by aggressive waters	Iron, resulting in turbidity and red-water complaints
Galvanized iron	Subject to galvanic corrosion of zinc by aggressive waters	Zinc and iron
Asbestos-cement	Good corrosion resistance; aggressive waters can leach calcium from cement	Asbestos fibers
Plastic	Resistant to corrosion	

Source: From Singley et al. (1984).

9.2.3.1 Types of corrosion. The major forms of corrosion are the following:

Uniform. Uniform corrosion occurs when there is a more or less equal amount of material loss and deposition of corrosion products across the surface of the pipe.

Pitting. Pitting occurs when a localized, nonuniform corrosion process forms holes in the pipe wall and can eventually lead to pipe failure.

Tuberculation. Tubercles are knoblike mounds that form in areas around pits where oxidation products of iron build up over time. Excessive tuberculation can significantly decrease a pipe's diameter, contribute to its roughness, and promote the growth of biofilm within its pore like structure.

Biological. Biological corrosion results from interactions between the pipe material and microorganisms attached to the pipe wall. Local changes in pH and dissolved oxygen in the vicinity of biofilm growth can enhance corrosive activity.

9.2.3.2 Factors affecting corrosion. Numerous physical, chemical, and biological factors can affect the occurrence and rate of corrosion. Some individual factors can either promote or inhibit corrosion, depending on other conditions. The most significant factors include the following

Flow velocity. High velocities increase the rate of mass transfer of materials to the pipe wall. This can enhance corrosion by increasing the rate of dissolved oxygen transfer or inhibit it by increasing the rate at which a protective scale can form. Stagnant conditions help promote tuberculation and pitting in iron pipes.

Temperature. In general, as with all chemical reactions, the rate of corrosion increases with increasing temperature. However, at higher temperatures, calcium carbonate has a greater tendency to precipitate and form a protective layer along the pipe wall.

pH. pH is inversely proportional to the concentration of hydrogen ions in water. Because these ions act as the electron acceptor in the corrosion reaction, lower pH helps promote corrosion. At pH values below 5, both copper and iron corrode rapidly and uniformly. Above pH 9, both metals usually are protected from corrosion (Singley et al., 1984). Between these pH values, pitting can occur if no protective film is laid down on the surface of the pipe wall. The formation and dissolution of such films are also affected by pH.

Alkalinity. Alkalinity is composed mostly of carbonate and bicarbonate ions. It provides water with buffering capacity to neutralize changes in pH. It also helps lay down a protective coating of metallic carbonate on the pipe wall and can help prevent the dissolution of calcium from cement-lined pipe or from asbestos-cement pipe.

Dissolved oxygen. Oxygen is considered to be the primary agent in electrochemical corrosion reactions. It is the substance that accepts the electrons given up by the metal at the anode of a corrosion cell as it corrodes. Oxygen also reacts with hydrogen released at the cathode to form water and prevent excess hydrogen buildup, which would otherwise retard further reaction. Oxygen will react with soluble ferrous metal produced by corrosion to form insoluble ferric hydroxide. This form of iron leads to the formation of tubercles and causes "red water" conditions.

Total dissolved solids. A high total of dissolved solids indicates a high ion concentration in water, which in turn increases conductivity, leading to higher flow of electrons and therefore helping to promote electrochemical corrosion.

Hardness. Hardness results from the presence of calcium and magnesium ions. Hard water is generally less corrosive than soft water because a protective lining of calcium carbonate is more likely to form on the pipe wall.

Bacteria. Bacteria present in biofilm attached to pipe walls can create local changes in pH and dissolved oxygen that help promote electrochemical corrosion. Iron-oxidizing bacteria can produce insoluble ferric hydroxide, which contributes to tuberculation. Sulfate-reducing bacteria produce hydrogen sulfide, which can be oxidized to produce sulfuric acid, which lowers pH and enhances corrosion.

9.2.3.3 Indicators of corrosion. Several efforts have been made to develop water-quality indexes that predict whether water will be corrosive. Most indexes are based on estimating the tendency for calcium carbonate to precipitate from solution and lay down a protective layer along the pipe wall, thus retarding corrosion. They usually consist of a simple formula involving water's pH, alkalinity, and calcium concentration. The more common indexes include the Langelier Saturation Index (LSI), the Aggressive Index (AI), the Ryznar Index (RI), and the Driving Force Index (DFI) (Singley, 1981). More direct means of testing for corrosion involve the use of pipe sections and coupons of new pipe or pipe taken from the field. Standard procedures are available for testing pipe coupons using visual observation, electron microscopy, weight-loss measurement, analysis of pipe scale or coatings using such techniques as x-ray diffraction, and the electrode method (Kirmeyer and Logsdon, 1983). The latter method uses electrodes immersed in water to measure voltage changes induced by the electrochemical corrosion (DeBerry et al., 1982).

9.2.3.4 Control of corrosion. Three general approaches are available for controlling corrosion: modify the water quality to make the water less corrosive, lay down a protective lining between the water and the pipe, or switch to pipe materials that are less prone to corrosion. Adjusting the pH is the most common form of corrosion control because it directly affects both electrochemical corrosion and the solubility of calcium carbonate, which can serve as a protective film. It also directly affects the dissolution of calcium from asbestos/cement pipes. Also available are a number of chemical inhibitors that help to develop a protective layer between the pipe wall and the water. Lime and soda ash can be used to promote deposition of a calcium carbonate film along the pipe. Inorganic phosphates and sodium silicates also lay down protective films through processes not yet completely understood. The optimal dosages and effectiveness of these inhibitors are system-specific and must be determined by field or pilot scale testing.

9.2.4 Biofilms

Biofilms play a major part in the microbial characterization of drinking water quality in distribution. These films or slimes become established in sediments, corrosion tubercles, static areas of slow water flow, dead-ends, standpipes, and storage tanks. Beyond the utility distribution system is the building- or home-plumbing network and associated attachment devices that also may harbor biofilm if not carefully suppressed.

9.2.4.1 Origins. Within the distribution system environment, there are many safe havens for biofilm development. Foremost among these sites are areas of corrosion and slow flow, where sediments accumulate and water movement is not sufficient to disrupt the development of colonization. It is in these sites within iron pipe where biologically mediated corrosion begins to develop. Soon, the area of corrosion expands as a result of microbial attack and is abetted by aggressive waters. At this point, corrosion begins to form tubercles at the site making the deposition better suited for a protected microbial habitat. Although cement pipe and polyvinyl chloride (PVC) pipe materials are not subject to this type of corrosion, the characteristics of water may change the surface structure of asbestos/cement mains, and biological activity creates pitting on the smooth inner surface of plastic pipe materials and splaying of cement. It is important to note that not all pipe sections will show evidence of deterioration even after years of service, the reason being the nature of the chemistry of water in the system and the continuous movement of water in high-demand areas (Geldreich and LeChevallier, in press). Water supply reservoirs and standpipes also may accumulate deposits of biofilm on side walls and in sediments on the bottom of the tank. These deposits are a particular nuisance in stratified static waters because of anaerobic growth of various organisms that cause taste and odors.

For organisms than can endure this harsh aquatic environment, growth proceeds slowly at first as the organisms adjust to the constraints of the pipe environment. In time, these microcolonies attract other organisms (such as *Legionella*), the more exacting nutritional needs of which may be found only in the by-products released in the metabolism of various organisms in the pioneering microbial community. Thus, a progressive diversity is brought into the biofilm as the site becomes populated with a variety of bacteria, protozoans, nematodes, and worms, as Geldreich (1996) observed.

9.2.4.2 Composition. Biofilms provide a variety of microenvironments for growth, including aerobic and anaerobic zones caused by limited oxygen diffusion (Characklis and Marshall, 1990). Within this complex structure is a community of diverse organisms embedded in a matrix of extracellular organic polymers adhering to moist surfaces. This matrix is interlaced with water channels that have been reported to constitute as much as 40 to 60 percent of the total volume of biofilm in a conceptual model proposed at the Center for Biofilm Engineering (1994).

9.2.4.3 Significance. When a water supply enters the distribution system, there will be changes in its microbial quality. The most dramatic changes are the result of contamination introduced by treatment failures, line breaks, illegal tap-ins, and less obvious cross-connections. Many of the more subtle day-by-day changes come from release of biofilm organisms into the bulk flow of water in the pipe network and its storage sites in the water distribution network.

Most of these water-quality changes occur in the population of heterotrophic bacteria. Of the wide spectrum of organisms that survive water treatment or are introduced to the distribution system from intrusions of contamination, some organisms deserve special attention. These organisms can interfere with detection of coliform, bacteria, create taste and odors, or become a health concern because some of them are known to be opportunistic pathogens.

With the growing awareness that some nonfecal coliform strains can colonize the distribution system, there is an urgent need to evaluate carefully the significance of total coliforms in a biofilm episode for public health risk. Although these occurrences of biofilm coliform may not be an immediate public health concern, they should not be ignored because the contamination suggests (1) existence of a habitat that could be used by pathogens, (2) possible leaks in the treatment barrier or distribution system, and (3) the accumulation of chlorine-demanding material that interferes with maintaining a disinfectant residual throughout the pipe network.

There is always a concern that biofilm occurrences could be hiding fecal contamination either from inadequate treatment or contamination of the distribution system. For this reason, any utility experiencing coliform biofilm in the system should intensify its monitoring program and search for evidence of fecal coliforms or *Escherichia coli* among the positive samples. If so, verified occurrence of these coliforms should call for a "boil water" order until repeat samples prove their disappearance. Any occurrence of fecal coliform bacteria or *E. coli* during a biofilm episode should not be brushed aside as an aberration in the laboratory results. These occurrences suggest very recent contamination because these organisms are not normally able to colonize biofilms permanently.

9.2.4.4 Treatment and control. Attempts to control biofilm in the distribution system have taken several directions. Perhaps the most drastic measures have involved the replacement of old pipe sections that did not respond to elevated levels of disinfectant and flushing. In *Halifax*, Nova Scotia, a 5-km section of new pipe was lined with cement to seal off *Klebsiella* colonization of wood forms left in the construction of a new pipe section (Martin et al., 1982). Avoiding the incipient releases of biofilm can be achieved through manipulation of water chemistry (pH, alkalinity, Langlier index) or the careful application of corrosion inhibitors not only to protect pipe materials but also to firm up the coating of sediments that harbor microbial communities.

Most often, the problem appears to be a reflection of poor system-flushing efforts to remove much of the accumulating pipe and storage tank sediments systematically. Increasing the concentration of free chlorine has not always been successful. LeChevallier et al. (1990) showed that disinfection of organisms on iron pipes was ineffective even when the organisms were exposed to 5 mg/L of free chlorine for several weeks. However, using 1 mg/L of free chlorine or monochloramine was effective for disinfection of biofilm on galvanized, copper or PVC. Apparently, a combination of factors involving corrosion rate, ratio of the molar concentration of chlorine and sulfate to bicarbonate (Larson index), selection of a chloramine residual, and the level of corrosion inhibitor applied could account for 75 percent of the variation in rates of biofilm disinfection for organisms grown on iron pipes, according to LeChevallier et al. (1993).

9.3 WATER-QUALITY MONITORING

Monitoring of distribution systems is a mechanism for identifying variations in water quality over time and space. The resulting database also can be used to understand the transformations that are occurring in the distribution system and in the calibration of mathematical models of the system. Monitoring can be classified as routine sampling and special studies. Routine monitoring is used to satisfy regulatory requirements and as an ongoing tool to assess the water quality throughout the distribution system. Special studies are generally more intensive efforts designed and implemented to answer specific water quality-questions about the distribution system.

9.3.1 Routine Monitoring

Routine monitoring must be designed, at a minimum, to satisfy regulatory requirements. More extensive monitoring can be performed to gather additional data on water quality in the system.

9.3.1.1 Regulatory requirements. In the United States, baseline regulatory requirements are set by the USEPA and are enforced by that agency or by the individual states that have been granted primacy by the agency. States may institute more stringent requirements if they deem them to be necessary. Although regulations historically have applied to the point at which water enters the distribution system, there is a trend toward the development of regulations at the point of use. Current regulations are summarized in Table 9.2 (Kirmeyer et al., 1999). In Europe, the European Community (EC) Directive on the Quality of Water for Human Consumption sets out a broad framework for drinking water quality throughout Europe. However, interpretation and implementation of the directive can be different from country to country.

9.3.1.2 Sampling methods. Samples can be collected and analyzed in two ways: grab samples and continuous samples. *Grab samples* are collected manually in the field and analyzed in the field or in the laboratory. *Continuous monitoring* is conducted by means of sensors and remote recording stations. Grab samples are more labor intensive and provide data only at the time of collection. Continuous monitoring requires a capital expense for the monitoring equipment and for equipment maintenance time, but it provides a continuous time-series profile of changes in water quality. A comprehensive monitoring plan can make use of both grab and continuous methods.

9.3.1.3 Sampling parameters. A routine monitoring program should be established to meet regulatory requirements and to collect additional water-quality information that is useful in the operation of a water system. Table 9.3 lists parameters that are frequently measured in a routine monitoring program (Kirmeyer et al., 1999).

9.3.2 Synoptic Monitoring

Synoptic monitoring studies are performed to address specific water-quality issues in distribution systems. Examples of special studies include the following:

- Measurement of disinfectant residuals throughout the system to determine residual loss in the distribution system and areas of low residual.

TABLE 9.2 U.S. Regulatory Limits for Finished Water Quality

Parameter	Sample Location	Regulatory Limit	Reference	Comments
Disinfectant residual	Entry point to distribution system	0.2 mg/L on a continuous basis	SWTR	Only applies to systems using surface water
Disinfectant residual or HPC bacteria count	Throughout distribution system	Detectable level of disinfectant residual or HPC bacteria count of 500 or less CFU per mL in 95% of samples collected each month for any 2 consecutive months	SWTR	Only applies to systems using surface water
Nitrite	Entry point to distribution system	1.0 mg/L as N	NPDWR (CFR 141.23)	Maximum contaminant level
Nitrate	Entry point to distribution system	10.0 mg/L as N	NPDWR (CFR 141.23)	Maximum contaminant level
TTHMs	Throughout distribution system	100 µg/L, running annual average based on quarterly samples	Total trihalomethane rule	Systems serving >10,000 people
Total coliform bacteria	Throughout distribution system	0 CFU in 95% of samples	TCR	Number of samples determined by population served
Lead	Customers' taps	0.015 mg/L	LCR	Action level at 90th percentile
Copper	Customers' taps	1.3 mg/L	LCR	Action level at 90th percentile
pH	Throughout distribution system	7.0 minimum pH units	LCR	Unless state determines otherwise

ABBREVIATIONS: SWTR: Surface Water Treatment Rule; NPDWR: National Primary Drinking Water Regulations; CFR: Code of Federal Regulations; TCR: Total Coliform Rule; LCR: Lead and Copper Rule; CFU: Colony forming units; HPC: Heterotrophic plate count; N: Nitrogen; TTHM: Total trihalomethanes

- Tracer studies to identify areas served by different sources and to assist in the calibration of a hydraulic network model.
- Identification of entry of contaminants into the distribution system.

Special studies are usually short-term, intensive activities aimed at developing a temporal and spatial understanding of the water quality in all or part of the distribution system. There are three phases in a sampling study: preparation, sample collection, and

TABLE 9.3 Monitoring of Water Quality Parameters

Parameter	Purpose	Sampling Procedure Used
Alkalinity	Indicates the potential buffering capacity against changes in pH	On-line ion-selective electrode or grab sample and laboratory analysis
Aluminum	Indicates potential coagulant overfeeding	On-line ion-selective electrode or grab sample and laboratory analysis
Ammonia, total and/or free	Indicates potential for nitrification	On-line ion-selective electrode or grab sample and laboratory analysis
Chlorine residual, total and/or free	Indicates protection from bacteria formation and provides early warning sign of water-quality deterioration; monitored at inlet and outlet to control rechlorination when practiced	On-line colorometric analyzer or grab sample and amperometric titration laboratory analysis
Coliform, total and/or fecal	Indicates presence of bacteria; provides early warning sign of water-quality deterioration	Grab sample and laboratory analysis
Conductivity, specific	Can quickly indicate relative changes in total dissolved solids e.g., alkalinity	On-line ion selective electrode or grab sample and laboratory analysis
Disinfection by-products	Represents potential for on-going chemical reactions and DBP formation	Grab sample and laboratory analysis
Heterotrophic bacteria	Indicates conformance to MCL; provides early warning sign of water-quality deterioration	Grab sample and laboratory analysis
Iron	Indicates potential corrosion reactions	On-line ion-selective electrode or grab sample and laboratory analysis
Nitrate	Indicates possibility of nitrification	On-line ion-selective electrode or grab sample and laboratory analysis
Nitrite	Indicates possibility of nitrification	On-line ion-selective electrode or grab sample and laboratory analysis
pH	Indicates changes from the water source. Indication of corrosion of concrete or an unlined new facility.	On-line ion-selective electrode or grab sample and laboratory analysis
Taste and odor	Evidence of water quality problem in progress	Grab sample and laboratory analysis
Temperature	Differences within storage facility indicate possible stratification and stagnant zones. Early warning sign of potential microbial problems.	On-line sensor
Turbidity	Provides early warning sign of water-quality deterioration	On-line turbidimeter sensor and analyzer

post-assessment of the data (Clark and Grayman, 1998). During the preparation stage, a detailed sampling plan should be developed. Frequently, a network model is applied to test the sampling locations and frequency under the operating conditions expected during the sampling study. The following issues should be considered in preparation of the sampling plan:

Sampling locations. Sampling locations should be selected to reflect both spatial diversity and locations of significant information. Easily accessible sites, such as dedicated sampling taps and hydrants, are preferable.

Sampling frequency. Temporal variability, flow velocities, and cost and availability of resources are all factors affecting the sampling frequency. Frequently, a crew will circulate among several sampling sites, taking samples at a prescribed frequency. The time required to sample at a single station and over the entire circuit can be estimated by a preliminary test of the circuit or by the times associated with each aspect of the task, such as flushing time, travel time, and so on.

System operation. The operation of a water system can have a significant impact on the movement of water through the system. During a sampling study, system operating conditions should be controlled (operated) to reflect normal operating conditions or a desired set of operations.

Preparation of sampling sites. Prior to actual sampling, the site should be prepared adequately. Preparation may include testing of hydrants, calculation of required flushing time, notification of owners, and marking the sites for easy identification.

Sample collection procedures. Procedures for collecting samples should be established, including required flushing times; methods for filling and marking sample containers, reagents or preservatives to be added to selected samples; methods for storing samples, and data-logging procedures. Bacteriological samples must be collected in sterile containers with the appropriate amount of dechlorinating agent

Analysis procedures. Samples taken in the field can be analyzed at the sampling site, at a field laboratory located in the sampling area, or in a centralized laboratory. Procedures to be followed for each type of analysis should be specified in the sampling plan.

Personnel organization and schedule. A detailed personnel schedule and logistical arrangements for the sampling study should be established in the sampling plan.

Safety issues. Many safety issues are associated with sampling studies. These are intensified by round-the-clock sampling and working in unfamiliar areas. The following safety-related concerns should be addressed in the sampling plan: notification of police and other governmental agencies; public notification (newspapers, television stations); notification of customers who may be directly affected; issuance of safety equipment, such as flashlights, vests, and so forth; use of marked vehicles and uniforms identifying the participants as official water utility employees or contractors; and issuance of official identification cards or letters explaining their participation in the study.

Data recording. An organized method for recording all data is required. Notes should be taken documenting all aspects of the study.

Equipment and supply needs. Equipment includes field sampling equipment (e.g., chlorine meter), safety equipment (vests, rain gear, flashlights), laboratory equipment, and the like. Expendable supplies include sampling containers, reagents, and marking pens. As part of the sampling plan, the needs and availability of equipment and supplies should be identified and alternative sources for equipment should be investigated. Because equipment malfunction or loss is possible, some redundancy in equipment should be planned for.

Training requirement. Training of sampling crews is essential and should be specified in the sampling plan.

Contingency plans. The old adage that "if something can go wrong, then it will" applies to field sampling studies and provides a good basis for contingency planning. Contingency planning should include consideration of equipment malfunction, illness of crew members, communication problems, severe weather, unexpected system operation, and customer complaints.

Communications. To coordinate actions during the study and to respond to unexpected events, a means of communication is needed. Alternatives include radios, cellular phones, walkie-talkies, or a person circulating in a vehicle among crews.

Calibration and review of analytical instruments. Field analytical equipment should be calibrated before and during the study.

9.4 WATER-QUALITY MODELING

It is difficult to use monitoring data alone to understand the fate and transformation of substances in drinking water as the water moves through a distribution system. Even medium-sized cities can have thousands of miles of pipes, making it impossible to achieve widespread monitoring. The flow pathways and travel times of water through these systems are highly variable because of the looped layout of the pipe network and the continuous changes in water usage over space and time. The common use of storage facilities out in the system makes things even more variable. At different times of the day, a location might be receiving relatively new water from the treatment works when storage tanks are being refilled or old water when storage tanks are being emptied. It usually is impractical to experiment on the entire distribution system by seeing how changes in pumping schedules, storage facility operations, or treatment methods affect the quality of water received by the consumer.

For these reasons, mathematical modeling of water-quality behavior in distribution systems has become an attractive supplement to monitoring. These models offer a cost-effective way to study the spatial and temporal variation of a number of water quality constituents, including.

- the fraction of water originating from a particular source,
- the age of water in the system,
- the concentration of a nonreactive tracer compound either added or removed from the system (e.g., fluoride or sodium),
- the concentration and loss rate of a secondary disinfectant (e.g., chlorine),
- the concentration and growth rate of disinfection by-products such as THMs, and
- the numbers and mass of attached and free-flowing bacteria in the system.

The models can be used to assist managers to perform a variety of water quality-related studies. Examples include the following:

- calibrating and testing hydraulic models of the system through the use of chemical tracers,
- locating and sizing storage facilities and modifying system operations to reduce the age of the water,

- modifying the design and operation of the system to provide a desired blend of waters from different sources,
- finding the best combination of pipe replacement, pipe relining, pipe cleaning, reduction in storage holding time, and location and injection rate at booster stations to maintain desired disinfectant levels throughout the system,
- assessing and minimizing the risk of consumer exposure to disinfectant by-products, and
- assessing the system's vulnerability to incidents of external contamination.

9.4.1 History

Although the use of mathematical models for hydraulic analysis of distribution systems dates back to the 1930s (Cross, 1936), water-quality models of distribution systems are a relatively recent development. Early work was limited to steady-state representations of networks. In a study of slurry flow in a pipe network, Wood (1980) presented an extension to a steady-state hydraulic model in which a series of simultaneous equations are solved for each node. A similar formulation was used later by Chun and Selznick (1985) in a 166-link representation of the Alameda County, California, Water District with three sources of water of differing hardness and by Metzger (1985) in studying blending, settling and flushing in distribution systems. In a generalization of this formulation, Males et al. (1985) used simultaneous equations to calculate the spatial distribution of concentration, travel times, costs, and any other variables that could be associated with links or nodes. This model, called SOLVER, was a component of the Water Supply Simulation Model, an integrated database management, modeling, and display system (Clark and Males, 1986).

Steady-state water-quality models proved to be useful tools for investigating the movement of a contaminant under constant conditions. However, the need for models that would represent the dynamics of contaminant movement led to the development of models that simulate the movement and transformation of contaminants in a distribution system under temporally varying conditions. Three such models were initially introduced at the American Water Works Association Distribution System Symposium in 1986 (Clark, et al., 1986; Hart et al., 1986; Liou and Kroon, 1986). Early applications of these models to trace contaminants and water age were reported by Grayman et al. (1988) and Kroon and Hunt (1989). More recent developments include the effects of both bulk and wall reactions in simulations of chlorine residual (Rossman et al., 1994, Vasconcelos et al., 1997), biofilm growth models (Servais et al., 1995), and THM formation models (Vasconcelos et al., 1996). An important development in the 1990s was the development of hydraulic and water quality models into graphically based user-friendly packages. Examples of commercial network modeling packages that contain water-quality modeling capabilities include CYBERNET, EPANET, H2ONET, and Stoner SynerGEE.

9.4.2 Governing Equations

A water distribution system consists of pipes, pumps, valves, fittings, and storage facilities that are used to convey water from source points to consumers. The actual physical system is modeled as a network of links that are connected at nodes in some specified branched or looped configuration. Links represent pipes, pumps, or valves. Nodes serve as junction, source, consumption, and storage points. A network water-quality model predicts how the concentration of a dissolved substance varies with time throughout the network under a known set of hydraulic conditions and source input patterns. Its governing equations rest on the principles of conservation of mass coupled with reaction kinetics. The following phenomena occurring in the distribution system are represented in a typical water-quality model (Rossman and Boulos; 1996; Rossman et al., 1994).

9.4.2.1 Advective transport in pipes. A dissolved substance will travel down the length of a pipe with the same average velocity as the carrier fluid, while at the same time reacting (either growing or decaying) at some given rate. Longitudinal dispersion is usually not an important transport mechanism under most operating conditions. This means there is no intermixing of mass between adjacent parcels of water traveling down a pipe. Advective transport within a pipe can be represented with the following equation:

$$\frac{\partial C_i}{\partial t} = -u_i \frac{\partial C_i}{\partial x} + r(C_i) \qquad (9.1)$$

where C_i = concentration (M/L³) in pipe i as a function of distance x and time t, u_i = flow velocity (L/T) in pipe i, and $r(C_i)$ = rate of reaction (M/L³/T) as a function of concentration.

9.4.2.2 Mixing at pipe junctions. At junctions receiving inflow from two or more pipes, the mixing of fluid is assumed to be complete and instantaneous. Thus, the concentration of a substance in water leaving the junction is simply the flow-weighted sum of the concentrations from the inflowing pipes. For a specific node k, one can write

$$C_{i|x=0} = \frac{\sum_{j \in I_k} Q_j C_{j|x=L_j} + Q_{k,ext} C_{k,ext}}{\sum_{j \in I_k} Q_j + Q_{k,ext}} \qquad (9.2)$$

where i = link with flow leaving node k, I_k = set of links with flow into k, L_j = length of link j, Q_j = flow (L³/T) in link j, $Q_{k,ext}$ = external source flow entering the network at node k, and $C_{k,ext}$ = concentration of the external flow entering at node k.

9.4.2.3 Mixing in storage facilities. Most water-quality models assume that the contents of storage facilities (tanks and reservoirs) are mixed completely. Thus, the concentration throughout the facility is a blend of the current contents and any entering water. At the same time, the internal concentration could be changing because of reactions. The following equation expresses these phenomena:

$$\frac{\partial (V_s C_s)}{\partial t} = \sum_{i \in I_s} Q_i C_{i|x=L_i} - \sum_{j \in O_s} Q_j C_s - r(C_s) \qquad (9.3)$$

where V_s = volume (L³) in storage at time t, C_s = concentration within the storage facility, Q_i = flow in inlet pipe, Q_j = flow in outlet pipe, I_s = set of links providing flow into the facility, and O_s = set of links withdrawing flow from the facility.

9.4.2.4 Bulk flow reactions. While a substance moves down a pipe or resides in storage, it can undergo reaction with constituents in the water column. The rate of reaction generally can be described as a power function of concentration:

$$r = kC^n \qquad (9.4)$$

where k = a reaction constant and n = the reaction order. Some examples of different reaction rate expressions are $r = -kC$ for chlorine decay (first-order decay), $r = k(C^* - C)$ for THM formation (first-order growth, where C^* = maximum THM formation possible), $r = 1$ for water age (zero-order growth), and $r = 0$ for conservative materials (e.g., fluoride).

9.4.2.5 Pipe wall reactions. While flowing through pipes, dissolved substances can be transported to the pipe wall and react with materials, such as corrosion products or biofilm, that are on or close to the wall. The amount of wall area available for reaction and the rate of mass transfer between the bulk fluid and the wall also will influence the overall rate of this reaction. The surface area per unit volume, which for a pipe equals 2 divided by the radius, determines the former factor. The latter factor can be represented by a mass-transfer coefficient, the value of which depends on the molecular diffusivity of the reactive species and on the Reynolds number of the flow (Rossman et al., 1994). For first-order kinetics, the rate of a pipe wall reaction can be expressed as

$$r = \frac{2k_w k_f C}{R(k_w + k_f)} \tag{9.5}$$

where k_w = wall reaction rate constant (L/T), k_f = mass transfer coefficient (L/T), and R = pipe radius (L). If a first-order reaction with rate constant k_b also is occurring in the bulk flow, then an overall rate constant k (T^{-1}) that incorporates both the bulk and wall reactions can be written as

$$k = k_b + \frac{2k_w k_f}{R(k_w + k_f)} \tag{9.6}$$

Note that even if k_b and k_w were the same throughout a system, the apparent rate k could still vary from one pipe to the next because of variations in pipe size and flow rate.

9.4.2.6 System of equations. When applied to a network as a whole, Eqs. (9.1), (9.2), and (9.3) represent a coupled set of differential/algebraic equations with time-varying coefficients that must be solved for C_i in each pipe i and C_s in each storage facility s. This solution is subject to the following set of externally imposed conditions:

- Initial conditions that specify C_i for all x in each pipe i and C_s in each storage facility s at time 0.
- Boundary conditions that specify values for $C_{k,ext}$ and $Q_{k,ext}$ for all time t at each node k which has external mass inputs.
- Hydraulic conditions that specify the volume V_s in each storage facility s and the flow Q_i in each link i at all times t.

9.4.3 Solution Methods

There are two general classes of water quality models—steady-state and dynamic.

9.4.3.1 Steady-state models. Steady-state models compute the spatial distribution of water quality throughout a pipe network under the assumptions that hydraulic conditions do not change and that storage does not affect water quality. The models can be derived from the general mass-conservation equations by setting all time derivatives to zero and requiring that all other coefficients be invariant with time. The resulting set of equations can be solved as a series of simultaneous equations (Males et al., 1985). Alternatively, a "marching out" solution can be used by performing a topological sort on the network (Boulos and Altman, 1993). This means finding a reordering of the nodes so that for any given node, the pipes with flow into the node are connected to nodes that appear earlier in the ordering. Although steady-state models are simple to set up and solve, the restrictive assumptions limit their applicability.

9.4.3.2 Dynamic models. Dynamic models of water quality in distribution systems take explicit account of how changes in flows through pipes and storage facilities occurring over an extended period of system operation affects water quality. These models thus provide a more realistic picture of system behavior. Solution methods for dynamic models can be classified spatially as either eulerian or lagrangian and temporally as either time-driven or event-driven. *Eulerian approaches* divide the pipe network into a series of fixed, interconnected control volumes and record changes at the boundaries or within these volumes as water flows through them. *Lagrangian models* track changes in a series of discrete parcels of water as they travel through the pipe network. *Time-driven simulations* update the state of the network at fixed time intervals. *Event-driven simulations* update the state of the system only at times when a change actually occurs, such as when a new parcel of water reaches the end of a pipe and mixes with water from other connecting pipes.

Each of these approaches assumes that a hydraulic model has determined the flow direction and velocity of flow in each pipe at specific intervals over an extended period. These intervals are referred to as hydraulic time steps and are typically 1 h for most applications. Within a hydraulic time step, the velocity within each pipe remains constant. Constituent transport and reaction proceed at smaller intervals of time known as the water-quality time step. Adjustments are made at the start of a new hydraulic time step to account for possible changes in flow velocity and direction. Brief descriptions of four different solution methods for dynamic models follow:

Finite difference method (FDM). FDM is an Eulerian approach that approximates the derivatives in Eq. (9.1) with their finite difference equivalents along a fixed grid of points in time and space (Islam et al., 1997).

Discrete volume method (DVM). DVM is an Eulerian approach that divides each pipe into a series of equally-sized, completely-mixed volume segments (Grayman et al., 1988; Rossman et al., 1993). At each successive water-quality time step, the concentration within each volume segment is first reacted and then transferred to the adjacent downstream segment. When the adjacent segment is a junction node, the mass and flow entering the node is added to any mass and flow already received from other pipes. After these reaction/transport steps are completed for all pipes, the resulting mixture concentration at each junction node is computed and released into the first segments of pipes with flow leaving the node.

Time-driven method (TDM). This Lagrangian method tracks the concentration and size of a series of nonoverlapping segments of water that fill each link of the network (Liou and Kroon, 1987). As time progresses, the size of the most upstream segment in a link increases as water enters the link while an equal loss in size of the most downstream segment occurs as water leaves the link. The size of the segments between the most upstream and downstream segments remains unchanged. This sequence of steps is repeated until the time when a new hydraulic condition occurs. The network is then resegmented to reflect changes in pipe travel times, mass is reapportioned from the old segmentation to the new one, and the computations are continued.

Event-driven method (EDM). EDM is a Lagrangian method that is similar to TDM, except that rather than updating the entire network at fixed time steps, individual link/node conditions are updated only at times when the leading segment in a link completely disappears through its downstream node (Boulos et al., 1995).

Comparisons of the alternative solution techniques (Rossman and Boulos, 1996) suggest that the Lagrangian TDM is the most efficient and versatile of the methods available for solving dynamic water-quality network models.

9.4.4 Data Requirements

Data requirements for water-quality models fall into the categories of hydraulic, water-quality, reaction-rate, and field data. A brief discussion of the requirements in each of these areas follows.

9.4.4.1 Hydraulic data. A water-quality model uses the flow solution of a hydraulic model as part of its input data. Steady-state models require only a single, steady-state flow value for each pipe. Dynamic models use a time history of flow in each pipe and of volume changes in each storage facility. These quantities are determined by making an extended-period hydraulic analysis of the system being studied. Most modeling software packages have the capacity to integrate the hydraulic and water-quality analyses together into a single operation. This relieves the analyst from having to supply flow data manually to the water-quality solver. Having a good hydraulic understanding of a network is essential for computing accurate water-quality results. A poorly calibrated hydraulic model will invariably lead to a poorly performing water-quality model.

9.4.4.2 Water-quality data. Dynamic models require a set of initial water-quality conditions to start the simulation. There are two basic approaches for establishing these conditions. One is to use the results from a field-monitoring survey. This approach is often used when calibrating the model to field observations. Sites in the model corresponding to sampled sites can have their initial quality set to the measured value. Initial conditions for other locations can be estimated by interpolating between the measured values. When using this method, it is important to get good estimates of quality conditions within storage facilities. Model results can be sensitive to these values, which can be slow to change during the simulation because of the usually slow replacement rate of water in storage. This approach cannot be used when modeling the age of water because there is no way to measure this parameter directly.

The other approach is to start the model simulation with arbitrary initial values and run it for a sufficiently long period under a repeating hydraulic loading pattern until the system's water-quality behavior settles into a periodic pattern. Note that the length of this pattern might be different from the length of the hydraulic pattern. Results from the last period would then be taken to represent the system's response to the imposed hydraulic loading. Good estimates of initial conditions in the storage facilities can reduce the time needed for the system to reach a dynamic equilibrium.

In addition to initial conditions, the water-quality model needs to know the quality of all external inflows into the system. These data can be obtained from existing source-monitoring records when simulating existing operations or could be set to specific values when investigating operational changes.

9.4.4.3 Reaction-rate data. The specific form of reaction-rate data needed to run a water-quality simulation depends on the constituent being modeled. It is essential that these data be developed on a site-specific basis since research has shown that reaction rates can differ by orders of magnitude for different water sources, treatment methods, and pipeline conditions.

First-order rate constants for chlorine decay in the bulk flow can be estimated by performing a bottle test in the laboratory. Water samples are stored in several amber bottles and kept at constant temperature. At several periods of time, a bottle is selected and analyzed for free chlorine. At the end of the test, the natural logarithms of the measured chlorine values are plotted against time. The rate constant is the slope of the straight line through these points. There is currently no similar direct test to estimate wall-reaction rate constants. Instead, one must rely on calibration against measured field data.

A similar bottle test can be used to estimate first-order growth rates for THMs. The test should be run long enough so that the THM concentration plateaus out to a constant level. This value becomes the estimate of the maximum potential of THM formation. A plot is then made of the natural logarithm of the difference in the formation potential and measured THM level versus time. The slope of the line through these points is the growth-rate constant.

9.4.5 Model Calibration

Model calibration is the process of adjusting characteristics and parameters of the model so that the model matches actual observed field data to some acceptable level. If a mathematical model exactly represented the actual physical processes and if we had perfect knowledge of all the parameters required by the model, calibration would not be necessary. Unfortunately, neither criterion listed above is generally met; thus, calibration is an important consideration in all studies.

Since water-quality models depend on hydraulic models to provide information on flows and velocities in pipes, an acceptably calibrated hydraulic model is a requirement for water-quality modeling. Frequently, hydraulic models are calibrated to match pressures measured in the field. Since calibration just for pressure does not guarantee that flows and velocity are accurately predicted (Grayman, 1998), additional hydraulic calibration may be required when performing water-quality modeling.

9.4.5.1 Calibration of conservative substances. In water-quality models of conservative substances and water age, the constituents travel with the flow and no transformations occur. At junctions, the constituent concentration or age is calculated based on a linear combination of the incoming flows. As a result, no intrinsic calibration is required or possible. Field results can be matched only by varying inflow concentrations or by the hydraulic parameters that control the flows in the system.

9.4.5.2 Calibration of nonconservative substances. The concentration of nonconservative substances changes over time as the substances travel through the distribution system because of reactions with other constituents in the water or through interaction with pipe walls and appurtenances. Laboratory and field data are generally required to establish both the form of the reaction and the reaction coefficients. For transformations that occur in the bulk water, bottle tests can be used to establish the transformation characteristics (see Sec. 9.4.4.3). For transformations that involve interaction with the distribution system itself, limited general information is available. The most common example is the pipe wall demand for chlorine residual. Pipe material, age, and condition are all factors that can affect this demand, and in situ field calibration studies are generally required to estimate it.

9.4.5.3 Uses for hydraulic calibration. Water-quality models can be used in association with field tracer studies as an alternative or supplementary method for calibrating network hydraulic models. Since water-quality models of distribution systems depend on hydraulic models to provide information on pipe flows, flow directions, and flow velocities, inaccuracies in the flow or velocity values provided by the hydraulic model will lead to inaccuracies in water-quality predictions.

Although most water-quality models can be used to represent both conservative and nonconservative substances, the use of conservative substances is more appropriate for calibration of hydraulic models. When modeling a conservative substance, there are essentially no water quality parameters that can be adjusted. In other words, if the hydraulic parameters are correct and the initial conditions and loading conditions for the substance are known accurately, the water-quality model should provide a good estimate of the concentration of the substance throughout the network. The use of the water-quality model and conservative tracer as a means of calibrating the hydraulic model is based on this relationship.

The calibration process using water-quality modeling can be summarized as follows:

1. A conservative tracer is identified for a distribution system. The tracer can be a chemical that is added to the flow at an appropriate location or, in the situation where there are multiple sources of water, can be a naturally occurring difference in the water sources, such as hardness. Chemicals that typically are used include fluoride, calcium chloride, sodium chloride, and lithium chloride. Selection of the tracer generally depends upon government regulations (e.g., some localities will not allow the use of fluoride), the availability and cost of the chemicals, the methods for adding the chemical to the system, and the measuring or analysis devices.

2. A controlled field experiment is performed in which either (1) the conservative tracer is injected into the system for a prescribed period of time (2) a conservative substance, such as fluoride, that normally is added is shut off for a prescribed period or (3) a naturally occurring substance that differs between sources is traced.

3. During the field experiment, the concentration of the tracer is measured at selected sites in the distribution system along with other parameters that are required by a hydraulic model, such as tank water levels, pump operations, flows, and so forth. In addition to the conservative tracer, other water-quality concentrations, such as chlorine residual, can be measured though these values are not generally used in the calibration process.

4. The model is then run with alternative hydraulic parameter values to determine the model parameters that result in the best representation of the field data. Perhaps traditional pressure and flow measurements are used to perform a first-step calibration. The water-quality model is then used to model the conservative tracer.

5. Good agreement between the predicted and observed tracer concentrations indicates a good calibration of the hydraulic model for the conditions being modeled. Significant deviations between the observed and modeled concentrations indicate that further calibration of the hydraulic model is required. Various statistical and directed search techniques can be used in conjunction with the conservative tracer data to aid the user in adjusting the hydraulic model parameters to achieve a better match with the observed concentrations.

REFERENCES

Boulos, P. F., and T. Altman, "Explicit calculation of water quality parameters in pipe distribution systems," *Journal of Civil Engineering Systems*, 10; 187–206, 1993.

Boulos, P. F., T. Altman, P. A. Jarrige, and F. Collevati, "Discrete simulation approach for network water quality models," *Journal of Water Resources Planning and Management, ASCE*, 121:49–60 1995.

Center for Biofilm Engineering, Biofilm Heterogeneity, Center for Biofilm Engineering News, Montana State University, 2(1):1–2, 1994.

Characklis, W. G., and K. C. Marshall, *Biofilms*, John Wiley & Sons, New York, 1990.

Chun, D. G. and H. L. Selznick, "Computer Modeling of Distribution System Water Quality," Proceedings of an ASCE Specialty Conference on Computer Applications in Water Resources, American Society of Civil Engineers, 448–456, New York, 1985.

Clark, R. M. and W. M. Grayman, *Modeling Water Quality in Drinking Water Distribution Systems*, American Water Works Association, Denver, CO, 1998.

Clark, R. M. and W. M. Grayman, R. M. Males and J, A. Coyle, "Predicting Water Quality in Distribution Systems, American Water Works Association, CO, 1998.

Cross, H., Analysis of Flow in Networks of Conduits or Conductors, Univ. of Ill. Eng. Experiment Station Bulletin 286, Urbana, Il, 1936.

DeBerry, D. W., J. R. Kidwell, arid D.A. Malish, Corrosion in Potable Water Systems, U.S. Environmental Protection Agency, Washington, DC, 1982.

Geldreich, E. E., *Microbial Quality of Water Supply in Distribution Systems*, CRC Press, Boca Raton, FL, 1996.

Geldreich, E. E., and M. W. LeChevallier, "Microbiological Quality Control in Distribution Systems," in *Water Quality and Treatment*, McGraw-Hill, New York (in press).

Grayman, W. M., R. M. Clark, and R. M. Males. "Modeling distribution system water quality: Dynamic approach," *Journal of Water Resources Planning and Management, ASCE*, 114:295–312, 1988.

Grayman, W. M., Use of Tracer Studies and Water Quality Models to Calibrate a Network Hydraulic Model. *Essential Hydraulics and Hydrology*, Haestad Press, Waterbury, CT, 1998.

Hart, F. L., J. L. Meader,. and S. N. Chiang, "CLNET - A Simulation Model for Tracing Chlorine Residuals in a Potable Water Distribution Network, Proceedings of the AWWA Distribution System Symposium, AWWA, Denver, CO, 193-203, 1986.

Kirmeryer, G. and G. S. Logsdon, "Principles of Internal Corrosion and Corrosion monitoring," Journal of the American Water Works Association, 75(2), 78-83, 1983.

Kirmeyer, G. J., L. Kirby, B. M. Murphy, P. F. Noran, K. Martel, T. W. Lund., J. L. Anderson, and R. Medhurst, Maintaining Water Quality in Finished Water Storage facilities, AWWA Research Foundation and American Water Works Association, Denver, CO. 1999.

Kroon, J. R. And W. A. Hunt, "Modeling Water Quality in the Distribution Network," Proceedings of the AWWA Distribution System Symposium, AWWA, Denver, CO, 1986.

Islam, M. R., M. H. Chaudhry, and R. M. Clark, "Inverse modeling of chlorine concentration in pipe networks under dynamic condition," *Journal of Environmental Engineering, ASCE*, 123:1033–1040, 1997.

LeChevallier, M. W., T. S. Babcock, and R. G. Lee, "Examination and Characterization of Distribution System Biofilms," *Applied Environmental Microbiology*, 53:2714–2724, 1987.

LeChevallier, M. W., C. D. Lowry, and R. G. Lee, "Disinfection of biofilms in a model distribution system," *Journal of the American Water Works Association*, 82(7):87–99, 1990.

LeChevallier, M. W., C. D. Lowry, R. G. Lee, and D. L. Gibbon "Examining the relationship between iron corrosion and the disinfection of biofilm bacteria," *Journal of the American Water Works Association*, 85(7):111–123, 1993.

Liou, C. P., and J. R. Kroon, "Modeling the propagation of waterborne substances in distribution networks," *Journal of the American Water Works Association*, 79(11), 54–58, 1987.

Liou, C. P., and J. R. Kroon, "Propagation and Distribution of Water borne Substances in Networks," Proceedings of the AWWA Distribution System Symposium, AWWA, Denver, CO, 231-242, 1986.

Males, R. M., R. M. Clark, P. J. Wehrman, and W. E. Gates, "Algorithm for mixing problems in water systems," *Journal of Hydraulics Division, ASCE*, 111:206–219, 1985.

Martin, R. S., W. H. Gates, R. S. Tobin, D. Grantham, R. Sumarah, P. Wolfe and P. Forestall et al., "Factors affecting coliform bacteria growth in distribution systems," *Journal of the American Water Works Association*, 64:34–37, 1982.

Metzger, I., "Water Quality Modeling of Distribution System," Proceedings of an ASCE Specialty Conference on Computer Applications in Water Resources, American Society of Civil Engineers, 422-429, New York, 1985.

Rossman, L. A. EPANET - Users Manual, EPA-600/R-94/057, U.S. Environmental Protection Agency, Risk Reduction Engineering Laboratory Cincinnati, OH, 1994.

Rossman, L. A., Boulos, P. F., and Altman, T., "Discrete volume-element method for network water-quality models," *Journal of Water Resources Planning and Management, ASCE*, 119:505–517, 1993.

Rossman, L. A., R. M. Clark, and W. M. Grayman, "Modeling chlorine residuals in drinking-water distribution systems," *Journal of Environmental Engineering, ASCE*, 120:803–820, 1994.

Rossman, L. A. and P. F. Boulos, (1996). "Numerical methods for modeling water quality in distribution systems: A comparison," *Journal of Water Resources Planning and Management*, ASCE, 122:137–146, 1996.

Servais, P., Laurent, P., Billen, G., and Gatel, D., "Development of a model of BDOC and bacterial biomass fluctuations in distribution systems," Rev. Sci. Eau, 8:427–462, 1995.

Singley, J. E. "The Search for a Corrosion Index," *Journal of the American Water Works Association*, 73(11): 579, 1981.

Singley, J. E., B. A. Beaudet, and P. H. Markey, Corrosion manual for Internal Corrosion of Water distribution Systems, EPA-570/9-84-001, U. S. Environmental Protection Agency, Office of Drinking Water, Washington, DC, 1984.

Vasconcelos, J. J., P. F. Boulos, W. M. Grayman, L. Kiene, O. Wable, P. Biswas, A. Bhari, L. A. Rossman, R. M. Clark, J. A. Goodrich et al., *Characterization and Modeling of Chlorine Decay in Distribution Systems*, American Water Works Association, Denver, CO, 1996.

Vasconcelos, J. J., L. A. Rossman, W. M. Grayman, P. F. Boulos, and R. M. Clark, "Kinetics of chlorine decay," *Journal of the American Water Works Association*, 89(7):54–65, 1997.

Wood, D. J., "Slurry Flow in pipe Networks", Journal of Hydraulics, ASCE, 106(1), 57-70, 1980.

CHAPTER 10
HYDRAULIC DESIGN OF WATER DISTRIBUTION STORAGE TANKS

Thomas M. Walski
Pennsylvania American Water Company
Wilkes-Barre, PA

10.1 INTRODUCTION

Water storage tanks are a commonly used facility in virtually all water distribution systems. Methods for sizing and locating these facilities to provide equalization and emergency storage have changed little over the years. Recent years, however, have seen an increased awareness of water-quality changes that can occur in storage tanks. Because water-quality problems are usually worse in tanks with little turnover, no longer it is safe to assume that a bigger tank is a better tank. There is increased emphasis on constructing the right-sized tank in the right location.

Water-quality concerns will not radically change tank design, but they do call for a reassessment of some design practices. This chapter summarizes the state-of-the-art in tank design.

10.2 BASIC CONCEPTS

Water distribution storage is provided to ensure the reliability of supply, maintain pressure, equalize pumping and treatment rates, reduce the size of transmission mains, and improve operational flexibility and efficiency. Numerous decisions must be made in the design of a storage tank, including size, location, type, and expected operation. This chapter focuses on the hydraulic aspects of design as opposed to structural, corrosion, safety, contamination, or instrumentation aspects. (In this chapter, the word "tanks" will be used to describe treated water-storage facilities, although the term "reservoirs" is preferred by some, whereas others use the word "reservoir" to describe only ground-level and buried tanks.)

The key considerations in the hydraulic design of water storage tanks are described in the following sections and procedures for design of the tanks are described in the remainder of the paper.

10.2.1 Equalization

One primary purpose for construction of storage facilities is *equalization*. Water utilities like to operate treatment plants at a relatively constant rate, and wells and pumping stations generally work best when pumped at a steady rate. However, water use in most utilities varies significantly over the course of the day. These variations in use can be met by continuously varying source production, continuously varying pumping rates, or filling and draining storage tanks. The process of filling and draining storage tanks is much easier operationally and is generally less expensive than other methods. Facilities serving portions of a distribution system with storage tanks generally need to be sized only to meet maximum daily demands, with storage tanks providing water during instantaneous peak demands.

10.2.2 Pressure Maintenance

To a great extent, the elevation of water stored in a tank determines the pressure in all pipes directly connected to the tank (i.e., not served through a pressure-reducing valve or pump). Ignoring headloss, which is usually small most of the time, the pressure can be estimated as

$$p = (H - z)w \qquad (10.1)$$

where p = pressure at elevation z, Pa (lb/ft^2), H = water level in tank, m (ft), z = elevation in distribution system, m (ft), and w = specific weight of water, N/m^3 (lb/ft^3). (When pressure is expressed in psi, w is 0.433.) The larger the tank volume, the more stable the pressures in the distribution system will be despite fluctuations in demand or changes in pump operation.

10.2.3 Fire Storage

If distribution storage tanks were not used, larger water transmission mains and larger treatment plant capacity would be required by most utilities to provide water needed for fire fighting. Especially for smaller systems, storage tanks are a much more economical and operationally reliable means for meeting the short-term large demands placed on a water-supply system during fire fighting.

10.2.4 Emergency Storage

In addition to fires, emergencies such as power outages, breaks in large water mains, problems at treatment plants, and unexpected shutdowns of water-supply facilities can cause failure of the water system if sufficient water is not available in storage. Storage tanks can meet demands during emergency situations. The extent to which emergency storage is needed in excess of fire storage depends on the reliability of the supply system. In addition to simply providing storage volume, tanks can provide a form of backup pressurization of the system in case of a loss of pumping capability (e.g., power outage, major pipe break). Such pressurization helps prevent contamination from cross connections if pumping should be lost.

10.2.5 Energy Consumption

To the extent that water is stored in a distribution at a higher level than the treatment plant, the energy in that water also is stored at that higher level. The existence of tanks enables utilities to store energy as well as water for later use. To the extent that equalization storage slows down the velocity (and friction losses) in the large transmission mains, the energy used to pump water is reduced by having distribution storage tanks that equalize pumping.

Most water utilities pay a demand charge, a capacity charge, or both to the electric utility based on peak rates of energy consumption. To the extent that availability of storage reduces peak energy usage, storage tanks can be helpful in reducing the demand charges for the utility. If time-of-day energy pricing is used, energy can be used during off-peak hours and that energy can be stored with the water in tanks to be used during peak hours.

10.2.6 Water Quality

Tanks may affect water quality in two general ways: (1) through chemical, physical, and biological processes that occur as water ages while stored in the tank and (2) through external contamination of water in tanks.

Sources of contamination may include the tank lining, sediments in the bottom of the tank, and animal or human contamination in either open reservoirs or through breeches in the tank roof, vents, and sides. These should be eliminated by proper design and maintenance. The greatest single change in tank design for water quality is the requirement to cover treated water-storage tanks.

Water ages in a storage facility because of detention times and differential aging caused by incomplete mixing. Common degradation mechanisms that may occur as a result of aging water include loss of disinfectant residual, which can result in microbial regrowth in the tank or distribution system, and formation of disinfection by-products, such as trihalomethanes.

There also is the potential for the escape of volatile organic compounds through the water's surface to the atmosphere, but this is generally considered to be inconsequential because of the low levels of volatile organic compounds that are mandated by the maximum contaminant levels and quiescent nature of the water surface in a tank. Another impact of tanks on water quality may be the decay of radon caused by volatilization and the short half-life of this contaminant. Water also can stratify in tall tanks, thus exacerbating water-quality problems.

Another impact of storage volume on water quality is the fact that storing water in tanks allows for a greater time interval between the occurrence of a distribution system problem (e.g. major pipe break or power outage) and the deterioration of water quality caused by cross-connections. (Water quality in storage tanks is discussed in greater detail in Chap. 11.)

10.2.7 Hydraulic Transient Control

Changes in velocity in water mains can result in hydraulic transients referred to as "waterhammer." (also see Chapter 6) These extremely high or low pressures caused by transients can be significantly dampened by storage tanks, especially those that "float-on-the-system." (See Sec. 10.3.1 for an explanation of this term.)

10.2.8 Aesthetics

Most people are neutral about storage tanks, considering them an acceptable part of the scenery. However, some individuals believe that tanks are unsightly and detract from views; thus, storage tanks are susceptible to the "NIMBY" syndrome (not in my back yard). Still others see tanks as a resource to promote their community or business. Tanks should be designed to satisfy the aesthetic considerations of stakeholders to the extent possible without sacrificing the purpose of the tank and efficiency of the system's operation.

10.3 DESIGN ISSUES

Design of distribution storage facilities involves resolving numerous issues and trade-offs. Some of these include location, levels, and volume, which are described in the following sections. The subsections immediately below describe some of the preliminary issues that must be resolved. Issues regarding the structural design of storage tanks are given for welded steel tanks in AWWA D100 and AWWA D-102 (AWWA, 1996, 1997a), bolted steel tanks in AWWA D103 (AWWA 1997b), prestressed concrete tanks in AWWA D110 and AWWA D115 (AWWA, 1995a, b) and pressure tanks in the appropriate ASME standard (ASME, 1998). A good overview of steel tank design is provided in AWWA Manual M42 (1998c).

10.3.1 Floating Versus Pumped Storage

For the purpose of this chapter, storage "*floating-on-the-system*" is defined as storage volumes located at elevations so that the hydraulic grade line outside the tank is virtually the same as the water level (or hydraulic grade line for pressure tanks) in the tank. In this type of storage, water can flow freely into and out of the tank. The converse of this is pumped storage, which refers to water that is stored below the hydraulic grade line in ground-level or buried tanks so that the water can leave the tank only by being pumped.

In many cases, storage floating-on-the-system is associated with higher capital but lower operating costs than is pumping from storage, and the trade-off must be made on a case-by-case basis. Capital costs are higher because the tank must be elevated or located at ground level on a hill somewhat removed from the service area. This is usually more costly than the capital costs of pumping equipment associated with the pumped tank. The energy cost associated with the pumped system is, of course, significantly higher.

The trade-offs between an elevated tank that floats on the system and a ground-level tank with pumping can best be illustrated by an example. Consider a 1.88 ML (500,000 gal) tank where 0.94 ML (250,000 gal) enter and leave the tank during the day. (Assume that labor and instrumentation costs are comparable between the alternatives.) The elevated tank (floating-on-the system) would cost $600,000 whereas the ground-level tank would cost $400,000 with the pumping equipment and controls costing $150,000. Based on initial cost, the pumped alternative is more economical in this case. However, the water level in the ground tank will normally be 30 m (100 ft) below the level of the water in the elevated tank. At an energy price of 5 cents per kilowatt-hour and a wire-to-water efficiency of 60 percent, the annual cost to operate the pumps is as follows:

Cost ($/yr) = [(62.4 lb/ft³)(0.25 Mgal/day) (1.54 ft³/Mgal/day)(100 ft)

$$\frac{(\$0.10/\text{kwh}) \ (8760 \ \text{h/yr})]}{[(0.6) \ (737 \ \text{ftlb/ks}]} = \$4760 \text{ per year.}$$

The extra cost for operations and maintenance labor and supplies may be $3000 per year. The present worth of this value at 5 percent interest over 20 years (present worth factor = 12.46) is given by present worth: (energy, labor) = ($4760 + $3000) (12.46) = $97,000. The total present worth of a ground tank = $400,000 + 150,000 + 97,000 = $647,000. The elevated tank ($600,000) will be slightly less expensive in this case. The costs, however, are highly site-specific and the comparison must be made on a case-by-case basis.

10.3.2 Ground Versus Elevated Tank

A decision related to whether the tank floats on the system is how the tank is constructed with respect to the ground. If the tank is constructed so that the bottom of the water is at or near ground level, the tank is referred to as a *"ground tank."* If the tank is significantly taller than it is wide, it is usually referred to as a *"standpipe."* Standpipes are usually constructed to float on the system, whereas ground-level tanks may float on the system if they are constructed at a sufficiently high elevation. In some areas, buried tanks are preferred because they are least susceptible to freezing. Figure 10.1 illustrates the different kinds of tanks.

In general, elevated tanks and standpipes are usually constructed of steel, buried tanks are usually constructed of concrete, and ground-level tanks are constructed of either steel or concrete. The risk of freezing increases as more of a tank's surface is exposed to cold weather. Heating of tanks has been addressed by Hodnett (1981).

Elevated tanks usually are the most expensive per unit volume, but they provide most of the storage at the desired elevation and are essential if storage that floats on the system is desired in flat areas. Terrain, aesthetics, seismic considerations, potential for freezing, land availability, budget and experiences with different types of tanks will influence a utility's choice of tank.

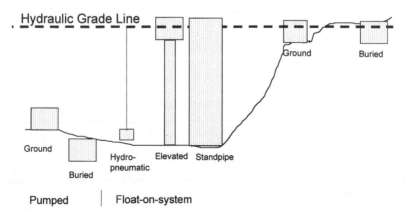

FIGURE 10.1 Tank terminology.

10.3.3 Effective Versus Total Storage

For tanks that float on the system, the tank will drain only if the hydraulic grade line outside the tank drops below the water level in the tank. When the water level in the tank drops too low, the pressures (i.e., hydraulic grade line) provided to customers at higher elevations in the service areas can drop below acceptable levels for customers at the highest elevation. Storage volume located at an elevation so that it can provide acceptable pressure is called "effective storage." Storage located below the minimum acceptable hydraulic grade line is not effective storage, although some individuals may consider that part of "total storage." A standpipe is an example of a tank in which only a small portion of the storage is effective storage.

Storage located below the effective storage zone, which still can provide a minimum pressure [say, 138 kPa (20 psi)], is sometimes referred to as "emergency storage" because a utility would only allow the level to drop to that level during an emergency.

This concept is illustrated in Fig. 10.2, which shows the highest customer located at elevation X requiring 241 kPa (35 psi) during non emergency situations and 138 kPa (20 psi) during emergency situations. The hydraulic grade line and tank water must be kept 25 m (81 ft) above elevation X during non emergency situations and 14.1 m (46 ft) above elevation X during emergencies neglecting headloss.

Water stored below the 14.1 m (46 ft) level can be referred to as ineffective storage (Walski et al., 1990). It does not help in equalization or fire fighting but does increase the detention time in tanks and hence contributes to the problem of disinfectant decay. Tall standpipes also can become stratified, which can contribute further to water-quality problems because the water in the top portion of the tank may have extremely long detention times. Kennedy et al. (1993) documented significantly lower chlorine concentrations in a tall tank. Standpipes with a great deal of ineffective storage should be discouraged for these reasons.

10.3.4 Private Versus Utility Owned Tanks

In most cases it is desirable for the utility to own all water distribution storage tanks. If a large customer needs a great deal of storage, it is usually better for all parties to have that customer contribute to the cost of a utility-owned tank. In some cases, however, a large customer (or a neighboring utility) may choose to construct its own tank. Once water enters the private system (or a neighboring utility's system), it should not be allowed back into the original utility's system unless the private system follows all precautions required to protect water quality. Issues regarding private water tanks are addressed in National Fire Protection Association Standard NFPA 22 (1998).

10.3.5 Pressurized Tanks

The tanks discussed thus far have been non pressurized tanks in which the water surface corresponds to the hydraulic grade line in the tank. If a tank is pressurized, the hydraulic grade line will be higher than the water surface. This can result in a ground-level tank that still effectively floats on the system. To allow the water level in the tank to fluctuate (so that the tank is not simply a wide spot in the pipe), some air is placed in the tank to expand or compress as the volume of water in the tank changes. (Tanks are only worthwhile if the volume of water in storage can change.) Such tanks are referred to as *hydropneumatic tanks*.

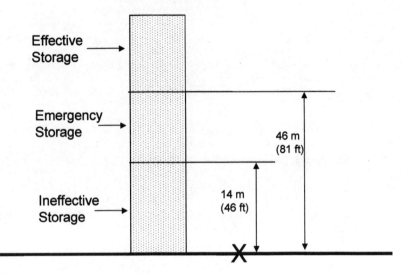

FIGURE 10.2 Definition of effective storage.

Pressurized tanks are much more expensive than non pressurized tanks. Therefore, their use is limited to small tanks that provide limited equalization and emergency storage and virtually no fire protection.

10.4 LOCATION

The location of storage provides an opportunity to make the most of a given volume of storage. However, because of restrictions on availability, terrain, and aesthetics, good storage sites may be difficult to find. Some considerations regarding location are described below.

10.4.1 Clearwell Storage

Clearwell storage at the downstream end of the water treatment plant or the outlet of a well is usually at ground level or in buried tanks that must be pumped. This type of storage can provide time to meet disinfection contact-time requirements. Usually, ground-level tanks are relatively inexpensive. However, because these tanks usually do not float on the system (except when the water treatment plant is on top of a sufficiently large hill), all the water must be pumped and standby power is required, especially if there is limited storage in the distribution system. A large pipe break near the plant also can completely eliminate the effectiveness of clearwell storage. In general, utilities should have some clearwell storage to provide contact time, but they should not rely solely on clearwell storage (unless they are small utilities with clearwell storage that float on the system).

10.4.2 Tanks Downstream of the Demand Center

Storage tanks are best placed on the downstream side of the largest demand from the source. An advantage of this is that if a pipe breaks near the source, or the break will not result in disconnecting all the storage from the customers. A second advantage is that if flow reaches the center of demand from more than one direction, the flow carried by any individual pipe will be lower and pipe sizes will generally be smaller, with associated cost savings. Of course, adequate capacity must be available to refill the tank in off-peak hours. Hydraulic grade lines with storage located opposite the peak demand center are shown in Fig. 10.3.

10.4.3 Multiple Tanks in the Pressure Zone

If there are to be multiple tanks in a pressure zone, the tanks should be placed roughly the same distance from the source or sources. If one tank is very close to the source and other tanks are farther away, it may be difficult to fill the remote tanks without shutting off (or overflowing) the closer tank. Because this is often a problem in systems that have evolved over many years, a tank that was on the fringe of the system years ago is now very close to the plant relative to new remote tanks in the growing service area. In most cases, use of control valves can enable multiple tanks to be used effectively. To make a system easier to operate, it may be desirable in some cases to abandon a small tank in an undesirable location when it needs maintenance.

When there are multiple tanks in a pressure zone, it is essential for all of them to have virtually the same overflow elevation. Otherwise, it may be impossible to fill the highest tank without overflowing or shutting off the lower tanks (thus causing water-quality problems in the lower tanks). Tanks are usually at consistent elevations when they are planned logically. However, when systems grow by annexation or regionalization, some

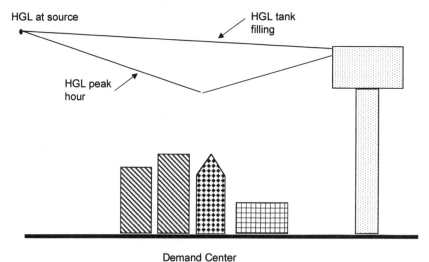

FIGURE 10.3 Hydraulic grade line for tank beyond demand center.

tanks may be at an inappropriate elevation and may need to be abandoned or modified, and pressure zone boundaries may need to be adjusted to operate effectively in the newly configured system.

10.4.4 Multiple Pressure-Zone Systems

In multiple pressure zones, the source is usually located in one of the lowest zones and the system is fed through pumps. In such cases, it is best to place sufficient storage volume in each zone so that each zone can operate almost independently. In this way, only water that is needed in the higher zones will be pumped to those zones. Pumping water to a higher zone only to have it run back down to a lower zone wastes energy and should be avoided if feasible.

In multiple pressure-zone systems where the source is located in the highest zone and lower zones are fed through pressure-reducing valves, most of the storage should be kept in the higher zones. In that way, stored water can feed virtually any zone. Tanks are justified in lower zones to ensure reliability in case of a pipe break and to reduce piping requirements to meet peak and localized large fire demands.

10.4.5 Other Siting Considerations

Because some individuals consider tanks to be an eyesore, it is important to select sites that minimize the visual impact of the tank by having it blend in with its surroundings or by making it attractive. Although the risk of structural failure of an elevated tank is small, setbacks from tanks should be such that the impacts of failure on neighbors will be minimal.

If a tank is planned for a site but is not to be constructed for several years, it is desirable to obtain the land for the tanks and the needed rights-of-way for pipes and roads well in advance. If possible, the zoning should be modified and land subdivision should be approved so that the tank is an accepted land use. In that way, construction will not be delayed because of zoning hearings and appeals.

10.5 TANK LEVELS

10.5.1 Setting Tank Overflow Levels

The most significant decision about a tank in terms of distribution system design is its overflow elevation. This elevation and the associated range (minimum normal day, bottom of tank) will determine the size and boundary of the pressure zone that can be served from the tank, the layout of transmission mains, and the head required at pumping stations. When a tank is being placed in an existing pressure zone, the overflow elevation and operating range should be consistent with existing tanks, as was discussed earlier.

If a tank is being designed for a new pressure zone, it is essential to select an overflow level that will still be acceptable when that pressure zone is built-out in the future and that will be consistent with tank overflows in neighboring utilities to the extent possible in case the systems may be combined into a regional system at some point in the future.

10.5.2 Identifying Tank Service Areas

Consider a tank with overflow elevation Z, as shown in Fig. 10.4. The bottom of its normal operating range is usually about 6 m (20 ft) below the overflow, and the tank bottom is about 12 m (40 ft) below the overflow. The highest customer that can be served is approximately 25 m (81 ft) below the normal low level (point H) or 14 m (46 ft) below the bottom of the emergency storage (point H0[1]).

The lowest customer that can be served is usually determined by customers who will receive excessive pressure 690–550 kPa (100–80 psi) when the tank is full and there is little head-loss in the system. In this example, 690 kPa (100 psi) is the maximum acceptable pressure, and the minimum elevation served is 70 m (231 ft) below the overflow (point L). Anyone above the shaded range will receive too little pressure, whereas anyone below the range will receive excessive pressure. Small pockets of customers with excessive pressure can be served through pressure-reducing valves, either in the system or on individual service lines. Larger groups of customers should be served to the extent possible from a lower-pressure zone to prevent the waste of energy.

10.5.3 Identifying Pressure Zones

Once the overflow elevation and the highest and lowest customers have been identified, it is best to locate all the area that falls between those elevations on a contour map and identify those areas by shading or coloring them. (It is acceptable for there to be some bands of elevations for which customers can be served from either the upper- or lower-pressure zone.) The exact location of the pressure-zone boundary should be consistent with long-term development plans, street layout and supply capability in the respective zones. Figure 10.5 shows the kind of map that can be prepared to identify which

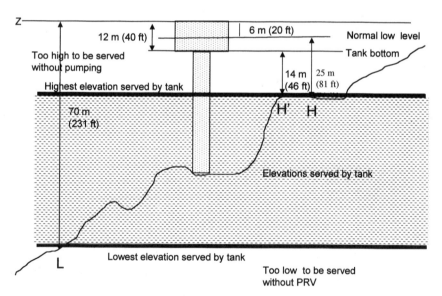

FIGURE 10.4 Elevation drawing showing a tank's service area.

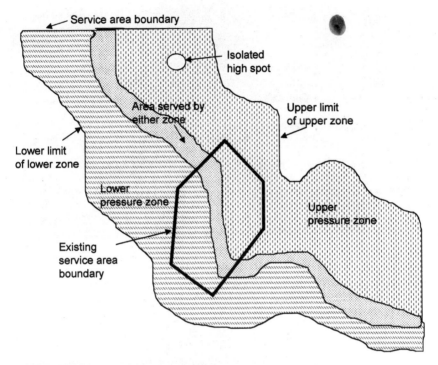

FIGURE 10.5 Plan view of a pressure zone layout.

customers can be served by tanks with various overflow elevations. The map should be prepared before land development occurs and should include land outside of the current service area that may be developed in the future. Isolated high and low points in a pressure zone can cause problems and should be identified in planning studies.

In general, pressure-zone hydraulic grade lines should differ by roughly 30 m (100 ft) from one pressure zone to the next. Significantly larger steps will result in some customers receiving excessively high or low pressure. Smaller steps between pressure zones result in an excessive number of tanks, pump stations, and pressure-reducing valves.

10.6 TANK VOLUME

10.6.1 Trade-offs in Tank Volume Design

Selecting the optimal tank volume involves trade-offs between improved reliability of the system provided by larger tanks and the higher costs and the disinfectant decay problems caused by loss of disinfectant residual in larger tanks. The issue is complicated further by the fact that there are substantial economies of scale in tank construction—doubling the volume of a tank only increases the cost by roughly 60 to 70 percent. Traditionally, the philosophy has been to build as big a tank as possible given long-term demands, budget limits, and site constraints. This can result in tanks that cause disinfectant residual problems.

Two overall approaches to sizing a tank are available: regulatory-driven design and functional design. Both are illustrated below.

10.6.2 Standards-Driven Sizing

Each state and province has its own standards for sizing tanks. For example, the "Ten State Standards" (Recommended Standards for Water Works, 1992) states the following:

> Fire flow requirements established by the appropriate state Insurance Services Office should be satisfied where fire protection is provided....
> The minimum storage capacity (or equivalent capacity) for systems not providing fire storage shall be equal to the average daily consumption. This requirement may be reduced when the source and treatment facilities have sufficient capacity with standby power to supplement peak demands in the system.

Another example is the Texas State Standards (Texas Department of Health, 1988), which devotes up to four pages of text to a description of volume sizing, depending on the size of the system and the nature of the source. The key points in the Texas standards for systems with more than 50 connections are the following: "Total storage capacity of 200 gallons per connection must be provided... Elevated storage in the amount of 100 gallons per connection is required for systems with over 2,500 connections." If more than 18,800 m^3 (5 million gal) of storage are required, utilities can substitute ground storage, pumping, and auxiliary power.

Hydropneumatic tanks can be sized on the basis of 20 gal per connection (with ground tank available) or 50 gal per connection (no ground tanks at source) (Texas Department of Health, 1988).

Sizing also can be determined on the basis of providing a reasonable number of pump starts: "The gross volume of the hydropneumatic tank, in gallons, should be at least ten times the capacity of the largest pump, rated in gallons per minute" (Ten State Standards, 1992).

Other states and provinces have variations on these standards. All these standards leave considerable discretion to the design engineer to provide storage and to regulators to accept the design.

10.6.3 Functional Design

Although the appropriate regulatory standards must be met, it also is helpful to examine why the volume is required. This involves summing up the storage required for each of the recognized purposes: (1) equalization, (2) fire protection, and (3) emergencies other than fires. Cesario (1995) referred to these three types of storage as supply, fire, and reserve, respectively. Each type is discussed in more detail below.

10.6.3.1 Equalization Storage. *Equalization storage* is used to enable the source and pumping facilities to operate at a predetermined rate, depending on the utility's preference. Some options for operating pumping facilities include the following:

1. Operate at a constant rate to simplify operation and reduce demand charges.

2. Adjust flows to roughly match demand and minimize use of storage.
3. Pump during off peak hours to take advantage of time of day energy pricing.
4. Match the demand exactly with variable speed pumps and have no storage.
5. Have a reasonable number of starts per unit time for hydropneumatic pumps.

A comparison of these pumping (or production) rates with a typical time-of-day demand pattern is shown in Fig. 10.6. The variable-speed pump alternative is not shown because it would correspond to the case where the demand and pumping are identical.

The amount of equalization storage required is given by the area between the demand and pumping curves on a peak day. The fraction of daily water production that must be stored depends on the individual community and the type of operation. Some typical values are summarized below.

Type of Operation	Equalization volume needed as a fraction of maximum daily demand
Constant pumping	0.10 – 0.25
Follow demand (constant speed)	0.05 – 0.15
Off peak pumping	0.25 – 0.50
Variable speed pumps	0

The higher values in the list are for systems with fairly peaked demands, and the lower values are for those with a flatter daily demand curve. For example, a utility with 7.5

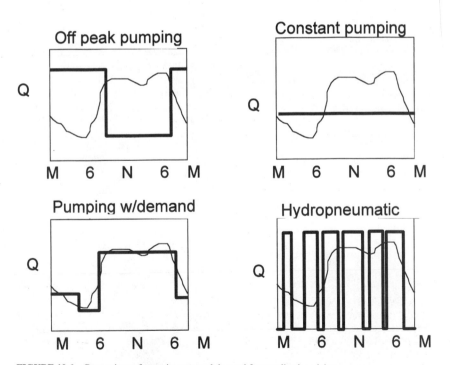

FIGURE 10.6 Comparison of pumping rate and demand for equalization sizing.

ML/day (2 Mgal/day) of maximum daily demand, would need roughly 1.5 ML (0.4 Mgal) of equalization storage (using a value of 0.2). Equalization storage can be checked using extended-period simulation models pipe networks.

10.6.3.2 Fire Storage. Fire storage requirements are based on the need for water to supplement the capacity of the water supply. If the capacity of the water supply is such that it can provide fire flow while still meeting maximum daily demand, no fire storage is required. This is sometimes the case in extremely large systems where fire demands are a tiny fraction of maximum daily demands.

The flow from fire storage required (in excess of equalization storage) can be given by

$$SSR = NFF + MDC - PC - ES - SS - FDS \qquad (10.2)$$

where SSR = storage supply required, NFF = needed fire flow, MDC = maximum daily consumption, PC = production capacity, ES = emergency supply, S = suction supply, and FDS = fire department supply.

All the above quantities are in flow units (volume per time), and the equation is based on the Fire Suppression Rating Schedule (Insurance Services Office, 1980). The storage supply required is water that must be delivered from storage. The needed fire flow is determined by the size and occupancy of the structure with the largest fire demand. The emergency supply is the water that can be brought into the system from connections with other systems. The suction supply is the supply that can be taken from nearby lakes and canals during the fire and cannot exceed the needed fire flow. The fire department supply is water that can be brought to the fire by trucks. The production capacity is either based on the capacity of treatment plant, the well capacity, or the pump capacity, depending on the system, as is shown in an example below. Except for the largest systems, it is usually safe to assume that only one major fire will occur at any given time.

The amount of water delivered to a fire can be analyzed best using pipe network models. Brock (1993) provided a graphical method for determining the amount of water that will flow from storage as opposed to the system source.

In large systems, where needed fire flow is small compared with treatment capacity, little fire storage is required. On the other hand, large amounts of fire storage, compared with equalization storage, are required for small- to medium-sized utilities with a structure that has a large needed fire flow.

Once the storage supply requirement is determined in flow units, the actual volume of fire storage must be established multiplying the requirement by the duration of the fire. For modest-sized fires, the duration is given below (AWWA, 1998):

Needed fire flow (gpm)	Needed fire flow (L/s)	Duration (h)
Less than 250	Less than 157	2
3000 – 3500	189 – 220	3
4000 – 12,000	251 – 755	4

In the United States, needed fire flows are usually rounded to the nearest 500 gpm (31.4 L/s) for flows in this range.

Multiplying the duration by the storage supply requirement gives the fire storage needed. This is compared with the available storage for fire protection, which is storage below the normal low level of equalization storage. For example, if a tank holds 0.5 ML (0.133 MG) and, at the normal low level, the stored volume is 0.3 ML (0.080 MG), only 0.3 ML (0.088 MG) is available for fire protection.

Hydraulic Design of Water Distribution Storage Tanks **10.15**

The calculations for a simple system are illustrated in Fig. 10.7. (The 0.5 ML is the existing volume in storage at normal low level.) The clearwell storage can be counted as available storage when one considers the plant as limiting, but the clearwell storage does

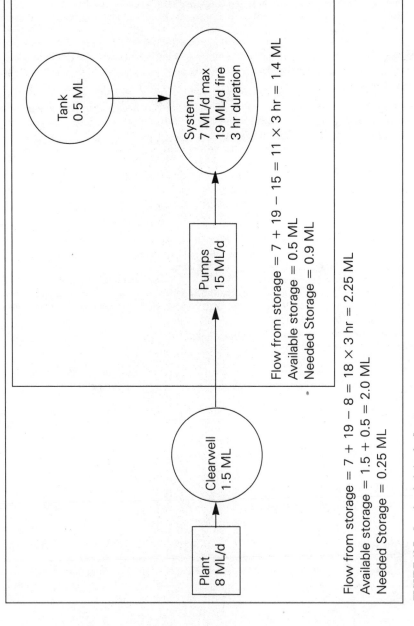

FIGURE 10.7 Sample calculation for fire storage.

not count toward available storage when the pump station is considered to be limiting. In this case, the plant is the limiting factor. When you look only at the overall system, you might conclude that only 0.25 ML of storage is needed. However, because the pumping system only can deliver 15 ML/day, it therefore cannot move water out of the clearwell fast enough and, actually, 0.9 ML of storage is needed.

Although the above calculation appears to be fairly simple, several decisions must be made. For example, maximum daily consumption (MDC) changes over time with growth of the utility. Selecting the year corresponding to the MDC can make a significant impact on volume required. Some would argue that MDC = PC (production capacity) is a good conservative assumption. Needed fire flows can be affected by use of fire resistant building materials, fire, walls and fire sprinkler systems in buildings. The extent to which these factors are required by the local building code can make a difference in storage requirements.

10.6.3.3 Emergency Storage. No formula exists for determining the amount of emergency storage required. The decision will have to be made on a judgment about the perceived vulnerability of the utility's water supply.

If a utility has several sources and treatment facilities with an auxiliary power supply (or power supplied from several sources), the need for emergency storage is small. Some storage should be available to handle a catastrophic pipe break that cannot be isolated easily.

If a utility has a single source without auxiliary power and a relatively unreliable distribution system, a significant volume of emergency storage is prudent.

10.6.3.4 Combination Equalization, Fire, and Emergency Storage. The volume of effective storage required should be based on a combination of equalization, fire, and emergency storage. Some engineers use the sum of the three types of storage, whereas others base design on the sum of equalization storage plus which is larger fire or emergency storage. The logic in such cases is that the fire is not likely to occur at the same time as a critical pipe break or power outage. The total storage can be summarized by equalization + maximum (fire, emergency).

The most economical tanks are constructed in standard sizes, so the number above is rounded (usually upward) to a standard size.

10.6.3.5 Summary of Functional Sizing. The results from an analysis of storage requirements, based on functional sizing should yield storage requirements similar to regulatory-based sizing. If there are significant differences, the utility and the regulatory agency need to work together to determine the appropriate storage requirement. No volume of storage can protect a utility from every possible emergency (e.g., several catastrophic fires and pipe breaks during a plant outage). However, the guidance above should provide for reasonable amounts of storage.

10.6.4 Staging Requirements

If a utility has a fairly slow rate of demand growth, then the volume required in storage should not increase dramatically over time. However, for fast growing utilities, there are significant questions concerning how tank construction should be staged.

For example, the typical question is whether a utility, which will ultimately need 1.88-ML (500,000 gal) of storage, should construct a 1.88-ML (500,000 gal) tank today or a

Hydraulic Design of Water Distribution Storage Tanks **10.17**

0.94-ML (250,000 gal) tank today and another 0.94-ML (250,000 gal) tank later. The key to the problem is the expected number of years the first, smaller tank will be adequate before the second is required. Suppose that the cost of the 1.8-ML tank is $500,000 and the cost of the smaller tank is $350,000. The present worth cost of the two smaller tanks for an interest rate of 7percent is:

$$\frac{\$350,000 + \$350,000}{(1+0.07)^n} \qquad (10.3)$$

where n = the number of years before the second tank is needed. If n is 5, the present worth is $600,000 and the single large tank is best. If n is 12 years, the alternatives are equal. If n is 20 years, the present worth is $440,000 and the two-tank alternative is superior. The break-even year will become larger as the interest rate decreases.

Economics is not the only consideration in the staging of storage tank construction. Having two properly located tanks in the system gives the utility more flexibility in its operations (e.g., if a tank must be taken off-line for painting and inspection). Figure 10.8 shows an example of two tanks located at a single site.

10.6.5 Useful Dead Storage

Some utilities may find that they need additional fire or emergency storage, but have a tall standpipe with only enough storage for equalization. One approach for converting the dead storage in the bottom of the tank into useful storage is to install emergency pumps that can withdraw water from the bottom during a fire or other emergency. In this way, the dead storage in the bottom of the tank can be used.

In such a situation, the tank should be refilled through a pressure-sustaining valve to prevent a localized drop in pressure when the tank is refilled. If the tank is the only storage in the system, backup power for the pump will be required, controls will be needed to

FIGURE 10.8 Use of two tanks at a site. (Photograph by T. M Walski).

prevent overpressurizing of the system, and a surge analysis will need to be performed to ensure that the pumping will not cause waterhammer.

10.7 OTHER DESIGN CONSIDERATIONS

10.7.1 Altitude Valves

Altitude valves are used to prevent tanks from overflowing by shutting off inflow to the tank when the water level in the tank approaches a high level. These valves are usually located in a vault at the base of a tank. Appropriate valving should be installed to enable the tank to function while the altitude valve is being serviced.

The two main types of altitude valves are single acting and double acting. The former type only allows water to flow into the tank; the latter type allows water to flow both into and out of the tank. When a single-acting valve is used, there must be a parallel line with a check valve that allows water to flow out of the tank. Although a double-acting valve is less expensive than a single-acting valve plus a check valve, there is usually a small risk that the double-acting valve may fail to open during an emergency. Therefore, a single-acting valve with a check valve is considered to be more reliable.

Although altitude valves provide some protection against overflow, they are not used when conditions that would cause an overflow are unlikely to occur: for example, when the tank level is monitored through a Supervisory Control And Date Acquisition (SCADA) system, with appropriate alarms and the impact of an overflow is minimal. Closing an altitude valve on the last tank in a pressure zone when pumps are running and demand is low could cause pressure to increase dramatically in some systems.

10.7.2 Cathodic Protection and Coatings

The presence of water and air in metal tanks causes the tanks to corrode rapidly if not protected adequately. Numerous types of paints and coatings are available. These coatings must be approved by the appropriate state regulatory agency and applied in accordance with AWWA D102, (AWWA, 1997a). Different types of paints are required for the interior and exterior of the tanks. Because new types of coatings are introduced and approved regularly, it is important for the design engineer to keep in contact with coating suppliers. Good inspection of the painting and frequent follow-up inspections are needed to prevent failure of the coating (Drisko, 1980; Dubcak, 1994; Knoy, 1983; Roetter, 1987).

Many metal tanks also are equipped with cathodic protection systems that further protect metal on the inside of the tank by offsetting the natural corrosion currents that set up when metal is in contact with an electrolyte (water). Cathodic protection of tanks involves passing direct current through a system of electrodes suspended in the water. Although benefits of cathodic protection have been documented, significant problems can develop in tanks in areas where formation of ice damages the cathodic protection system. Installation of cathodic protection systems is described in AWWA D104 (AWWA 1998a).

10.7.3 Overflows and Vents

Tanks should have an overflow pipe capable of handling the maximum potential overflow volume from the tank (e.g., failure of altitude valves with pumps running). The pipe should have an air gap at its discharge and a check valve that can prevent birds and insects from entering the pipe. Depending on the applicable environmental regulations and the potential for flooding downstream, a detention basin and erosion protection may be required at the tank.

The draining and filling of tanks also requires that a large volume of air enters and leaves the tank during each cycle. These vents should be screened to prevent birds and insects from entering the tank. In climates subject to freezing, a frostproof design is needed to prevent ice from blocking the vents.

REFERENCES

ASME, *Boiler and Pressure Vessel Code*, 1998, American Society of Mechanical Engineers, New York.

AWWA, *Circular Prestressed Concrete Water Tanks with Circumferential Tendons, AWWA D115-95*, American Water Works Association, Denver, CO, 1995a.

AWWA, *Wire and Strand Wound Circular Prestressed Concrete Water Tanks, AWWA D110-95*, American Water Works Association, Denver, CO, 1995b.

AWWA, *Welded Steel Tanks for Water Storage, AWWA D100-96*, American Water Works Association, Denver, CO, 1996.

AWWA, *Coating Steel Water Storage Tanks, AWWA D102-97*, American Water Works Association, Denver, CO, 1997a.

AWWA, *Factory Coated Bolted Steel Tanks for Water Storage, AWWA D103-97*, American Water Works Association, Denver, CO, 1997b.

AWWA, *Automatically Controlled, Impressed Current Cathodic Protection for the Interior of Steel Water Tanks, AWWA C104-98*, American Water Works Association, Denver, CO, 1998a.

AWWA, *Distribution Requirements for Fire Protection, AWWA Manual M-31*, American Water Works Association, Denver, CO, 1998b.

AWWA, *Steel Water Storage Tanks, AWWA Manual M-42*, American Water Works Association, Denver, CO, 1998c.

Brock, P.D., *Fire Protection Hydraulics and Water Supply Analysis*, Fire Protection Publications, University of Oklahoma, Norman, OK, 1993.

Cesario, A. L., *Modeling Analysis and Design of Water Distribution Systems,* American Water Works Association, Denver, CO, 1995.

Drisko, R.W., "Methods of Quality Control of Coatings," *AWWA Distribution Symposium,* Los Angeles, CA, 1980, p. 103.

Dubcak, T.O.,"Marriage Between Paint and Steel Tanks–The First Anniversary is Crucial," *AWWA Annual Convention*, New York, p. 569, 1994.

Great Lakes and Upper Mississippi River Board of State Public Health and Environmental Managers, "Recommended Standards for Water Works", Albany, NY, 1992.

Hodnett, R. M., "Water Storage Facilities and Suction Supplies," *Fire Protection Handbook*, National Fire Protection Association, Quincy, MA, 1981.

Kennedy, M. S., et al., "Assessing the Effects of Storage Tank Design on Water Quality," *Journal of American Water Works Association*, 85, (7): 78, 1993.

Knoy, E. C., "What You Should Know About Tank Maintenance," *AWWA Distribution Symposium*, Birmingham, AL, 1983, p. 205.

NFPA, *Standards for Water Tanks for Private Fire Protection, NFPA 22*, National Fire Protection Association, Quincy, MA, 1998.

Insurance Services Office, "Fire Suppression Rating Schedule," ISO Commercial Risk Services, Duluth, Ga, 1980.

Roetter, S., "Rehabilitation and Maintenance of Steel Water Storage Tanks," T. M. Walski (Ed.), in American Society of Civil Engineering, 1987.

Texas Department of Health, *Rules and Regulations for Public Water Systems*, Austin, TX, 1988.

Walski, T. M., J. Gessler, and J. A. W. Sjostrom, *Water Distribution Systems: Simulation and Sizing*, Lewis Publishers, Chelsea, MI, 1990.

CHAPTER 11
QUALITY OF WATER IN STORAGE

Walter M. Grayman
Consulting Engineer
Cincinnati, OH

Gregory J. Kirmeyer
Economic and Engineering Services, Inc.
Bellevue, WA

11.1 INTRODUCTION

Storage facilities are an essential component of water distribution systems. Traditionally, the facilities have been designed and operated to meet hydraulic requirements: providing emergency storage, equalizing pressure, and balancing water use throughout the day. However, the design and operation of storage facilities also can negatively affect water quality, and these potential impacts should be considered in designing, operating, and maintaining distribution system tanks and reservoirs. This topic has been addressed in depth in two recently published research reports sponsored by the AWWA Research Foundation (Grayman et al., 1999; Kirmeyer et al., 1999).

11.1.1 Overview

Treated water storage tanks and reservoirs are an integral part of most drinking-water distribution systems. They are used to equalize pumping requirements and operating pressures and to provide emergency water for fire fighting and pumping outages. These tanks come in a variety of sizes, can be ground level or elevated, and can operate in either

---

The information in this chapter draw heavily on work performed by teams led by the authors on two recently completed projects of the American Water Works Association Research Foundation (AWWARF) titled "Water Quality Modeling of Distribution System Storage Facilities" and "Maintaining Water Quality in Finished Water Storage Facilities." The Foundation financial support and the foresight in sponsoring these important research topics is gratefully acknowledged. Lewis Rossman, Rolf Deininger, Clifford Arnold, Charlotte Smith, James Smith, and Rita Schnipke were the investigators on the first project, and Lynn Kirby, Brian M. Murphy, Paul F. Noran, Kathy Martel, Theodore W. Lund, Jerry L. Anderson, and Richard Medhurst were the investigators on the second project. Comments and views presented may not necessarily reflect the views of AWWARF officers, directors, affiliates or agents.

a simultaneous inflow/outflow mode or in a fill-and-draw mode. Distribution storage will be defined as facilities that are not part of the treatment or contact time requirements.

From a water-quality standpoint, however, their impact is mainly negative. Long detention times spent in storage can cause water to lose its disinfectant residual through bulk water decay, can promote microbial regrowth in the distribution system, and can cause harmful disinfectant by-products to increase, especially in the presence of free chlorine residual. Poor mixing can worsen these impacts by creating pockets of water with even longer residence times than normal. Thus, the long detention times can be attributed to two problems—underutilization (lack of use) and poor mixing. Furthermore, improper maintenance can provide a pathway to the introduction of contaminants into the storage facility or can facilitate their growth.

11.1.2 Definitions

Distribution system storage tanks and reservoirs. Storage facilities located in or part of the distribution system as opposed to storage that is part of the treatment process.

Detention time. Average length of time that water is resident in a storage facility.

Standpipe. A tank that is generally cylindrical in shape with a height greater than its diameter.

CFD (computational fluid dynamics). A mathematical modeling system for simulating the dynamics of motion in a liquid or gas.

Systems model. A class of models in which physical processes are represented by conceptual, empirical relationships.

Clearwell. A unit process in a treatment plant generally used to provide contact time or mixing at the end of the treatment trains.

11.2 WATER QUALITY PROBLEMS

Three main categories of problems occur in storage facilities—chemical, microbiological, and physical. Many problems fit into more than one category. These potential water-quality problems, possible causative factors, and potential methods for improvements are discussed in this section.

11.2.1 Chemical Problems

Several problems associated with finished water-storage facilities are caused by or are the result of a chemical reaction. These include, but are not limited to, loss of disinfectant residual, formation of disinfection by-products, development of taste and odor, increase in pH, corrosion, buildup of iron and manganese, occurrence of hydrogen sulfide, and leachate from internal coatings. Loss of disinfectant residual and formation of disinfection by-products are the most common chemical problems, and loss of disinfectant residual can lead to other microbiological problems discussed below.

11.2.1.1 Loss of disinfectant residual. The loss of disinfectant residual is a chemical process resulting in the decrease of the secondary disinfectant, generally either free chlorine or total chlorine. It is probably the most common water-quality concern and is a function of time and rate of disinfectant decay (loss). The rate of loss can be affected by microbiological activity, temperature, nitrification, exposure to ultraviolet light (sun), and amount and type of disinfectant-demanding compounds present, such as organic and inorganic compounds. Since the volume of water in a storage facility normally is large compared to the amount of exposed surface area of the container, the effect of the wall and floor on disinfectant decay are normally not significant. Thus, disinfectant decay in storage facilities normally can be attributed to bulk decay rather than to wall effects. A finished water-storage facility can affect the loss of disinfectant by increasing the amount of time the water remains in the distribution system before serving the customer.

The actual detention time of the stored water can be excessively long, approaching weeks or even months in facilities with poor turnover rates. A long detention time can allow the disinfectant residual to be completely depleted, thereby not protecting the finished water from possible microbial regrowth that might occur in the distribution system downstream of the storage facility. The result could adversely affect the quality of water provided to the customer and compliance with drinking water regulations.

11.2.1.2 Formation of disinfection by-products. Disinfection by-products (DBPs) are formed when the disinfectant used reacts chemically with the organic material in the treated water. The types of DBPs formed vary with the type of primary disinfectant used. Generally, DBPs are higher in sources of surface water than sources of ground water because of the higher levels of natural organic matter. Organic material also can enter open finished water facilities directly or can be increased internally by growth of algae. The greatest focus has been on regulating chlorinated DBPs, such as trihalomethane compounds (THMs). Free chlorine tends to produce much higher DBP levels compared with combined chlorine. Thus, all other things being equal, utilities practicing chloramination for secondary disinfection typically have lower levels of THMs and haloacetic acids (HAAs) in the distribution system.

According to Règli (1993), the main factors influencing formation of DBPs include contact time, chlorine dose and residual, temperature, pH, concentrations of precursors and bromide ions.

A finished water-storage facility can affect four of these six factors. As indicated above, contact time (detention time) in storage facilities is increased and longer contact times normally increase the formation of DBPs. Rechlorination systems within storage facilities can expose the water to higher chlorine residuals, increasing the potential for more DBPs. Higher water temperatures resulting from steel tanks during summer seasons can increase DBPs because chemical reactions proceed faster and go further at higher temperatures. Finally, pH levels can increase in some storage facilities caused by the leaching of hydroxides and carbonates from concrete surfaces, thereby increasing the potential for formation of THMs.

11.2.1.3 Development of taste and odor. Burlingame and Anselme (1995) reported on the sources of tastes and odors in the distribution system, including emissions from construction materials, external contaminants, biological activity, disinfectant residuals, and DBPs.

Construction materials used in finished water storage facilities have caused taste and odor problems. Burlingame and Anselme (1995) cited numerous examples of odoriferous organic solvents leaching from reservoir linings. For example, complaints of solvent-like odors in one water distribution system were found to be coming from organic contaminants in bitumastic and epoxy reservoir linings.

Anaerobic biological activity can develop in stagnant areas of the distribution system, producing sulfur compounds, such as hydrogen sulfide gas. Accumulated material that settles on the bottom of finished water storage facilities could create conditions conducive to anaerobic activity. The loss of chlorine residual and subsequent biological growth can theoretically lead to off flavors. This situation caused customer complaints at one set of standpipes in Philadelphia (Burlingame and Brock, 1985).

Burlingame and Anselme (1995) summarized research by Montiel (1987) in Paris, where a strong musty taste in the water was linked to fungi that transformed chlorophenols to chloroanisoles. "Chlorine introduced into a properly treated surface water increases the taste threshold of the water, depending on the chlorine dose and the contact time." Thus, rechlorination must be done with care. DBPs, such as THMs, can produce medicinal tastes, especially when iodine or bromine are present or algae growth occurs.

Burlingame and Anselme (1995) presented the issue of shelf life on taste and odor. Irrespective of construction materials, the taste or odor of water may change as it stands. Under some conditions, chlorine odors can be reduced; however, the loss of chlorine can unmask underlying tastes and odor.

Cross-connections are another source of external contaminants causing tastes and odors. Burlingame and Anselme (1995) cited several examples from an American water system. Cross-connections between a storage facility's drain or overflow pipes could lead to similar contamination problems. Utilities must be diligent in checking to ensure that no cross-connections occur associated with the storage facilities.

11.2.1.4 Increase in pH. Lack of pH stability within the distribution system can contribute to water-quality degradation. A stable pH is essential for developing and maintaining effective corrosion control passivating layers. In addition, the higher pH levels that can occur during storage can increase the potential for higher levels of THMs. pH levels in storage facilities can change in several ways. If the level carbon dioxide (CO_2) in source water is high-as it is in many groundwaters—CO_2 may evolve in the storage facility and the pH will rise. Rechlorination can change pH levels either up or down, depending on the chemical used in the rechlorination process.

Concrete storage facilities, especially new ones, can increase the pH of the water in contact with the walls and floor. Long-term exposure to the walls, such as a long detention time in a storage facility, can affect the magnitude of the increase in pH. Decreasing this detention time can help to alleviate the problem. Operationally, a utility can fluctuate the water level more or provide an internal coating over the areas in contact with the water.

11.2.1.5 Corrosion. Customers may notice that the water is reddish if there is a problem with iron uptake from the exposed metal surfaces of storage facilities. The amount of reddish water is dependent on many factors, including the pH, alkalinity and temperature of the water, whether there is proper cathodic protection, the integrity of any internal coating, and the flow of the water. Prevention of internal corrosion of the storage facility surfaces may require a properly installed, calibrated, and maintained cathodic protection system and an internal coating that is inspected and maintained properly.

11.2.1.6 Buildup of iron and manganese. Dissolved iron and manganese in the source water can precipitate out of solution and settle in storage facilities, particularly if there is a long detention time or an oxidant, such as chlorine or oxygen, is present. If an improper dose of sequestering agent is used or detention times are long, the iron and manganese may still come out of solution in a storage facility or in the distribution mains. Sediment that accumulates in water mains can be stirred up when the water velocity increases because of water main or hydrant breaks, fire demand, or flow testing of fire hydrants (AWWA, 1986a). This sediment can enter the reservoir, where it resettles. To help prevent the metals from building up to a point where they adversely affect the quality of the water, a regularly scheduled cleaning program to remove sediment buildup from storage facilities may be beneficial. Iron and manganese and other precipitates that settle in a finished water storage facility may be reintroduced to the finished water during rapid discharge from the facility.

11.2.1.7 Occurrence of hydrogen sulfide. Hydrogen sulfide gas is an aesthetic concern because of its rotten egg odor. It can occur naturally in groundwaters with little to no dissolved oxygen, or it has a greater tendency to form in distribution systems when the following conditions are present: high levels of sulfate ions, sulfate-reducing bacteria, excess electrons, and low or no dissolved oxygen. The water's detention time in the distribution and storage system is a contributing factor, affecting the reaction time and the resultant hydrogen sulfide concentration (Pettie, 1990). A storage facility can greatly increase the overall detention time of the distribution system. An aeration or ventilation system at a storage facility also may improve conditions.

11.2.1.8 Leachate from internal coatings. Chemicals contained in protective coatings and linings may leach out into the finished water over time, depending on the chemical composition, the rate of migration, and the water temperature. After a coating has been applied to a finished water storage facility, volatile organic compounds could be introduced to the water if sufficient curing time is not allowed. An American Water Works Association Research Foundation (AWWARF) study of five organic coatings used in potable water distribution systems determined that drinking water contaminants at levels above 1 µg/L are common during the first 30 days in service (Alben et al., 1989). Organic paints and coatings often contain nutrients that can support bacterial growth. Organic concentrations from leachate decreased at a faster rate when in contact with water, compared to prolonged air drying at the same temperature. The AWWARF study also looked at the effect of chlorine on leachate concentrations and found that chlorinated by-products in leachate contributed minimally to the total concentrations of leachate compounds in the water.

11.2.2 Microbiological Problems

Microorganisms can be introduced into storage facilities from biofilms in distribution system mains, new or repaired water mains or storage facilities that were not properly disinfected, inadequately treated water, surface or ground-water infiltration into ground storage facilities, and cross-connections.

Additionally, microorganisms can enter from outside sources, such as uncovered reservoirs; poorly constructed or inadequately maintained storage facility covers, roofs, or sidewall joints; and faulty vents, hatches, and other penetrations. For example, several documented cases of salmonellosis were caused by drinking water reservoirs contaminated with bird droppings (Ongerth, 1971; Smith and Burlingame, 1994). Salmonellosis is a serious disease and can cause death. Other sources of microbial contamination include the atmosphere through the air-water interface in finished water storage facilities and through water main breaks (Antoun et al., 1995).

11.2.2.1 Bacterial regrowth. Bacterial regrowth and biofilms are typically more of a problem in piping systems than in storage facilities because of the greater surface-area-to-water-volume ratio when compared to storage facilities. However, after storage, water may have less disinfectant residual and be warmer; both factors increase the propensity for regrowth and biofilms in the piping system. Smith et al. (1990) identified the following factors that provide a favorable environment for microbial growth in the distribution system: the seasonal variation in water temperature, the availability of growth-promoting nutrients and minerals, the occurrence of corrosion products in the distribution system, distribution system disinfection practices, and distribution system hydrodynamics (flow and velocity).

Bacterial growth is common on tank surfaces and in other noncirculating zones of a tank. The presence of microorganisms can be a regulatory concern and can contribute to increased chlorine demand, red water problems, lowered levels of dissolved oxygen, taste and odor problems, and nitrification. Factors that provide optimum conditions for microorganisms to multiply include long water-detention times, adequate nutrient levels, and warm temperatures. These biofilm slimes can make it difficult to maneuver safely during reservoir inspections and cleaning and may require extraordinary methods to remove them. Shoenen (1992) reported that bacterial regrowth in service reservoirs was largely system-dependent. His study revealed that growth in the service reservoir was caused by water movement, leading to resuspension of bacteria adhering to the walls of the water chamber, rather than the influence of dissolved substrates in the water body.

Organic paints and coatings often contain nutrients that can support bacterial growth (Schatz, 1992; Sonntag, 1986). This process can lead to biocorrosion of the reservoir's structure (Boudouresque, 1988) or to increased porosity of the walls, thus creating spaces for bacterial colonies. All materials used in distribution systems are not necessarily intended to contact water directly; however, in some instances, pores in linings and fittings have enabled these materials to come into direct contact with water and release chemical agents that may be nutritive to bacteria. Bellen et al. (1993) refined laboratory test methodologies that utilities can use to measure the potential of materials in contact with potable water to support biological growth.

11.2.2.2 Nitrification. Nitrification is a two-step process caused by bacteria that includes the conversion of ammonia to nitrite, then nitrite to nitrate. Nitrification can occur in drinking-water systems with a natural presence of ammonia or in systems that add ammonia as part of the chloramine disinfection treatment process (Kirmeyer et al., 1993). Excess free ammonia can be present as the chloramine residual dies off or if the chlorine-to-ammonia ratio is too low. An estimated two-thirds of medium and large systems in the United States that chloraminate experience nitrification to some degree (Kirmeyer et al., 1995).

Certain water-quality and system conditions can result in nitrification in the distribution system. Important water-quality factors include pH, temperature, chloramine residual, ammonia concentration, chlorine-to-ammonia-nitrogen ratio, and concentrations of organic compounds. Important system factors include system detention time, reservoir design and operation, sediment and tuberculation in piping, biofilm, and the absence of sunlight.

Nitrification in the distribution system or in a storage facility can have several effects on water quality. For example, it may degrade chloramine residuals, consume dissolved oxygen, slightly decrease pH, increase heterotrophic bacterial populations, increase nitrite and nitrate levels, increase concentrations of organic nitrogen, and reduce ammonia concentrations in the distribution system.

From a storage-facility standpoint, one method of reducing nitrification would be to increase the turnover rate to decrease overall detention time. Another method is to remove the facility from service and practice breakpoint chlorination. Operationally, changing the chlorine-to-ammonia ratio also may help to reduce the occurrence of nitrification by reducing the available ammonia.

11.2.2.3 Worms and Insects. It is possible for worms or insects to enter the distribution system through finished water-storage facilities, cross-connections, dead-ends, or from stirred-up sediment in the bottom of distribution mains. Midge fly larvae, which look like small worms, can enter a storage facility that is not equipped with secure insect screening on vents and openings. Worms or insects can develop at dead-ends using accumulated organic material as a food supply (AWWA, 1986b).

11.2.3 Physical Problems

A water-quality problem may be related to a physical occurrence, such as accumulation of sediment and direct entry of contaminants into storage facilities. These physical problems can lead to other chemical or microbiological problems if not addressed by either a scheduled maintenance program or treatment of source waters. Uncovered distribution storage facilities present special contamination concerns that must be addressed by utilities.

11.2.3.1 Sediment buildup. In addition to iron and manganese buildup in storage facilities because of chemical precipitation, other suspended material present in the source water may settle out in storage facilities. Particulates may be introduced to the distribution system if the water treatment facility is not working properly. Sediment frequently accumulates in storage tanks where the velocities are minimal. Sediment can be resuspended because of surges in flow, and sediment is a potential cause of water-quality degradation by contributing to the water's chlorine demand. One utility traced elevated levels of coliform bacteria in reservoir water to accumulated sediment (Beuhler et al., 1994). In another study, the authors speculated that a taste and odor problem was linked to the bottom sediment, where substrate is abundant for diverse biological growth (Burlingame and Brock, 1985).

Block et al. (1996) collected sediment samples from finished water-storage facilities and water mains and determined their composition. The sediment samples varied in their appearance, with colors ranging from reddish to yellow-brown to dark brown. Their contents were mainly grains or flocs or a mixture of the two. The samples of reservoir sediment had the following composition: insolubles (18 percent), iron oxides (19 percent), volatile solids (19 percent of total), aluminum hydroxides (15 percent), calcium carbonates (10 percent), and unknown (16 percent). Block et al. (1996) reported that the reservoir and pipe sediment samples were heavily colonized by microorganisms.

One method of preventing the accumulated sediments from reentering the distribution system once settling in a storage facility has occurred is to install a riser pipe on the outlet. Avoiding flow situations that would scour the bottom surface or changing the inlet-outlet configuration to improve flow patterns also may keep the sediments from reentering the distribution system. In the case of an improperly operated water treatment plant, a plant evaluation to determine more effective treatment operations may be beneficial. Cleaning the storage facilities to minimize sediment buildup also is important. Flushing of the distribution system to remove accumulated sediments before they reach the storage facility is in order as well.

11.2.3.2 Entry of contaminants. The Water Industry Data Base statistics from 1992 estimated that 300 uncovered reservoirs were in use in the United States as finished water-storage facilities at that time. These uncovered reservoirs provide the largest opportunity for entry of contaminants into the distribution system. Open reservoirs or storage facilities potentially are subject to contamination from bird droppings and other animal excrement that have the potential to transmit disease-causing organisms to the finished water.

Microorganisms also can be introduced into open reservoirs from windblown dust, debris, and algae. Algae proliferate in open reservoirs with adequate sunlight and nutrients. Organic matter, such as leaves and pollen, also are a concern in open reservoirs. Covering an open reservoir or replacing it with a covered storage facility can reduce or eliminate the potential for direct entry of contaminants. Properly designed and maintained rigid covers afford the best protection; however, flexible membrane covers are considered to be a cost-effective alternative by many utilities.

Covered storage facilities are generally protected much better than open reservoirs; however, they can be susceptible to airborne microorganisms entering through access hatches, roofs, vents, and other penetrations. Several hundred cases of illness and five deaths were reported in a Missouri community that had a system where bird droppings and avian remains were found inside a closed storage tank (Atkinson, 1995). Maintaining screens, locking all access hatches, repairing holes in covers, and conducting frequent inspections will help to minimize the introduction of contaminants into covered storage facilities.

11.2.3.3 Temperature. The temperature of the water can change in storage facilities and can increase or decrease depending on the type of structure and how it is operated. This can lead to stratification in a tank that can inhibit mixing. This topic is addressed in Sec. 11.3.2.

11.3 MIXING AND AGING IN STORAGE FACILITIES

The water-quality changes that can occur within finished water-storage tanks and reservoirs are influenced significantly by the degree of mixing and subsequent residence time that water experiences within these facilities. Inflow and outflow and movement of fluid within these structures determines mixing and residence times.

11.3.1 Ideal Flow Regimes

There are two theoretical ways in which water can flow through a storage tank or reservoir: in a completely mixed state or in a completely unmixed state. These two ideal flow regimes, termed mixed and plug flow, are illustrated in Fig. 11.1. In mixed flow, water entering the reservoir mixes instantaneously and completely with water in the reservoir, resulting in a uniform mixture at all times. The composition of water discharging from the reservoir is the same as the uniform composition in the reservoir at the time of discharge. In plug flow, water moves through the reservoir without mixing with the water in the reservoir. This produces a first-in, first-out ordering.

Many factors affect the actual mixing processes in a reservoir, thus resulting in flow that is neither fully mixed nor plug flow. Factors that can lead to nonideal flow conditions include thermal effects, interaction between inlet and outlets resulting in short-circuiting, stagnant zones in which flow is incomplete, and small-scale eddies.

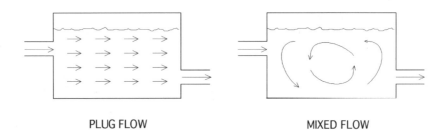

PLUG FLOW   MIXED FLOW

FIGURE 11.1 Ideal flow regimes.

Unit processes, such as chlorine contact chambers and clearwells at water treatment plants, are generally designed for plug-flow operation to ensure that all parcels of water experience approximately the same residence time. Distribution system storage facilities may be designed to encourage mixed flow or plug flow. A major factor influencing the selection of a flow regime is the minimization of disinfectant loss. Since the rate of disinfectant loss with respect to time is concentration-dependent (Fair et al., 1948), the higher the concentration of chlorine, the faster it decays because of reactions with other constituents in the water. When reaction kinetics exhibit this type of behavior, a tank operated under plug flow will have more disinfectant loss than it would have under mixed flow (Grayman et al., 1999).

Plug flow is best achieved in long, narrow, shallow basins or through the use of baffling, diffusers, or stilling chambers to reduce flow momentum. Near-complete mixing has been observed in many storage tanks through normal jet mixing without the use of special structural additions or mixing devices (Baur and Eisenbart, 1982; Boulos et al., 1996; Grayman and Clark, 1993; Hammer and Marotz, 1986; Kennedy et al., 1993; Poggenburg et al., 1981).

11.3.2 Jet Mixing

When water enters a storage tank, a jet of water is formed. Jet flow can be classified as either laminar or turbulent. For circular jets, the Reynolds number remains constant throughout the jet structure and is equal to that of the flow exiting the inlet pipe. Fully turbulent jets have been characterized as those with Reynolds numbers above 3000, whereas fully laminar jets have values below 1000 (McNaughton and Sinclair, 1966). For a turbulent jet, as the water moves from the inlet, the diameter of the jet expands because of entrainment of the surrounding water into the jet. When the jet reaches a free surface or a boundary, the flow direction changes, resulting in further mixing. This behavior is illustrated in Fig. 11.2 in a series of diagrams adapted from Okita and Oyama (1963). Turbulent jet mixing serves as the primary mixing mechanism in a storage tank. For a laminar jet, the entering fluid and the ambient fluid remain separate entities, with mixing occurring only by the much slower mechanism of molecular diffusion.

Thermal forces also affect the mixing of a jet. When the incoming water is warmer or cooler than the tank contents, then the resulting jet is buoyant and the temperature differences and associated density differences provide another source of motion. A positively buoyant jet (an inflow temperature greater than the ambient temperature) will tend to rise, whereas in a negatively buoyant jet (an inflow temperature less than the ambient temperature), vertical movement is inhibited. Temperature differences between the entering water and the ambient water may result from heat exchange through the tank walls, with the air space above the water surface, or through variations in inflow temperatures because of operational changes in the system. The behavior of a buoyant jet and its effectiveness as a mixing device is governed by the interplay of the two driving forces: momentum and buoyancy.

11.3.3 Mixing Times

Jet mixing has commonly been used in the chemical processing and oil and gas industries as an effective means of mixing the contents of large storage tanks. Experience in those industries can be applied to mixing in drinking-water storage facilities. In the oil and gas field, Fossett and Prosser (1949) studied the effect of inlet diameter and orientation on mixing in tanks of differing diameter and height. In that study and others, the performance

11.10 Chapter Eleven

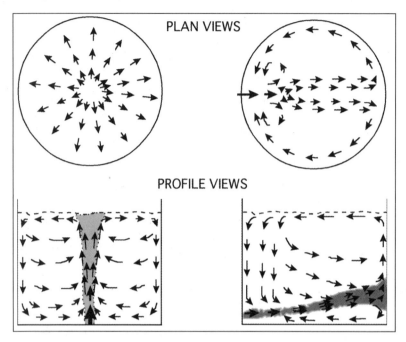

FIGURE 11.2 Views of jet mixing (Based on Okita and Oyama, 1963).

of jet mixers is generally measured by determining a "mixing time." The mixing time is the time needed for a known amount of added tracer material to reach a specific degree of uniformity in the tank. The result of several studies that have related mixing times to tank dimensions and flow rates are discussed in Fossett (1973), Hyman (1962), Rossman and Grayman (1999) and Simpson (1975) and are summarized in Table 11.1. Each formula indicates increased mixing times as the tank size increases and decreased mixing times as the momentum flux (flow $\times$ velocity) increases.

11.3.4 Stratification

Water in a tank can exchange heat with the walls and air space of a storage facility, which in turn can exchange heat with the ambient environment. The resulting cooling or heating can create a temperature difference between the contents of a tank and the tank's inflow. Temperature differences also can result from abrupt changes in the temperature of a tank's inflow caused by temperature changes in the source water or by changes in the source of the inflow. The temperature difference between a tank's contents and inflow creates a corresponding density difference, which in turn causes buoyancy forces to develop during the inflow period. Negative buoyancy occurs when the inflow is colder and denser than the tank's contents, whereas positive buoyancy occurs for the opposite case. An inflow jet with excessive buoyancy will cease to function as an effective mixing device and will lead to stable stratified conditions within some portion of the tank.

TABLE 11.1 Mixing Time Formulas for Storage Tanks

Investigators	Formula	Constant	Remarks
Fossett and Prosser (1949)	$\dfrac{KD^2}{M^{1/2}}$	$K = 8$	D = tank diameter M = inlet momentum = velocity × flow
Van de Vusse (1959)	$\dfrac{KD^2}{M^{1/2}}$	$K = 9$	
Okita and Oyama (1963)	$\dfrac{KH^{1/2}D^{3/2}}{M^{1/2}}$	$K = 4.9$ Re > 5000	H = water level
Fox and Gex (1956)	$\dfrac{KH^{1/2}D}{Re^{1/6}M^{2/3}g^{1/6}}$	K depends on units	Re = Reynolds number g = gravitational constant
Rossman and Grayman (1999)	$KV^{2/3} / M^{1/2}$	$K = 10.2$	V = tank volume

For jet discharges into unconfined or semiconfined bodies of water, such as lakes and rivers, the densimetric Froude number (Fr) is a measure of stability (Fischer et al., 1979). Fr is the ratio of its inertial forces to its buoyancy forces and is computed by:

$$Fr = \frac{u}{(g'd)^{1/2}} \quad (11.1)$$

where u = discharge flow velocity, d = discharge pipe diameter, $g' = g(\rho_f - \rho_a)/\rho_a$ g = acceleration of gravity, ρ_f = density of discharge water, and ρ_a = density of ambient water.

Fischer et al. (1979) showed experimentally that the terminal height of rise of a vertical, negatively buoyant jet discharging into an unbounded region is proportional to the product of d and Fr, where the proportionality constant is between 0.5 and 0.7. Lee and Jirka (1981) demonstrates that for vertical positively buoyant jets discharging into environments of unbounded horizontal extent, stratified conditions can be avoided when Fr > 4.6 H/d, where H is the water depth and d is the diameter of the inlet. Rossman and Grayman (1999) used scale modeling to study stratification effects in a tank. In a formulation analogous to the general form of the Jirka and Lee equation, Fr > C H/d, they developed values for C for tanks with both horizontal and vertical jets that are either positively and negatively buoyant. Based on the above-referenced studies, values of C for the various conditions are compared in Table 11.2, which was derived through and is applicable to scale models of tanks. Its applicability to full-scale tanks has not been confirmed.

11.3.5 Aging

Deterioration in water quality is frequently associated with the age of the water. Loss of disinfectant residual, formation of disinfectant by-products, and bacterial regrowth can all result from aging of water. As a result, an implicit objective in both the design and operation of distribution system storage facilities is the minimization of detention time and the avoidance of parcels of water that remain in the storage facility for long periods. The allowable detention time should depend on the quality of the water, its reactivity, the

TABLE 11.2 Classification Coefficients for Stratification

Inlet Orientation	Inflow Buoyancy	Coefficient from Tank Experiments	Coefficient for Unconfined Conditions
Vertical	Negative	0.8	0.5–0.7
Vertical	Positive	1.5	4.6
Horizontal	Negative	1.5	—
Horizontal	Positive	0.8	—

type of disinfectant used, and the travel time before and after the water's entry into the storage facility.

The mean detention time within a reservoir is dependent on the inflow and outflow pattern and the volume of water in the reservoir. For a storage facility operating in fill-and-draw mode, the detention time can be estimated by dividing the duration of an average fill-and-draw cycle by the fraction of the water that is exchanged during the cycle. Mathematically, this can be stated by the following equation:

$$\text{Average detention time} = [0.5 + (V / \Delta V)] (\tau_d + \tau_f) \qquad (11.2)$$

where τ_f = fill time, τ_d = draw time, V = volume of water at start of the fill period, ΔV = change in water volume during the fill period.

The results of the application of this equation are presented in Fig. 11.3. As an example, for a reservoir that has one fill-and-draw cycle per day and in which 20 percent of the volume is exchanged, the average detention time is 5.5 days.

11.4 MONITORING AND SAMPLING

Monitoring and sampling for water quality in storage facilities is an important activity that can help the utility confirm high water quality, anticipate water-quality problems, supplement data from inspections, help plan maintenance, and assist in the utility's regulatory compliance program. The monitoring program is considered to be an adjunct to good design, operation, inspection, and maintenance.

11.4.1 Routine Monitoring

Drinking-water regulations apply to the distribution and storage facilities, and the focus is on public health parameters. The water-quality parameters of most regulatory interest include those presented in Table 11.3. As indicated, the emphasis is on maintaining a disinfectant residual in the distribution system; on maintaining low levels of microbiological activity as indicated by heterotrophic plate counts and coliform; and keeping disinfection by-products at low levels as well. The regulations do not actually require sampling within the storage facility; however, since the storage facility is such an integral part of the distribution system, routine sampling within the facility is highly advised. In addition to water-quality regulations, most states address disinfection after cleaning and maintenance, and they address planning and design features related to sizing, siting,

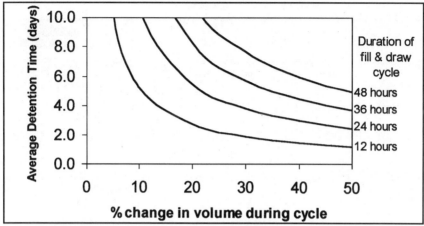

FIGURE 11.3 Average detention time in fill and draw reservoirs.

penetrations, coatings, linings, and so forth through reference to other codes and manuals. Key state references include AWWA's standards and manuals, the Ten States Standards (1997) and the National Sanitation Foundation (NSF) standards. In addition to the drinking water, there are environmental standards that apply to the discharge of chlorinated waters into storm sewers and surface waters that must be adhered to.

11.4.1.1 Typical parameters of water quality. Water quality parameters to be monitored in a finished water-storage facility are driven by a utility's desire and need for sampling. Sampling a storage facility can provide information on treatment, disinfection, the distribution system, the facility's operation, and seasonal impacts on water quality. When developing a monitoring program, a utility should build on a minimum program of chlorine residual, heterotrophic plate count (HPC), and coliform. Not all parameters need to be monitored at all facilities nor necessarily throughout the year. It is essential that a utility evaluate the needs of each facility when developing a monitoring program. Table 11.4 lists some monitored water-quality parameters and the sampling options available.

Beyond the listed water-quality parameters for a storage facility's bulk water, a utility should evaluate three special monitoring issues: nitrification, sediments, and biofilms. Monitoring for the occurrence of nitrification is a bulk-water issue. Sediments and biofilms are factors within the storage facility that are not addressed by bulk-water parameters but may affect bulk-water quality.

11.4.1.2 Parameters of nitrification monitoring. For utilities practicing chloramination for secondary disinfection, Kirmeyer et al. (1995) recommended monitoring strategies to anticipate and identify nitrification in the distribution system. Each utility practicing chloramination should develop a specific monitoring program to identify occurrences of nitrification in finished water-storage facilities. Table 11.5 lists the minimum, recommended, and additional sampling parameters for determining nitrification in the distribution system.

11.4.1.3 Parameters of sediment monitoring. Storage facility sediments are typically outside of the bulk-water column but can affect bulk-water quality. There are no regulations which pertain directly to monitoring of sediments. Suspended materials, such as iron

TABLE 11.3 Water Quality Parameters and Associated Regulations for Storage Facilities

Parameter	Sample location	Regulatory Limit	Reference	Comments
Disinfectant residual	Entry point to distribution system	0.2 mg/L on a continuous basis	United States SWTR	Only applies to systems using surface water supplies.
			UK	No UK maximum or minimum limit on disinfectant residual. Residual must be sufficient to insure appropriate bacteriological quality.
Disinfectant residual or HPC bacteria count	Throughout distribution system	Detectable level of disinfectant residual or HPC bacteria count of 500 or less CFU mL in 95% of samples collected each month for any two consecutive months	United States SWTR	Only applies to systems using surface water supplies. In U.S, *Legionella* is also regulated by a treatment technique.
	Storage facility		United Kingdom	UK storage facilities are required to be sampled weekly for HPC.
TTHMs	Throughout distribution system	100µg/L, running annual average based on quarterly samples	Total Trihalomethane Rule	Systems serving >10,000 people
		100µg/L rolling 3 month average	United Kingdom Prescribed Concentration or Value (PCV)	If fewer than four samples collected annually, no single TTHM sum allowed above 100 µg/L.
Total coliform bacteria	Throughout distribution system	0 CFU in 95% of samples	U.S. Total Coliform Rule	Number of samples determined by population served.
	Storage facility		United Kingdom	UK storage facilities are required to be sampled weekly.

TABLE 11.4 Water Quality Parameters for Finished Water Storage Facilities

Parameter	Purpose	Sampling procedure used
Alkalinity	Indicates the potential buffering capacity	On-line ion–selective electrode or grab sample and laboratory analysis
Aluminum	Indicates potential coagulant overfeeding	On-line ion–selective electrode or grab sample and laboratory analysis
Ammonia, total or free	Indicates potential for nitrification	On-line ion–selective electrode or grab sample and laboratory analysis
Chlorine residual, total or free	Indicates protection from bacterial growth and provides early warning sign of water-quality deterioration; control rechlorination when practiced	On-line colorimetric analyzer or grab sample and laboratory analysis
Coliform, total and/or fecal	Indicates presence of indicator bacteria	Grab sample and laboratory analysis
Conductivity, specific	Can quickly indicate relative changes in total dissolved solids e.g., alkalinity	On-line ion–selective electrode or grab sample and laboratory analysis
Disinfection by products	Represents potential for ongoing chemical reactions and DBP formation	Grab sample and laboratory analysis
Heterotrophic bacteria	Indicates conformance to MCL; provides early warning sign of water-quality deterioration	Grab sample and laboratory analysis
Iron	Indicates potential corrosion reactions	On-line ion–selective electrode or grab sample and laboratory analysis
Nitrate	Indicates possibility of nitrification	On-line ion–selective electrode or grab sample and laboratory analysis
Nitrite	Indicates possibility of nitrification	On-line ion–selective electrode or grab sample and laboratory analysis
pH	Indicates changes from the water source. Indication of corrosion of concrete or an unlined new facility.	On-line ion–selective electrode or grab sample and laboratory analysis
Taste and odor	Evidence of water-quality problem in progress	Grab sample and laboratory analysis
Temperature	Differences within storage facility indicate possible stratification and stagnant zones. Early warning sign of potential microbial problems	On-line sensor
Turbidity	Provides early warning sign of water-quality deterioration	On-line turbidimeter sensor and analyzer

TABLE 11.5 Nitrification Monitoring Parameters

Minimum	Recommended	Additional
Heterotrophic plate count	Heterotrophic plate counts	Dissolved oxygen
Nitrite	Free and total ammonia	Alkalinity
Nitrate	Nitrite	pH
Ammonia	Nitrate	Total organic carbon
Chlorine species	Chlorine dose	Dissolved organic carbon
	Chloramine residual	Assimilable organic carbon (AOC)
	Temperature	Ammonia–oxidizing bacteria counts

and manganese; turbidity; suspended solids; and precipitated calcium hardness entering storage may settle in storage facilities in which the flow velocities are minimal. Resuspension of sediment caused by flow surges is a potential source of water-quality degradation and may contribute to the water's disinfectant demand. Though not a regulatory issue, sediment sampling is of use to a utility because it can provide information on causes of water-quality problems, such as taste and odor or recurring bacterial counts. Monitoring for sediment buildup can therefore be used to avoid water-quality problems caused by sediments. Determining the rate of sediment accumulation in a facility can be a useful tool in determining the time between cleanings. The analysis of sediments may provide information useful to treatment operations, such as excess aluminum residuals. Table 11.6 lists some parameters that utilities may consider using to evaluate sediment samples.

When a utility decides to sample sediments, it will need to develop a method of collecting them. Sediment samples are difficult to collect while the facility is in service. The best opportunity for sampling is after drainage and before cleaning activities. One suggested method of sampling is the use of a sediment gauge. Operating on the principle of a rain gauge, the sediment gauge is a calibrated cylinder fastened to the bottom of the

TABLE 11.6 Sample Parameters for Sediment Monitoring

Parameter	Indicated by presence of
Iron oxide	Distribution system corrosion
Aluminum hydroxides	Excess aluminum due to afterfloc
Calcium carbonates	Supersaturation of minerals in hard waters
Manganese	Source water problem
Heterotrophic plate counts	Possible source of taste and odor problems
	Potential source of recurring bacterial counts
Depth of sediment	Rate of accumulation
	Possible source of disinfectant residual loss due to resuspension
Gross microbial examination	System-cross connection, poor hydraulic circulation, failed facility screening

storage facility. The gauge provides a known sample area, a representative sampling of the surrounding facility floor, calibrated depth measurements, and undisturbed samples after drainage of the facility.

11.4.1.4 Parameters of biofilm monitoring. Continuous and grab sampling of a finished water-storage facility for analysis of bacteria, such as coliform and HPC, provide determination of organisms in the bulk water. A negative test of a bulk-water sample does not necessarily mean the facility is free of microorganisms. Biofilms can be a potential source of taste and odor problems, disinfectant residual loss, and recurring bacterial problems. Like sediment monitoring, biofilm monitoring is not required by regulation, but including it in a sampling program may help utilities deal proactively with potential problems or identify causative elements more quickly when problems occur.

Biofilms occur at the water-tank interface. The area of potential for biofilm growth includes all facility walls from the sediment line to a point of maximum water level. Several specific areas should be considered for biofilm monitoring: surfaces always submerged during operation, the zone in which the water level fluctuates during operation, and in zones of possible stagnation.

Biofilm sampling can be combined with collection of sediment after drainage and before cleaning procedures. Utilities will again need to develop a sample-collection method that best suits their facility and analytical capabilities. One possible approach is a coupon method. In this method, a coupon of standardized size is coated and treated in a fashion identical to the interior coating of the tank. The coupon is stabilized flush with the tank wall at the sampling location and left in place during operation. During the next cleaning period, the coupon is removed and transported to the laboratory for analysis. This method provides a standardized surface area, replicates tank wall conditions, and simplifies collection.

11.4.2 Sampling Methods and Equipment

Sampling methods and equipment are intertwined. Deciding on the method of sampling will determine the equipment needed to complete the job. Two types of samples can be collected: grab samples and continuous samples. For the purposes of this discussion, the distinction between the two is defined as follows: If the sample is collected and analyzed manually, it is a grab sample; if sampling is conducted by means of sensors and remote recording stations, it is considered to be continuous sampling or monitoring. Grab samples are more labor intensive and, depending on the facility, may be difficult to collect. They provide a snapshot of the state of the water quality at the time of collection. Continuous sampling/monitoring requires greater up-front equipment cost and equipment-maintenance time. This method provides a continuous time series profile of changes in water quality. The materials used in grab or continuous monitoring equipment should be considered carefully because materials can either promote or inhibit microbial growth and can release metals or chemicals in to the water being sampled. A comprehensive monitoring plan can make use of both grab and continuous methods to round out the facility's monitoring program.

Grab samples require access to the stored water either by means of installed sample taps or through the access hatches. Sampling through the access hatches can be performed using a common lake- or tank-sampling device (cleaned and disinfected). These sampling devices are designed to be engaged and closed at predetermined depths within the storage facility, enabling sampling at varying depths from a single entry point. To ensure that a consistent sampling depth is maintained between sampling periods and among sampling teams, a calibrated cord should be used when lowering the sampler to a prescribed depth.

A second collection option for grab samples is the use of sample taps. Sample taps can provide samples from several fixed depths and eliminate the time and safety concerns that may be involved with sampling through the hatch of an elevated storage facility, a ground-level tank, or a standpipe. If not installed with the original storage facility, taps can be retrofitted and can be designed for either gravity or pumped flow. Taps can be continuously flowing, which reduces the need for extended flushing times before sampling, or be on-off taps where flushing would be required. Continuously flowing taps are more prone to biofilm growth. Dedicated sample taps designed to allow periodic chlorination with backflushing to a drain are preferred. Ideally, sampling taps should be located indoors (e.g., in sheds and pump houses), enclosed for sanitary reasons, and secured for sampling only. Continuous maintenance of clean and reliable taps is beneficial and helps ensure accurate monitoring results (DuFresne et al. 1997).

Continuous or on-line sampling is automated monitoring that uses sensors, probes, and data-recording and transmitting equipment. Several water-quality parameters that must be analyzed on-site for accuracy if collected by grab sample can be measured with on-line instrumentation. Parameters commonly monitored on-line by utilities include chlorine residual, pH, temperature, turbidity, and sometimes, particle counts. With continuous monitoring equipment, these parameters can be integrated with hydraulic conditions of flow rate, water levels, and demand within the storage facility.

When using on-line monitoring, a continuous sample stream must be obtained from the desired location within the storage facility. The sample lines and sensors should be placed to facilitate access for cleaning and maintenance. Utilities considering on-line monitoring should be aware that the ion-selective electrodes used for analysis require regular maintenance; if regular maintenance is impossible, grab sampling and laboratory analysis are preferable.

The benefit of on-line sampling goes beyond the simple continuous recording of data. One example is the integration of on-line chlorine analyzers and rechlorination system controls to provide immediate booster chlorination as needed. When the level of chlorine reaches a predetermined level within the storage facility, the residual analyzer signals the chlorination system to start, stop, or trim the chlorine feed.

Data collected by continuous monitoring equipment can be gathered by one of two methods. The first option is for personnel to visit the site and collect hard copy data records or download electronic data to a portable computer. The second option is the use of telemetry equipment to transmit the information from the remote site to a centralized control center. Telemetry enables utility personnel to evaluate conditions for numerous storage facilities concurrently and continuously and to modify operations or treatment in real time. This type of integrated network of data acquisition, transmission, and system control is known as Supervisory Control and Data Acquisition or SCADA.

No single monitoring method is likely to address all of a utility's monitoring needs. Each utility system, and possibly each storage facility within the system, has unique features associated with it. Utilities could benefit by reviewing each finished water storage facility for access, monitoring needs, availability of personnel, and applicability of a particular monitoring method when establishing or evaluating monitoring methods.

11.4.3 Monitoring Frequency and Location of Samples

Once a utility has determined what parameters to monitor and what method of collection will be used, two questions remain: Where to collect samples? and How often should samples be collected?

Routine monitoring is normally performed to detect water-quality trends during operation that are of regulatory concern. Parameters included are those that are required by regulation or can show variation over a longer period of time and may be indicative of larger problems. Routine monitoring sites are likely to be (1) at the inlet to establish a baseline on water quality entering the storage facility, (2) at the outlet to determine what changes in quality may have occurred during storage and (3) at one or more locations internally determined to be "representative" of the water in the facility.

Documenting the operation of all storage facilities before and during monitoring is important and is particularly relevant for facilities with common inlets and outlets. Sampling regimes and timing of sampling during a filling or drawing cycle need to reflect the operation.

A single sampling site may fulfill regulatory requirements; however, what becomes apparent is that a single sample per facility cannot represent conditions throughout storage. Multiple sampling sites will provide valuable information when attempting to identify the presence or location of water-quality problems in a storage facility. Smith and Burlingame (1994) found that direct monitoring of tank effluent may not detect all potential water-quality problems. Their results showed that although effluent samples could result in zero bacteria counts, microorganisms could still be present as biofilms on tank surfaces, in tank sediment, or in the water. In another study, Boulos et al. (1995) measured free chlorine residual, water temperature, and reservoir inflow rates and water levels and used a fluoride tracer to measure detention time in the storage facility. The authors determined that the reservoir had two distinct mixing zones: a well-mixed zone shaped like an annular ring from the reservoir edge toward the center and an inner cylinder with a slower downward flow rate.

Such variation has significant implications for monitoring strategies, suggesting that characterization of the quality of water delivered is a more complex problem, possibly requiring more sampling, than has previously been believed. These studies show the importance of identifying system-specific conditions before developing a water-quality monitoring program. As when determining the sampling methodology, a utility should review each storage facility for its unique sampling requirements. For each facility being sampled, the utility should be aware of the following before using routine sampling:

Configuration of the inlet/outlet piping and facility. This will affect the circulation and mixing characteristics of the water.

Mode of operation when sampling common inlet/outlet facilities. Samples during both filling and drawing need to be represented.

Flow patterns. A baffled facility that approaches a plug-flow situation requires a different type of sampling program than does a mixed facility. A last-in, first-out situation within a stand-pipe can result in misrepresentative results.

Location of stagnant zones. Knowledge of inlet/outlet configurations and flow patterns will assist in the identification of tank zones that have the potential for stagnation.

Temperature stratification. Temperature stratification can impede mixing throughout the depth of the tank and create separate, possibly stagnant zones at the top and bottom of the tank.

A review and understanding of these variations within storage facilities will provide the necessary information when deciding on the locations and frequency of both routine and diagnostic sampling.

Weekly, monthly, or quarterly samples may prove sufficient to meet current regulations but may lack the detail necessary to profile the true condition of the stored water or the causes of water-quality degradation. A properly implemented sampling program can alert a utility to acute changes in water quality, provide diagnostic information on how and where degradation is occurring, and determine whether operational practices (or the facility itself) are contributing factors. An effective monitoring schedule is designed with forethought based on water-quality, sediment, and biofilm parameters; seasonal temperature variations; patterns of use changes in operations; type of disinfectant; potential for nitrification; regulatory requirements; and the frequency of inspection and cleaning. The monitoring strategy should be reviewed and modified on the basis of collected data parameters.

11.4.4 Special Studies

Routine monitoring generally provides "snapshots" of water quality at different times at a single location within a tank (usually at the outlet). However, to understand more fully the temporal and spatial variation of water quality in such a facility and to observe the mixing and flow characteristics, more intensive studies can be conducted.

11.4.4.1 Intensive studies of water quality and tracers. A variety of automated and manual techniques are available for monitoring water quality. These include grab samples that are analyzed in the field or in a laboratory, automated samplers that extract a sample at a preset interval and perform some form of analysis, and in situ electronic sensors that measure and log water-quality values at preselected intervals. For all of these techniques, quality assurance/quality control procedures must be followed to develop a dependable database. Parameters that are frequently measured in these studies include disinfectant residual, pH, and temperature.

Mixing and flow patterns can be studied by adding a conservative tracer to the inflow of a tank or reservoir and measuring the resulting concentrations over time at various sites within the facility and in the outflow. When a tracer study is performed for determining the CT (chlorine residual times contact time) value of a clearwell or treatment-plant storage facility, measurements are taken only in the reservoir outflow. However, when studying a distribution system's reservoirs and tanks, the tracer concentrations should be measured at available locations within the facility in addition to the outflow to assess the internal mixing patterns. Tracers should be conservative (i.e., they should not decay over time) and must conform to local, state, and federal regulations. Fluoride, calcium chloride, sodium chloride and lithium chloride, are the usual candidates for such studies.

Figure 11.4 illustrates the type of information that can be collected in an intensive water-quality and tracer study. In this 2-MG elevated tank, concentrations of chlorine and a calcium chloride tracer were measured in the inlet and outlet and at sampling taps at different elevations. The results indicated a relatively well-mixed tank.

11.4.4.2 Temperature monitoring. Temperature variations in a tank or reservoir can affect the mixing in the facility and result in stratification. Water temperatures can be measured manually from a sample tap, or water can be drawn from different depths using a pump or sampling apparatus. However, to determine long-term trends in temperature at different depths within a tank or reservoir, an apparatus composed of a series of thermistors and a data logger can be used (Fig. 11.5). Thermistors are positioned at different fixed depths or are attached to floats to measure temperatures at fixed preset depths below the water's surface and are connected to a data logger that can be set to take readings at preset intervals, such as hourly. A computer can be attached to the data logger at any time to download the collected data. Figure 11.6 illustrates the type of data that can

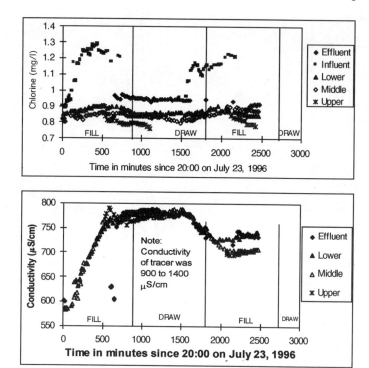

FIGURE 11.4 Variation in chlorine concentration and tracer concentration during a monitoring study.

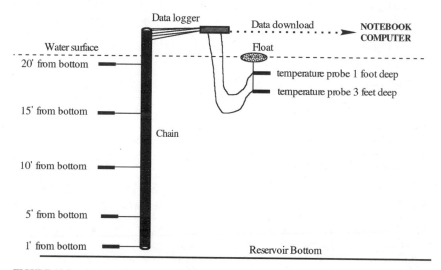

FIGURE 11.5 Automated temperature measurement and logger device.

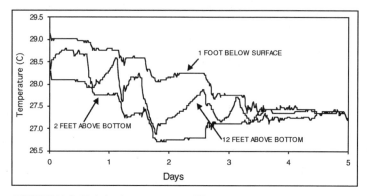

FIGURE 11.6 Variation in temperature at different depths in a reservoir.

be collected using the automated device. In this 25-MG reservoir, temperatures varied considerably between the top and bottom of the reservoir and at different times during the recording period.

11.5 MODELING

Modeling can provide information on what will happen in an existing, modified, or proposed facility under a range of operating situations. There are two primary types of models: physical scale models and mathematical models. Physical scale models are constructed from such materials as wood or plastic. Dyes or chemicals are used to trace the movement of water through the model. In mathematical models, equations are written to simulate the behavior of water in a tank or reservoir. These models range from detailed representations of the hydraulic mixing phenomena in the facility called computational fluid dynamics (CFD) models to simplified conceptual representations of the mixing behavior called systems models. Information collected during monitoring studies can be used to calibrate and confirm both types of models.

11.5.1 Scale Models

The history of scale modeling of hydraulic structures dates back more than a century. The experimental station of the U.S. Army Corps of Engineers at Vicksburg, Mississippi, and the laboratories of the U.S. Bureau of Reclamation in Denver, Colorado (Water and Power Resources Service, 1980), have constructed and tested large-scale models of dams, reservoirs, spillways, harbors, and canals. Scale modeling of mixing within storage vessels has been carried out in the chemical process industries (Fossett, 1973; Germeles, 1975; McNaughton and Sinclair; 1966; and Simpson, 1966) and in the drinking water field (Crozes, 1999; Grayman et al., 1996; Rossman and Grayman, 1999).

11.5.1.1 Principles of similitude. A physical scale model is a reduced-size representation of the prototype system of interest. Hydraulic models are usually designed to look like their prototypes, but the important factor is that they operate in a manner similar to the prototype. Various similitude criteria must be followed so that the results of the model mimic the real-world system.

The laws governing models are based on a consideration of the forces of gravity, buoyancy, inertia, viscosity and surface tension. Dimensionless groups used in hydraulic experimentation are mainly the Froude, Reynolds, and Weber numbers. The Froude number (Fr) represents the ratio of the inertial force to the gravity force, the Reynolds number (Re) represents the ratio of the inertial force to the viscous force, and the Weber number (W) is the ratio of the inertial to the surface-tension force:

$$Fr = \frac{U}{\sqrt{Lg}} \tag{11.3}$$

$$Re = \frac{UL\rho}{\mu} \tag{11.4}$$

and

$$W = \frac{U^2 L\rho}{\sigma} \tag{11.5}$$

where U = characteristic velocity of system, L = length characteristic, diameters, or depth, g = acceleration of gravity, ρ = density of fluid, μ = viscosity of fluid, and σ = surface tension of fluid.

Similitude exists between two geometrically similar systems when the Re, Fr, and W are the same for the model and the prototype. When using physical scale models to predict the behavior of a prototype system, it is seldom possible to achieve simultaneous equality of the various force ratios. The scaling laws are then based on the predominant force, and strict dynamic similarity is not attained. The Re is generally used in modeling of flows in pipes. The Fr is the governing factor in flows with a free surface because gravitational forces are predominant. Hydraulic structures, such as spillways, weirs, and water reservoirs, are modeled according to Froude's law.

The modeling of the flow in a distribution system reservoir, has been based primarily on Froude's law. The following are the fundamental relationships:

Length scale=L

Velocity scale=$L^{0.5}$

Area scale=L^2

Flow scale=$L^{2.5}$

Time scale=$L^{0.5}$

Volume scale=L^3

Once a length scale is chosen, all other scales are fixed. For a model constructed with a length scale of 1:100, the width, length, and height of the model will each be 1/100th of the dimension of the actual structure, and the time scale will be $(1/100)^{0.5}$ or 1/10 of the real-time value. Thus, in the operation of the model, 6 min would correspond to 60 min in the real world.

11.5.1.2 Construction of a model. To minimize the impacts of model scale on the mixing processes, it is desirable to create a model that is as large as possible. The literature on reservoir studies mentions scale models that range from 1:10 to 1:100, with the majority of the historical scale models being in the range of 1:30–1:50. Frequently, small-scale models are built to explore general mixing behavior, where as larger models are used in final design situations. Figure 11.7 illustrates a small-scale model experimental setup.

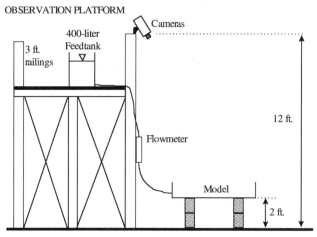

FIGURE 11.7 Example small scale model setup.

Materials that are typically used in model construction are wood, fiberglass, clear acrylic, sheet metal, and cement. For regularly shaped reservoirs (cylinders, rectangles), preconstructed small-scale vessels are often available from plastics companies, aquarium supply companies, farm supply companies, or companies in other related fields. Larger-scale models and models of irregularly shaped reservoirs must be custom-fabricated specifically for the studies being undertaken. The type of tracer that is used frequently dictates the type of construction material that must be used. When visible dyes are used as tracers, a transparent model is required unless viewing from the top is considered sufficient.

11.5.1.3 Types of tracers. In most reservoir studies, the objective is to study the movement of the water from the inflow as it mixes within the vessel and then exits the facility in the outflow. This is accomplished by adding a tracer to the inflow to the model and observing its movement in the vessel. Under some circumstances, such as when studying distributions of residence times, it may be sufficient to observe tracer only in the outflow, but for most distribution system reservoir studies, knowledge of the tracer's movement and mixing within the reservoir is necessary.

Options for tracers are chemicals that are traced with instrumentation or dyes that are observed visually or recorded by a camera. Tracers can be added continuously to the inflow by pump or from an elevated feed tank or can be injected as a slug directly into the feed line to the model.

Chemical tracers that are typically used in scale models include fluoride, calcium chloride, sodium chloride, and lithium chloride. Several German studies have used sodium nitrate. Concentrations of the tracer can be measured accurately using ion-selective electrodes or conductivity probes mounted in the model and attached to external meters. Data can be read manually from meters or captured automatically using a data logger.

When dyes are used as tracers, commercially available dyes with a density comparable to water should be used. Video cameras, standard still cameras, or digital cameras can be used to record results of the experiments. Figure 11.8 depicts a visible dye used to trace the movement through a scale model of a reservoir. When transparent models are used, results can be recorded from all perspectives. For opaque models, results can be recorded

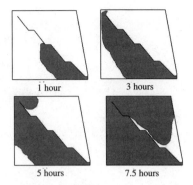

FIGURE 11.8 Movement of tracer dye as predicted by a scale model.

only from above, thus limiting the analysis of vertical mixing patterns. The resulting pictures can be analyzed visually to achieve a qualitative understanding of the mixing patterns. Alternatively, the variation of the intensity of the dye in the model can be analyzed through "pixel" analysis on the computer to estimate variations in concentration. For fluorescent dyes (such as rhodamine WT), a fluorometer can be used to measure directly the concentration of the tracer in the outflow. However, since a sample must be extracted for analysis in the fluorometer, use of this technique for measuring tracer concentrations in the model requires a carefully designed extraction method to avoid affecting the mixing patterns in the model.

11.5.1.4 Temperature modeling. Temperature can have a significant effect on the mixing patterns in tanks and reservoirs. For scale modeling studies of isothermal (constant temperature) situations, care must be taken to ensure that the temperature of the water in the tank and the temperature of the inflow are the same and that they remain relatively constant during the experiment. When the impacts of temperature are being studied in a scale model, methods of controlling and measuring temperature are required. For experiments of relatively short duration (< 1 h) in which the impacts of temperature differences between the tank and the inflow are being studied, it will generally suffice if the temperatures of the water in the tank and in the feed reservoir are established at the beginning of the experiment. For experiments of longer duration, more sophisticated methods of controlling the ambient temperature in the laboratory and the temperature of the feed reservoir are required.

11.5.2 Computational Fluid Dynamics

Computational fluid dynamics (CFD) is a modeling technique that has gained much popularity as an engineering evaluation tool over the past 20 years. The technique is used to study the dynamics of motion associated with both liquids and gases. Applications are common in industries as diverse as the auto industry and the aerospace industry and in medical research.

Its introduction into the drinking-water field is relatively recent. Gatel et al. (1996) and Grayman et al. (1996, 1999) have described its use for studying mixing in reservoirs. Crozes et al. (1998), Hannoun (1995), and Murrer et al. (1995) applied CFD models in the design and operation of clearwells and contact tanks. In water treatment studies, Jones et al. (1996) investigated the use of CFD in studying static mixers for coagulation, and Ta et al. (1996) modeled dissolved air flotation tanks.

11.5.2.1 Mathematical formulations of CFD Models. A physical object in a CFD model can be represented in three basic processes: the mathematical representation, the numerical representation of the mathematical model, and the computational method for solving the numerical representation.

CFD is a technology based on modeling the physics of fluid motion from first principles. Five coupled, nonlinear partial differential equations describe the basic conservation properties: conservation of mass, conservation of momentum (three equations for three spatial components), and conservation of energy.

Other physical processes occurring in fluid systems typically are not treated from first principles of physics, but instead with a heuristic, parameterized model. Examples of such processes include turbulence, mixing of multiple phases, and phase changes caused by flow.

The mathematical formulation for the fluid motions can be transformed into a numerical representation in a number of different ways. Standard methods implemented in commercial software map the partial differential equations into a set of finite arithmetic operations applied to a grid (or mesh) of nodes and elements that approximate the fluid volume of interest. Figure 11.9 shows a mesh construction for a cylindrical reservoir with a finer mesh in the vicinity of the inlet and outlet. Some numerical algorithms transport the physics in a local fashion, relying on properly chosen time steps for numerical stability (an explicit, time-marching algorithm). Other algorithms enforce the conservation principles of the physics over the entire domain of the vessel in a globally consistent way at any given time (an implicit algorithm).

Numerical methods for CFD are in a constant state of evolution. Over the last 40 years, improvements in general-purpose algorithms toward solving the partial differential equations in CFD have resulted in gains in performance greater (on a regular basis) than improvements in the speed of computer hardware over the same period.

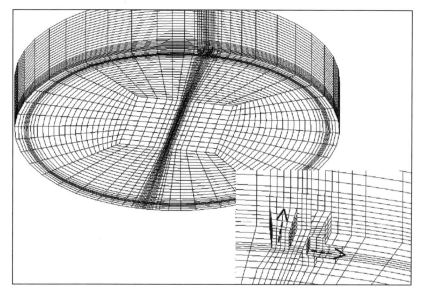

FIGURE 11.9 Mesh for a cylindrical tank with a tangential inlet and nearby outlet. (Detail for inlet and outlet shown in box).

Quality of Water in Storage   **11.27**

TABLE 11.7 Sample Commercial Computational Fluid Dynamics (CFD)

CFD Package	Company	Website
Fidap, Fluent	Fluent, Inc.	www.fluent.com
Phoenics	CHAM	www.cham.co.uk
FLOW-3D	Flow Science, Inc.	www.flow3d.com
FIRE	AVL List	www.avl.com/htlm/296.htm
CFX4 and CFX5	AEA Technology	www.aeat.co.uk
CF Design/TK and HydroTank	Blue Ridge Numerics	www.brni.com
CDF-ACE	CDF Research Corp.	www.cfdrc.com

Computational techniques and software systems for solving the numerical formulation of a fluid system are now readily available outside the research community that originally dominated the field. The detailed techniques vary somewhat from case to case, but all have the same general goal: to integrate the system of nonlinear equations consistent with the appropriate set of boundary conditions. Most software systems perform this integration on a mesh of "finite elements" or "cells." The techniques that dominate the field are categorized into two groups: finite element methods and finite volume methods (also known as finite difference methods). Over time, both approaches have borrowed the best characteristics from the other, thus their differences are no longer as clear cut as in the past.

11.5.2.2 Application of CFD models.   CFD modeling is an advanced form of modeling that has a relatively high cost in terms of software, hardware, and training. Many commercial CFD packages that handle a wide range of situations are available. These software packages require a significant investment in training before they can be used productively. Table 11.7 contains a list of popular commercial CFD packages that have been used in the simulation of reservoirs.

Outputs from CFD applications include graphical depictions of the spatial variation of concentrations of a tracer or disinfectant in a reservoir (Fig. 11.10), time histories of concentrations in the outflow of a reservoir, or statistics on the variability of concentrations in the reservoir.

FIGURE 11.10   Simulation of a vertical jet in a standpipe using a CFD model.

11.5.3 Systems Models

Systems models are a class of models in which physical processes (e.g. the mixing phenomena in the tank or reservoir) are represented by highly conceptual, empirical relationships. This type of model is also frequently referred to as a black-box model or an input-output model. Because such models do not use detailed mathematical equations to describe the movement of water within the tank, they rely heavily upon field data and past experience to define the parameters that control the behavior of the model.

11.5.3.1 Background. Systems-type models are used in a variety of applications in the field of water resources. In the field of hydrology, watersheds are often represented by unit-hydrograph models (Sherman, 1932) or as a series of cascading reservoirs (Nash, 1967). In the chemical engineering field, systems models commonly are used (Levenspiel, 1972) to represent the behavior of chemical reactors. In the drinking-water field, residence time models using formulations such as N-tanks-in-series have been applied successfully to disinfection contactors (Clark, 1996).

11.5.3.2 Elemental systems models. Tanks can be represented by simple systems models that assume a theoretical flow process for the tank. The most commonly used systems-type model of a tank is the mixed flow model or continuously stirred tank reactor (CSTR), which assumes the tank is instantaneously and completely mixed at all times. Network distribution system water-quality models use the CSTR representation of storage tanks. On the other hand, contact basins or clearwells that are designed to provide sufficient contact time for disinfectants are frequently represented as simple plug-flow reactors, which assume a "first in, first out" pistonlike behavior. A third basic system model is a short-circuiting model, which assumes a "last-in, first-out" behavior. These three basic models are considered to be elemental systems models of tanks and reservoirs.

Figure 11.11 illustrates the behavior of the three separate elemental models when applied to tanks of similar size and inflow characteristics. In this example, a 1-MG tank is represented. A 12-h fill and 12-h draw period is simulated with equal fill-and-draw rates resulting in one-third of the water being exchanged per day. The three parameters that are simulated include a conservative tracer with a concentration of 1 mg/L, water age, and chlorine that enters the tank at a concentration of 1 mg/L and decays at an exponential rate with a rate constant of -0.5/day.

11.5.3.3 Compartment models. In situations where the elemental systems models are not adequate, a more complex systems model can be constructed by combining the elemental building blocks into a more sophisticated representation. These are called "compartment" models because the elemental blocks are combined to form a series of interacting compartments (Grayman and Clark, 1993). Figure 11.12 illustrates several compartment models, all of which can be constructed to model a nonreactive (conservative) constituent, a reacting substance that decays or grows with time, and water age.

11.5.3.4 Application of systems models. Systems models of tanks and reservoirs can be applied in two broad areas: (1) as a conceptual and approximate representation of the tank or reservoir's mixing behavior and (2) as a truly calibrated representation of the expected behavior of an existing or future storage facility.

As a conceptual model, a systems model provides a useful mechanism for testing the impacts of alternative flow regimes. For example, the loss of disinfectant in a reservoir can easily be compared for the case of plug flow versus completely mixed flow.

Systems models also can be used to represent the behavior of an actual or proposed storage facility. However, like any model, the systems model must be calibrated and tested

Quality of Water in Storage   **11.29**

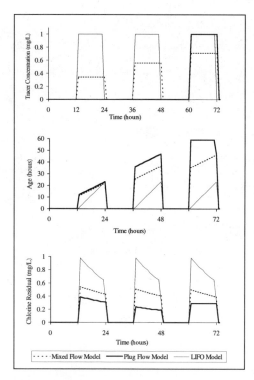

FIGURE 11.11 Behavior of alternative elemental models of tanks.

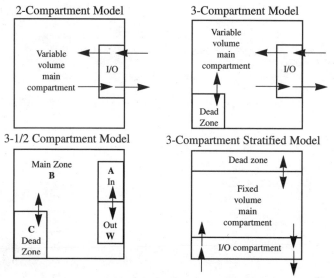

FIGURE 11.12 Examples of compartment models of tanks.

properly before its application. Because systems models do not attempt to mimic the actual physical processes that are occurring in the real-world system, the parameters that control the behavior of the models are conceptual, macro parameters, such as the volume of the compartments in the multi-compartment models. As a result, the calibration and testing process is both more difficult and more important than is the case with more physically based models.

11.6 DESIGN AND OPERATIONAL ISSUES

The general laws of fluid mechanics govern the movement of water and the mixing processes in a tank. Understanding fluid mechanics of a tank leads to improved design and operation of storage facilities to meet water-quality objectives.

11.6.1 Water-Quality Design Objectives

The primary water-quality objectives in the design and operation of a distribution system reservoir are to (1) minimize the detention time in the facility to reduce disinfectant loss and biological regrowth and (2) avoid stagnant zones where parcels of water may be significantly older. This differs from the design of water treatment plant facilities such as chlorine contact chambers and clearwells at water treatment plants where the objective is to minimize short circuiting and, more generally, to ensure that all parcels of water experience approximately the same residence time. These two different sets of objectives lead to significantly different design criteria.

11.6.2 Modes of Operation: Simultaneous Inflow-Outflow Versus Fill and Draw

There are two basic modes of operation for reservoirs: (1) simultaneous inflow and outflow through separate inlets and outlets; and (2) fill and draw in which water levels vary in the reservoir as water alternately enters and leaves the reservoir through combined inlet-outlets or through separate inlets and outlets. The mode of operation of a specific reservoir is generally controlled by both its design and its relationship to the water distribution system.

In the simultaneous inflow-outflow mode of operation, inflow and outflow rates may differ, resulting in changes in the reservoir's water level, or may be relatively equal and constant over time. Under a fill-and-draw operation, the operation may be dominated by repeating diurnal patterns, such as filling during the low-demand nighttime period and drawing during the higher demand periods, or may be more random, resulting in brief irregular periods of fill and draw. The mode of operation can affect the mixing characteristics of the reservoir and the resulting water quality.

11.6.3 Flow Regimes: Complete Mix Versus Plug Flow

Water can be made to flow through a storage tank or reservoir in two distinct ways: in a completely mixed state or in a plug-flow manner. Although idealized mixed flow or plug flow are never fully achieved in actual situations, reservoir design can result in approximations of either flow regime. Design of a reservoir to achieve a particular flow regime should reflect the objectives of the storage facility and the physical configuration and mechanisms required to achieve the regime.

11.6.3.1 Effects of flow regime on loss of disinfectant in reservoirs. Because the rate of disinfectant decay is concentration-dependent, when comparing two tanks of the same size and inflow-outflow characteristics, the one operating under plug-flow conditions will generally lose more disinfectant than the one operating under mixed-flow conditions. As illustrated in Fig. 11.13, the difference in loss of disinfectant between the two regimes grows with increasing disinfectant reactivity, and residence time. This figure applies to reservoirs operating with a simultaneous inflow and outflow of equal magnitude. Similar relationships hold for reservoirs operating in a fill and draw mode.

11.6.3.2 Mixed flow. Mixing a fluid requires a source of energy input. In distribution system storage facilities, this energy is normally introduced from the facility's inflow. As the water enters the facility, jet flow occurs, the ambient water is entrained into the jet, and circulation patterns are formed that result in mixing. To have efficient mixing, the path of the jet must be long enough to allow for the entrainment and mixing process to develop. Therefore, the inlet jet should not be pointed directly toward nearby impediments, such as a wall, the reservoir bottom, or deflectors. For entrainment to occur the jet also must be a turbulent rather than laminar jet. Though turbulent jets are the norm in most storage facilities, the following equations can be used to determine whether a jet is turbulent or laminar. Fully turbulent jets have been characterized as those with Reynolds numbers above 3000. To ensure turbulent jet flow, the following relationship between inflow (Q) and inlet diameter (d) must hold:

Classical (English) units (Q in gal/min and d in Ft): $Q/d > 11.5$ at 20°C and $Q/d > 17.3$ at 5°C.

Metric units (Q in L/s and d in m): $Q/d > 2.4$ at 20°C and $Q/d > 3.6$ at 5°C.

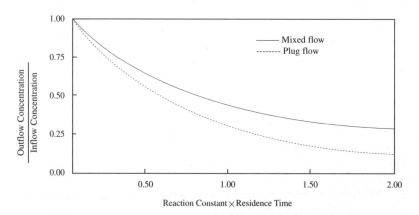

FIGURE 11.13 Disinfectant loss in mixed and plug flow tanks with simultaneous inflow and outflow.

For a tank operating in a fill-and-draw mode, mixing occurs primarily during the fill cycle; as a result, if the tank is relatively well mixed at the end of the fill cycle, mixing problems are unlikely. For a wide range of tank and reservoir designs, experimentation has shown that the mixing time depends primarily on the volume of water in the facility and the momentum flux (flow × velocity). Since momentum flux depends on the rate of flow and the diameter of the inlet, the following relationship has been developed for cylindrical tanks under fill-and-draw operation:

$$\text{Mixing time (seconds)} = 9\ V^{2/3}\ (d\ /\ Q) \quad (11.6)$$

where V = volume of water in tank at start of fill in feet^3 (m^3), Q = inflow rate in ft^3/s (m^3/s), and d = inlet diameter in ft (m). With the requirement that the fill time must exceed the mixing time, Eq. (11.6) can be restated in terms of the required change in water volume during the fill cycle as a fraction of the volume at the start of fill:

$$\Delta V/V > 9\ d/V^{1/3} \quad (11.7)$$

where ΔV = change in water volume during the fill period (Rossman and Grayman, 1999).

All units should be in consistent length dimensions (ft or m). The relationship given in Eq. (11.7) is shown graphically in Fig. 11.14 using reservoir volume in millions of gallons.

For reservoirs operating in simultaneous inflow-outflow mode, in addition to momentum flux and reservoir volume, mixing also depends on the juxtaposition of the inlet and outlet and the relative inflow an outflow rates. As a result, advanced models (scale models or CFD models) are frequently needed to determine the effects of specific designs on the mixing characteristics in the facility.

11.6.3.3 Plug flow. Plug flow is best achieved in long, narrow, shallow basins through the use of baffling to create long serpentine patterns or through the use of diffusers or stilling chambers to reduce flow momentum. Internal baffles are placed in vessels to direct

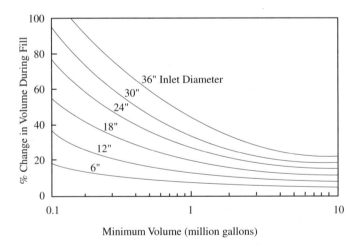

FIGURE 11.14 Minimum percent turnover needed for mixing during a fill cycle.

and control the flow through the facility to encourage an orderly movement of flow through the storage facility, to minimize short circuiting between inlets and outlets, and to reduce the formation of dead zones. Plug flow is achieved most easily in reservoirs that operate in a simultaneous inflow-outflow mode with minimal variations in flow rates over time. Variations in the rate of flow introduce disorder in the flow patterns. Plug flow is virtually impossible to achieve in reservoirs operating in a fill-and-draw mode because of the changes in flow rate and flow reversals.

11.6.3.4 Recommendations. Although a goal of plug flow is generally set for contact chambers at treatment plants to avoid short-circuiting and to encourage a constant travel time through the facility, the choice of mixed flow or plug flow for distribution system reservoirs depends on several factors.

In the case of reservoirs operating in fill-and-draw mode, plug flow cannot be achieved because of variations in flow and flow reversals. Therefore, such reservoirs should be designed to encourage "complete" mixing through adequate inflow momentum, which generally can be achieved through proper placement and sizing of the inlet to assure the formation of a turbulent jet. Overly large inlet diameters; inlets pointed directly at nearby reservoir walls, the reservoir floor, or the water surface; and the use of baffles should be avoided.

For reservoirs operating in a simultaneous inflow-outflow mode, trade-offs exist between mixed flow and plug flow. Plug flow results in a greater loss of disinfectant than does mixed flow, and the effects are most significant in cases of long detention time and highly reactive water. However, under some circumstances, plug flow affords better control in reducing stagnant zones. Use of scale and CFD models is encouraged to examine the impacts of alternative designs on movement of water in the proposed facilities.

11.6.4 Stratification in Reservoirs

Whenever there is a temperature difference between the contents of a tank and its inflow, the potential for poor mixing and stratification exists. Temperature differences result in a buoyant jet. Positive buoyancy (the temperature of the inflow is higher than the temperature of the ambient water) causes the inflow to rise toward the water surface, where as negative buoyancy (the temperature of the inflow is less than that of the ambient water) causes the inflow to sink toward the bottom. An inflow jet with excessive buoyancy (either positive or negative) relative to its momentum will lead to ineffective mixing and cause stable stratified conditions within the tank.

The critical temperature difference, ΔT in °C, which can lead to stratification, can be estimated based on the following equation:

$$|\Delta T| = C\, Q^2\, /\, (d^3\, H^2), \tag{11.8}$$

where C is a coefficient dependent on configuration of the inlet type of buoyancy and diameter of the tank; Q = inflow rate (ft^3/s, L/m), H = depth of water (ft, m), and d = inlet diameter (ft, m).

Based on this relationship, tall tanks and tanks with large-diameter inlets have a greater tendency toward stratification. If significant temperature differences are experienced, increasing the inflow rate is an effective strategy for reducing the likelihood of stratification. Because temperature differences frequently are not known and there is considerable uncertainty in the value for C, Eq. (11.8) can be used best to provide a qualitative measure of the propensity toward stratification. Smaller values of ΔT indicate a greater likelihood for the formation of stratified conditions.

11.7 INSPECTION AND MAINTENANCE ISSUES

Finished water-storage facilities require regular inspection and maintenance to increase their useful life and to keep them clean so that they do not degrade water quality. Each utility should develop written standard operating procedures for each of its storage facilities and should develop and maintain a good record-keeping system to document practices and facilitate troubleshooting.

11.7.1 Inspections

Inspections can take many forms and can include routine, periodic, and comprehensive procedures. Both interior and exterior inspections are needed to assure that physical integrity, security, and high water quality are maintained. The type and frequency of the inspection is driven by the type of storage facility, its susceptibility to vandalism, its age and condition, the time since its last cleaning or maintenance, and its history of water quality, plus other local criteria. Routine inspections can usually be conducted from the ground during normal or weekly tasks. Periodic inspections are normally more rigorous, and may require climbing the tank, and may be conducted every 3 or 4 months. Comprehensive inspections are major undertakings that often require the facility to be removed from service or drained. Comprehensive inspections should be conducted by a multidisciplinary team of experts representing water-quality, maintenance, operations, and engineering and possibly outside specialists. To properly inspect surfaces, cleaning often needs to occur first so that the condition of the surface can be ascertained.

The need for comprehensive inspections is generally recognized by the water industry; however, the frequency is not well defined. The recently withdrawn *AWWA Standard D101* recommended a 5-year maximum inspection interval. Although many states require disinfection when a storage facility is returned to service per AWWA Standard C652-92, and refer to the Ten States Standards (1997) for design features, they do not mention inspection frequencies. The frequency is normally left to the discretion of the utility.

In Europe, the frequency of inspections varies by country. The guidelines in the United Kingdom specify a minimum of once every 10 years. In Switzerland and France, reservoirs must be completely drained, cleaned, and disinfected once a year. No consistent guidelines are provided on the necessary frequency of inspections, possibly because conditions vary so widely throughout the utility industry.

As a good management practice, all three types of inspections should be conducted. With regard to comprehensive inspections, Kirmeyer et al. (1999) recommends that they be conducted every 3–5 years for structural condition and possibly more often for water quality. However, the actual frequency should be based on water-quality monitoring and the condition of the facility and can be as frequent as once per year. Because uncovered reservoirs have unique problems, consideration should be given to comprehensive inspection once or twice per year. Reservoirs with flexible membrane covers require specific routine, periodic, and comprehensive inspections to ensure their integrity and to assess the deterioration of the cover.

The methods of inspections include visual, float down, and wet. The most popular method is to drain, clean, and inspect the facility visually. Underwater or wet inspections with either a diver or a remotely operated vehicle are increasing dramatically and account for an estimated one-third of commercial inspections. The obvious advantage of wet inspection is that the reservoir does not have to be drained. Underwater inspections should be conducted by specially trained teams, including commercial divers certified in necessary safety requirements. Float-down inspections are used the least.

The most common problems found by commercial inspectors are lack of bug screens on vents and overflows, cathodic protection systems not operating or not adjusted properly, hatches not locked, presence of lead paint in the interior, or exterior, and presence of paints not approved by the National Sanitation Foundation (NSF). Another issue of concern to the utility would be accumulation of sediment.

Keeping records of and thoroughly documenting inspections cannot be emphasized enough. Consistent forms and checklists help to ensure that the same points are inspected and evaluated each time. Quantitative measurements also are important to document the condition of paints and coatings and the facility's structural integrity. Photographs with adequate field notes can be a valuable tool, as can videotaping with audio explanation.

11.7.2 Maintenance

Maintenance activities can be divided into maintenance planning, which includes preventive and predictive activities, and emergency maintenance. Preventive activities are performed on a regularly scheduled basis to extend life, prevent failure of structures and equipment, and preclude water-quality problems. Predictive activities include the use of technology and other methods to collect and analyze data to identify present conditions, forecast failure, and act before more costly repairs are required. *Emergency maintenance* is unplanned, is performed as a result of a natural or manmade disaster, and is disruptive in nature.

Maintenance of storage facilities is conducted on a system-specific basis. Maintenance activities include cleaning, painting, and repairing of structures to maintain serviceability, and so on. A review of state regulations indicated that much discretion is left to the utility but many states recommend adhering to AWWA Standards, ANSI/NSF Standard 61, the Ten State Standards (1997), and other standards for maintenance, disinfection procedures, and approval of coatings. Most states do not recommend a cleaning frequency; however, some guidelines were "as often as necessary" or "at reasonable intervals."

Most utilities that have regular cleaning programs have a cleaning interval between 2 and 5 years. Kirmeyer et al (1999) suggests that covered facilities be cleaned a minimum of every 3–5 years, or more often if needed on the basis of inspections and water-quality monitoring. It also suggests that uncovered storage facilities be cleaned once or twice a year (Kirmeyer et al, 1999). Floating covers require a specific preventive maintenance program that should include water-quality monitoring, periodic inspection, operational guidelines, and evaluation of materials.

Cleaning procedures can be divided into out-of-service and in-service methods. The traditional out-of-service method includes draining and cleaning with high-pressure hoses, brooms, shovels, and so for. In-service cleaning is becoming more popular and involves the use of vacuum devices operated by either a diver or a remotely operated vehicle. The benefits and drawbacks of each method should be weighed. The obvious advantages to in-service cleaning are that its quicker and does not require taking the facility off-line for extended periods.

Coatings and linings are important components related to the protection and life of structures and to water quality. Special precautions should be used when selecting approved coatings, surface preparation, and curing conditions. NSF-approved materials should be used. Finally, cathodic protection systems must be maintained properly because malfunctioning systems can increase corrosion and subsequent problems.

REFERENCES

Alben, K., A. Bruchet, and E. Shpirt, *Leachate from Organic Coating Materials Used in Potable Water Distribution Systems,* American Water Works Research Foundation and American Water Works Association, Denver, CO, 1989.

Antoun, E. N., R. J. Theiss, J. E. Dyksen, and D. J. Hiltebrand, "Role of Distribution Systems in Achieving Compliance With Safe Drinking Water Act," *Proceedings of the 1995 Annual AWWA Conference,* American Water Works Association, Denver, CO, 275-280, 1995.

Atkinson, R., "A Matter of Public Health: Contamination of Water Storage Tanks," *Missouri Municipal Review,* P 22, 1995.

AWWA D101, *AWWA Standard for Inspecting and Repairing Steel Water Tanks, Standpipes, Reservoirs, and Elevated Tanks for Water Storage,* American Water Works Association, Denver, CO, 1986a.

AWWA, *Maintaining Distribution System Water Quality,* American Water Works Association, Denver, CO, 1986b.

AWWA D130, *AWWA Standard for Flexible-Membrane Lining and Floating-Cover Materials for Potable-Water Storage,* American Water Works Association, Denver, CO, 1996.

Baur, A., and K. Eisenbart, "Mixing in Reservoirs with Different Geometry and Inlet/Outlet Configurations", *GWF Wasser/Abwasser,* 123: 487–491, 1982.

Bellen, G. E., S. H. Abrishami, P. M. Colucci, and C. J. Tremel, *Methods for Assessing the Biological Growth Support Potential of Water Contact Materials,* American Water Works Association, Research Foundation, Denver, CO, 1993.

Beuhler, M. D., D. A. Foust, and R.W. Mann, "Monitoring to Identify Causative Factors of Degradation of Water Quality: What to Look For," *Proceedings of the 1994 Annual AWWA Conference,* American Water Works Association; Denver, CO, 1994.

Block, J. V., V. Gauthier, C. Rosin, L. Mathieu, J. M. Portal, P. Chaix, and D. Gatel, "Characterization of the Loose Deposits in Drinking Water Distribution Systems," In *Proceeding of the AWWA Distribution System Symposium.,* American Water Works Association, Denver, CO, 1996.

Boudouresque, P., "Dégradation et Corrosion Internes des Réseaux de Distribution D'eau," *Water Supply,* 6:15–18, 1988.

Boulos, P.F., W. M. Grayman, R. W. Bowcock, J. W. Clapp, L. Rossman, R. Clark, R. A. Deininger, and A. K. Dhingra, "Comprehensive Sampling Study of Storage Reservoir Water Quality: Characterization of Chlorine Dynamics in Terms of Contact Time and Loss of Residual," In *Proceeding of the 1995 Annual AWWA Conference,* American Water Works Association, 815-839 Denver, CO, 1995.

Boulos, P. F., W. M. Grayman, R. W. Bowcock, J. W. Clapp, L. A. Rossman, R. M. Clark, R. A. Deininger, and A. K. Dhingra, "Hydraulic Mixing and Free Chlorine Residual in Reservoirs," *Journal of the American Water Works Association,* 88(7): 48–59, 1996.

Burlingame, G. A., and C. Anselme., "Distribution System Tastes and Odors," in *Advances in Taste-and-Odor Treatment and Control,* American Water Works Association Research Foundation; Denver, CO, 1995.

Burlingame, G. A., and G. L. Brock., "Water Quality Deterioration in Treated-Water Storage Tanks," in *Proceedings of the 1985 Annual AWWA Conference,* American Water Works Association, Denver, CO, 1985.

Clark, M. M., *Transport Modeling for Environmental Engineers and Scientists.* John Wiley & Sons, New York, 1996.

Crozes, G. F., J. P. Hagstrom, M. M. Clark, J. Ducoste, and C. Burns, *Improving Clearwell Design for CT Compliance,* American Water Works Association Research Foundation, Denver, CO, 1999.

Dufresne L., G. Burlingame, C. Cochrane, L. Maley, S. Shahid, and M. Toch, "Eliminating 'Noise' in Distribution System Coliform Monitoring," presented at the American Water Works Association Water Quality Technical Conference, Denver, CO, 1997.

Fair, G. M., J. C. Morris, S. L. Chang, I. Weil, and R. P. Burden, "The Behavior of Chlorine as a Water Disinfectant," *Journal of the American Water Works Association*, 40(10): 1051–1061, 1948.

Fischer, H. B., E. J. List, R. C. Y. Koh, J. Imberger, and N. H. Brooks, *Mixing in Inland and Coastal Waters*, Academic Press, New York,1979.

Fossett, H., and L. E. Prosser, "The Application of Free Jets to the Mixing of Fluids in Bulk" *Institute Mechanical. Engineering.*, 160:224, 1949.

Fossett, H. "Some Observations on the Time Factor in Mixing Processes," in *Fluid Mechanics of Mixing*, American Society of Civil Engineers, 1973.

Fox, E.A. and V.E. Gex, "Single-phase Blending of Liquids", *Journal of the American Institute of Chemical Engineering*, 2, 539-544, 1956.

Gatel, D. C. Henriet, T. Braekman, P. Servais, A. Maret, and J. Cavard. "Impact of Reservoirs on Drinking Water Quality" *Proceedings of the AWWA Water Quality Technology Conference*, Boston, MA, AWWA, 1996.

Geldreich, E. E., *Microbial Quality of Water Supply in Distribution Systems*, Lewis Publishers, Boca Raton, FL, 1996.

Germeles, A. E., "Forced Plumes and Mixing of Liquids in Tanks," *Journal of Fluid Mechanics*, 71:601–623, 1975.

Grayman, W. M., and R. M. Clark, "Using Computer Models to Determine the Effect of Storage on Water Quality," *Journal of the American Water Works Association*, AWWA, 85(7):67–77, 1993.

Grayman, W. M., R. A. Deininger, A. Green, P. F. Boulos, R. W. Bowcock, and C. C. Godwin, "Water Quality and Mixing Models for Tanks and Reservoirs," *Journal of the American Water Works Association* 88(7): 60–73, 1996.

Grayman, W.M., L.A. Rossman, C. Arnold, R.A. Deininger, C. Smith, J.F. Smith, R. Schnipke, Water Quality Modeling of Distribution System Storage Facilities, American Water Works Association Research Foundation and American Water Works Association, Denver, CO, 1999.

Hammer, D., and G. Marotz, "Improvement of Water Exchange in Drinking Water Reservoirs by Hydraulic Measures." *Wasserwirtschaft*, 76:145–150, 1986.

Hannoun, I. A., "Application of Computational Fluid Dynamics to the Optimization of a Water Plant Clearwell," *Proceedings. of the AWWA Computer Conference; Computers in the Water Industry*, 1-15. American Water Works Association. 1995.

Hyman, D., "Mixing and Agitation," in T.B. Drew, J. W. Hoppes Jr., and T. Vermeulen, eds., *Advances in Chemical Engineering* Vol. 3, Academic Press, New York, 1962.

Jones, S. C., A. Amirtharajah, and L. N. Sankar, "Static Mixes for Coagulation in Water. Treatment: A Computational Fluid Dynamics Model", *Proceedings of the Annual AWWA Conference*, American Water Works Association, Toronto, Ontario, 1996.

Kennedy, M. S., S. Moegling, S. Sarikelle, and K. Suravallop, "Assessing the Effects of Storage Tank Design on Water Quality." *Journal of the American Water Works Association* AWWA, 85(7):78-88, 1993.

Kirmeyer, G. J., G. W. Foust, G. L. Pierson, J. J. Simmler, and M. W. LeChevallier, *Optimizing Chloramine Treatment,* American Water Works Association Research Foundation, Denver, CO.,1993.

Kirmeyer, G. K., L. H. Odell, J. Jacangelo, A. Wilczak, and R. Wolfe, *Nitrification Occurrence and Control in Chloraminated Water Systems*, American Water Works Association Research Foundation and American Water Works Association, Denver, CO, 1995.

Kirmeyer, G. J., L. Kirby, B. M. Murphy, P. F. Noran, K. Martel, T. W. Lund, J. L. Anderson and R. Medhurst. *Maintaining Water Quality in Finished Water Storage Facilities*, American Water Works Association Research Foundation and American Water Works Association, Denver, CO, 1999.

Lee, J. H., and G. H Jirka, "Vertical Round Buoyant Jet in Shallow Water," *Journal of Hydraulic Eng.*, ASCE, 107: 1651–1675, 1981.

Levenspiel, O., *Chemical Reaction Engineering*, John Wiley & Sons, New York, 1972.

McNaughton, K. J., and C. G. Sinclair, "Submerged Jets in Short Cylindrical Flow Vessels", *Journal of Fluid Mech.*, 25(2): 367–375, 1966.

Montiel, A., J. Ouvrard, S. Rigal, and G. Bousquet, "Etude de L'origine et du Mécanisme de Formation de Composés Sapides Responsables de Goûts de Moisi Dans les Eaux Distribuées," TSM, 82(2):73 83, 1987.

Murrer, J., J. Gunstead, and S. Lo, "The Development of an Ozon Contact Tank Simulation Model," *Ozone: Science and Engineering*, 17: 607–617, 1995.

Nash, J. E.,"The Form of the Instantaneous Unit Hydrograph", IASH, Pub. 45, 3:114–121, 1957.

Okita, N., and Y. Oyama, "Mixing Characteristics in jet Mixing", *Japanese Chemical Engineering*, 1(1):94 101. 1963.

Ongerth, H. J., "Quality Control in Distribution Systems," *Water Quality and Treatment*, McGraw-Hill, New York, 1971.

Pettie, R., "Air, Other Gases Make Powerful Cross Connections," *OPFLOW*, 16(1):1, 3–4, 1990.

Poggenburg, W. J. Schubert, J. Uhlenberg, "Planning and Construction of a 16MG Reservoir for Düsseldorf," GWF Wasser/Abwasser, 122:451–459, 1981.

Pontius, F. W., "Microbial-Disinfection By-Products Rule on Expedited Schedule," *OPFLOW*, 23(6):7–8, 1997.

Regli, S., "Intent of the D/DBP Rule and the Regulatory Framework," *Proceedings of the AWWA Satellite Teleconference: Disinfectants and Disinfection By-Products: Understanding the Proposed D/DBP Rule: Participant Guide*, American Water Works Associations, Denver, CO, 1993.

Rossman, L. A., and W.M. Grayman, "Scale Model Studies of Mixing in Drinking Water Storage Tanks," *Journal of Environmental Engineering, ASCE*, 125(8): 755–761, 1999.

Schatz, O., "Kriterien für die Instandsetzung von Wasserbehälter in Bau- und Betrieblicher Sicht," *3R International*, 31(1/2):46–51, 1992.

Sherman, L. K., "Streamflow from Rainfall by the Unit-Graph Method," *Engineering News Record*, 108(April 7):501–505, 1932.

Shoenen, D., "Recolonization of Drinking Water: Experimental Studies of Three Different Waters Under Various Test Conditions," GWF-Wasser/Abwasser, 133:73–186, 1992.

Simpson, L. L., "Industrial Turbulent Mixing," In R. S. Brodkey, ed., *Turbulence in Mixing Operations*, Academic Press, New York,1975.

Smith, C., and G. Burlingame, "Microbial Problems in Treated Water Storage Tanks," Proceedings of the 1994 Annual AWWA Conference, American Water Works Association, Denver, CO, pg. 107-111, 1994.

Smith, D. B., A. F. Hess, and S. A. Hubbs, "Survey of Distribution System Coliform Occurrences in the United States," *Proceeding of the Eighteenth Annual AWWA Water Quality Technology Conference*, American Water Works Association, Denver, CO, pg 1103-1116, 1990.

Sonntag, H. G., "Experience With Bacterial Growth" in Waterworks Systems, *Water Supply*, 4:195–197, 1986.

Ta, C. T., A. Eades, and A. J. Rachwal, "Practical Methods for Validating a Computational Fluid Dynamics-Based Dissolved Air Flotation Method, "*Proceedings of the AWWA Water Quality Technology Conference*," American Water Works Association; 1996.

Van de Vusse, J.G., "Vergleichende Ruhrversuche zum Mischen Ioslicher Flussigkeiten in einem 12000 - m cubed – Behalter" (A comparison of different mixing experiments for the mixing of soluble fluids in a 12,000 cubic meter tank), *Chemie Ingenieur Technik*, Vol. 31, 583-587, 1959.

Water and Power Resources Service, *Hydraulic Laboratory Techniques*, Denver, CO, U.S. Department of the Interior, 1980.

CHAPTER 12
COMPUTER MODELS/EPANET

Lewis A. Rossman
*U.S. Environmental Protection Agency,
National Risk Management Research Laboratory
Cincinnati, OH*

12.1 INTRODUCTION

Pipe network flow analysis was among the first civil engineering applications programmed for solution on the early commercial mainframe computers in the 1960s. Since that time, advancements in analytical techniques and computing power have enabled us to solve systems with tens of thousands of pipes in seconds using desktop personal computers. This chapter discusses how modern-day computer models are used to analyze the hydraulic and water-quality behavior of distribution systems. It covers how computer models are applied to actual systems, what the internals of the models consist of, and the capabilities and operation of one particular model in the public domain, EPANET. The chapter focuses only on models that analyze successive periods of steady flow through a general arrangement of connected pipes, pumps, valves, and storage facilities. Other, more specialized computer models, such as programs for surge analysis, are not addressed.

12.1.1 Need for Computer Models

The classical pipe-network flow problem asks what the flows and pressures are in a network of pipes subject to a known set of inflows and outflows. Two sets of equations are needed to solve this problem. The first set requires conservation of flow to be satisfied at each pipe junction. The second specifies a nonlinear relation between flow and headloss in each pipe, such as the Hazen-Williams or Darcy-Weisbach equation. Whenever a network contains loops or more than one fixed-head source, these equations form a coupled set of nonlinear equations. Such equations can be solved only by using iterative methods, which for all but the smallest-sized problems require the aid of a computer. Because most distribution systems of interest are looped, computer models have become a necessity for analyzing their behavior.

Computer models also provide other advantages that enhance distribution system modeling. These include the following:

- Systematic organization, editing, and error checking of the input data required by the model
- Aid in viewing model output, such as color-coded maps, time-series plots, histograms, contour plots, and goal-specific queries
- Linkages to other software, such as databases, spreadsheets, computer aided design (CAD) programs, and geographic information systems (GIS)
- Ability to perform other kinds of network analyses, such as optimal pipe sizing, optimal pump scheduling, automated calibration, and water-quality modeling

12.1.2 Uses of Computer Models

Cesario (1991) discusses a number of different ways that network computer models are used in planning, engineering, operations, and management of water utilities. Some examples include the following.

1. Network models are run to analyze what capital improvements will be needed to serve additional customers and maintain existing services in future years. They also can help a utility prepare for planned outages of specific system components, such as reservoirs and pump stations.
2. Network models are used to locate and size specific network components, such as new mains, storage tanks, pumping stations, and regulator valves.
3. Pump scheduling, tank turnover analysis, energy optimization, and operator training are some ways in which network models can be used to improve system operations.
4. Extensions to hydraulic models allow them to analyze a host of questions related to water quality. They can determine how water from different sources blends together throughout a system, how operational changes can reduce the time that water spends in the system, and what steps can be taken to maintain adequate disinfectant residuals without excessive levels of disinfection by-product formation throughout the system.
5. Fire-flow studies are used to determine if adequate flow and pressure are available for fire-fighting purposes, as required for fire insurance ratings.
6. Vulnerability studies are used to test a system's susceptibility to unforeseen occurrences, such as loss of power, major main breaks, extended drought periods, and intrusion of waterborne contamination.

12.1.3 History of Computer Models

The groundwork for computer modeling of distribution systems was laid by the numerical method developed by Hardy Cross in the 1930s for analyzing looped pipe networks (Cross, 1936). The first mainframe programs for pipe-network analysis that appeared in the 1960s were based on this method (Adams, 1961), but these were soon replaced with codes that used the more powerful Newton-Raphson method for solving the nonlinear equations of pipe flow (Dillingham, 1967; Martin and Peters, 1963; Shamir and Howard, 1968).

The 1970s saw a number of new advancements in network solution techniques. New, more powerful solution algorithms were discovered (Epp and Fowler, 1970; Hamam and

Brameller, 1971; Wood and Charles, 1972), techniques for modeling such nonpipe elements as pumps and valves were developed (Chandrashekar, 1980; Jeppson and Davis, 1976). Ways were found to implement the solution algorithms more efficiently (Chandrashekar and Stewart, 1975; Gay et al., 1978). The extension from single-time-period to multitime-(or extended) period analysis was made (Rao and Bree, 1977).

The 1980s were marked by the migration of mainframe codes to personal desktop computers (Charles Howard and Associates; 1984; Wood, 1980). They also saw the addition of water-quality modeling to network analysis packages (Clark et al., 1988; Kroon, 1990). In the 90s, the emphasis has been on graphical user interfaces (Rossman, 1993) and the integration with CAD programs and water utility databases (Haestad Methods, 1998).

12.2 USE OF A COMPUTER MODEL

A water distribution system model actually has two parts (Walski, 1983): the computer program that makes the calculations and the data describing the physical components of the water system, customer demands, and operational characteristics. The steps involved in the modeling process can be summarized as follows:

1. Determine the kinds of questions the model will be used to answer.
2. Represent the real-world components of the distribution system in terms that the computer model can work with.
3. Gather the data needed to characterize the components included in the model.
4. Determine water use throughout the modeled network within each time period being analyzed.
5. Characterize how the distribution system is operated over the period of time being analyzed.
6. Calibrate the model against observations made in the field.
7. Run the model to answer the questions identified in step 1 and document the results.

12.2.1 Network Representation

12.2.1.1 Network components. Computer models require a real-life distribution system to be conceptualized as a collection of links connected together at their end points, which are called nodes. Water flows along links and enters or leaves the system at nodes. The actual physical components of a distribution system must be represented in terms of these constructs. One particular scheme for accomplishing this is shown in Fig. 12.1. In this scheme, links consist of pipes, pumps, or control valves. Pipes convey water from one point to another, pumps raise the hydraulic head of water, and control valves maintain specific pressure or flow conditions. Other types of valves, such as shutoff or check valves, are considered to be properties of pipes. Nodes consist of pipe junctions, reservoirs, and tanks. *Junctions* are nodes where links connect together and where water consumption occurs. *Reservoir nodes* represent fixed-head boundaries, such as lakes, groundwater aquifers, treatment plant clearwells, or connections to parts of a system not being modeled. Tanks are storage facilities, the volume and water level of which can change over an extended period of system operation.

12.4 Chapter Twelve

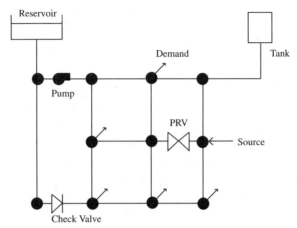

FIGURE 12.1 Node-link representation of a distribution system.

12.2.1.2 Network skeletonization. When building a network model, one must decide which pipes to include in the model. The process of representing only selected pipes within a model is called skeletonization. At one end of the spectrum, a transmission-mains model might include only the major pipelines connecting points of water entry to pump stations, storage tanks, control valves, and major consumers. At the other end, an all-mains or street-level model includes every pipe, short of lateral connections to individual homes. Where a model lies along this spectrum depends in part on the kind of questions the model is being used to answer. For example, a highly skeletonized model may be sufficient for capital improvement planning or for pump-scheduling studies. Such a model might not be suitable for water-quality modeling or for fire-flow analysis, where more localized impacts are of interest.

One usually skeletonizes a network by first deciding on the smallest diameter of pipe to include in the model. To these pipes are added those that connect to large water users, major facilities, and particular points of interest, such as monitoring locations. Additional pipes might be included to close loops that are judged to be important. The advantages of a skeletonized model are reduced data-handling requirements and easier comprehension of model output. Disadvantages include the need to use engineering judgment about which pipes to include and difficulties in aggregating demand from individual water users to the nodes contained in the model. All-mains models provide more accurate depiction of true system behavior at the expense of having to supply more descriptive data and producing output that is more difficult to understand. It has become increasingly easier to develop and use all-mains models as more utilities have used computerized asset management systems and as the user interfaces and data-handling capabilities of network modeling packages have become more sophisticated.

12.2.2 Compilation of Data

Table 12.1 lists the minimum set of properties that must be supplied for the various components included in a network model. Further explanation of some of these items is provided below.

TABLE 12.1 Minimum Set of Properties Needed to Model Network Components

Component	Properties
Junctions	ID label
	Elevation
	Demand
	Demand pattern
Reservoirs	ID label
	Elevation
Tanks	ID label
	Bottom elevation
	Initial water level
	Water level-volume curve
Pipes	ID label
	Start node label
	End node label
	Diameter
	Length
	Roughness coefficient
Pumps	ID label
	Start node label
	End node label
	Head-discharge curve
Valves	ID label
	Start node label
	End node label
	Type (PRV, PSV, FCV, etc.)
	Pressure/flow setting

12.2.2.1 ID labels. Each node and link must be assigned a unique number or label so that it can be identified during processing. Character labels provide more flexibility than numbers do because it is easier to include useful information, such as pressure zones or place names in the label.

12.2.2.2 Nodal elevations. It is important to obtain accurate elevations at system boundary points, such as reservoirs and storage tanks, and at locations where pressure measurements are made for calibration purposes. Each foot of error in elevation will introduce almost half a pound per square inch (psi) of error in pressure estimates.

12.2.2.3 Pipe diameters. Historical information on pipe diameters only reflects the size of the pipe at the time of installation. The diameter of unlined iron pipe can be significantly reduced over time because of tuberculation caused by corrosion. The effect of this reduction on computing flows and headlosses is usually lumped together with modifications made to the pipe's roughness coefficient during model calibration. Although this works well enough for hydraulic modeling, use of an incorrect pipe diameter could lead to difficulties when trying to model water quality. The behavior of a water-quality model can be affected by the travel time of water through a pipe, which is a function of the pipe's diameter. Thus, a well-calibrated hydraulic model might not perform as well in modeling water-quality behavior.

12.2.2.4 Pipe roughness. A pipe's roughness coefficient represents the contribution of irregular wall surfaces to headloss caused by friction. The type of coefficient used depends on the headloss formula being used. For the Hazen-Williams formula, the coefficient is a dimensionless quantity known as the C-factor, which decreases in value with increasing surface roughness. The Darcy-Weisbach formula uses a coefficient with units of length that represents the height of roughness elements along the pipe wall. Thus, its value increases with increasing surface roughness. Tables are available that provide typical values for both new and older pipe of different materials (Lamont, 1981, as reproduced in Walski, 1984). Values from these tables can be used as starting points when building a network model. They should ultimately be refined through both field testing and model calibration.

12.2.2.5 Pump curves. Pumps add energy to water to lift it from lower to higher levels. For a fixed rotational speed, a pump has a unique relationship between the head it can deliver and the flow rate it can supply. The curve showing this relationship is known as the head-discharge curve, and it typically has a concave downward shape. The set of curves showing head, efficiency, and power as a function of flow rate are known as a pump's characteristic curves. They are usually provided by the pump manufacturer. However, since pump characteristics can change with time, pumping tests should be performed periodically to assess the pump's actual performance.

Another method used to model pump performance when pump curves are not available is to assume that the pump operates at a constant horsepower rating. This approach should be used with caution, particularly with extended-period simulations because the resulting head-discharge combinations produced for the pump can sometimes become totally unrealistic.

Sometimes, it is more convenient to replace a pump connected directly to a supply reservoir with a single node that is assigned a negative demand equal to the pump's discharge. This can be especially useful in studies where pump performance is not an issue but where it is important to model inflows to the system accurately.

12.2.3 Estimation of Demand

Water demands, or consumption rates, for a distribution system are analogous to the loads placed on a structure. Both play a major role in determining the behavior of their respective systems. Average demands can be estimated and assigned to network junctions in several ways. In order of increasing level of detail and accuracy, they are by category of land use, type and number of dwellings, meter routes, and individual meter billing records. These four methods are illustrated in Fig. 12.2. Special attention must be paid to large water users, such as certain industries, commercial establishments, universities, and hospitals. Unaccounted for water, which can be as much as 10 to 20 percent of total demand, is usually distributed uniformly across all the junctions in the network.

Average rates of water use should be adjusted to reflect the season and time of day for which the model is being run. Seasonal adjustment factors can be based on average system-production rates recorded at different times of the year. Diurnal adjustment factors can be found by performing a system water budget over a 24-h time span. This process computes total system consumption in each hour of the day as the difference between the amount of water entering the system from all external inflows and the net amount of water added to storage. The latter quantity, which can be either positive or negative, can be

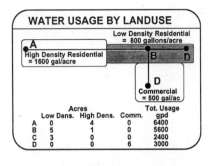
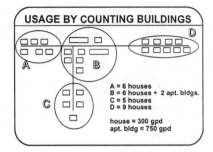
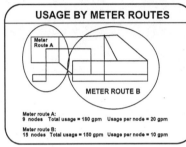
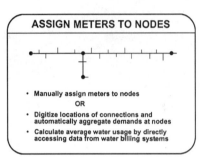

FIGURE 12.2 Methods for estimating water demands.

determined from changes in water levels in the storage tank. The adjustment factor for any hour of the day equals the consumption in that hour divided by the average daily consumption.

12.2.4 Operating Characteristics

Additional information needed to run a network model includes the status of all pumps and valves, the initial water levels in all storage tanks, and, for water-quality analyses, the initial water quality at all nodes. When making an extended-period analysis, the model also needs to know how pumps and valves are controlled throughout the simulation period. This information might be represented through a fixed time schedule of pump/valve openings and closings or through a set of rules that describe what conditions (e.g., tank water levels or nodal pressures) will cause a pump or valve to change status.

12.2.5 Reaction-Rate Information

When modeling the fate and transport of reactive substances within a distribution system, one must be able to characterize their rate of reaction. Typically, one assumes that reaction rates are proportional to the amount of substance present: in other words, that reaction rates are first-order. Other reaction orders can be used as well (e.g., Vasconcelos et al., 1996). First-order reaction rates can be modeled using a single parameter called a reaction-rate coefficient. Rate coefficients can be different for reactions occurring in the bulk flow, at the pipe wall, and within storage tanks. Although global values for these coefficients may be convenient to use, localized conditions, such as differences in pipe material and age, may require the use of different coefficients on a pipe-by-pipe basis. The need for such detail can be determined only from field monitoring and model calibration efforts.

12.2.6 Model Calibration

Calibration is the process of making adjustments to model inputs so that the model output reproduces observed measurements to a reasonable degree of accuracy. Adjustable model inputs primarily include pipe-roughness coefficients and nodal demands. For water-quality models, they include initial water-quality conditions and reaction-rate coefficients. Observable model outputs are pressures, flows, tank water levels, and water-quality predictions.

One can perform two levels of calibration. One level serves as a reality check that the model is producing reasonable, but not necessarily highly accurate results. The modeler should check for the following problematic behavior:

- Unreasonably low (e.g., negative) or high pressures.
- Pumps operating outside of their allowable range or being shut down for this reason.
- Pumps cycling on/off in an unreasonable fashion.
- Tanks that continuously keep filling or emptying.
- Nodes disconnected from any source because of closed pipes, pumps, or valves.

Any of these conditions indicates that there was a problem in representing some aspect of the system to the computer.

The second level of calibration involves adjustments to model input parameters that match best with field observations. This requires the collection of field data, preferably under more than one operating condition. When collecting these data, priority should be given to measuring conditions at the system boundaries. This would include flow rates and pressures at supply points or at interzone connections and water levels in storage tanks. For water-quality models, one would want to have constituent concentrations measured at these points as well. Selection of additional sampling points within the system depends on what use is being made of the model. Avoid selecting locations that provide redundant information. If possible, try to include readings from any installed flowmeters because computed flows tend to show more response to changes in input parameters than do pressures. More detailed discussions of model calibration can be found in Chaps. 9 and 14.

12.3 COMPUTER MODEL INTERNALS

The computer code that solves a network model typically includes the following functions:

- input processing,
- topological processing,
- hydraulic solution algorithm,
- linear equation solver,
- extended period algorithm,
- water-quality algorithm, and
- output processing.

12.3.1 Input Processing

The input processor takes a description of a network model that is meaningful to a human analyst and converts it into an internal representation that can be used by the computer. The network description typically is encoded into a text file with some specified format for storing the type of information shown in Table 12.1. The computer model's input processor parses and interprets the contents of this file and assigns each piece of data to the correct internal data structure. The better computer codes obey the principle of not requiring the user to supply information that the computer can determine for itself. Thus, there should be no need for the user to specify how many nodes or links are contained in the network (the computer can count for itself as the data for each object is read). Nor should the user have to provide information on network connectivity, such as identifying closed loops or determining which nodes are linked to a given node through a link (it is enough to know which two nodes make up the end points of a link).

12.3.2 Topological Processing

The computer needs to determine certain topological relations between the nodes and links in a model so that its solution algorithms can be implemented. First among these is a determination of which nodes are directly connected to other nodes via links. This can be determined by creating an adjacency list for each node. The *adjacency list* is a linked list of data structures. Each element of the list contains three items: the index of the link that connects to the node in question, the index of the node on the other end of the link, and a pointer to the next item in the list. An array is used to store the address of the first element of the list for each node.

A second type of topological processing is required for hydraulic solution methods that use Kirchoff's law, which requires that the sum of headlosses and gains around each closed loop in a network be zero. An efficient means is needed to determine a set of basic or fundamental loops. The number of such loops in a network with NN nodes and NL links is $NL - NN + 1$. Osiadacz (1987) described a number of ways available to generate a set of loops, all of them based on finding a *spanning tree* for the network. A spanning tree is a collection of links that connects every junction node in the network back to a reservoir or tank node without forming any loops. The spanning tree also is useful for determining an initial set of flows that satisfies nodal continuity.

12.3.3 Hydraulic Solution Algorithms

The basic equations required to solve for flows and hydraulic heads (from which pressures can be obtained) in a network are

$$\sum_i Q_{ij} - \sum_k Q_{jk} = D_j \text{ for each node } j \qquad (12.1)$$

and

$$H_i - H_j = aQ_{ij}|Q_{ij}^{b-1}| \text{ for each link connecting nodes } i \text{ and } j \qquad (12.2)$$

where Q_{ij} = flow in link connecting nodes i and j (positive if flow is from i to j, otherwise negative), H_j = head at node j, and D_j = demand at node j. Equation (12.1) is the nodal continuity relation and Eq. (12.2) is the flow-headloss relationship, where a and b are coefficients. For the Hazen-Williams formula, and

$$a = \frac{10.69L}{C^{1.85}d^{4.87}} \quad (12.3)$$

$$b = 1.85 \quad (12.4)$$

where $L=$ pipe length (m), $d =$ pipe diameter (m), and $C=$a roughness coefficient.

Computer models can reduce these equations into a simpler system to be solved in four different ways: (1) the node method (H equations), (2) the flow method (Q equations), (3) the loop method (ΔQ equations), and (4) the gradient or node-loop method (H-Q equations).The Newton-Raphson technique is used with each method to solve the resulting system of nonlinear equations by means of an iterative solution of a system of linear equations. Well-behaved systems typically will converge in under four to six iterations. More information about the computational details of these methods can be found in Jeppson (1976), Osiadacz (1987), and Salgado et al. (1988). Although the original Hardy Cross method cannot compete with the numerical efficiency of these procedures, it still appears in sample computer programs for pipe flow analysis found in many textbooks (e.g., Clark et al., 1977).

Table 12.2 compares some pertinent characteristics of the four different methods. The flow method results in the largest number of equations and the loop method has the smallest. The node method solves for heads, and the flow and loop methods solve for flows. Once either heads or flows are known, it is a straightforward matter to determine the complementary set of unknowns using the flow-headloss relations. The gradient and node-loop methods are unique in that both heads and flows are determined in a recursive fashion during the Newton-Raphson iterations, with each new set of flows serving as a feedback signal used to update the next new set of heads.

Both the flow and loop methods require that a set of fundamental loops must be identified. The loop method also requires that an initial estimate of link flows must be found to satisfy continuity. The node method is known to have convergence problems on some networks that have low resistance links connected to high-resistance ones and

TABLE 12.2 Characteristics of Different Hydraulic Solution Methods

	Node Method	Flow Method	Loop Method	Gradient/Node Loop Method
Number of equations	NJ	NJ + NL	NL−NJ	NJ
Variable solved for	Head	Flow	Flow adjustment	Head and flow
Requires loop generation	No	Yes	Yes	No
Requires initial flow solution	No	No	Yes	No
Convergence properties	Poor to good	Good	Good	Good
Symmetric coefficient matrix	Yes	No	Yes	Yes
Relative degree of sparsity	High	Medium	Low	High

Abbreviations: NJ = number of junctions; NP = number of links.

pumps with steep head-discharge curves (Salgado et al., 1988). The node, loop, and gradient/node-loop methods all produce symmetric coefficient matrices for the system of linear equations resulting from the Newton-Raphson procedure. Symmetric matrices require less computer memory for storage and permit the use of more efficient solution techniques. The node and gradient/node-loop methods have an easier time accommodating nonpipe elements-particularly check valves, regulating valves, and closed pipes-than do the loop-based methods. Special techniques are required by the loop-based methods to handle situations in which a pipe can be closed and its flow becomes zero.

12.3.4 Linear-Equation Solver

Most of the work in computing network hydraulics involves solving the system of simultaneous linear equations that results at each iteration of the Newton-Raphson method. Indeed, improvements in this aspect of the computer model are likely to have more impact on the size and speed at which network problems can be solved than will the choice of solution algorithm. The coefficient matrices associated with the various solution methods are moderately to highly sparse, meaning that most entries are zeroes. For example, the number of nonzero coefficients in a row of the matrix derived from either the node or the gradient/node-loop methods represents the number of links connected to a particular node. Because pipe networks rarely have more than five links connecting to a node, a system with 1000 nodes will have at most only 5000 nonzero entries in the 1,000,000 elements of its coefficient matrix. Having to store and carry all the zero-value elements through the process solution can impose a significant computational penalty.

Direct methods for solving sets of linear equations use a series of elementary operations (e.g., addition and multiplication) on the elements of the coefficient matrix to transform it into a triangular form so that the equations can be solved via a sequence of simple substitutions. This process, known as *factorization*, adds additional nonzero entries to the transformed coefficient matrix, called fill-ins, which in turn increases the computational burden. Simply by rearranging the rows and columns of the matrix, the number of fill-ins can often be reduced significantly. For the node and gradient/node–loop methods, such a rearrangement is equivalent to simply reordering (or renumbering) the nodes in the network. The techniques used to minimize fill-in and to store and operate on only the nonzero elements of the coefficient matrix are known as sparse matrix methods (see George and Liu, 1981, and Pissanetzky, 1984, for further discussions). Needless to say, without the use of such methods, it would be impossible to analyze large pipe networks in reasonable amounts of time using personal computers.

12.3.5 Extended-Period Solver

Simulating the behavior of a distribution system over an extended period of time is known as extended-period simulation (EPS), which allows the modeler to capture the effects that changes in customer demands and tank water levels have on system performance. EPS also is a prerequisite for performing meaningful water-quality analysis. The method used to perform EPS is to integrate the differential equation that represents the change in head at storage tanks with respect to time, using steady-state network analysis to compute network flows as a function of tank heads. The equations to be solved for each storage tank s can be expressed as

$$\frac{dV_s}{dt} = Q_s \qquad (12.5)$$

and

$$H_s = E_s + h(V_s) \tag{12.6}$$

where V_s = volume of water in storage tank s, t = time, Q_s = net flow into (+) or out of (−) tank s, H_s = head (water surface elevation) in tank s, E_s = elevation of bottom of tank s, and $h(V_s)$ = water level as a function of water volume in tank s.

A number of different methods are available to perform this integration. The simplest method is known as the Euler method, which replaces the dV_s/dt term in Eq. (12.5) with its forward difference approximation so that

$$V_s(t + \Delta t) = V_s(t) + Q_s(t)\Delta t \tag{12.7}$$

and

$$H_s(t + \Delta t) = E_s + h(V_s(t + \Delta t)) \tag{12.8}$$

where $X(t)$ denotes the value of X at time t. Under this scheme, the tank levels $H_s(t)$ and nodal demands existing at time t are used to solve a network flow analysis that produces a set of net flows $Q_s(t)$ into the tanks at time t. Equations (12.7) and (12.8) are used to determine new tank levels after a period of time Δt. Then a new steady state analysis is run for time $t + \Delta t$, using the new tank levels as well as new demands and operating conditions that apply to this new time period. The simulation proceeds in this fashion from one time period to the next.

Under most conditions, the Euler method produces acceptable results because over a typical time period of say, 1 h, demands and tank levels do not change dramatically. Other integration schemes, such as the predictor-corrector method (Rao and Bree, 1977), are available to accommodate more rapid changes in conditions. These methods require additional flow solutions to be made at intermediate time steps. As mentioned above, EPS uses a succession of steady-state flow solutions that does not account for either inertial or compressibility effects in pipe flow. Hence, it cannot be called a truly dynamic simulation approach. However, for the kinds of model uses described in Sec. 12.1.2, it has proved to work well enough in practice.

12.3.6 Water-Quality Algorithms

The numerical methods for tracking the propagation and fate of water-quality constituents in distribution systems were reviewed in Chap. 9. All of these methods require as input the link flows computed over each time step of an extended-period hydraulic analysis. The time steps used in a water-quality analysis typically are much shorter than those used for extended-period hydraulic analysis (e.g., 5 min. instead of 1 h). Because all the methods use some form of subdiscretization of the network's links, a substantial amount of computer memory can be required for water-quality analysis. Thus, most models compute water-quality conditions after a hydraulic analysis has already been performed rather than attempt to implement the two procedures simultaneously.

12.3.7 Output Processing

Output processing conveys the results computed by the models in a format that is useful and informative to the model's users. The amount of information generated by these models can be enormous—flows, pressures, heads, and water quality at thousands of nodes and links for dozens or even hundreds of time periods. Such huge amounts of data require that selective output reporting be used or that disk files be used to archive it. Several issues and options arise in designing effective output processors. These include the following:

1. *Time steps.* Extended-period hydraulic simulations usually proceed at some fixed time step set by the model's user. However, new solutions can be generated at intermediate times, such as when tanks are taken off-line because they become empty or full or when pumps open and close because certain tank levels are reached. A decision must be made regarding whether such intermediate results are saved and made available to the model's user. A compromise strategy is to keep a separate log of changes in system status that records when intermediate solutions occur and what system components changed state but does not provide complete reporting of such solutions to the user.

2. *Reporting options.* Input options should be available to allow the user to state which nodes and links should be reported and in which time periods. A common technique used in water-quality analyses to establish repeating periodic conditions over a prescribed operating period is to run the model in repeating fashion for dozens or even hundreds of such periods until this state is achieved. Clearly, there is no need to save results generated during the transient start-up phase of such simulations. Another type of reporting option allows the user to request that reports be made only when a certain condition is encountered, such as the pressure at a node being below a set level or the head loss in a pipe exceeding a certain level.

3. *Binary output files.* To warehouse the potentially enormous amount of output data that can be generated by a network model, the use of binary files is a must. These provide much quicker data access and more compact data storage than do formatted text files. Creation of such files provides great flexibility in using postprocessing software to examine the data in a multitude of different ways. As alternatives to simple binary files, output results also can be saved in formats used by commercial spreadsheet and database programs, which also are usually binary in nature.

4. *Error reporting.* Output processing should report all error and warning conditions encountered in processing a network's input data and in running an analysis. Noteworthy conditions that warrant such reporting include failure of a hydraulic analysis to converge to a solution, pumps operating outside the limits of their head-discharge curves, occurrence of negative pressures, and nodes disconnected from any source of water as a result of link closures.

12.4 EPANET PROGRAM

12.4.1 Background

EPANET is a public-domain, water-distribution-system modeling package developed by the U.S. Environmental Protection Agency's Water Supply and Water Resources Division. It performs extended-period simulation of hydraulic and water-quality behavior within pressurized pipe networks and is designed to be a research tool that improves our understanding of the movement and fate of drinking-water constituents within distribution systems. EPANET first appeared in 1993 (Rossman, 1993), and a new version is slated for release in 1999. The program can be downloaded from the World Wide Web at *http://www.epa.gov/ORD/NRMRL/wswrd/epanet.html.*

12.4.2 Program Features

EPANET models a distribution system using the same objects as described in Sec. 12.2.1.1. These objects include the following:

- Junctions (where pipes connect and water consumption occurs)
- Reservoirs (which represent fixed-head boundaries)
- Tanks (which are variable-volume storage facilities)
- Pipes (which can contain either shutoff valves or check valves)
- Pumps (which can include fixed-speed, variable-speed, and constant-horsepower pumps)
- Control valves (which can include pressure-reducing valves, pressure-sustaining valves, flow control valves, and throttle-control valves)

In addition to these physical objects, the following informational objects also can be used to represent a distribution system:

- Time patterns (sets of multipliers used to model diurnal water demands)
- Curves (*x-y* data used to represent head-discharge curves for pumps and water level-volume curves for tanks)
- Operational controls (rules that change link status depending on such conditions as tank levels, nodal pressures, and time)
- Hydraulic analysis options (choice of headloss formula, flow units, viscosity, and specific gravity)
- Water-quality options (choice of type of water-quality analysis, type of reaction mechanism, and global reaction rate coefficients)
- Time parameters (simulation duration, time steps for hydraulic and water-quality analyses, and time interval at which output results are reported)

In addition to steady-state or extended-period hydraulic analysis, EPANET can be used to run the following kinds of water-quality analyses: (1) tracking the propagation of a nonreactive constituent, such as one that would be used in a tracer study or for reconstruction of a contamination event, (2) determining what percentage of water from a particular source is received by each location in a network, (3) estimating the age of water received at various locations in the network, (4) modeling the fate of chlorine and chloramines, which decay with time and can react both in the water phase and at the pipe wall, and (5) modeling the growth of certain disinfection by-products, such as trihalomethanes, which grow with time up to a limiting value.

EPANET consists of two modules. One is a network solver that performs the hydraulic and water-quality simulation; the other is a graphical user interface that serves as a front-and-back end for the network solver. In the program's normal mode of operation, the user interacts directly with the graphical interface with the solver's presence being transparent. The solver also can be run as a stand-alone executable, receiving its input from a text file and writing its results to a formatted text-report file or an unformatted, binary-output file. The solver also exists as a library of callable functions that third-party developers can use in custom applications.

12.4.3 User Interface

EPANET's graphical user interface is responsible for constructing the layout of the network to be simulated, editing the properties of the network's components and its simulation options, calling on the solver module to simulate the behavior of the network, and accessing results from the solver to display to the user in a variety of formats. It was written specifically for the Windows 95/98/NT platform using Inprise's Delphi language (an object-oriented version of Pascal).

Figure 12.3 depicts the user interface as it might appear when editing a network. The Network Map provides a schematic layout of the distribution system that gives the user a visual sense of where components are located and how they connect together. Because the nodes and links can be colored according to the value of a particular variable, such as pressure or flow, the Network Map also offers a holistic view of how quantities vary spatially across the network. The toolbar that appears above and to the right of the map allows the user to add new components visually using point-and-click with the mouse. Existing items on the map also can be selected and then moved, edited, or deleted. The toolbar also includes icons for panning and zooming in or out on the map as well.

The Browser window is the central control panel for EPANET. It is used to (1) select specific network objects, (2) add, delete, or edit network objects including such nonvisual objects as time patterns, operating rules, and simulation options, (3) select which variable

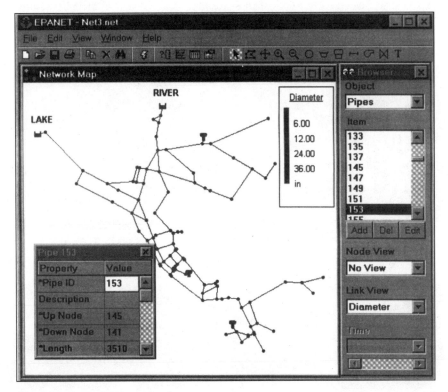

FIGURE 12.3 EPANET's user interface.

is viewed via color-coding on the map, and (4) select which time period is viewed on the map for extended-period simulations.

The third, smaller window shown in Fig. 12.3 is the Property Editor, which is used to change the properties of the item currently selected on the Network Map and in the Browser. Other, more specialized property editors exist for editing the network's nonvisual data objects, such as time patterns, curves, and operating rules.

The Network Map, the Browser, and the Property Editor are all connected to one another in the sense that any selection or change made in one is carried over to the others. For example, if the user clicks on a specific node on the map, that node becomes the current object shown in the Browser and in the Property Editor. If the user changes the diameter of a pipe in the Property Editor and diameter is the current variable selected in the Browser for viewing, the pipe will be redrawn on the map with its color changed.

EPANET's user interface keeps all the data that describes a network in its own object-oriented internal database. When the user wants to run an analysis, the program writes these data to a text file, which is then passed on to the solver module for processing. The solver makes it computations and writes its results to an unformatted binary file. The user interface then accesses this file to display selected results back to the user on request.

Figure 12.4 provides some examples of the kinds of output views that EPANET's user interface can generate after a successful simulation has been made. The top left window shows the results of a query made to the Network Map, asking it to identify all nodes where the pressure was below 50 psi. The top right window is a tabular display of link results at hour 6 of the simulation, where a filter was applied to list only those links where the headloss per 1000 ft was above 1.0. The bottom left window depicts a time-series plot for pressure at two different locations in the network. Finally, the bottom right window displays a contour plot of pressure throughout the network at hour 6.

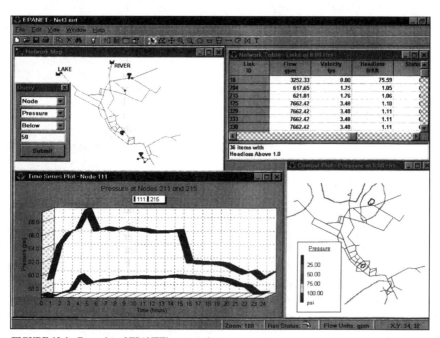

FIGURE 12.4 Examples of EPANET's output views.

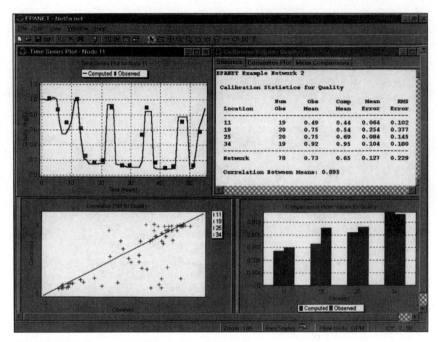

FIGURE 12.5 Example of EPANET's calibration report.

EPANET also provides tools to help aid in calibration of the network. These tools are illustrated in Fig. 12.5 for the case where a 54-h. fluoride tracer study was made in a particular network. The top left window shows a time series plot for fluoride at a specific node where both the simulated and measured values are shown for comparison. The window to its right depicts a calibration report. It compares errors between observed and computed fluoride values at all measuring locations. The two lower plots are different views of the same data. One compares measured and computed results for all samples at each location; the other compares the mean values of the computed and observed samples for each location.

12.4.4 Solver Module

EPANET's solver program is written in American National Standards Institute standard C with separate code modules for input processing, hydraulic analysis, water-quality analysis, sparse matrix/linear-equation analysis, and report generation. This modular approach facilitates making modifications to the program's features and computational procedures. The data-flow diagram for the solver is shown in Fig. 12.6. The processing steps depicted in the diagram can be summarized as follows:

1. The input processor module of the solver receives a description of the network being simulated from an external input file (.INP). This description is written in an easily understood Problem Description Language that will be discussed in more detail below. The file's contents are parsed, interpreted, and stored in a shared memory area.

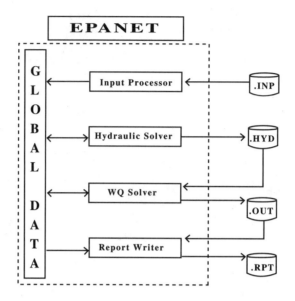

FIGURE 12.6 Data flow diagram for EPANET's solver.

2. The hydraulics module carries out a complete, extended-period hydraulic simulation, with the results obtained at every time step written to an external, unformatted (binary) hydraulics file (.HYD). Some of these time steps might represent intermediate points in time when system conditions change because tanks become full or empty or pumps turn on or off because of level controls or timed operation.

3. If a water-quality simulation was requested, the water-quality module accesses the flow data from the hydraulics file as it computes substance transport and reaction throughout the network over each hydraulic time step. During this process, it writes both the formerly computed hydraulic results as well as its water-quality results for each pre-set reporting interval to an unformatted (binary) output file (.OUT). If no water-quality analysis was called for, then the hydraulic results stored in the .HYD file are simply written out to the binary output file at uniform reporting intervals.

4. If requested by the input file, a report writer module reads back the results of the computed simulation from the binary output file (.OUT) for each reporting period and writes out selected values (as instructed by the user) to a formatted report file (.RPT). Any error or warning messages generated during the run also are written to this file.

When called by the Windows user interface, the solver skips Step 4 because the interface itself is used to generate output reports. The input file fed to the solver is written using a Problem Description Language (PDL) that makes it easily readable and self-documenting. Excerpts from such a file are shown in Fig. 12.7. Each category of input data is placed in a separate section identified by a keyword in brackets. Comment lines, beginning with a semicolon, can be placed throughout the file. The properties for multiple network objects of the same type, such as junctions and pipes, are entered in a columnar

```
[TITLE]
Sample Pipe Network
[JUNCTIONS]
;       Elev.   Demand
;ID     ft.     gpm
; ---------------------------
1       1090
101     1090
2       1122
3       1138
4       1157    500
5       1180
<etc.>

[TANKS]
;       Elev.   Init.   Min.    Max.    Diam.
ID      ft.     Level   Level   Level   ft.
; ---------------------------------------------
17      910
18      810     10      0       20      50

[PIPES]
;       From    To      Length  Diam.
;ID     Node    Node    ft.     In.     C-factor
; ---------------------------------------------
1       1       2       1500    12      130
2       2       3       1000    8       130
3       3       4       1200    10      120
4       4       5       2000    10      120
<etc.>

[PUMPS]
;       From    To      Head    Flow
;ID     Node    Node    ft.     gpm
; ---------------------------------------
171     17      101     456     2700

[VALVES]
;       From    To              Setting
;ID     Node    Node    Type    psi
; ---------------------------------------
25      21      23      PRV     75

[OPTIONS]
UNITS           GPM
HEADLOSS        H-W
QUALITY
```

FIGURE 12.7 Excerpts from an EPANET input data file.

format that saves space and enhances readability. Singular properties, such as analysis options, are entered in keyword-value format. When working with the Windows user interface, the existence of the PDL input file is invisible to the user, although the user can generate such a file so that a stand-alone, human-readable version of a network's data can be made available.

The solver uses the gradient method with extensions, as described in Salgado et al. (1988), to solve the network hydraulic equations that result at each time step of the simulation. A special procedure is implemented on top of this to keep track of changes in status that can occur in pumps and valves as the iterations of the gradient method unfold. The minimum-degree node reordering method of George and Liu (1981) is used to minimize fill-ins in the system of linear equations solved via Cholesky factorization. The coefficient matrix for these equations is stored in a sparse row-wise matrix format (Pissanetzky, 1984). Simple Euler integration is used to update water levels in storage tanks between hydraulic time steps.

The water-quality solution method used by the solver is based on a time-driven, lagrangian transport scheme described in Rossman and Boulos (1997). Reactions in both the bulk fluid and at the pipe wall can be modeled using the approach described in Rossman et al. (1994). General nth-order reactions with a limiting growth or decay potential can be modeled in the bulk phase, whereas either zero or first-order, mass-transfer-limited reactions can be modeled at the pipe wall. Storage tanks can be modeled as completely mixed, plug-flow, or two-compartment reactors.

12.4.4 Programmer's Toolkit

The functions in EPANET's network solver have been compiled into a library of routines that can be called from other applications. The toolkit functions permit a programmer to (1) open an EPANET input file, read its contents, and initialize all necessary data structures, (2) modify the value of selected network objects, such as nodal demands, pipe diameters, and roughness coefficients, (3) run repeated hydraulic and water-quality simulations using sets of modified parameters, (4) retrieve the value of selected simulation results, and (5) generate a custom report on simulation results.

The toolkit allows network modelers to incorporate state-of-the-art hydraulic and water-quality analysis routines into their own custom applications without having to worry about the details of programming these capabilities on their own. The toolkit should prove useful for developing specialized applications, such as optimization models or parameter estimation models, that require running many network analyses with modified input parameters. It also can simplify adding analysis capabilities to integrated network-modeling environments based on CAD, GIS, and database packages.

12.5 CONCLUSION

Computer models of distribution systems have achieved a remarkable degree of power and sophistication over the past 30 years. Indeed, it is probably unlikely that any more major advances can be made in solving the basic network-flow problem. This is not to say that network modeling is a dead issue. Several key themes now pervade the water industry's view of network modeling and will continue to challenge developers of computer models in the years to come.

Foremost among these themes is systems integration. Network models do not exist in a vacuum. They rely on many different sources of data to supply the information needed to run them, and their results can be used by a variety of different groups within a water utility. The issue involves being able to link network models efficiently and productively to other corporate information systems, such as engineering CAD systems, GIS, customer billing records, customer complaint records, water-quality monitoring records, pipe replacement and repair records, and real-time Supervisory Control and Data Acquisition (SCADA) systems. The topic continues to be one of high interest within the water industry, as evidenced by the recent papers of Kroon (1997) and Lerner and DiSera (1997).

A second ongoing theme is the embedding of network simulation models in other types of computer models aimed primarily at questions of system optimization and control. Examples include optimization models for pipe sizing (Dandy et al., 1996), optimal calibration programs (Lingireddy and Ormsbee, 1998), neural network models of system operation (Swiercz, 1994), and models for optimally locating and operating satellite chlorination stations (Boccelli et al., 1998). Developing libraries of network simulation functions using an open architecture, such as the EPANET Programmer's Toolkit, will help to simplify the task of embedding simulation capabilities in other codes and perhaps even encourage the creation of new and more innovative applications.

A third piece of unfinished business for network computer models involves enhancements to their water-quality modeling capabilities. Advancements continue to be made as our understanding of the chemical and biological behavior of treated water within the pipe environment improves. Whereas the first generation of models could address only nonreactive substances or substances following first-order decay reactions, recent improvements have added capabilities to model pipe-wall reactions (Rossman et al., 1994), production of trihalomethane (Vasconcelos et al., 1996), and bacterial growth (Piriou et al., 1996). Further advances are needed to model the transport and fate of alternative disinfectants, such as chloramines; to account for the effect that blending of water from different sources has on reaction kinetics; and to track the movement and fate of particulates.

Computer modeling of distribution systems has become a mainstay of the water industry. It is a classic example of how research and development carried out mainly at universities and government laboratories has been transferred into practical, useful tools for everyday practitioners. Distribution system modeling has reached a level of maturity and reliability that make it a valuable asset to any water utility.

REFERENCES

Adams, R. W., "Distribution Analysis by Electronic Computer," *Institute of Water Engineers*, 15:415–428, 1961.

Bhave, P. R., *Analysis of Flow in Water Distribution Networks*, Technomic, Lancaster, PA, 1991.

Boccelli, D. L., et al., "Optimal Scheduling Model for Booster Disinfection in Water Distribution Networks," *ASCE Journal, Water Resources, Planning, and Management*, 124:99, 1988.

Cesario, A. L., "Network Analysis from Planning, Engineering, Operations, and Management Perspectives," *Journal of the American Water Works Association*, 83:38–42, 1991.

Chandrashekar, M., "Extended set of Components in Pipe Networks," *ASCE Journal of the Hydraulic Division*, 106(HY1):133, 1980.

Chandrashekar, M., and Stewart, K. H., "Sparsity Oriented Analysis of Large Pipe Networks," *ASCE Journal of the Hydraulic Division*, 101(HY4):341, 1975.

Charles Howard and Associates, *Water Distribution Network Analysis: SPP8 Users Manual*, Charles Howard and Associates, Victoria, B.C., 1984.

Clark, R. M., W. M. Garyman, and R. M. Males, "Contaminant Propagation in Distribution Systems," *ASCE Journal of Environmental Engineering*, 114:929–943, 1988.

Clark, J. W., W. Viessman, Jr., and M.J. Hammer, *Water Supply and Pollution Control,* 3rd ed., IEP, New York, 1977.

Cross, H., *Analysis of Flow in Networks of Conduits or Conductors,* University of Illinois Engineering Experiment Station Bulletin 286, Urbana, IL, 1936.

Dandy, G. C., A. R., Simpson, and L. J. Murphy, "An Improved Genetic Algorithm for Pipe Network Optimization," *Water Resources Research,* 32:449–458, 1996.

Dillingham, J. H., "Computer Analysis of Water Distribution Systems, Parts 1, 2, 4," *Water and Sewage Works,* 114(1):1, 114(2):43, 114(4):141, 1967.

Epp, R., and A.G. Fowler, "Efficient Code for Steady State Flows in Networks," *ASCE Journal of the Hydraulic Division,* 96(HY1):43, 1970.

Gay, R. K. L., et al., "Node Reordering Algorithms for Water Network Analysis," *International Journal of Numerical Methods in Engineering,* 12:1241–1259, 1978.

George, A., and, J. W-H. Liu,. *Computer Solution of Large Sparse Positive Definite Systems,* Prentice-Hall, Englewood Cliffs, NJ, 1981.

Haestad Methods, *CYBERNET 3.1 Users Manual,* Haestad Press, Waterbury, CT, 1998.

Hamam, Y. M., and A. Brameller, "Hybrid Method for the Solution of Piping Networks," *Proceedings of the IEE,* 113:1607–1612, 1971.

Jeppson, R. W., *Analysis of Flow in Pipe Networks,* Ann Arbor Science, Ann Arbor, MI, 1976.

Jeppson, R. W., and A.L. Davis, "Pressure Reducing Valves in Pipe Network Analysis," ASCE *Journal of the Hydraulic Division,* 102(HY7):987, 1976.

Kroon, J. R., "An Innovation in Distribution System Water Quality Modeling," *Waterworld News,* July/August, 1990.

Kroon, J. R., "What is 'Integration'? Embedding vs. Interoperation," *1997 Computer Conference Proceedings,* American Water Works Association, Denver, CO, 1997.

Lamont, P. A., "Common Pipe Flow Formulas Compared with the Theory of Roughness," *Journal of the American Water Works Association,* 73(5):274, 1981.

Lerner, N. B., and D. P. DiSera, "Secrets to Successful Systems Integration," *Proceedings of the 1997 Computer Conference,* American Water Works Association, Denver, CO, 1997.

Lingireddy, S., and L. E. Ormsbee, "Neural Networks in Optimal Calibration of Water Distribution Systems," in I. Flood and N. Kartam, eds., *Artificial Neural Networks for Civil Engineers: Advanced Features and Applications,* American Society of Civil Engineers, Reston, VA, 1998.

Martin, D. W. and G. Peters, "The Application of Newton's Method of Network Analysis by Digital Computer," *Journal of the Institute of Water Engineers,* 17(2):115, 1963.

Osiadacz, A. J., *Simulation and Analysis of Gas Networks,* E. & F. N. Spon, London, UK, 1987.

Piriou, P. H., S. Dukan, Y. Levi, and P. A. Jarrige, "PICCOBIO: A New Model for Predicting Bacterial Growth in Drinking Water Distribution Systems", Proceedings of the 1996 Water Quality Technology Conference: Part I, American Water Works Association, Denver, CO, 1996.

Pissanetzky, S., *Sparse Matrix Technology,* Academic Press, London, UK, 1984.

Rao, H. S., and D.W. Bree, "Extended Period Simulation of Water Systems—Part A," *ASCE Journal of the Hydraulic Division,* 103(HY2):97, 77.

Rossman, L. A., "The EPANET Water Quality Model" in B. Coulbeck, ed., *Integrated Computer Applications in Water Supply,* Vol. 2, Research Studies Press Ltd., Somerset, England, 1993.

Rossman, L. A., EPANET Users Manual, EPA–600/R–94/057, Risk Reduction Engineering Laboratory, U.S. Environmental Protection Agency, Cincinnati, OH, 1994.

Rossman, L. A., and P. F. Boulos, "Numerical Methods for Modeling Water Quality in Distribution Systems: A Comparison," *ASCE Journal of Water Resources Planning and Management,* 122:137–146, 1996.

Rossman, L. A., P. F., Boulos, and T. Altman, "Discrete Volume Element Method for Network Water-Quality Models," *ASCE Journal Water Resources Planning and Management,* 119:505–517, 1993.

Rossman, L. A., R. M., Clark, and W. M. Grayman, "Modeling Chlorine Residuals in Drinking Water Distribution Systems," *ASCE Journal Environmental Engineering,* 120:803–820, 1994.

Salgado, R., E. Todini, and P. E. O'Connell, "Extending the Gradient Method to Include Pressure Regulating Valves in Pipe Networks," B. Coulbeck and C. H. Orr, eds. *Computer Applications in Water Supply: Vol. 1, Systems Analysis and Simulation,* Research Studies Press, 1988.

Shamir, U., and C. D. D. Howard, "Water Distribution Systems Analysis," *ASCE Journal of the Hydraulic Division,* 94(HY1):219, 1968.

Simon, A. L., *Hydraulics,* 3rd. ed., John Wiley Sons, New York, 1986.

Swiercz, M., "Using Neural Networks to Simplify Mathematical Models of Water Distribution Networks: A Case Study," First International Symposium on Mathematical Models in Automation and Robotics, Miedzyzdroje, Poland, September 1–3, 1994.

Vasconcelos, J. J., et al., *Characterization and Modeling of Chlorine Decay in Distribution Systems,* American Water Works Association Research Foundation, Denver, CO, 1996.

Walski, T.M. "Using Water Distribution System Models," *Journal of the American Water Works Association,* 75:58–63, 1983.

Walski, T. M. *Analysis of Water Distribution Systems,* Van Nostrand Reinhold, New York, 1984.

Wood, D. J., *Computer Analysis of Flow in Pipe Networks Including Extended Period Simulations (KYPIPE),* Office of Continuing Education, University of Kentucky, Lexington, 1980.

Wood, D. J. and C. Charles. "Hydraulic Network Analysis Using Linear Theory", *ASCE Journal of the Hydraulic Division,* 98(HY):1157, 1992.

CHAPTER 13
WATER QUALITY MODELING-CASE STUDIES

Robert Clark
U. S. Environmental Protection Agency
National Risk Management Research Laboratory
Water Supply and Water Resources Division
Cincinnati, OH

13.1 INTRODUCTION

The Safe Drinking Water Act of 1974 and its Amendments of 1986 requires the U.S. Environmental Protection Agency (EPA) to establish goals for maximum levels for each contaminant that may have an adverse effect on people's health. Each goal must be set at a level at which no known or anticipated adverse effects on health occur, allowing for an adequate margin of safety (Clark et al., 1987). Maximum contaminant levels, which are the enforceable standards, must be set as close to the goals as feasible.

The Act has posed a major challenge to the United States drinking water industry because of the large number of regulations being implemented over a short time frame. The Safe Drinking Water Act and its Amendments of 1996 added even more complexity to the set of national regulations affecting the drinking water industry. Although most of the regulations promulgated under the 1996 Act and amendments have focused on treated water, there is substantial evidence that water quality can deteriorate between the treatment plant and the point of consumption. Factors that can influence the quality of water in distribution systems include the chemical and biological quality of source water; the effectiveness and efficiency of treatment processes; the adequacy of treatment facilities, storage facilities, and the distribution system; the age, type, design, and maintenance of the distribution network; and the quality of treated water (Clark and Coyle, 1990). Initially, these regulations were promulgated with little understanding of the effect that the distribution system can have on water quality.

However, the 1996 Act and its amendments also has been interpreted as meaning that some maximum contaminant levels, shall be met at the consumer's tap, which in turn has forced the inclusion of the entire distribution system when considering compliance with a

---
The author wishes to thank Jean Lillie and Steven Waltrip for their assistance in preparing this chapter.

number of the maximum contaminant levels, rules, and regulations. Consequently, there is growing awareness of the possibility that the quality of drinking water can deteriorate between the treatment plant and the consumer. For example, the regulations, emphasizing system monitoring include the Surface Water Treatment Rule, the Total Coliform Rule, the Lead and Copper Rule, and the Trihalomethane Regulation. The first two rules specify treatment and monitoring requirements that must be met by all public water suppliers. The first rule requires a detectable disinfectant residual to be maintained at representative locations in the distribution system to provide protection from microbial contamination. The rule pertaining to coliform bacteria requires regulation of bacteria that are used as "surrogate" organisms to indicate whether contamination of the system is occurring. Monitoring for compliance with the Lead and Copper Rule is based entirely on samples taken at the consumer's tap.

The recent Stage 1 Disinfectants and Disinfection By-Products Rule has lowered the standard for trihalomethanes from 0.1 mg/L to 0.08 mg/L. This standard applies to all community water supplies in the United States. Monitoring and compliance is required at selected points in the distribution system. Some of the regulations may provide contradictory guidance, however. For example, the rules concerning surface water treatment and coliform recommend the use of chlorine to minimize risk from microbiological contamination. However, chlorine or other disinfectants interact with natural organic matter in treated water to form disinfection by-products. Raising the pH of treated water will help to control corrosion but may increase the formation of trihalomethanes.

Distribution systems are extremely complex and difficult to study in the field. Therefore, interest has been growing in the use of hydraulic and water-quality models as a mechanism for evaluating the various factors that influence the deterioration of water quality in drinking-water distribution systems. This chapter presents a number of case studies that illustrate the use of this important tool.

13.2 DESIGN OF DISTRIBUTION SYSTEMS IN THE UNITED STATES

Distribution systems in the United States are frequently designed to ensure hydraulic reliability, which includes adequate quantity and pressure of water for fire flow as well to meet domestic and industrial demand. To meet these goals, large amounts of storage are usually incorporated into system designs, resulting in long residence times, which in turn may contribute to deterioration of water quality. Many water distribution systems are approaching 100 years old, and an estimated 26 percent of the distribution system piping is unlined cast iron and steel and is in poor condition. At current rates for replacement of distribution system components, a utility will replace a pipe every 200 years (Kirmeyer et al., 1994).

Conservative design philosophies, aging water supply infrastructure, and increasingly stringent drinking-water standards are resulting in concerns about the viability of drinking water systems in the United States. Questions have been raised concerning the structural integrity of these systems as well as their ability to maintain water quality from the treatment plant to the consumer.

Many distribution systems serve communities from multiple sources, such as a combination of wells, surface sources, or both. A factor that is considered infrequently, and may influence water quality in a distribution system is the effect of mixing of water from these different sources. The mixing of waters from different sources that takes place within a distribution system is a function of complex system hydraulics (Clark et al., 1988a, 1991a, 1991b).

It is difficult to study the problems of system design and the effects of long residence times in full-scale systems. Constructing specially designed pipe loops is one approach to simulating full-scale systems; however, properly configured and calibrated mathematical hydraulic models can be used to study water quality effectively in situ. Such models also can be used to assess various operational and design decisions, determine the impacts resulting from the inadvertent introduction of a contaminant into the distribution system, and assist in the design of systems to improve water quality.

13.3 WATER QUALITY IN NETWORKS

Figures 13.1 and 13.2 illustrate some transformations that take place in the bulk–water phase and at the pipe wall as water moves through a network. Cross-connections, failures at the treatment barrier, and transformations in the bulk phase can all degrade water quality. Corrosion, leaching of pipe material, and biofilm formation and scour can occur at the pipe wall to degrade water quality. Contaminant's may be conservative or may experience decay or growth as they move through the system. Many investigators have attempted to understand the possible deterioration of water quality once it enters the distribution system. Changes in bacteriological quality may cause aesthetic problems involving taste and odor, discolored water, slime growths, and economic problems, including corrosion of pipes and biodeterioration of materials (Water Research Center, 1976). Levels of bacteria tend to increase during distribution and are influenced by a number of factors, such as bacterial quality of the finished water entering the system, temperature, residence time, presence or absence of a disinfectant residual, construction materials, and availability of nutrients for growth (Geldreich et al., 1972; LeChevallier et al., 1987 Maul et al., 1985a, 1985b).

The relationship of bacteriological quality to turbidity and particle counts in distribution water was studied by McCoy and Olson (1986). The authors selected an upstream and a downstream sampling site in each of three distribution systems (two surface water supplies and a ground water supply) and sampled them twice per month for 1 year. Turbidity was found to be related in a linear manner to total particle concentration but not to the number of bacterial cells. Degradation of bacterial water quality was shown to be the result of unpredictable intermittent events that occurred within the system.

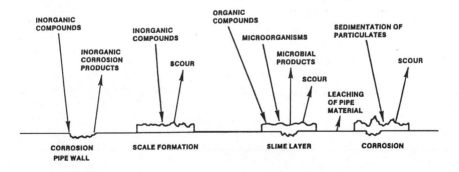

FIGURE 13.1 Water distribution pipe.

13.4 Chapter Thirteen

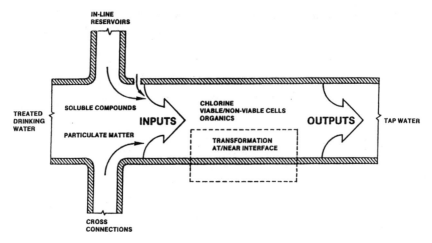

FIGURE 13.2 Transformation at pipe wall.

13.4 HYDRAULIC AND WATER-QUALITY MODELS

Cross (1936) proposed the use of mathematical methods to analyze the flow in networks, and such methods have been in use for more than half a century. Computer-based models for performing this type of analysis were first developed in the 1950s and 1960s and were greatly expanded and made more available in the 1970s and 1980s. Currently, dozens of such models are readily available on computers ranging from microcomputers (Wood, 1980a) to supercomputers (Sarikelle et al., 1989). Figure 13.3 illustrates the evolution of hydraulic and water-quality models since the 1950s.

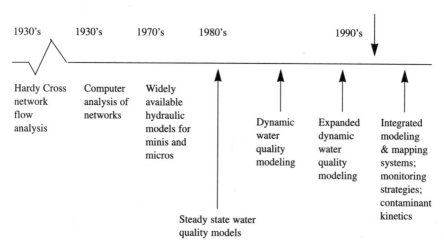

FIGURE 13.3 Historic development of water distribution system modeling.

Hydraulic models developed to simulate flow and pressures in a distribution system either under steady-state conditions or under time-varying demand and operational conditions are generally referred to as extended-period simulation (EPS) models. Hydraulic models also may incorporate optimization components that aid the user in selecting system parameters resulting in the best match between the system's observed performance and model results (Gessler and Walski, 1985). The theory and application of such hydraulic models is thoroughly explained in many widely available references (AWWA, 1989; Gessler and Walski, 1985).

In addition, hydraulic models are required to provide flow information used in distribution system water-quality models. Hydraulic models and water-quality models can be tightly bundled into a single entity, or a stand-alone hydraulic model can be used to generate a file containing hydraulic flow conditions that is then used by a stand-alone water-quality model. The usefulness and reliability of a water-quality model depends totally on the proper hydraulic characterization of the network.

13.4.1 Steady-State-Water-Quality Models

The use of models to determine the spatial pattern of water quality in a distribution system resulting from sources of differing quality was suggested by Wood (1980b) in a study of slurry flow in a pipe network. He presented an extension to a steady-state hydraulic model in which a series of simultaneous equations are solved for each node. A similar formulation was used later by Chun and Selznick (1985) in a 166-link representation of the Alameda County, California, Water District with three sources of water of differing hardness. Metzger (1985) proposed a similar approach.

In a generalization of this formulation, Males et al., (1985) used simultaneous equations to calculate the spatial distribution of variables that could be associated with links and nodes, such as concentration, travel times, costs, and other variables. This model, called SOLVER, was a component of the Water Supply Simulation Model, an integrated database management, modeling, and display system that was used to model water quality in networks (Clark, 1993a; Clark and Goodrich, 1993; Clark and Males, 1986).

An alternative steady-state "marching out" solution was introduced by Clark et al., (1988b) for calculating spatial patterns of concentrations, travel times, and the percentage of flow from sources. In this approach, links are ordered, hydraulically starting with source nodes and progressing through the network until all nodes and links are addressed (Grayman, et al., 1988b).

Wood and Ormsbee (1989) investigated alternative methodologies for predicting water quality and for determining the source of delivered flow under steady-state conditions and found an iterative cyclic procedure to be both effective and efficient. This procedure was similar to the marching out solution described previously for networks that Wood and Ormsbee identified as source-dependent (networks where the nodes can be sequenced hydraulically starting from sources). However, for non-source-dependent networks, which are rare, their algorithm iterates until a unique solution is found.

13.4.2 Dynamic Water-Quality Models

Although steady-state water-quality models proved to be useful tools, the need for models that would represent the dynamics of contaminant movement was recognized. Thus, in the mid-1980s, several models that simulated the movement and transformation of contaminants in a distribution system under temporally varying conditions were developed and applied. Three such models were initially introduced at the American

Water Works Association Distribution Systems Symposium in 1986 (Clark et al., 1986; Hart et al., 1986; Liou and Kroon, 1986). Grayman et al., (1988a) developed and applied a water quality simulation model that used flows previously generated by a hydraulic model and a numerical scheme to route conservative and nonconservative contaminants through a network. In this model, each pipe link was represented as a series of "sublinks" and "subnodes" with the length of each sublink selected to approximate the distance that a contaminant would travel during each time step. The number of sublinks varied with the velocity of flow in a link (Grayman et al., 1988b). Kroon and Hunt (1989) developed a similar model that originally was implemented on a minicomputer (Liou and Kroon, 1986). This model is tied directly to a hydraulic model and generates both tabular and graphical output displaying the spread of contaminants through a network. Hart (1991) developed a model using the GASP IV simulation language.

13.5 EARLY APPLICATIONS OF WATER–QUALITY MODELING

One of the first projects to investigate the feasibility of modeling water quality in drinking-water distribution systems was conducted under a cooperative agreement initiated between the North Penn Water Authority in Lansdale, Pennsylvania, and the U.S. EPA. The project focused on the mixing of water from multiple sources and investigated the feasibility for development and application of a steady-state water-quality model. As the study progressed, it became obvious that the dynamic nature of both demand patterns and variations in water quality required the development of a dynamic water-quality model. In addition, techniques for semicontinuous monitoring of volatile organic contaminants were explored (Clark et al., 1988b). The concept of water-quality modeling was extended to the South Central Connecticut Regional Water Authority in a study that also documented the possible negative impact of storage tanks on water quality in distribution systems. A water-quality model called the Dynamic Water Quality Model (DWQM) resulted from these studies. The DWQM was later applied to a waterborne disease outbreak in Cabool, Missouri (see Sec. 13.5.3).

13.5.1 North Penn Study

At the time of the North Penn Water Authority EPA study, the authority served 14,500 customers in 10 municipalities and supplied an average of 5 million gallons (mgd) of water per day (Clark et al., 1988b). Water sources included a 1 mgd-treated surface source of water purchased from the Keystone Water Company and 4 mgd from 40 wells operated by the authority. Figure 13.4 is a schematic representation of the 225 mi of pipe in the North Penn distribution system, showing the location of wells, the Keystone "tie-in," and the three pressure zones: Souderton Zone, Lansdale Low Zone, and Hillcrest Zone.

Surface water entered the North Penn system at the Keystone tie-in. The rate of flow into the system was determined by the elevation of the tank in the Keystone system and by a throttling valve at the tie-in. Flow was monitored continuously and was relatively constant. Water flowed into the Lansdale low-pressure zone and, from there, entered the Lawn Avenue tank and was then pumped into the Souderton Zone. Additional water from the Hillcrest pressure zone entered the Lansdale system at the Office Hillcrest transfer point; this water was derived solely from wells in the Hillcrest zone. Except for unusual and extreme circumstances, such as fire or main breaks, water did not flow from the Souderton zone into the Lansdale Low Zone nor from the Lansdale Zone into Hillcrest.

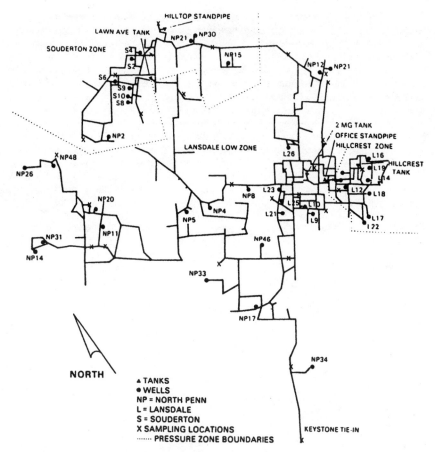

FIGURE 13.4 Distribution system of the North Penn Water Authority.

Distinct chemical characteristics were associated with the Keystone water compared with the well waters. Keystone water contained total trihalomethanes (THMs) at significantly higher levels than did the well water. Certain wells showed the occurrence of trichloroethylene, cis-1, 2-dichloroethylene, or both. Inorganic chemicals also varied from well to well and between the wells and Keystone.

13.5.1.1 Network modeling. The North Penn distribution system was modeled in a network representation consisting of 528 links and 456 nodes, and water demands for modeling represented conditions during May-July 1984. The network hydraulic model used was developed by the U.S. Army Corps of Engineers, which contained provisions for both steady-state and quasi-dynamic hydraulic modeling or extended-period simulation (Gessler and Walski, 1985). The model was used to study the overall sensitivity of the system to well pumpage, demand, and other factors resulting in the development of a number of typical flow scenarios. It revealed that significant portions of the system were subject to flow reversals.

13.8 Chapter Thirteen

13.5.1.2 Variations in water quality data To investigate the nature of variability in water quality under dynamic conditions within the system, a sampling program was conducted at six sites selected according to spatial variations determined from historical data and modeling results. Figure 13.5 shows the various sampling points used in the field study, and Figure 13.6 depicts the results of the intensive sampling program using THMs as a tracer. Laboratory and field evaluations demonstrated that the THMs in the Keystone water had reached their formation potential and were relatively stable and that any THMs formed from well sources were relatively minor. Figure 13.6 also depicts the variation in hardness of these same points (hardness was associated primarily with flow from the wells). At the Mainland sampling point, a flushing of water back and forth between the surface source and the well sources can be seen. The peaks of the THMs at Mainland were approximately 12 h out of phase with the peaks from the wells, indicating that water flow at this point was affected by the sources of surface and ground water. These results illustrated the problems involved in attempting to predict a dynamic situation using a steady-state approach and the dynamic nature of water movement in the distribution system (Clark et al.,1988b).

13.5.1.3 Development of a dynamic water-quality algorithm. As a consequence of the lesson learned in the North Penn Study DWQM was developed that used a numerical routing solution to trace water quality through the network. Demands and inflows (both values and concentrations) were assumed to be constant over a user-defined period, and a quasi-dynamic externally generated hydraulic solution was used for each period. Flow and

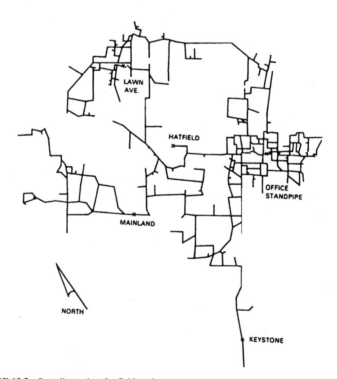

FIGURE 13.5 Sampling points for field study.

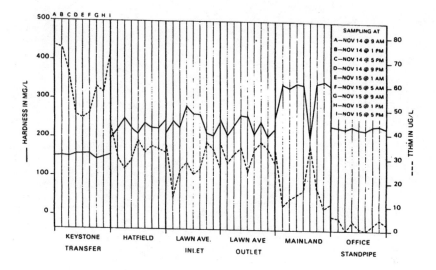

FIGURE 13.6 Sampling results from six sampling stations in North Penn Water Authority.

velocity for each link was known from the hydraulic solution for each time period, which was evenly divided into an integer number of computational time steps. Each link was then divided into sublinks by a series of evenly spaced subnodes (though the distance between subnodes varied from link to link or for a link at different time periods), so that the travel time from a subnode (or node) to the adjacent subnode (or node) was approximately equal to Δt.

The solution algorithm used in the DWQM operated sequentially by time period. During a time period, all external forces affecting water quality were assumed to remain constant (e.g., demand, well pumpage, tank head).

The DWQM was used to simulate a 34-h period corresponding to conditions present during the pilot-level sampling program conducted on November 14-15, 1985. Parameters of the model were adjusted so that predicted tank levels and flows at selected sites represented those measured during the sampling period. A comparison of measured and modeled hydraulic conditions at three locations are shown in Fig. 13.7 (Clark et al., 1988a;1988c).

For chloroform, THM, and hardness, the predicted concentrations compared favorably with the observed values at the three selected sampling stations. Fig. 13.8, which is a plot of (THMs), illustrates these results. In each case, there were some differences in the timing of peak or minimum values. When the spatial variation of predicted concentrations of THMs was compared to the historical average THM level, the same general patterns were apparent. In addition, the predicted patterns bracketed the pattern corresponding to the long-term historical average, a result that would be expected since the two selected times corresponded to the extreme spatial patterns during the sampling period.

13.5.2 South Central Connecticut Regional Water Authority

The North Penn case study provided an excellent test-bed for development of a dynamic water-quality model (Clark and Coyle, 1990). To extend the North Penn application, the

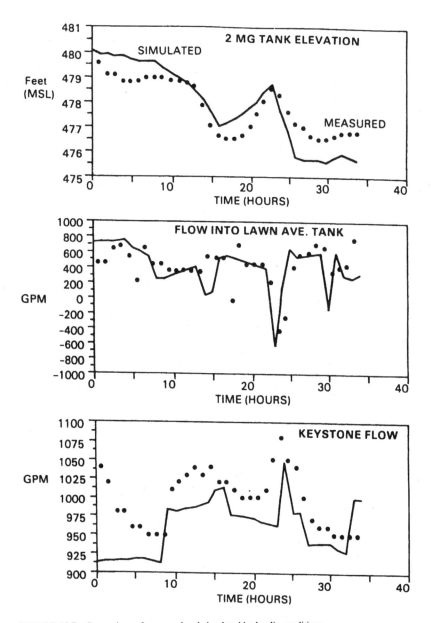

FIGURE 13.7 Comparison of measured and simulated hydraulic conditions.

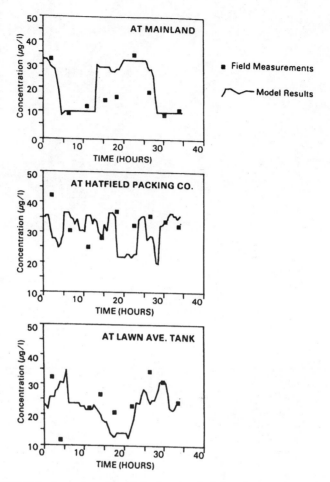

FIGURE 13.8 Comparison of modeled and measured trihalomethanes at selected stations.

U.S. EPA initiated another cooperative agreement with the University of Michigan. In conjunction with the EPA, the university initiated a program with the South Central Connecticut Regional Water Authority (SCCRWA) to test the previously developed modeling concepts including field studies to verify and calibrate the model (Clark and Goodrich, 1993; Clark et al., 1988a).

At the time of the study, SCCRWA supplied water to approximately 95,000 customers (380,000 individuals) in 12 municipalities in the Greater New Haven area. The service area was divided into 16 separate pressure/distribution zones (Fig. 13.9). Average production was 50 mgd with a safe yield of approximately 74 mgd. Water sources included four sources of surface water (Lake Gaillard, Lake Saltonstall, Lake Whitney, and the West River System). Five well fields served as sources (North Cheshire, South Cheshire, Mt. Carmel, North Sleeping Giant, and South Sleeping Giant). Approximately

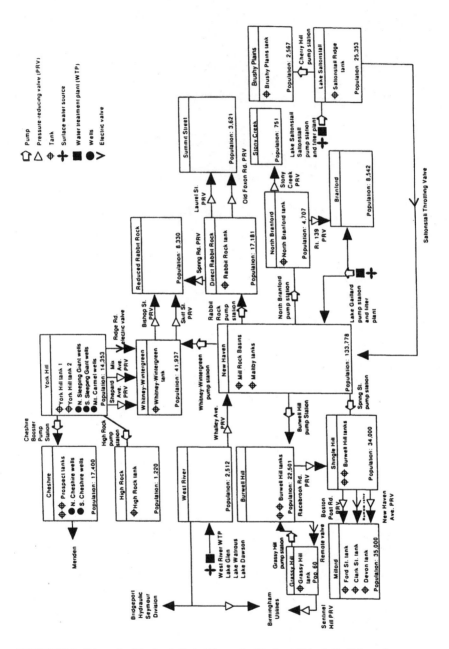

FIGURE 13.9 Schematic of the South Central Connecticut Regional Water Authority's service area.

80 percent of the water in use in the system came from surface sources and the remaining 20 percent came from wells. All water was treated with chlorination, filtration, and addition of a phosphate-corrosion inhibitor. The system included 22 pumping stations, 23 storage tanks, and approximately 1300 mi of water mains.

13.5.2.1 System modeling. Preliminary efforts to develop and validate a model for the SCCRWA were concentrated on the Cheshire service area (Clark, 1993a; Clark et al., 1993b). This area was relatively isolated and provided a prototype for modeling the remainder of the system. To validate the model, an extensive study was planned in which the fluoride feed at the North Well Field was turned off and the propagation of the fluoride feed water was tracked through the system (Clark et al., 1991b).

Prior to the water-quality modeling effort and the related field study, extensive hydraulic analyses were conducted on the system. For the preliminary modeling effort described here, the full system network was represented by approximately 520 nodes and 700 links. In most cases, the system was represented by "skeletonizing" the network (i.e., selectively choosing pipes based on their size and perceived impact as transmission mains). In a few cases, "surrogate" pipes were used to represent the effects of several pipes. For example, a single 35.6-cm (14-in) pipe can be used to represent the effects of a pair of parallel 20.3-cm (8-in) and 30.5-cm (12-in) pipes. In these model applications, the hydraulic and water-quality models were applied to the Cheshire system-simulating the proposed fluoride study. The DWQM developed in the North Penn study was applied to the Cheshire system to simulate the propagation of fluoride feed water and also to select sampling locations for a field study.

13.5.2.2 Design of the field study. Based on the simulation results and the objectives of the proposed field study, a field sampling scheme was designed. Fluoride was selected as the tracer because it was added regularly to the water at a concentration of approximately 1 mg/L, as required by the State Department of Health Services. Fluoride does not dissipate from the water, it is easily tested for, and turning off the fluoride feed can be done with no health or aesthetic effects. By tracing the changes in the fluoride concentration in the distribution system, accurate travel times could be determined for the water in the distribution system by relating the time the fluoride feed was shut down at the well fields to the time it dissipated at the sampling points (Skov et al., 1991). The sampling sites are shown in Fig. 13.10.

13.5.2.3 Results from the field study. The model predictions and the sampling results were extremely close. Figures 13.11 and 13.12 compare the model predictions and the field sampling results for nodes 37, 88, 182, 323, 537, and 570 (tank).

The behavior of the tanks was of special interest. During the early portion of the sampling period, variations in tank levels were held to a minimum [<0.91 m (3 ft)]. After 2 days, little change in fluoride concentrations was found in the tank and, as a result, the water level was then allowed to vary approximately 0.28 m (8 ft). The wider range in tank-level variation had the effect of turning the water over relatively rapidly. Even with the rapid turnover, it took nearly 10 days to replace old water with new water in the tanks completely. It was clear from this analysis that tanks could have a detrimental effect on water quality, particularly as water aged in the tank.

On August 13 to 15, 1991, another sampling program at the Cherry Hill/Brushy Plains Service Area was initiated with the goal of validating the previously discussed simulation results. The purpose of this sampling program was to gather information to characterize the variation of water quality in the service area and to study the impact of tank operation on water quality.

13.14 Chapter Thirteen

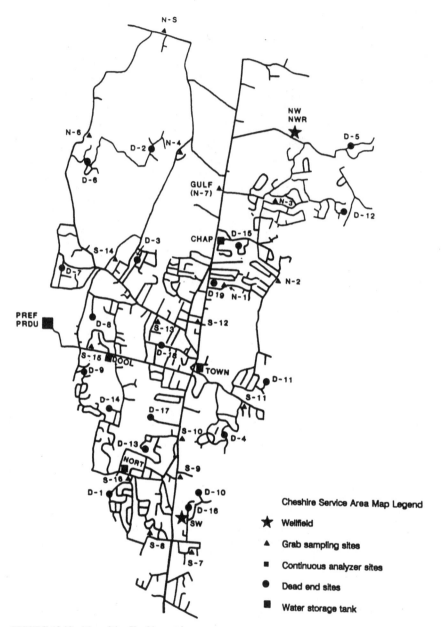

FIGURE 13.10 Map of the Cheshire service area.

Water Quality Modeling-Case Studies **13.15**

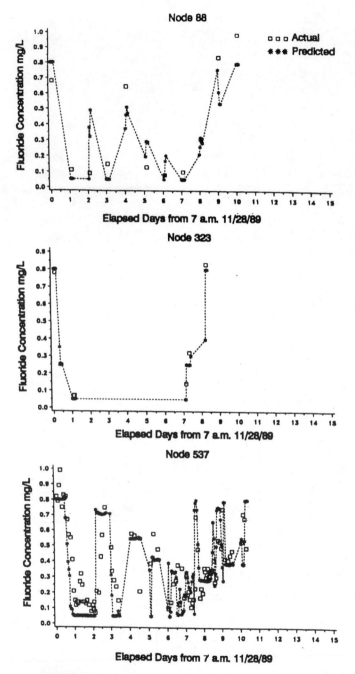

FIGURE 13.11 Actual versus predicted fluoride concentrations for nodes 88, 323, and 537.

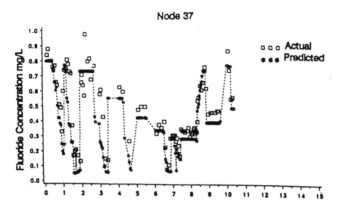
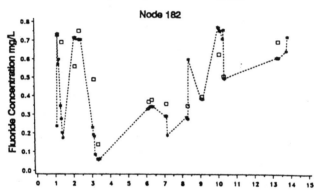
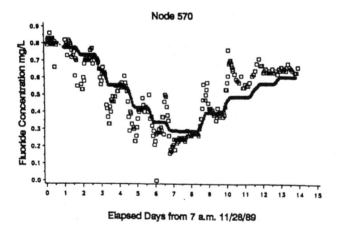

FIGURE 13.12 Actual versus predicted fluoride concentrations for nodes 37, 182, and 570.

13.5.2.4 Verification study. The Cherry Hill/Brushy Plains Service Area covered approximately 2 mi^2 in the Town of Branford in the eastern portion of the SCCRWA service area (Clark et al., 1993b). This service area was almost entirely residential, containing both single-family homes and apartment/condominium units. Average water use during the sampling period was 1700 m^3/day (0.46 mgd). The water distribution system was composed of 20.3-cm (8-in) and 30.48-cm (12-in) mains, as shown in the schematic in Fig. 13.13. The terrain in the Cherry Hill/Brushy Plains Service Area was generally moderately sloping, with elevations varying from approximately 15.2 m (50 ft) mean sea level (MSl) to 70.1 m (230 ft) MSl.

Cherry Hill/Brushy Plains received its water from the Saltonstall system. Water was pumped from the Saltonstall system into Brushy Plains by the Cherry Hill Pump Station. Within the service area, storage was provided by the Brushy Plains tank. The pump station contained two 10.2-cm (4-in) centrifugal pumps with a total capacity of 5300 m^3/day (1.4 mgd). The operation of the pumps was controlled by water elevation in the tank. Built in 1957, the tank had a capacity of 3800 m^3/day (1.0 mgd). It had a diameter of 15.2 m (50 ft), a bottom elevation of 58.8 m (193 ft) MSL, and a height (to the overflow) of 80.2 m (263 ft) MSL. During normal operation, the pumps were set to go on when the water level in the tank dropped to 15.2 m (56 ft) and to turn off when the water level reached 19.8 m (65 ft).

13.5.2.5 Presampling procedures. As had occurred during the study of the Chesire service area, the WADISO hydraulic model and the DWQM were applied to establish flow patterns within the service area. In addition, during the periods of May 21–22, July 1–3, July 8–10, and July 30–August 1, 1991, chlorine residuals were monitored at the tank and operational patterns (pump records), and variations in tank water level were studied. On the basis of these model runs and sampling data, a sampling strategy was adopted that involved turning the fluoride off at the Saltonstall Treatment Facility and sampling for both fluoride and chlorine in the Cherry Hill/Brushy Plains Service Area. The intention was to use defluorided water as a conservative tracer for the movement of flow through the system and to calibrate the DWQM. The DWQM and a chlorine decay model based on hydrodynamic principles was used to model the dynamics of this substance. Seven sampling sites in the distribution system, in addition to sampling sites at the pump station and tank, were identified as shown in Fig. 13.13.

13.5.2.6 Analysis of sampling results. The WADISO hydraulic model and DWQM model were used to simulate the Cherry Hill/Brushy Plains Service Area for a 53-h period from 9:00 am on August 13 to 3:00 pm on August 15, 1991. A skeletonization was developed representing the Cherry Hill/Brushy Plains distribution system. This skeletonization, as shown in Fig. 13.13, included all 30.4-cm (12-in) mains, major 20.3-cm (8-in) mains and loops, and pipes that connected to the sampling sites. Pipe lengths were scaled from mappage, actual pipe diameters used, and, in the absence of any other information, a HazenWilliams roughness coefficient of 100 was assumed for all pipes.

Figures 13.14 and 13.15 show the results of the fluoride sampling study and the modeling efforts at each sampling node. From these results, it is clear that the modeling effort matches the sampling efforts well, with the exception of the dead-ends. The shaded areas at the top of the individual plots in Figs. 13.14 and 13.15 show the on-and-off cycle for the pumps. Clearly, the pump cycles influence water quality heavily at several sampling points. For example, at node 11, during the pumps-on cycle the fluoridated water was pumped into the system. When the system was being fed from the tank (pumps-off), the system was receiving water that had reached an equilibrium concentration of fluoride before the stoppage of the fluoride feeders.

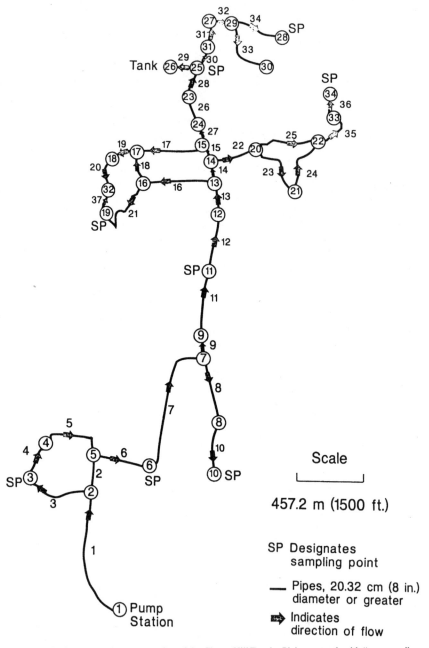

FIGURE 13.13 Link-node representation of the Cherry Hill/Brushy Plains network with "pumps on" scenario.

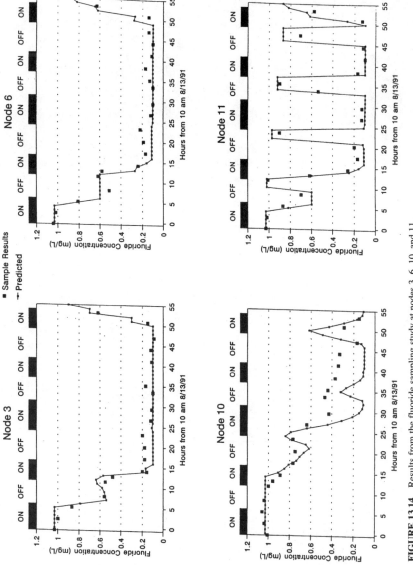

FIGURE 13.14 Results from the fluoride sampling study at nodes 3, 6, 10, and 11.

13.20 Chapter Thirteen

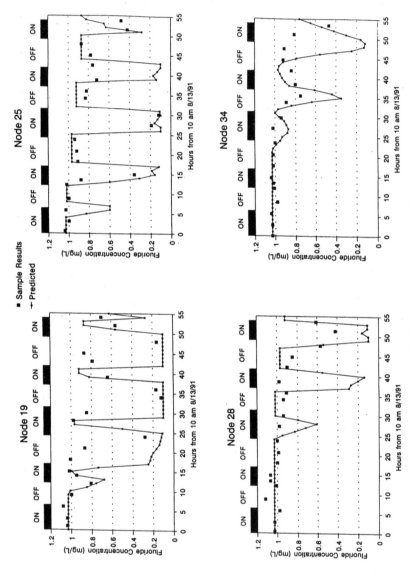

FIGURE 13.15 Results from Fluoride Sampling Study at Nodes 19, 25, 28, 35.

13.5.2.7 Modeling of chlorine residual. As mentioned above the two scenarios evaluated during the sampling study were with the pumps-on and pumps-off condition. Table 13.1 shows the length of pipe in feet, average velocity in feet per second, residence time in days, and the average chlorine residual at the beginning and ending nodes for various links in the system with the pumps on.

Using the upstream and downstream chlorine concentration and the residence times in the link, the chlorine decay coefficient was calculated for each link. Chlorine demand was calculated according to a first-order assumption, as defined by Eq. (13.1):

$$C = C_0 e^{-kt} \tag{13.1}$$

where C = the concentration at time t, C_0 = initial chlorine concentration, k = decay rate in min^{-1}, and t = time in min.

A bench study was conducted in which chlorine demand for the raw water was calculated using Cherry Hill/Brushy Plains water, and the chlorine decay rate was calculated as 0.55 day^{-1}. This decay rate might be regarded as the bulk decay rate or the decay rate of chlorine in the treated water. As can be seen by the ratio shown in the "Ratio of pipe to Bulk Decay" column in Table 13.1, the total system demand was much higher than just the bulk decay rate. This additional demand most likely was caused by pipe wall demand, biofilm, and tubercles and was highly significant. The results raised the possibility that some system components ultimately may have to be replaced to meet water-quality goals. The links resulting in dead-ends (which also had the lowest average velocity during the pumps-on scenario) also exhibited the highest chlorine demand. Furthermore, it is clear from the "Decay Coefficient" column in Table 13.1 that a single first-order decay rate would not predict chlorine residual adequately.

TABLE 13.1 Hydraulic Conditions During "Pumps On" Scenario

Link Beginning/Ending Nodes	Length* (ft)	Residence Time (days)	Chlorine Concentration in mg/L (upstream downstream)	Decay Coefficient (days^{-1})	Ratio of Pipe to Bulk Decay
1/3	3700	0.0414	1.08/1.00	1.86	3.38
1/6	4400	0.0321	1.08/1.00	2.40	4.36
6/11	3800	0.0286	1.00/0.98	0.71	1.29
6/10	4900	0.8049	1.00/0.36	1.27	2.30
11/19	5400	0.1634	0.98/0.16	11.09	20.16
11/25	4350	0.0424	0/98/0.94	0.98	1.78
11/34	64000	1.3714	0.98/0.12	1.53	2.78
25/28	2400	1.4937	0.94/0.16	1.19	2.16
Tank	—	3.0000	0.94/0.16	0.59	1.07

* To convert from ft to m multiply by 0.3048.

13.5.3 Case Study of Cabool, Missouri

A recommendation from the 1991 EPA/American Water Works Association Research Foundation conference (Clark et al, 1991b) on water-quality modeling was to develop water-quality modeling techniques that could be applied to outbreaks of waterborne disease. The first opportunity to attempt this type of application arose as a result of an outbreak that occurred between December 15, 1989, and January 20, 1990, in Cabool, Missouri (population 2090) (Geldreich et al., 1992). During the outbreak, 243 residents and visitors to Cabool experienced diarrhea (85 bloody), and six deaths occurred. The illnesses and deaths were attributed to Escherichia coli serotype 0157:H7. At the time of the outbreak, the water source was untreated groundwater. Shortly after the outbreak, the EPA was invited to send a team to conduct a research study with the goal of determining the underlying cause of the outbreak.

Exceptionally cold weather before the outbreak contributed to two major breaks in the water system lines and required the replacement of 43 water meters throughout the area. The sewage collection lines in Cabool were situated, for the most part, away from the drinking water distribution lines, but they did cross or were close to water lines in several locations. At the time of the outbreak, storm water drained via open ditches along the sides of the streets and roads. During heavy rainfalls, sewage was observed to overflow manhole covers and to overflow streets in several locations, parking lots, and residential foundations.

The DWQM was applied to examine the movement of water and contaminants in the system. Steady-state scenarios were examined, and a dynamic analysis of the movement of water and contaminants associated with the meters of replacement and the aforementioned breaks was conducted. When typical demand patterns were developed from available meter usage for each service connection, it was found that the water demand was 65 percent of the average well production, indicating inaccurate meters, unmetered uses, and a high loss of water in the system.

The modeling effort revealed that the pattern of illness was consistent with patterns of water movement in the distribution system, assuming two water-line breaks. Therefore, the conclusion was that some disturbance in the system, possibly the two line breaks or the 43 meter replacements, allowed contamination to enter the water system. Analysis showed that the simulated movement of contaminants accounted for 85 percent of the infected population.

13.6 EVOLUTION OF WATER QUALITY MODELING

On the basis of the results from the case studies just described and from other studies, it had become obvious that water-quality modeling had the potential to provide insight into the factors that degrade water quality in networks. It also became increasingly obvious that despite the treatment investments being forced by regulations in the 1996 Safe Drinking Water Act and amendments, the potential still existed for deterioration of water quality in the network itself. This realization led to the development of several public- and private-sector water-quality models. Only the public-sector models will be discussed here.

EPANET, developed by Rossman (1994) and Rossman et al., (1994), was based on mass transfer concepts. Another approach to the propagation of contaminants developed by Biswas et al., (1993) used a steady-state transport equation that accounted for the simultaneous advective transport of chlorine in the axial direction, diffusion in the radial direction, and consumption by first-order reaction in the bulk-liquid phase. Islam (1995) and Islam et al., (1997) developed a model called QUALNET, which predicted the temporal and spatial distribution of chlorine in a pipe network under slowly varying unsteady-flow conditions. Boulos et al., (1995) proposed a technique called the Event Driven-Method, which is based on a "next event" scheduling approach and can significantly reduce computing times.

Several different types of numerical methods have been proposed to solve these types of models, including the eulerian Finite-Difference Method, the eulerian Discrete Volume Method, the lagrangian Time-Driven Method, and the lagrangian Event-Driven Method.

The Finite-Difference Method approximates derivatives with finite-difference equivalents along a fixed grid of points in time and space. Islam et al., (1997) used this technique.

The Discrete Volume Method divides each pipe into a series of equal-sized completely mixed volume segments. At the end of each successive water-quality time step, the concentration within each volume segment is, first, reacted and then transferred to the adjacent downstream segment. This approach was used in the early DWQM studies.

The Time-Driven Method tracks the concentration and size of a nonoverlapping segment of water that fills each link of a network. As time progresses, the size of the most upstream segment in a link increases as water enters the link, whereas an equal loss in size of the most downstream segment occurs as water leaves the link. The size of these segments remains unchanged.

The Event-Driven Method is similar in nature to the Time-Driven Method, but rather than update an entire network at fixed time steps, individual link/node conditions are updated only when the leading segment in a link disappears completely through this downstream node.

The development of the EPANET hydraulic model has satisfied the need for a comprehensive public-sector model and has been a key component in providing the basis for water-quality modeling in many utilities throughout the United States. EPANET is based on the extended-period simulation approach to solving hydraulic behavior of a network. In addition, it is designed to be a research tool for modeling the movement and fate of drinking-water constituents within distribution systems. EPANET calculates all flows in cubic feet per second (cfs) and has an option for accepting flow units in gallons per minute (gpm), million gallons per day (mgd), or liters per second (L/s).

EPANET uses the Hazen-Williams formula, the Darcy-Weisbach formula, or the Chezy-Manning formula to calculate the head loss in pipes. It also models pumps, valves, and minor loss. To model water quality within distribution systems, the concentration of a particular substance must be calculated as it moves through the system from various points of entry (e.g., treatment plants) and on to water users. This movement is based on three principles: (1) conservation of mass within differential lengths of pipe, (2) complete and instantaneous mixing of the water entering pipe junctions, and (3) appropriate kinetic expressions for the growth or decay of the substance as it flows through pipes and storage facilities.

13.7 MODELING PROPAGATION OF CONTAMINANTS

Maintenance of chlorine or other disinfectant residuals is generally considered to be a water-quality goal in the United States, and most American water systems attempt to maintain a detectable residual throughout the distribution system. Before leaving the treatment plant, water is usually chlorinated in a final disinfection step and then is stored in a clear well or basin. When the water is discharged from the clear well, it is transported throughout the distribution system and home plumbing to the consumer. It is presumed that a detectable chlorine residual will minimize the potential for waterborne disease and biofilm growth in the system.

As dissolved chlorine travels through the pipes in the network, it reacts with natural organic matter in the bulk water and with biofilm and tubercles on the pipe walls or with the pipe wall material itself (Clark et al., 1993a). This reaction results in a decrease in chlorine residual and a corresponding increase in disinfection by-products, depending on the residence time in the network pipes and holding time in storage facilities. Understanding these reactions will help water-utility managers deliver high-quality drinking water and meet regulatory requirements under the 1996 Safe Drinking Water Act and amendments.

Water-quality modeling has the potential to provide insight into the factors that influence the variables affecting changes in water quality in distribution systems. Understanding the factors that influence the formation of TTHMs and maintenance of chlorine residuals is of particular interest (Clark et al., 1996c). EPANET has proved to be especially useful for modeling both formation of TTHMs and the propagation and maintenance of chlorine residuals.

Among the first studies to address these issues using EPANET was one conducted by the EPA in collaboration with the North Marin Water District in California (Clark et al., 1994). Another recently completed study conducted jointly by the EPA and the American Water Works Association Research Foundation (AWWARF) examined these same issues (Vasconcelos et al., 1996). This study evaluated various types of models to describe both the formation of TTHMs and loss of chlorine residual.

13.7.1 Case Study of the North Marin Water District

The North Marin district serves a suburban population of 53,000 people who live in or near Novato, California. It used two sources of water: Stafford Lake, and the North Marin Aqueduct. The aqueduct is a year-round source, but Stafford Lake is in use only during the warm summer months, when precipitation is virtually nonexistent and demand is high. Novato, the largest population center in the North Marin service area, is located in a warm inland coastal valley with a mean annual rainfall of 68.58 cm (27 in). Virtually no precipitation occurs during the growing season from May through September. Eighty-five percent of total water use is residential, and the service area contains 13,200 single-family detached homes, which accounted for 65 percent of all water use (Clark et al., 1994).

The water quality of the two sources differed greatly. Stafford Lake water had a high humic content and was treated with conventional treatment and prechlorination doses of between 5.5 and 6.0 mg/L. The treated water had a residual of 0.5 mg/L when it left the treatment plant clearwell. The potential for formation of THMs in the Stafford Lake water was high. The North Marin Aqueduct water was derived from a Raney Well Field along the Russian River. Technically, groundwater, the source water, was likely to contain a high proportion of naturally filtered water. Aqueduct water was disinfected only and was low in precursor material with a correspondingly low potential for formation of THMs. Both sources carried a residual chlorine level of approximately 0.5 mg/L when the water entered the system.

Figure 13.16 is an overall schematic showing the entire North Marin service area. Figure 13.17 shows the distribution network for zone 1, which was the major focus of the study. Figure 13.17 shows both sources, a schematic of the major pipes in the service area's distribution system, the major tanks and pumps, and the sampling points used in the study. As mentioned above, depending on the time of year and the time of day, water entered the system from either one source or both sources. The North Marin Aqueduct source operated year-round, 24 h per day. The Stafford Lake source operated only during the peak demand period from 6:00 to 10:00 pm and generally operated for 16 h per day. Table 13.2 summarizes the characteristics of North Marin Water use.

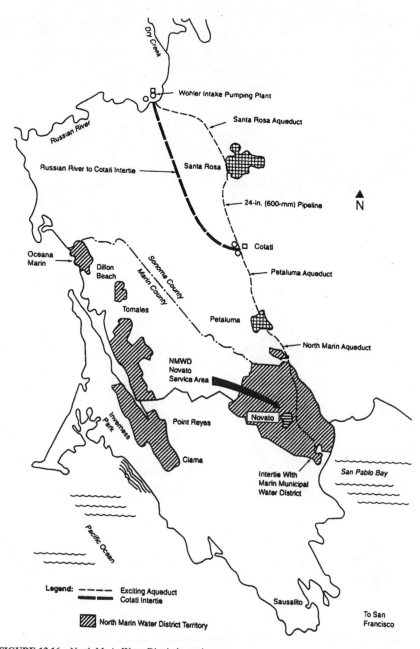

FIGURE 13.16 North Marin Water District's service area.

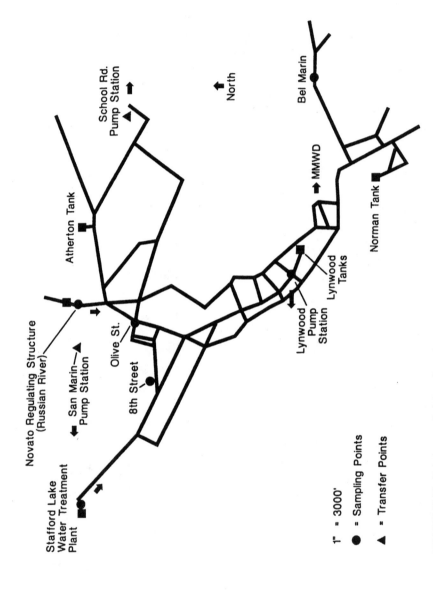

FIGURE 13.17 Existing zone 1 piping schematic.

EPANET was used to model the system hydraulics, including the relative flow from each source, TTHMs, and propagation of chlorine residual (Rossman et al., 1994). The model was based on an earlier representation of the network made by Montgomery Watson, Inc., for North Marin and was calibrated based on a comparison of simulated versus actual tank levels for the May 27–29, 1992, period of operation.

The dynamic nature of the system led to variable flow conditions and variable water-quality in the network. Flow directions frequently reverse within a given portion of the network during a typical operating day. Figure 13.18 shows the predicted percentage of water from Stafford Lake at the various sampling points used during the 2 day study. The consequences of these variable flow patterns for water quality are discussed in the following sections.

TABLE 13.2 Characteristics of Water Use in the North Marin Water District

Service area	259.0 km² (100 mi²)
Principal service center: greater Novato area	
Population	53,000
Character	Suburban (near San Francisco)
Normal rainfall	69.1 cm (27.2 in/y)
Normal reference	
evapotranspiration (ETO):	111.76 cm (44 in/y)
Applied water requirement	
for cool season grasses	70.61 cm (27.8 in/y)
Water supply	Surface
	North Marin Aqueduct 83%
	Stafford Lake 17%
All accounts metered?	Yes
Overall average per capita use	510.98 L (135 gal)*
Distribution of metered water use	
Residential	84.7%
Commercial	15.3%
Annual residential Use*	
Single family (67.3%)	545.04 L (144 gal)
Townhouse/condominium (12.7%)	435.28 L (115 gal)
Mobile home (16.6%)	291.44 L (77 gal)
Apartment (3.4%)	299.02 L (79 gal)
Unaccounted–for water and water loss	5.7%

*Gallons per capita per day (based on avg. household density of 2.7 persons/dwelling).

13.28 Chapter Thirteen

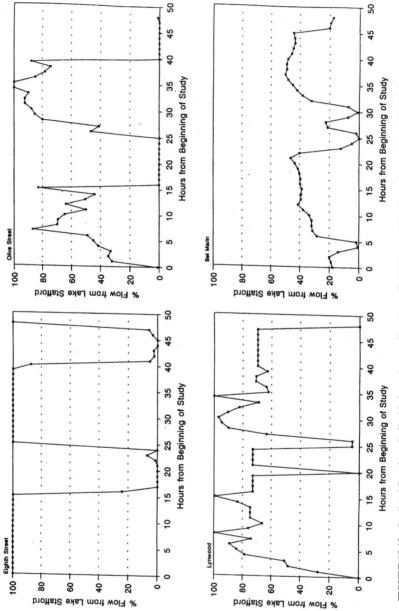

FIGURE 13.18 Predicted flow from Stafford Lake to the sampling sites at Eighth Street, Olive Street, Lynwood, and Bel Marine.

13.7.1.1 Water-quality study.
To characterize the water quality in the North Marin Water District, the US EPA designed a sampling protocol and sent a team of investigators to work with the district staff for the period May 27–29, 1992. Figure 13.19 shows the time formation curves for THMs and chlorine demand for the two source waters. These figures illustrate the significant difference in water quality from the two sources. For example, at the "Eighth Street" sampling point, chloroform levels vary from 38.4 to 120.1 µg/L over the 2 day period. This variability is caused by the penetration of water from the two different sources. As can be seen from Fig. 13.18, a sample taken at the "Eighth Street"

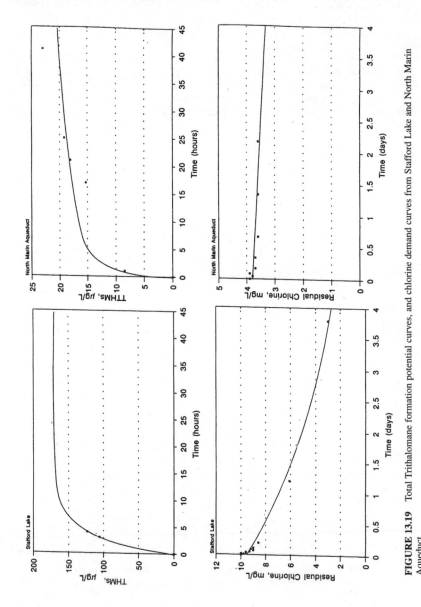

FIGURE 13.19 Total Trithalomane formation potential curves, and chlorine demand curves from Stafford Lake and North Marin Aqueduct.

13.30 Chapter Thirteen

sampling site may consist of water from the Aqueduct or Stafford Lake or a blend of both. However, over the sampling period, THM levels were relatively constant at the two sources. Given the extreme differences in water quality from the two sources and the variability in percent ages of source water penetrating within the network, it is reasonable to assume that mixing and blending of water is a major factor affecting water quality.

13.7.1.2 Modeling of total trihalomethane formation. To test this mixing hypothesis, the average values for THMs at each source were used. These levels of THMs were treated as conservative, and levels at each sampling station were calculated on the basis of the percentage of water from a source at that point over time. This assumption was verified with field and bench testing. For example, the average THM levels for Stafford Lake water and for Aqueduct water were 151.1 and 20.7 g/L, respectively. If, at a given place in the system at a given time, 50 percent of the flow is from Stafford Lake and 50 percent is from the Aqueduct, the predicted THM values at that point would be 85.9 µg/L. Figure 13.20 shows the results of this assumption at the various sampling points. Some differences between actual and predicted values in the figures are no doubt the result of the calibration problems discussed above.

As part of the study, it was observed that ultraviolet absorbency and THMs were closely related. To establish that ultraviolet absorbance is in fact a good predictor for THMs, a

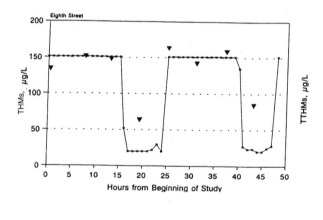

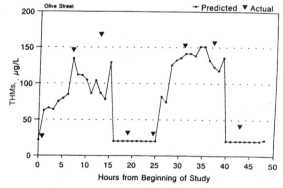

FIGURE 13.20 Predicted versus actual total trithalomethanes at the Eighth Street, Olive Street, Lynwood, and Bel Marin sampling sites.

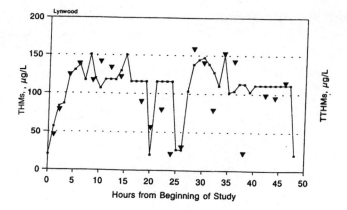
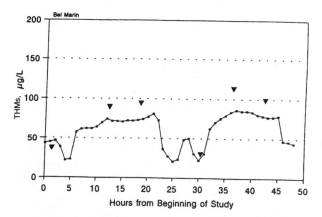

FIGURE 13.20 Predicted versus actual total trithalomethanes at the Eighth Street, Olive Street, Lynwood, and Bel Marin sampling sites.

regression relationship between ultraviolet absorbance and THMs was established using the data from the two sources. Data from the first day of the study was used to develop a regression model between THMs and ultraviolet absorbance as follows:

$$\text{THM} = 2.056 + 1648.2 * \text{UV} \tag{13.2}$$

Where THM = total trihalomethanes (μg/L) and UV = ultraviolet absorbance in the 250-nm range (cm^{-1}) with a p-value = 0.0001, r = 0.9875, and mean square error (MSE) = 11.1. The assumptions of the model (constant variance and normality of error terms) were checked and deemed to be reasonable. Figure 13.21 shows the data used to develop the regression model. Figure 13.22 shows the use of Eq. (13.2) to predict THMs at the various sampling points for the second day of the study. As can be seen, the predictions were relatively good.

Figure 13.23 shows simulated versus actual ultraviolet absorbance data at the four sampling sites based on the same mixing assumptions used to predict THM levels.

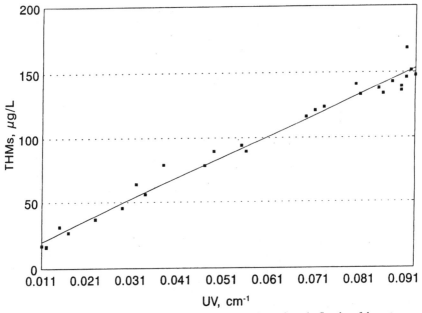

FIGURE 13.21 Ultraviolet absorbance versus total trithalomethanes from the first day of the water-quality study.

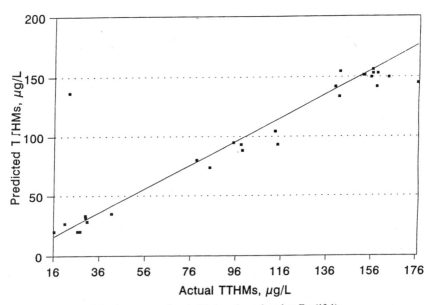

FIGURE 13.22 Predicted versus actual total trithalomethanes based on Eq. (13.1).

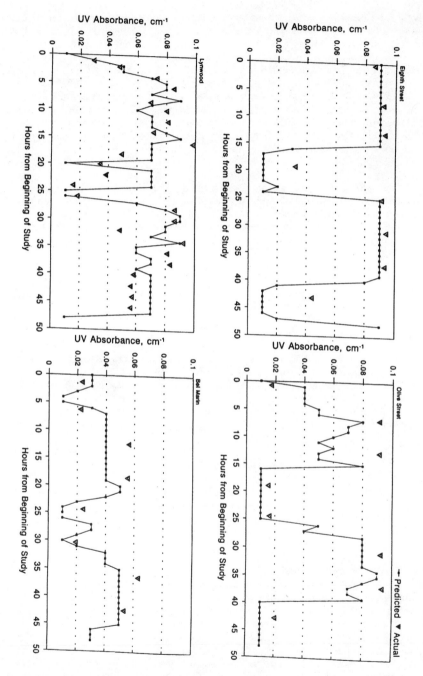

FIGURE 13.23 Predicted versus actual ultraviolet absorbance at the Eighth Street, Olive Street, Lynwood, and Bel Marin sampling sites.

13.7.1.3 Chlorine demand. Both Stafford Lake water and the North Marin Aqueduct water generally maintained a chlorine residual level of 0.5 mg/L as the treated water entered the system. Figure 13.19 illustrates chlorine demand for the two sources. As can be seen, the Stafford Lake water had a much higher chlorine demand than did the Aqueduct water. To predict chlorine demand at the various sampling points, a first-order decay relationship was assumed, as shown in Eq. (13.1) (Clark et al., 1995).

In EPANET, chlorine decay is represented by decay in the bulk phase and by decay in the pipe wall. Based on bulk-water calculations, the first-order decay coefficients or bulk demand for the Stafford Lake and the Aqueduct sources were 0.31 and 0.03 day-1, respectively. Using EPANET and the previously assumed hydraulic conditions, the chlorine residuals were estimated. Figure 13.24 shows these predictions based on the four sampling points.

13.7.1.4 Effect of system demand It is evident that the pipes in the distribution network can exhibit a demand for chlorine. This demand may come from tubercles, biofilm, and perhaps the pipewall material itself (Clark et al., 1995). A comparison between chlorine residuals using the first-order assumptions predicted from EPANET versus actual chlorine residuals provided an excellent illustration of this point.

Clearly, the demand in the system went beyond just bulk-water decay. Because EPANET has the capacity to incorporate a wall demand factor in addition to the bulk demand factors for chlorine, the system was simulated again using the bulk demand for the two sources, and trial and error was used to estimate wall demands for four sections of the network, as shown in Fig. 13.25. Wall demand, or k_w, as defined in EPANET, is given in feet per day. One might think of k_w as a penetration velocity for the disinfectant.

When chlorine residuals were reestimated at the four sampling points (Fig. 13.26), wall demand obviously played a major role in chlorine residual loss in the North Marin system. However, Fig. 13.26 still shows some large differences between predicted and actual chlorine residuals during the middle portion of the study.

The wall demand effect may be the result of the source or the age of the system. For example, as illustrated by Fig. 13.25, the maximum wall demand was found in the areas served primarily by Stafford Lake. However, those pipes also were the oldest in the system.

13.7.2 Complement to the North Marin study

As a complement to the North Marin study, the EPA and AWWARF (Vasconcelos et al., 1996) initiated a project that involved case studies of several utilities and was intended to gain a better understanding of the kinetic relationships describing chlorine decay and formation of THMs in water distribution systems.

Several kinetic models were evaluated, tested, and validated using data collected in these field sampling studies based on the EPANET distribution network model as the framework for analysis.

Another goal was to provide information and guidelines for conducting water-quality sampling and modeling studies by water utilities. The specific objectives of the EPA/AWWARF study were to (1) establish protocols for obtaining data to calibrate predictive models for chlorine decay and THM formation, (2) evaluate alternative kinetic models for chlorine decay and THM formation in distribution systems, (3) develop a better understanding of the factors influencing the reactivity of chlorine in distribution and transmission pipes, and (4) make these results available for use in system operations and in planning of similar water-quality studies. The utilities that participated in this study were selected to represent a wide range of source-water qualities. In addition, systems that

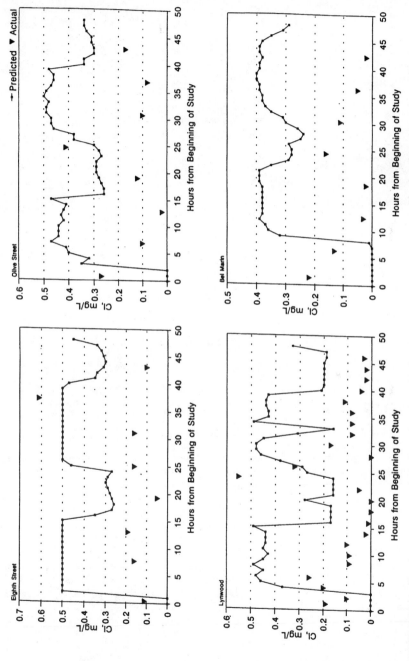

FIGURE 13.24 Predicted versus actual chlorine residual, assuming first-order decay at the eighth Street, Olive Street, Lynwood, and Bel Marin sampling sites.

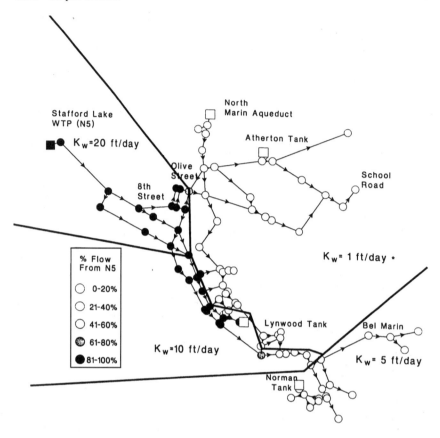

FIGURE 13.25 North Marin hydraulic calibrations for hour 1200 incorporating well demand.

had small isolated zones or long isolated lines were included, when possible. Systems having significant potential for formation of TTHMs also were selected as part of the study.

13.7.3 Waterborne Outbreak in Gideon, Missouri

The DWQM was applied to the Cabool outbreak, as mentioned above. With the increasing sophistication of water-quality propagation modeling, it has now become possible to apply these types of models to outbreaks of waterborne disease even more easily than was possible during the Cabool outbreak. Therefore, when an outbreak occurred in 1993 in Gideon, Missouri, EPANET was used to analyze propagation of contaminants in the system.

In December 1993, six to nine cases of diarrhea were reported at a local nursing home in Gideon, Missouri, raising the possibility of a waterborne outbreak (Clark et al., 1996a). After an initial investigation by the Missouri Department of Health, the Missouri

Water Quality Modeling-Case Studies 13.37

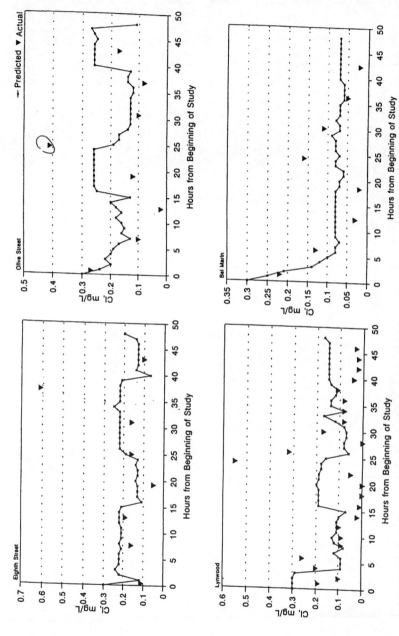

FIGURE 13.26 Predicted versus actual chlorine residual, assuming wall demand factors at the Eighth Street, Olive Street, Lynwood, and Bel Marin sampling sites.

Department of National Resources was contacted, and water samples were taken at various points in the system between December 17 and 21, 1993. Several samples were positive and yielded one to six total coliforms per 100 ml and a few samples were positive for fecal coliform. Several other samples yielded results that were too numerous to count for coliforms and also were positive for fecal coliform. Original speculation regarding the cause of the outbreak focused on a water tank situated on private property. The tank was constructed in 1930 and appeared to be heavily rusted and in an obvious state of disrepair. This privately owned tank, connected via a backflow–prevention valve to the city water system, was used primarily for fire protection at the Cotton Compress, a local cotton-baling facility.

The municipal system had two elevated tanks. One tank was a 189-m^3 (50,000-gal) elevated tank; the other was a 378-m^3 (100,000-gal) elevated tank. The privately owned, tank was located on the Cotton Compress property. It had a volume of 378-m^3 (100,000-gal). Both 378-m^3 (100,000-gal) tanks had broad flat roofs, whereas the smaller municipal tank had a much steeper pitch.

On January 14, 1994, an EPA field team, in conjunction with the Centers for Disease Control and Prevention (CDC) and the State of Missouri, initiated a field investigation that included a sanitary survey and microbiological analyses of samples collected on site. A system evaluation was conducted in which EPANET was used to develop various scenarios to explain possible transport of contaminants in the Gideon system (Table 13.3).

13.7.3.1 Description of the system. Gideon, Missouri, is located in Anderson Township in New Madrid County, which is in the southeastern part of the State. The topography is flat and the predominant crop is cotton. In 1990, the population of Gideon was 1104, with a median income of $14,654 (25 percent of the population was below the poverty level).

TABLE 13.3 Assumptions Used in Analysis of the Gideon, Missouri, Distribution System

Item	Value
No. of homes in Gideon	429
No. of residents in Gideon	1104
Persons/households in Gideon	2.6
Average daily consumption*	492 m^3 (130,000 gal)
189–m^3 (50,000-gal) municipal tank (T200)	
Height	8.8 m (29 ft)
Diameter	5.5 m (18 ft)
378–m^3 (100,000-gal) municipal tank (T300)	
Height	7.5 m (24.5 ft)
Diameter	9.1 m (30 ft)
378–m^3 (100,000-gal) Cotton Compress tank (T400)	
Height	10.1 m (33 ft)
Diameter	7.3 m (7.3 ft)

*Drinking, bathing, washing, cooking, lawn watering, etc.

The unemployment rate was 11.3 percent. The major employers in Gideon are a nursing home with 68 residents and a staff of 62 and the Gideon schools with 444 students (kindergarten through 12th grade).

The Gideon municipal water system was originally constructed in the mid-1930s and obtains water from two adjacent 396-m (1300-ft) deep wells. The well waters were not disinfected at the time of the outbreak. The distribution system consists primarily of small-diameter [5-, 10-. and 15-cm (2-, 4-, and 6-in)] unlined steel and cast iron pipes. Tuberculation and corrosion are a major problem in the distribution pipes. Raw water temperatures are unusually high for a ground water supply [14 °C (58 °F)] because the system overlies a geologically active fault. Under low-flow or static conditions, the water pressure is close to 3.5 kg(Kilograms) /cm^2 (50 psi). However, under high-flow or flushing conditions, the pressure drops dramatically, as will be discussed below. These sharp pressure drops were evidence of major problems in the Gideon distribution system. In the Cotton Compress yards, water was used for equipment washing, in rest rooms, and for consumption. The pressure gradient between the Gideon system and the Cotton Compress system was such that the private storage tank overflowed when the municipal tanks were filling. To prevent this from occurring, a valve was installed in the influent line to the Cotton Compress tank. This same pressure differential kept water in the Cotton Compress tank unless there was a sudden demand in the warehouse area. The entire Cotton Compress water system was isolated from the Gideon system by a backflow-prevention valve. There were no residential water meters in the Gideon system, and residents paid a flat service rate of $11.50 per month for both water and sewage service. The municipal sewage system operated by a gravity flow with two lift stations and, as of December 31, 1993, served 429 households.

13.7.3.2 Identification of the outbreak. On November 29, 1993, the Missouri Department of Health became aware of two high school students from Gideon who were hospitalized with culture-confirmed salmonellosis (Clark et al., 1996d). Within 2 days, five additional patients living in Gideon were hospitalized with salmonellosis (one student, one child from a day care facility, two nursing home residents, and one visitor to the nursing home). The State Public Health Laboratories identified the isolates as dulcitol-negative *Salmonella*, and the CDC laboratories identified the organism as *Salmonella* serovar *typhimurium*. Interviews conducted by the health department suggested that the majority of patients had no exposure to food in common. All the ill individuals had consumed municipal water.

The Missouri Department of Natural Resources was informed that the health department suspected a water-supply link to the outbreak. Water samples collected by the natural resources departments on December 16 were positive for fecal coliform. On December 18, the city of Gideon, as required by the department, issued a "boil water" order. Signs were posted at city hall and in the grocery store, and two area radio stations announced the boil-water order.

Several water samples collected by the Department of Natural Resources on December 20 also were found to be positive for fecal coliform. On December 23, the department placed a chlorinator on line at the city well, and nine samples were collected by both Missouri departments from various sites in the distribution system. None of the samples contained chlorine, but one sample collected from a fire hydrant was positive for dulcitol-negative-.*Salmonella* serovar *typhimurium*. The health department had informed the CDC about the outbreak in Gideon in early December and requested information about dulcitol-negative *Salmonella* serovar *typhimurium* On December 17, the health department informed the CDC that contaminated municipal water was the suspected cause of the outbreak and on December 22, invited the CDC to participate in the investigation. A flyer explaining the boil-water order, jointly produced by State's Departments of Health and

Natural Resources, was placed in the mailboxes of all the homes in Gideon on December 29 and the privately owned water tower was physically disconnected from the municipal system on December 30. The natural resources department mandated that Gideon permanently chlorinate its water system. At the end of the study, the EPA provided input to the natural resources department on the criteria necessary to lift the boil–water order (Angulo et al., 1997).

Through January 8, 1994, the Department of Health had identified 31 cases of laboratory-confirmed salmonellosis associated with the Gideon outbreak. The State Public Health Laboratories identified 21 of these isolates as dulcitol-negative *Salmonella* serovar *typhimurium*. Fifteen of the 31 culture-confirmed patients were hospitalized (including two patients hospitalized for other causes who developed diarrhea while in the hospital). The patients were admitted to 10 different hospitals. Two of the patients had positive blood cultures, and seven nursing home residents exhibiting diarrhea illness died, four of whom were culture confirmed (the other three were not cultured). All the culture-confirmed patients were exposed to Gideon municipal water.

Ten culture-confirmed patients did not reside in Gideon, but all of them traveled to Gideon frequently either to attend school (eight patients), use a day care center in town (one patient), or work at the nursing home (one patient). The earliest onset of symptoms in a culture-confirmed case was on November 17 (this patient was last exposed to Gideon water on November 16). A CDC survey indicated that approximately 44 percent of the 1104 residents of Gideon, or almost 500 people, were affected with diarrhea between November 11 and December 27, 1993. Nonresidents who drank Gideon's water during the outbreak experienced an attack rate of 28 percent (Angulo et al., 1997).

13.7.3.3 Possible causes. The investigation clearly implicated consumption of Gideon's municipal water as the source of the outbreak of Salmonellosis. Speculation focused on a sequential flushing program conducted on November 10 involving all 50 hydrants in the system. The program began in the morning and continued through the entire day. Each hydrant was flushed for 15 minutes at an approximate rate of 2.8 m^3/min (750 gal/min). It was observed that the pump at well 5 was operating at full capacity during the flushing program (approximately 12 h), indicating that the municipal tanks were discharging during this period. The flushing program was conducted in response to complaints about taste and odor.

It was hypothesized that taste and odor problems may have resulted from a thermal inversion, which may have occurred because of a sharp temperature drop the day before the complaint. If stagnant or contaminated water were floating on the top of a tank, a thermal inversion could have caused this water to be mixed throughout the tank and to be discharged into the system, thus resulting in taste and odor complaints (Fennel et al., 1974). As a consequence, the utility initiated a citywide flushing program. Turbulence in the tank from the flushing program could have stirred up the tank sediments, which were transported into the distribution system. It is likely that the bulk water, the sediments, or both were contaminated with *Salmonella* serovar *typhimurium*.

During the EPA's field visit, a large number of pigeons were observed roosting on the roof of the 378-m^3 (100,000-gal) municipal tank. Shortly after the outbreak, a tank inspector found holes at the top of the Cotton Compress tank, rust on the tank, and rust, sediment and bird feathers floating in the water. According to the inspector, the water in the tank looked black and was so turbid he could not see the bottom. Another inspection, conducted after the EPA's field study, confirmed the disrepair of the Cotton Compress tank and also found the 378-m^3 (100,000-gal) municipal tank in such a state of disrepair that bird droppings could, in the inspector's opinion, have entered the stored water. Bird feathers were in the vicinity or in the tank openings of both the Cotton Compress tank and the 378-m^3 (100,000-gal) municipal tank.

It was initially speculated that the backflow valve between the Cotton Compress and the municipal system might have failed during the flushing program. After the outbreak, the valve was excavated and found to be working properly. Because the private tank was drained accidently after the outbreak during an inspection, it was impossible to sample water in the bowel of the tank. However, sediment in the Cotton Compress tank contained dulcitol-negative *Salmonella* serovar *typhimurium* as did samples found in a hydrant sample and in culture-confirmed patients. The *Salmonella* found in a hydrant matched the serovar of the patient isolate when analyzed by the CDC laboratory comparing DNA fragments using pulse field gel electrophoresis. The isolate from the tank sediment, however, did not provide an exact match with the other two isolates. No *Salmonella* isolates were found elsewhere in the system.

13.7.3.4 Evaluation of the System.

The purpose of the system evaluation was to study the effects of distribution system's design and operations, demand, and hydraulic characteristics on the possible propagation of contaminants in the system. Given the evidence from the survey and the results from the valve inspection at the Cotton Compress, the conclusion was that the most likely source of contamination was bird droppings in the large municipal tank. Therefore, the analysis concentrated on propagation of water from that municipal tank in conjunction with the flushing program. This did not rule out other possible sources of contamination, such as cross-connections.

The system's layout, demand information, pump characteristic curves, tank geometry, flushing program, and so on, and other information needed for the modeling effort were obtained from maps and demographic information and from numerous discussions with consulting engineers and city and natural resource department officials. EPANET, was used to conduct the contaminant propagation study (Rossman, 1994).

13.7.3.5 Performance of the System.

EPANET was calibrated by simulating flushing at the hydrants shown in Table 13.4, assuming a discharge of 2.8 m³/min (750 gpm) for 15 min. The "C" factors were adjusted until the headloss in the model matched headlosses observed in the field. (Table 13.4).

The hydraulic scenario was initiated by "running" the model for 48 h. The water level reached 122 m (400.59 ft) in the Cotton Compress tank, 122 m (400.63 ft) in the 378 m³ (100, 000-gal) tank (T300), and 122 m (400.66 ft) in the (50,000-gal) tank (T200). At 8 am on the third day, the simulated flushing program was initiated by sequentially imposing a 2.8 m³/min (750 gpm) demand on each of 50 hydrants, for 15 min. The entire process consumed 12.5 h. Using the TRACE option in EPANET, the percentages of water from both municipal tanks were calculated at each node over a

TABLE 13.4 Pressure Test Result

Hydrant Number	Pressure			
	Static		Dynamic	
	psi	kilograms/cm²	psi	kilograms/cm²
4	58	0.22	7	0.026
9	53	0.20	8	0.033
49	50	0.19	18	0.068

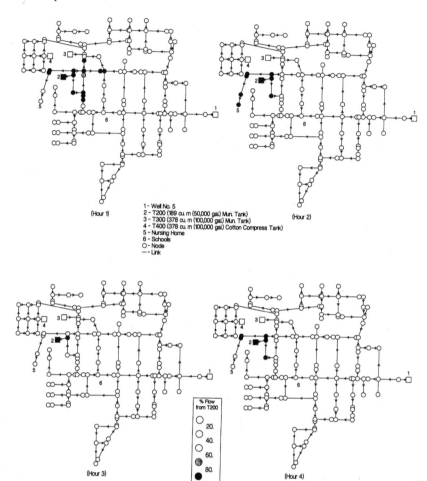

FIGURE 13.27 Movement of water from T200 during hours 1 through 4 of a 72-h simulation period (T400 valve closed).

period of 72 h. (Fig. 13.27). On the basis of the findings from excavating the backflow-prevention valve, the impact of flows from the Cotton Compress tank were not considered in the simulation.

The simulation indicated that the pump operated at more than 3 m^3/min (800 gal/min) during the flushing program and then reverted to cyclic operation thereafter. The elevation for both municipal tanks fluctuated, and both tanks discharged during the flushing program. At the end of the flushing period, nearly 25% of the water from the large municipal tank passed through the small municipal tank, where it was again discharged into the system.

Pressure drops during the flushing program were simulated at the hydrants used for calibration. The model predicted dramatic pressure drops during the flushing program. It was believed that, based on the information available, these results replicated the conditions that existed during the flushing program closely enough to provide the basis for an analysis of water movement in the system.

13.7.3.6 Propagation of the contaminant. Data from the simulation study, the microbiological surveillance data, and the outbreak data could be used to provide insight into the nature of both general contamination problems in the system and into the outbreak itself. The patterns of water movement showed that the majority of the special samples that were positive for coliform and fecal coliform occurred at points that lie within the zone of influence of the small and large tanks. During both the flushing program and large parts of normal operation, these areas are served predominately by tank water, which might lead us to believe that the tanks were the source of the fecal contamination since there were positive fecal coliform samples before chlorination.

Data from the early cases, in combination with the water movement data, were used to infer the source of the outbreak. Using data supplied by the CDC and the water movement simulations, an overlay of the areas served by the small and large tanks during the first 6 h of the flushing period and the earliest recorded cases was created (Fig. 13.28). As can

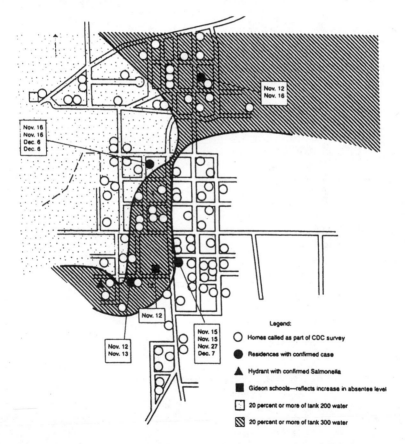

FIGURE 13.28 Comparison of early confirmed cases and a Salmonella-positive sample versus penetration of tank water during the first 6 h of a flushing program.

be seen in Fig. 13.28, the earliest recorded cases and the hydrant sample positive for *Salmonella* were found in the area that was served primarily by the large tank, but was outside the small tank area of influence, during the flushing period. One can conclude that during the first 6 h of the flushing period, the water, that reached the homes with confirmed cases and the Gideon School was almost totally from the large tanks. Therefore, it was logical to conclude that these sites should experience the first signs of the outbreak, which makes a strong circumstantial case for the large tank as the source of contamination.

Based on the results of the sampling program conducted by the Departments of Health and Natural Resources, it is likely that the contamination had been occurring over a period of time, which is consistent with the possibility of bird contamination. If the cause had been a single event, the contaminant would most likely have been "pulled" through the system during the flushing program.

13.8 CURRENT TRENDS IN WATER–QUALITY MODELING

Distribution system water-quality modeling has evolved from the basic models available in the late 1980s and early 1990s to the full-featured models currently in use. This new generation of models contains sophisticated tank-mixing models, GIS capability, and flexible user-interface features.

Several case studies presented in this section demonstrate the extended capability of water-quality models. The recent development of models for analyzing water quality in storage tanks is discussed.

13.8.1 Study in Cholet, France

Cholet (population 60,000) is a municipality situated in the western part of France. Its major sources of water are the reservoirs of Moulin de Ribou and Verdon (Heraud et al., 1997). The largest treatment plant in the municipality treats water from the Moulin de Robou reservoir. Production is approximately 30,000 m^3/day. Treated water is discharged into the distribution system, which consists of more than 280 km of pipes and two tanks.

The Piccolo hydraulic model, which is proprietary and was developed by the Research Center of the Lyonnaise des Eaux Groupe, was used to analyze flow in the network. The model was used to demonstrate that the main system was, in fact, made up of two hydraulically independent subsystems. A unique feature of the study was the incorporation of a continuous on-line chlorine residual monitor as part of the water-quality modeling effort.

13.8.2 Case Study in Southington, Connecticut

The Southington, Connecticut, water supply system has a distribution system with 299 km (186 mi) of pipe and nine wells that are capable of pumping more than 0.2965 m^3/day (4700 gpm) (Aral and Masila, 1997; Aral et al., 1996). Three municipal reservoirs also are incorporated into the system. The ground water was contaminated by volatile organic chemicals during the 1970s. EPANET, in conjunction with a GIS system, was used to simulate four exposure scenarios that represented pumping conditions for 1970, 1974, and 1979. The study concluded that (1) exposure to contamination by volatile organic chemicals can exhibit significant spatial variation from one census block to another even when census blocks are adjacent to each other within a specified radius, (2) the use of peak demand conditions may not yield the maximum exposure, and (3) hydraulic and water-quality modeling is a superior mechanism for quantifying the exposure of populations to past contamination.

13.8.3 Mixing in Storage Tanks

Storage tanks are the most visible component of a water distribution system. However, they are least understood in terms of their impact on water quality. Although tanks play a major role in providing hydraulic reliability for fire-fighting needs and for providing reliable service, they may serve as vessels for complex chemical and biological changes that may cause the deterioration of water quality. Grayman and Clark (1993) conducted studies indicating that water quality degrades as the result of long residence time in tanks. These studies highlighted the importance of the design, location, and operation of tanks with regard to water quality. Mau et al., (1995) developed compartmental models to represent the different mixing conditions in tanks assuming steady-state conditions. Clark, et al., (1996b) extended this compartmentalization approach by assuming non-steady-state conditions.

A collaborative study between the AWWARF and the EPA was initiated to study the effects of tanks on water quality (Vasconcelos et al., 1997). As part of this study, Boulos, et al., (1996) published the results of an extensive study of the Ed Hauk reservoir in Azusa, California. This tank has a capacity of 14,782.6 m^3 (4 million gal) and was built to provide operational and emergency storage and to maintain contact time for water leaving the Azusa treatment plant. The normal mode of operation, unlike many system storage tanks, is simultaneous inflow and outflow. The study placed primary emphasis on providing an understanding of the hydraulic mixing and free chlorine residual in the reservoir. The reservoir was found to be mixed completely, with two exceptions. Short-circuiting was found to exist between the inlet and outlet and the presence of a stagnant zone in the center core of the reservoir. The result was the possibility of stratification or partitioning in the reservoir.

The storage tank models in the early versions of water-quality distribution models, such as EPANET, were relatively simple. Generally, they assumed complete mixing or used two-compartment models, even though the actual mixing regime might be much more complex. As part of an AWWARF/EPA study, Grayman et al., (1996) examined three approaches to describing the behavior of tanks. One was the use of physical models, the second was a simplified systems model that emphasized the input and output of the reservoir or tank, and the third was a computational fluid dynamics model based on mathematical equations. They authors found that each approach has some advantages.

13.9 SUMMARY AND CONCLUSIONS

There is growing recognition that water quality can deteriorate significantly between the treatment plant and the consumer. Factors that can cause deterioration include the quality of the source water, the type of treatment process used, the storage facilities, and the age, type design, and maintenance of the distribution system.

Distribution systems in the United Stated are designed to ensure adequate fire flow and pressure as well as to satisfy domestic and industrial demand. Hydraulic and water-quality models are growing in acceptance as a mechanism for analyzing the flow in networks and for determining the factors that contribute to the deterioration of water quality. The North Penn Water Authority study was among the first applications of water-quality modeling. This study demonstrated the dynamic nature of quality in drinking water systems. The DWQM was developed and applied to the South Central Connecticut Regional Water Authority. That model required an external hydraulic model. The South Central study demonstrated the potential of storage tanks to degrade water quality.

The development of EPANET marked the next generation of water-quality models. This integrated hydraulic and water-quality model has been applied in a number of water utilities. For example, it has been applied to the tracking of THMs in the North Marin Water Authority and to an outbreak of waterborne disease in Gideon, Missouri. It has recently been applied to study exposure to volatile organic chemicals in Southington, Connecticut. As a consequence of a study conducted by AWWARF and the EPA, computational fluid dynamics and physical models have now been developed to model water quality in tanks.

Finally, water-quality models have become useful for studying water quality in networks. Current trends indicate that these models will become increasingly sophisticated and user friendly.

REFERENCES

AWWA, "Distribution Network Analysis for Water Utilities," AWWA Manual M-32, *American Water Works Association,* Denver, CO, 1989.

Angulo, F. J., S. Tippen, D. J. Sharp, B. J. Payne, C. Collier, J. E. Hill, T. J. Barrett, R. M. Clark, E. E. Geldriech, H. D. Donnell, and D. L. Swerdlow, "A Community Waterborne Outbreak of Salmonellosis and the Effectiveness of a Boil Water Order," *American Journal of Public Health,* 87:580–584, 1997.

Aral, M. M., and M. L. Maslia, "Exposure Assessment Using Simulation and GIS," *Proceedings of the 1997 CSCE/ASCE Environmental Engineering Conference,* Edmonton, Alberta, July 22–26, 885–892, 1997.

Aral, M. M., M. L. Maslia, G. V. Ulirsch, and J. J. Reyes, "Estimating Exposure to Volatile Organic Compounds from Municipal Water-Supply Systems: Use of a Better Computational Model," Archives of Environmental Health, 51:300–309, 1996.

Biswas, P., C. Lu, and R. M. Clark, "A Model for Chlorine Concentration Decay in Drinking Water Distribution Pipes," *Water Research,* 27:1715–1724, 1993.

Boulos, P. F., T. Altman, P. A. Jarrige, and F. C. Collevati, "Discrete Simulation Approach for Network-Water-Quality Models," *ASCE Journal of Water Resources Planning and Management,* 121(1):49–60, 1995.

Boulos, P. F., W. M. Grayman, R. W. Bowcock, J. W. Clapp, L. A. Rossman, R. M. Clark, R. A. Deininger, and A. K. Dhingra, "Hydraulic Mixing and Free Chlorine Residual in Reservoirs," *Journal of the American Water Works Association,* 88(7):48–59, 1996.

Chun, D. G., and H. L. Selznick, "Computer Modeling of Distribution System Water Quality," *ASCE Special Conference on Computational Application in Water Resources,* Buffalo, NY, June 1985, pp.

Clark, R. M., "Applying Water Quality Models," in M. H. Chaudry and L. M. Mays eds., *(Computer Modeling of Free-Surface and Pressurized Flow,)* Kluwer Dordrecht, Netherlands, 581–612, 1993a.

Clark, R. M., "Development of Water Quality Models," In M. H. Chaudry and L. M., Mays, *(eds., Computer Modeling of Free-Surface and Pressurized Flows,)* Kluwer Academic Publishers, Dordrecht, Netherlands, 553-580, 1993b.

Clark, R. M., and J. A. Coyle, "Measuring and Modeling Variations in Distribution System Water Quality," *Journal of the American Water Works Association,* 82(8): 1990.

Clark, R. M., and J. A. Goodrich, "Water Quality Modeling in Distribution Systems," in *Strategies and Technologies for Meeting SDDWA Requirements*, Technomics Publishing, Lancaster, PA, 344–359, 1993.

Clark, R. M., and R. M. Males, "Developing and Applying the Water Supply Simulation Model," *Journal of the American Water Works Association*, 78(8), 1986.

Clark, R. M., W. M. Grayman, R. M. Males, and J. A. Coyle, "Predicting Water Quality in Distribution Systems," *Proceedings of the AWWA Distribution System Symposium*, Minneapolis, MN, September 1986.

Clark, R. M., J. Q. Adams, and R. M. Miltner, "Cost and Performance Modeling for Regulatory Decision Making," Water, 28(3):20–27, 1987.

Clark, R. M., W. M. Grayman, and R. M. Males, "Contaminant Propagation in Distribution Systems," *ASCE Journal of Environmental Engineering*, 114: 1988a.

Clark, R. M., J. A. Coyle, W. M. Grayman, and R. M. Males, "Development, Application, and Calibration of Models for Predicting Water Quality in Distribution Systems," *Proceedings of the AWWA Water Quality and Technology Conference*, St. Louis, MO, 1988b.

Clark, R. M., W. M. Grayman, R. M. Males, and J. Coyle, "Modeling Contaminant Propagation in Drinking Water Distribution Systems," *Journal of Water Supply Research and Technology—Aqua*, (3):137–151, 1988c.

Clark, R. M., W. M. Grayman, J. A. Goodrich, R. A. Deininger, and A. F. Hess, "Field Testing Distribution Water Quality Models," *Journal of the American Water Works Association*, 84(7): 1991a.

Clark, R. M., W. M. Grayman, and J. A. Goodrich, "Water Quality Modeling: Its Regulatory Implications," Proceedings of the American Water Works Association Research Foundations, Environmental Protection Agency Conference on Water Quality Modeling in Distribution Systems, Cincinnati, OH, 1991b.

Clark, R. M., J. A. Goodrich, and L. J. Wymer, "Effect of the Distribution System on Drinking Water Quality," *Journal of Water Supply Research and Technology—Aqua*, 42(1):03–38, 1993a.

Clark, R. M., W. M. Grayman, R. M. Males, and A. F. Hess, "Modeling Contaminant Propagation in Drinking Water Distribution Systems," *ASCE Journal of Environmental Engineering*, 119:349–364, 1993b.

Clark, R. M., G. Smalley, J. A. Goodrich, R. Tull, L. A. Rossman, J. T. Vasconcelos, and P. F. Boulos, "Managing Water Quality in Distribution Systems: Simulating TTHM and Chlorine Residual Propagation," *Journal of Water Supply Research and Technology—Aqua*, 43(4):182–191, 1994.

Clark, R. M., L. A. Rossman, and L. G. Wymer, "Modeling Distribution System Water Quality: Regulatory Implications," *ASCE Journal of Water Resources Planning and Management*, 121(6):423–428, 1995.

Clark, R. M., E. E. Geldreich, K. R. Fox, E. W. Rice, C. H. Johnson, J. A. Goodrich, J. A. Barnick, F. Abdesaken, J. E. Hill, and F. J. Angulo, "A Waterborne Salmonella typhimurium Outbreak in Gideon, Missouri: Results from a Field Investigation," *International Journal of Environmental Health Research*, 6:187–193, 1996a.

Clark, R. M., F. Abdesaken, P. F. Boulos, and R. Mau, "Mixing in Distribution System Storage Tanks: Its effect on Water Quality," *ASCE Journal of Environmental Engineering*, 122:814–821, 1996b.

Clark, R. M., H. Pourmoghaddas, L. G. Wymer, and R. C. Dressman, "Modeling the Kinetics of Chlorination By-product Formation: The Effects of Bromide," *Journal of Water Supply Research and Technology—Aqua*, 45(1):1–8, 1996c.

Clark, R. M., E. E. Geldreich, K. R. Fox, E. W. Rice, C. H. Johnson, J. A. Goodrich, J. A. Barnick, and F. Abdesaken, "Tracking a *Salmonella* serovar *typhimurium* Outbreak in Gideon, Missouri: Role of Contaminant Propagation Modeling," *Journal of Water Supply Research and Technology—Aqua*, 45:171–183, 1996d.

Cross, H., "Analysis of Flow in Networks of Conduits or Conductors," University of Illinois Engineering Experiment Station Bulletin 286, Urbana, IL, 1936.

Fennel, H., D. B. James, and J. Morris, "Pollution of a Storage Reservoir by Roosting Gulls," *Journal of the Society of Water Treatment Examiners*, 23(5):24, 1974.

Geldreich, E. E., H. D. Nash, D. J. Reasoner, and R. H. Taylor, "The Necessity of Controlling Bacterial Populations in Potable Water: Community Water Supply," *Journal of the American Water Works Association*, 64:596–602, 1972.

Geldreich, E. E., K. R. Fox, J. A. Goodrich, E. W. Rice, R. M. Clark, and D. L. Swerdlow, "Searching for a Water Supply Connection in the Cabool, Missouri Disease Outbreak of *Escherichia coli* 0157:H7," *Water Research*, 84:49–55, 1992.

Gessler, J., and T. M. Walski, *Water Distribution System Optimization*, TREL-85-11, WES, U.S. Army Corps of Engineers, Vicksburg, MS, 1985.

Grayman, W. M., and R. M. Clark, "Using Computers to Determine the Effect of Storage on Water Quality," *Journal of the American Water Works Association*, 85(7):67–77, 1993.

Grayman, W. M., R. M. Clark, and R. M. Males, "A Set of Models to Predict Water Quality in Distribution Systems," *Proceedings of the International Symposium on Computer Modeling of Water Distribution Systems*, Lexington, KY, May 1988a.

Grayman, W. M., Grayman, W. M., R. M. Clark, and R. M. Males, "Modeling Distribution-System Water Quality: Dynamic Approach," *ASCE Journal of Water Resources Planning and Management*, 114(3): 1998b.

Grayman, W. M., R. A. Deininger, A. Green, P. F. Boulos, Bowcock, and C. C. Godwin," Water Quality and Mixing Models for Tanks and Reservoirs," *Journal of the American Water Works Association*, 88(7):60–73, 1996.

Hart, F. L., "Applications of the NET Software Package," *Proceedings of the AWWARF/EPA Conference on Water Quality Modeling in Distribution Systems*, Cincinnati, OH, February 1991.

Hart, F. L., J. L. Meader, and S. N. Chiang, "CLNET—A Simulation Model for Tracing Chlorine Residuals in a Potable Water Distribution Network," *Proceedings of the AWWA Symposium on Distribution Systems*, Minneapolis, MN, 1986.

Heraud, J., L. Kiene, M. Detay, and Y. Levy, "Optimized Modeling of Chlorine Residual in a Drinking Water Distribution System with a Combination of On-line Sensors," *Journal of Water Supply Research and Technology—Aqua*, 46(2):59–70, 1997.

Islam, M. R., "Modeling of Chlorine Concentration in Unsteady Flows in Pipe Networks," Ph.D. dissertation, *Washington State University*, Pullman, May 1995.

Islam, M. R., M. H. Chaudhry, and R. M. Clark, "Inverse Modeling of Chlorine Concentration in Pipe Networks Under Dynamic Conditions," ASCE *Journal of Environmental Engineering*, 123:1033–1040, 1997.

Kirmeyer, G. J., W. Richards, and C. D. Smith, "An Assessment of the Condition of North American Water Distribution Systems and Associated Research Needs," *American Water Works Association Research Foundation*, Denver, CO, 1994.

Kroon, J. R., and W. A. Hunt, "Modeling Water Quality in the Distribution Network," *AWWA Water Quality Technical Conference*, Philadelphia, PA, November 1989.

LeChevallier, M. W., T. M. Babcock, and R. G. Lee, "Examination and Characterization of Distribution System Biofilms," *Applied and Environmental Microbiology*, 53:2714–2724, 1987.

Liou, C. P., and J. R. Kroon, "Propagation and Distribution of Waterborne Substances in Networks," *Proceedings of the AWWA Distribution System Symposium, Minneapolis*, MN, September 1986,

Males, R. M., R. M. Clark, P. J. Wehrman, and W. E. Gates, "Algorithm for Mixing Problems in Water Systems," *ASCE Journal of Hydraulic Engineering*, 111(2): 1985.

Mau, R. P. Boulos, R. Clark, W. Grayman, R. Tekippe, and R. Trussell, "Explicit Mathematical Models of Distribution System Storage Water Quality," *ASCE Journal of Hydraulic Engineering*, 121(10):699–709, 1995.

Maul, A., A. H. El-Shaarawi, and J. C. Block, "Heterotrophic Bacteria in Water Distribution Systems, I. Spatial and Temporal Variation," *Science of the Total Environment*, 44:201–214, 1985a.

Maul, A., A. H. El-Shaarawi, and J. C. Block, "Heterotrophic Bacteria in Water Distribution Systems II. Sampling Design for Monitoring," *The Science of the Total Environment*, 44:215–224, 1985b.

McCoy, W. F., and B. H. Olson, "Relationship Among Turbidity Particle Counts and Bacteriological Quality Within Water Distribution Lines," *Water Research*, 20:1023–1029, 1986.

Metzger, I, "Water Quality Modeling of Distribution Systems," *Proceeding of ASCE Special Conference on Computational Application in Water Research*, Buffalo, NY, June 1985.

Rossman, L. A., *EPANET Users Manual*, Drinking Water Research Division, USEPA, Cincinnati, OH, 1994.

Rossman, L. A., R. M. Clark, and W. M. Grayman, "Modeling Chlorine Residuals in Drinking Water Distribution Systems," *ASCE Journal of Environmental Engineering*, 120:803–820, 1994.

Sarikelle, S., Y Chuang, and G. A. Loesch, "Analysis of Water Distribution Systems on a Supercomputer," *Proceeding of AWWA Comp. Spec. Conf.*, Denver, CO, 1989.

Skov, K. R., A. F. Hess, and D. B. Smith, "Field Sampling Procedures for Calibration of a Water Distribution System Hydraulic Model," in Water Quality Modeling In Distribution Systems, *American Water Works Research Foundation/ U.S EPA*, Denver, CO, 1991.

Vasconcelos, J. J., P. F. Boulos, W. M. Grayman, L. Kiene, O. Wable, P. Biswas, A. Bhari, L. A. Rossman, R. M. Clark, and J. A. Goodrich, "Characterization and Modeling of Chlorine Decay in Distribution Systems," Report No. 90705, *American Water Works Association Research Foundation*, Denver, CO, 1996.

Vasconcelos, J. J., L. A. Rossman, W. M. Grayman, P. F. Boulos, and R. M. Clark, "Kinetics of Chlorine Decay," *Journal of the American Water Works Association*, 89(7):54–65, 1997.

Water Research Centre, "Deterioration of Bacteriological Quality of Water During Distribution," *Notes on Water Research*, No. 6, 1976.

Wood, D. J., "Computer Analysis of Flow in Pipe Networks," Department of Civil Engineering, University of Kentucky, Lexington, 1980a.

Wood, D. J., "Slurry Flow in Pipe Networks," *ASCE Journal of Hydraulic Engineering*, 106: 1980b.

Wood, D. J., and L. E. Ormsbee, "Supply Identification for Water Distribution Systems," *Journal of the American Water Works Association*, 81(7): 1989.

CHAPTER 14
CALIBRATION OF HYDRAULIC NETWORK MODELS

Lindell E. Ormsbee and Srinivasa Lingireddy
Department of Civil Engineering
University of Kentucky,
Lexington, KY

14.1 INTRODUCTION

Computer models for analyzing and designing water distribution systems have been available since the mid-1960s. Since then, however, many advances have been made with regard to the sophistication and application of this technology. A primary reason for the growth and use of computer models has been the availability and widespread use of the microcomputer. With the advent of this technology, water utilities and engineers have been able to analyze the status and operations of the existing system as well as to investigate the impacts of proposed changes (Ormsbee and Chase, 1988). The validity of these models, however, depends largely on the accuracy of the input data.

14.1.1 Network Characterization

Before an actual water distribution system can be modeled or simulated with a computer program, the physical system must be represented in a form that can be analyzed by a computer. This normally requires that the water distribution system first be represented by using node-link characterization (Fig. 14.1). In this case, the links represent individual pipe sections and the nodes represent points in the system where two or more pipes (links) join together or where water is being input or withdrawn from the system.

14.1.2 Network Data Requirements

Data associated with each link will include a pipe identification number, pipe length, pipe diameter, and pipe roughness. Data associated with each junction node will include a junction identification number, junction elevation, and junction demand. Although it is

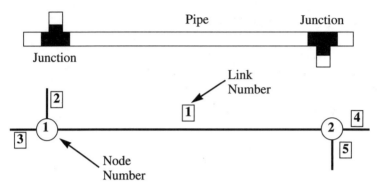

FIGURE 14.1 Node-link characterization.

recognized that water leaves the system in a time-varying fashion through various service connections along the length of a pipe segment, it is generally acceptable in modeling to lump half the demands along a line to the upstream node and the other half of the demands to the downstream node as shown in Fig. 14.2.

In addition to the network pipe and node data, physical data for use in describing all tanks, reservoirs, pumps, and valves also must be obtained. Physical data for all tanks and reservoirs normally includes information on tank geometry as well as the initial water levels. Physical data for all pumps normally include either the value of the average useful horsepower or data for use in describing the pump flow/head characteristics curve. Once this necessary data for the network model has been obtained, the data should be entered into the computer in a format compatible with the selected computer model.

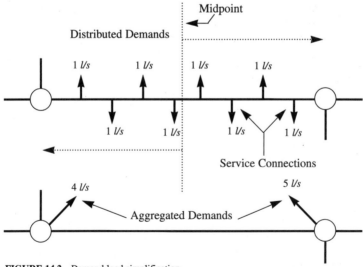

FIGURE 14.2 Demand load simplification.

14.1.3 Model Parameters

Once the data for the computer network model has been assembled and encoded, the associated model parameters should then be determined before actual application of the model. In general, the primary parameters associated with a hydraulic network model include pipe roughness and nodal demands. Because obtaining economic and reliable measurements of both parameters is difficult, final model values are normally determined through the process of model calibration. Model calibration involves the adjustment of the primary network model parameters (i.e., pipe roughness coefficients and nodal demands) until the model results closely approximate actual observed conditions, as measured from field data. In general, a network-model calibration effort should encompass the following seven basic steps (Fig. 3). Each step is discussed in detail in the following sections.

14.2 IDENTIFY THE INTENDED USE OF THE MODEL

Before calibrating a hydraulic network model, it is important to identify its intended use (e.g., pipe sizing for master planning, operational studies, design projects, rehabilitation studies, water-quality studies) and the associated type of hydraulic analysis (steady-state versus extended-period). Usually, the type of analysis is directly related to the intended use. For example, water-quality and operational studies require an extended-period analysis, whereas some planning or design studies can be performed using a study state analysis (Walski, 1995). In the latter, the model predicts system pressures and flows at an instant in time under a specific set of operating conditions and demands (e.g., average or maximum daily demands). This is analogous to photographing the system at a specific point in time. In extended-period analysis, the model predicts system pressures and flows over an extended period (typically 24 hours). This is analogous to developing a movie of the system's performance.

Both the intended use of the model and the associated type of analysis provide some guidance about the type and quality of collected field data and the desired level of agreement between observed and predicted flows and pressures (Walski, 1995). Models for steady-state applications can be calibrated using multiple static flow and pressure observations collected at different times of day under varying operating conditions. On the other hand, models for extended-period applications require field data collected over an extended period (e.g., 1.7 days).

In general, a higher level of model calibration is required for water-quality analysis or an operational study than for a general planning study. For example, determining ground evaluations using a topographic map may be adequate for one type of study, whereas another type of study may require an actual field survey. This of course may depend on the contour interval of the map used. Such considerations obviously influence the methods used to collect the necessary model data and the subsequent calibration steps. For example, if one is working in a fairly steep terrain (e.g. greater than 20 foot contour intervals), one may decided to use a GPS unit for determining key elevations other than simply interpolating between contours.

14.3 DETERMINE ESTIMATES OF THE MODEL PARAMETERS

The second step in calibrating a hydraulic network model is to determine initial estimates of the primary model parameters. Although most models will have some degree of uncertainty associated with several model parameters, the two parameters that normally have the greatest degree of uncertainty are the pipe roughness coefficients and the demands to be assigned to each junction node.

14.3.1 Pipe Roughness Values

Initial estimates of pipe-roughness values can be obtained using average values in the literature or values directly from field measurements. Various researchers and pipe manufacturers have developed tables that provide estimates of pipe roughness as a function of various pipe characteristics, such as pipe material, pipe diameter, and pipe age (Lamont, 1981). One such typical table is shown in Table 14.1 (Wood, 1991). Although such tables can be useful for new pipes, their specific applicability to older pipes decreases significantly as the pipes age as a result of the effects of such factors as tuberculation, water chemistry, and the like. As a result, initial estimates of pipe roughness for all pipes other than relatively new ones normally should come directly from field testing. Even when new pipes are being used, it is helpful to verify the roughness values in the field since the roughness coefficient used in the model actually may represent a composite of several secondary factors such as fitting losses and system skeletonization.

14.3.1.1 Chart the pipe roughness. A customized roughness nomograph for a particular water distribution system can be developed using the process illustrated in Figs. 14.4.A-C To obtain initial estimates of pipe roughness through field testing, it is best to divide the water distribution system into homogeneous zones based on the age and material of the associated pipes (Fig. 14.4A). Next, several pipes of different diameters should be tested

TABLE 14.1 Typical Hazen-William Pipe Roughness Factors

Pipe Material	Age (years)	Diameter	C Factor
Cast iron	New	All sizes	130
	5	>380 mm (15in)	120
		>100 mm (4in)	118
	10	>600 mm (24in)	113
		>300 mm (12in)	111
		>100 mm (4in)	107
	20	>600 mm (24in)	100
		>300 mm (12in)	96
		>100 mm (4in)	89
	30	>760 mm (30in)	90
		>400 mm (16in)	87
		>100 mm (4in)	75
	40	>760 mm (30in)	83
		>400 mm (16in)	80
		>100 mm (4in)	64
Ductile iron	New		140
Polyvinyl chloride	Average		140
Asbestos cement	Average		140
Wood stave	Average		120

Calibration of Hydraulic Network Models **14.5**

Identify the intended use of the model
Determine initial estimates of the model parameters
Collect calibration data
Evaluate the model results
Perform the macro-level calibration
Perform the sensitivity analysis
Perform the micro-level calibration

FIGURE 14.3 Seven basic steps for network model calibration.

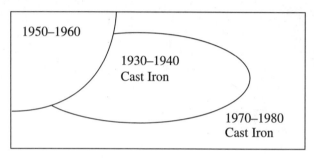

FIGURE 14.4A Subdivided network into homogeneous zones of like age and material.

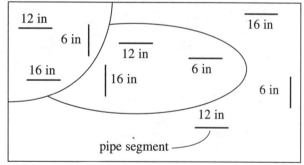

FIGURE 14.4B Select representative pipes from each zone.

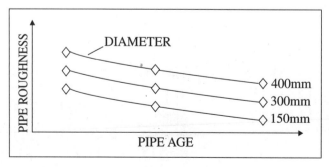

FIGURE 14.4C Plot associated roughness as a function of pipe diameter and age.

14.6 Chapter Fourteen

in each zone to obtain individual estimates of pipe roughness (Fig. 14.4B). Once a customized roughness nomograph is constructed (Fig. 14.4C), it can be used to assign values of pipe roughness for the rest of the pipes in the system.

14.3.1.2 Field test the pipe roughness. Pipe roughness values can be estimated in the field by selecting a straight section of pipe that contains a minimum of three fire hydrants (Figure 14.5A). When the line has been selected, pipe roughness can be estimated using one of two methods (Walski, 1984): (1) the parallel-pipe method (Fig. 14.5B) or (2) the two-hydrant method (Figure 14.5C). In each method, the length and diameter of the test pipe are determined first. Next, the test pipe is isolated, and the flow and pressure drop are measured either by using a differential-pressure gauge or two separate pressure gauges. Pipe roughness can then be approximated by a direct application of either the Hazen-Williams equation or the Darcy-Weisbach equation. In general, the parallel-pipe method

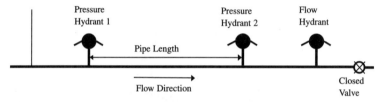

FIGURE 14.5A Pipe roughness test configuration.

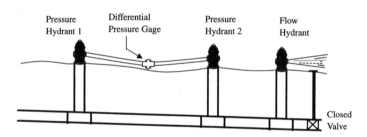

FIGURE 14.5B Parallel pipe method.

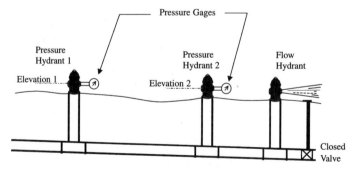

FIGURE 14.5C Two gage method.

is preferable for short runs and for determining minor losses around valves and fittings. For long runs of pipe, the two-gage method is generally preferred. Also if the water in the parallel pipe heats up or if a small leak accurs in the parallel line, it can lead to errors in the associated headloss measurements (Walski, 1985).

Parallel-pipe method. The steps involved in the application of the parallel pipe method are summarized as follows:

1. Measure the length of pipe between the two upstream hydrants (L_p) in meters.
2. Determine the diameter of the pipe (D_p) in millimeters. In general, this should simply be the nominal diameter of the pipe. It is recognized that the actual diameter may differ from this diameter because of variations in wall thickness or the buildup of tuberculation in the pipe. However, the normal calibration practice is to incorporate the influences of variations in pipe diameter via the roughness coefficient. It should be recognized, however, that although such an approach should not significantly influence the distribution of flow or headloss throughout the system, it may have a significant influence on pipe velocity, which in turn could influence the results of a water-quality analysis.
3. Connect the two upstream hydrants with a pair of parallel pipes, (typically a pair of fire hoses) with a differential pressure device located in between (Figure 14.5B). The differential pressure device can be a differential pressure gauge, an electronic transducer, or a manometer. Walski (1984) recommended the use of an air-filled manometer because of its simplicity, reliability, durability, and low cost. (*Note*: When connecting the two hoses to the differential pressure device, make certain that there is no flow through the hoses. If there is a leak in the hoses, the computed headloss for the pipe will be in error by an amount equal to the headloss through the hose.)
4. Open both hydrants and check all connections to ensure there are no leaks in the configuration.
5. Close the valve downstream of the last hydrant, then open the smaller nozzle on the flow hydrant to generate a constant flow through the isolated section of pipe. Make certain the discharge has reached equilibrium condition before taking flow and pressure measurements.
6. Determine the discharge Q_p (L/s) from the smaller nozzle in the downstream hydrant. This is normally accomplished by measuring the discharge pressure P_d of the stream leaving the hydrant nozzle using either a hand-held or nozzle-mounted pitot. Once the discharge pressure P_d (in kPa) is determined, it can be converted to discharge (Q_p) using the following relationship:

$$Q_p = \frac{C_d D_n^2 P_d^{0.5}}{900.3} \qquad (14.1)$$

where D_n is the nozzle diameter in millimeters and C_d is the nozzle discharge coefficient, which is a function of the type of nozzle (Fig. 14.6). (*Note*: When working with larger mains, sometimes you can't get enough water out of the smaller nozzles to get a good pressure drop. In such cases you may need to use the larger nozzle).
7. After calculating the discharge, determine the in-line flow velocity V_p (m/s) where

$$V_p = \frac{Q_p}{(\pi D_p^2/4)^2} \qquad (14.2)$$

8. After the flow through the hydrant has been determined, measure the pressure drop D_p through the isolated section of pipe by reading the differential pressure gauge. Convert

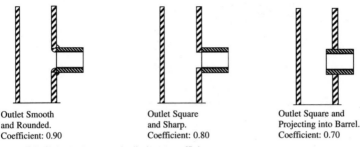

| Outlet Smooth and Rounded. | Outlet Square and Sharp. | Outlet Square and Projecting into Barrel. |
| Coefficient: 0.90 | Coefficient: 0.80 | Coefficient: 0.70 |

FIGURE 14.6 Hydrant nozzle discharge coefficients.

the measured pressure drop in units of meters (H_p) and divide by the pipe length L_p to yield the hydraulic gradient or friction slope S_p:

$$S_p = \frac{H_p}{L_p} \quad (14.3)$$

9. Once these four measured quantities have been obtained, the HazenWilliams roughness factor (C_p) can then be determined using the HazenWilliams equation as follows:

$$C_p = \frac{218 V_p}{D_p^{0.63} S_p^{0.34}} \quad (14.4)$$

To calculate the actual pipe roughness e, it is necessary to calculate the friction factor f using the Darcy-Weisbach equation as follows (Walski, 1984):

$$f = \frac{g S_p D_p}{500 \, V_p^2} \quad (14.5)$$

where g = gravitational acceleration constant (9.81 m/s²).

Once the friction factor has been calculated, the Reynolds number (Re) must be determined. Assuming a standard water temperature of 20°C (68°F), the Re is

$$Re = 993 \, V_p D_p \quad (14.6)$$

When the friction factor f and the Re have been determined, they can be inserted into the Colebrook-White formula to give the pipe roughness e (mm) as

$$e = 3.7 D_p \left[\exp(-1.16 \sqrt{f}) - \frac{2.51}{R\sqrt{f}} \right] \quad (14.7)$$

Two-hydrant method. The two-hydrant method is basically identical to the parallel-pipe method, with the exception that the pressure drop across the pipe is measured using a pair of static pressure gauges (Fig. 14.5C). In this case, the total headloss through the pipe is the difference between the hydraulic grades at both hydrants. To obtain the hydraulic grade at each hydrant, the observed pressure head (m) must be added to the elevation of the reference point (the hydrant nozzle). For the two-hydrant method, the headloss through the test section H_p (m) can be calculated using the following equation:

$$H_p = \frac{(P_2 - P_1)}{9.81} + (Z_2 - Z_1) \qquad (14.8)$$

where P_1 is the pressure reading at the upstream gauge (kPa), Z_1 is the elevation of the upstream gauge (m), P_2 is the pressure reading at the downstream gauge (kPa), and Z_2 is the elevation of the downstream gauge (m).

The difference in elevation between the two gauges should generally be determined using a transit or a level. As a result, one should make certain to select two upstream hydrants that can be seen from a common point. This will minimize the number of turning points required to determine the differences in elevation between the nozzles of the two hydrants. As an alternative to the use of a differential survey, topographic maps can sometimes be used to obtain estimates of hydrant elevations. However, topographic maps usually should not be used to estimate the elevation differences unless the contour interval is 1 m or less. One hydraulic alternative to measuring the elevations directly is to simply measure the static pressure readings (kPa) at both hydrants before the test and convert the observed pressure difference to the associated elevation difference (m) using the relations Z1 − Z2 = [P2(static) − P1(static)]/9.81.

General suggestions. Hydrant pressures for use in pipe-roughness tests are normally measured with a Bourdon tube gauge, which can be mounted to one of the hydrant's discharge nozzles using a lightweight hydrant cap. Bourdon tube gauges come in various grades (i.e., 2A, A, and B), depending on their relative measurement error. In most cases, a grade A gauge (1 percent error) is sufficient for fire-flow tests. For maximum accuracy, one should choose a gauge graded in 5-kPa (1-psi) increments, with a maximum reading less than 20 percent above the expected maximum pressure (McEnroe et al., 1989). In addition, it is a good idea to use pressure snubbers to eliminate the transient effects in the pressure gauges. A pressure snubber is a small valve that is placed between the pressure gauge and the hydrant cap which acts as a surge inhibitor (Walski, 1984).

Before conducting a pipe roughness test, it is always a good idea to make a visual survey of the test area. When surveying the area, make certain that there is adequate drainage away from the flow hydrant. In addition, make certain that you select a hydrant nozzle that will not discharge into oncoming traffic. Also, when working with hydrants in close proximity to traffic, it is a good idea to put up traffic signs and use traffic cones to provide a measure of safety during the test. As a further safety precaution, ensure that all personnel are wearing highly visible clothing. It also is a good idea to equip testing personnel with radios or walkie-talkies to help coordinate the test.

While the methods outlined previously work fairly well with smaller lines (i.e. less than 16in in diameter), their efficiency decreases as you deal with larger lines. Normally, opening hydrants just doesn't generate enough flow for meaningful head-loss determination. For such larger lines you typically have to run conduct the headloos tests over very much longer runs of pipe and use either plant or pump station flow meters or change in tank level to determine flow (Walski, 1999).

14.3.2 Distribution of Nodal Demands

The second major parameter determined in calibration analysis is the average demand (steady-state analysis) or temporally varying demand (extended-period analysis) to be assigned to each junction node. Initial average estimates of nodal demands can be obtained by identifying a region of influence associated with each junction node, identifying the types of demand units in the service area, and multiplying the number of each type by an associated demand factor. Alternatively, the estimate can be obtained by identifying the area associated with each type of land use in the service area, then multiplying the area of each type by an associated demand factor. In either case, the sum of these products will provide an estimate of the demand at the junction node.

14.3.2.1 Spatial distribution of demands. Initial estimates of nodal demands can be developed using various approaches depending on the nature of the data each utility has on file and how precise they want to be. One way to determine such demands is by employing the following strategy.

1. Determine the total system demand for the day to be used in model calibration, (*TD*). The total system demand may be obtained by performing a mass balance analysis for the system by determining the net difference between the total volume of flow which enters the system (from both pumping stations and tanks) and the total volume that leaves the system (through pressure reducing valves (PRVs) and tanks).

2. Use meter records for the day and try to assign all major metered demands (e.g., MD_j, where j = junction node number) by distributing the observed demands among the various junction nodes serving the metered area. The remaining demand will be defined as the total residual demand (*TRD*) and can be obtained by subtracting the sum of the metered demands from the total system demand:

$$TRD = TD - \Sigma \, MD_j \qquad (14.9)$$

3. Determine the demand service area associated with each junction node. The most common method of influence delineation is to simply bisect each pipe connected to the reference node, as shown in Fig. 14.7A.

4. Once the service areas associated with the remaining junction nodes have been determined, an initial estimate of the demand at each node should be made. This can be accomplished by identifying the number of different types of demand units within the service area, then multiplying the number of each type by an associated demand factor (Fig. 14.7B). Alternatively, the estimate can be obtained by identifying the area associated with each different type of land use within the service area, then multiplying the area of each type by an associated unit area demand factor (Fig. 14.7C). In either case, the sum of these products will represent an estimate of the demand at the junction node. Although in theory the first approach should be more accurate, the latter approach can be expected to be more expedient. Estimates of unit demand factors are normally available from various water resource handbooks (Cesario, 1995). Estimates of unit area demand factors can normally be constructed for different land use categories by weighted results from repeated applications of the unit demand approach.

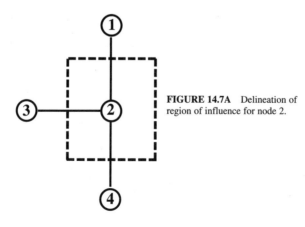

FIGURE 14.7A Delineation of region of influence for node 2.

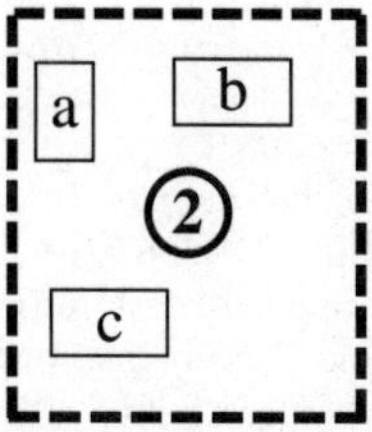

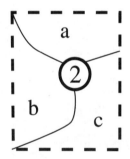

Type of Establishment	Units	Average Annual Demand (gpd/unit)	Maximum Daily Demand (gpd/unit)
a. metered residential	gpcd	70.00	140.00
b. garden apartment	gpd/unit	213.00	272.00
c. car wash	gpd/ft²	4.78	10.3

FIGURE 14.7B Demand assignment using individual units.

Type of Land Use (gpd/acre)	Unit Demand (acres)	Area (acres)	Total Demand (gpd)
a. metered residential	700	5	3500
b. garden apartament	600	4	2400
c. car wash	160,000	1	160000

FIGURE 14.7C Demand assignment using land use units.

5. Once an initial estimate of the demand has been obtained for each junction node j, (IED_j), a revised estimated demand (RED_j) can be obtained using the following equation:

$$RED_j = IED_j * TRD / \Sigma\, IED_j \qquad (14.10)$$

6. Finally, with the revised demands obtained for each junction node, the final estimate of nodal demand can be achieved by adding together both the normalized demand and the metered demand (assuming there is one) associated with each junction node:

$$D_j = RED_j + MD_j \tag{14.11}$$

14.3.2.2 Temporal distribution of demands. Time-varying estimates of model demands for use in extended-period analysis can be made in one of two ways, depending on the structure of the hydraulic model. Some models allow the user to subdivide the demands at each junction node into different use categories, which can then be modified separately over time using demand factors for water-use categories. Other models require an aggregate-use category for each node. In the latter case, spatial-temporal variations of nodal demands are obtained by lumping nodes of a given type into separate groups, which can then be modified uniformly using nodal demand factors. Initial estimates of either water-use category demand factors or nodal-demand factors can be obtained by examining historical meter records for various water-use categories and by performing incremental mass-balance calculations for the distribution system. The resulting set of temporal demand factors can then be fine-tuned through subsequent calibration of the model.

14.4 COLLECT CALIBRATION DATA

After model parameters have been estimated, the accuracy of the model parameters can be assessed. This is done by executing the computer model using the estimated parametric values and observed boundary conditions and by comparing the model results with the results from actual field observations. Data from fire-flow tests, pump-station flowmeter readings, and tank telemetric data are used most commonly in such tests.

In collecting data for model calibration, it is very important to recognize the significant impact of measurement errors. For example, with regard to calibrating pipe roughness, the C factor may expressed as:

$$C = k(V + error)/(h + error)^{0.54} \tag{14.12}$$

If the magnitude of V and h are on the same order of magnitude as the associated measurement errors (for V and h) then the collected data will be essentially useless for model calibration. That is to say, virtually any value of C will provide a "reasonable" degree of model calibration (Walski, 1986). However, one can hardly expect a model to accurately predict flows and pressures for a high stress situation (i.e. large flows and velocities) if the model was calibrated using data from times when the velocities in the pipes were less than the measurement error (e.g. less than 1 ft/s). The only way to minimize this problem is to either insure that the measurement errors are reduced or the velocity or headloss values are significantly greater than the associated measurement error. This latter condition can normally be met either using data from fire flow tests or by collecting flow or pressure reading during periods of high stress (e.g., peak hour demand periods).

14.4.1 Fire-Flow Tests

Fire-flow tests are useful for collecting both discharge and pressure data for use in calibrating hydraulic network models. Such tests are normally conducted using both a normal pressure gauge (to measure both static and dynamic heads) and a pitot gauge (to calculate discharge). In performing a fire-flow test, at least two separate hydrants are selected for use in the data collection effort. One hydrant is identified as the pressure or residual hydrant and the other hydrant is identified as the flow hydrant. The general steps for performing a fire flow test can be summarized as follows (McEnroe et al., 1989):

1. Place a pressure gauge on the residual hydrant and measure the static pressure.
2. Determine which of the discharge hydrant's outlets can be flowed with the least amount of adverse impact (flooding, traffic disruption, and so on).
3. Make certain that the discharge hydrant is initially closed to avoid injury.
4. Remove the hydrant cap from the nozzle of the discharge hydrant to be flowed.
5. Measure the inside diameter of the nozzle and determine the type of nozzle (i.e, rounded, square edge, or protruding) to determine the appropriate discharge coefficient. (Fig. 14.6).
6. Take the necessary steps to minimize erosion or the impact of traffic during the test.
7. Flow the hydrant briefly to flush sediment from the its lateral and barrel.
8. If using a clamp-on pitot tube, attach the tube to the nozzle to be flowed, then slowly open the hydrant. If using a hand held pitot tube, slowly open the hydrant and then place the pitot tube in the center of the discharge stream, being careful to align it directly into the flow.
9. Once an equilibrium flow condition has been established, make simultaneous pressure readings from both the pitot tube and the pressure gauge at the residual hydrant.
10. Once the readings are completed, close the discharge hydrant, remove the equipment from both hydrants, and replace the hydrant caps.

To obtain sufficient data for an adequate model calibration, data from several fire flow tests must to collect be collected. Before conducting each test, it also is important to collect the associated system boundary condition data. This includes information on tank levels, pump status, and so forth. To obtain an adequate model calibration, it is normally desirable for the difference between the static and dynamic pressure readings measured from the residual hydrant to be at least 35 kPa (5 psi), preferably with a drop of 140 kpa (20 psi) (Walski, 1990a). In the event that the discharge hydrant does not allow sufficient discharge to cause such a drop, it may be necessary to identify, instrument, and open additional discharge hydrants.

In some instances, it may also be beneficial to use more than one residual hydrant (one near the flowed hydrant and one off the major main from the source). The information gathered from such additional hydrants can sometimes be very useful in tracking down closed valves (Walski, (1999).

14.4.2 Telemetric Data

In addition to static test data, data collected over an extended period (typically 24 h) can be useful when calibrating network models. The most common type of data will include flow-rate data, tank water-level data, and pressure data. Depending on the level of instrumentation and telemetry associated with the system, much of the data may already have been collected as part of the normal operations. For example, most systems collect and record tank levels and average pump station discharges on an hourly basis. These data are especially useful to verify the distribution of demands among the various junction nodes. If such data are available, they should be checked for accuracy before using them in the calibration effort. If such data are not readily available, the modeler may have to install temporary pressure gauges or flowmeters to obtain the data. In the absence of flowmeters in lines to tanks, inflow or discharge flow rates can be inferred from incremental readings of the tank level.

14.4.3 Water-Quality Data

In recent years, both conservative and nonconservative constituents have been used as tracers to determine the travel time through various parts of a water distribution system (Cesario, et al., 1996; Grayman, 1998; Kennedy et al., 1991). The most common type of tracer for such applications is fluoride. By controlling the injection rate at a source, typically the water treatment plant, a pulse can be induced into the flow that can then be monitored elsewhere in the system. The relative travel time from the source to the sampling point can be determined. The measured travel time thus provides another data point for use in calibrating a hydraulic network model.

Alternatively, the water distribution system can be modeled using a water-quality model such as EPANET (Rossman, 1994). In this case, the water quality-model is used to predict tracer concentrations at various points in the system. Since the result of all water-quality results depend on the underlying hydraulic results, deviations between the observed and predicted concentrations can thus provide a secondary means of evaluating the adequacy of the underlying hydraulic model. For additional insights into water-quality modeling and the use of such models in calibration, refer to Chap.9.

14.5 EVALUATE THE RESULTS OF THE MODEL

In using fire-flow data, the model is used to simulate the discharge from one or more fire hydrants by assigning the observed hydrant flows as nodal demands within the model. The flows and pressures predicted by the model are then compared with the corresponding observed values in an attempt to assess the accuracy of the model. In using telemetric data, the model is used to simulate the variation of tank water levels and system pressures by simulating the operating conditions for the day over which the field data was collected. The predicted tank water levels are then compared with the observed values in an attempt to assess the model's accuracy. In using water-quality data, the travel times (or constituent concentrations) are compared with model predictions in an attempt to assess the model's accuracy.

The accuracy of the model can be evaluated using a variety of criteria. The most common criterion is absolute pressure difference (normally measured in psi) or relative pressure difference (measured as the ratio of the absolute pressure difference to the average pressure difference across the system). In most cases, a relative pressure difference criterion is usually preferred. For extended-period simulations, comparisons are normally made between the predicted and observed tank water levels. To a certain extent, the desired level of model calibration will be related to the intended use of the model. For example, a higher level of model calibration will normally be required for analysis of water quality or an operational study rather than use of the model in a general planning study. Ultimately, the model should be calibrated to the extent that the associated application decisions will not be affected significantly. In the context of a design application, the model should normally be calibrated to such an extent that the resulting design values (e.g., pipe diameters and tank and pump sizes or locations) will be the same as if the exact parameter values were used. Determining such thresholds often requires the application of model sensitivity analysis (Walski, 1995).

Because of the issue of model application, deriving a single set of criteria for a universal model calibration is difficult. From the authors' perspective, a maximum deviation of the state variable (i.e., pressure grade, water level, flow rate) of less than 10 percent is generally satisfactory for most planning applications, whereas while a maximum deviation of less than 5 percent is highly desirable for most design, operation, or water quality applications. Although no such general set of criteria has been officially developed for the United States, a set of "Performance Criteria" has been developed by the Sewers and Water Mains Committee of the Water Authorities Association (1989) in the United Kingdom. For steady-state models, the criteria are as follows:

1. Flows agree to 5 percent of measured flow when flows are more than 10 percent of total demand, and to 10 percent of measured flow when flows are less than 10 percent of total demand.
2. Pressures agree to 0.5 m (1.6 ft) or 5 percent of headloss for 85 percents of test measurements, to 0.75m (2.31 ft) or 7.5 percent of headloss for 95 percent of test measurements, and to 2 m (6.2 ft) or 15 percent of headloss for 100 percent of test measurements.

For extended-period simulation, the criteria require that three separate steady-state calibrations must be performed for different time periods and that the average volumetric difference between measured and predicted reservoir storage must be within 5 percent. Additional details can be obtained directly from the Water Authorities Asociation's report (1989).

Deviations between the results of the model application and the field observations can be caused by several factors, including (1) erroneous model parameters (e.g. pipe-roughness values and nodal demand distribution), (2) erroneous network data (e.g. pipe-diameters or lengths), (3) incorrect network geometry (e.g. pipes connected to the wrong nodes), (4) incorrect pressure zone boundary definitions, (5) errors in boundary conditions (e.g. incorrect PRV value settings, tank water levels, pump curves), (6) errors in historical operating records (e.g. pumps starting and stopping at incorrect times), (7) measurement equipment errors (e.g. pressure gauges not properly calibrated), and (8) measurement errors (e.g. reading the wrong values from instruments). It is hoped that the last two sources of errors can be eliminated, or minimized at least, by developing and implementing a careful data-collection effort. Eliminating the remaining errors frequently requires the iterative application of the last three steps of the model calibration process—macro-level calibration, sensitivity, and micro-level calibration. Each of these steps is described in the following sections.

14.6 PERFORM A MACRO-LEVEL CALIBRATION OF THE MODEL

In the event that one or more of the measured state variable values differ from the modeled values by an amount that is deemed to be excessive (i.e., greater than 30 percent), the cause of the difference is likely to extend beyond errors in the estimates for either the pipe-roughness values or the nodal demands. Although such differences have many possible causes, they may include (1) closed or partially closed valves, (2) inaccurate pump curves or tank telemetry data, (3) incorrect pipe sizes (e.g., 6 in instead of 16 in), (4) incorrect pipe lengths, (5) incorrect network geometry, and (6) incorrect pressure zone boundaries, (Walski, 1990a).

The only way to address such errors adequately is to review the data associated with the model systematically to ensure the model's accuracy. In most cases, some data will be less reliable than others. This observation provides a logical place to begin an attempt to identify the problem. Model sensitivity analysis provides another means of identifying the source of the discrepancy. For example, if one suspects that a valve is closed, this assumption can be modeled by simply closing the line in the model and evaluating the resulting pressures. Potential errors in pump curves can sometimes be minimized by simulating the pumps with negative inflows set equal to observed pump discharges (Cruickshank and Long, 1992). This of course assumes that the error in the observed flow rates (and the induced head) are less that the errors introduced by using the pump curves. In any case, only after the model results and the observed conditions are within some reasonable degree of correlation (usually less than a 20 percent error) should the final step of micro-level calibration be attempted.

14.7 PERFORM A SENSITIVITY ANALYSIS

Before attempting a micro-level calibration, it is helpful to perform a sensitivity analysis of the model to identify the most likely source of model error. This analysis can be accomplished by varying the different model parameters by different amounts, then measuring the associated effect. For example, many current network models have as an analysis option the capability to make multiple simulations in which global adjustment factors can be applied to pipe-roughness values or nodal-demand values. By examining such results, the user can begin to identify which parameters have the most significant impact on the model results and thereby identify potential parameters for subsequent fine-tuning through micro-level calibration.

14.8 PERFORM A MICRO-LEVEL CALIBRATION OF THE MODEL

After the model results and the field observations are in reasonable agreement, a micro-level model calibration should be performed. As discussed previously, the two parameters adjusted during this final calibration phase normally will include pipe roughness and nodal demands. In many cases, it may be useful to break the micro calibration into two separate steps: steady-state calibration, and extended-period calibration. In a steady-state calibration, the model parameters are adjusted to match pressures and flow rates associated with static observations. The normal source of such data is fire-flow tests. In an extended-period calibration, the model parameters are adjusted to match time-varying pressures and flows as well as tank water-level trajectories. In most cases the steady state calibration is more sensitive to changes in pipe roughness, whereas the extended-period calibration is more sensitive to changes in the distribution of demands. As a result, one potential calibration strategy would be to fine-tune the pipe-roughness parameter values using the results from fire-flow tests and then try to fine-tune the distribution of demands using the flow-pressure-water level telemetric data.

Historically, most attempts at model calibration have typically used an empirical or a trial-and-error approach. However, such an approach can be extremely time-consuming and frustrating when dealing with most typical water systems. The level of frustration will, of course, depend to some degree on the modeler's expertise, the size of the system, and the quantity and quality of the field data. Some of the frustration can be minimized by breaking complicated systems into smaller parts and calibrating the model parameters using an incremental approach. Calibration of multitank systems can sometimes be facilitated by collecting multiple data sets with all but one of the tanks closed (Cruickshank and Long, 1992). In recent years, several researchers have proposed different algorithms for use in automatically calibrating hydraulic network models. These techniques have been based on the use of analytical equations (Walski, 1983), simulation models (Boulos and Ormsbee, 1991; Gofman and Rodeh, 1981; Ormsbee and Wood, 1986; Rahal et al., 1980) and optimization methods (Coulbeck, 1984; Lansey and Basnet, 1991; Meredith, 1983; Ormsbee, 1989; and Ormsbee et al., 1992).

14.8.1 Analytical Approaches

In general, techniques based on analytical equations require significant simplification of the network through skeletonization and the use of equivalent pipes. As a result, such techniques may only get the user close to the correct results. Conversely, both simulation and optimization approaches take advantage of using a complete model.

14.8.2 Simulation Approaches

Simulation techniques are based on the idea of solving for one or more calibration factors through the addition of one or more network equations. The additional equation or equations are used to define an additional observed boundary condition (such as fire-flow discharge head). With the addition of an extra equation, an additional unknown can be determined explicitly.

The primary disadvantage of simulation approaches is that they can handle only one set of boundary conditions at a time. For example, in applying a simulation approach to a system with three different sets of observations—all of which were obtained under different boundary conditions (e. g.) different tank levels or pump statuses-three different results can be expected. Attempts to obtain a single calibration result will require one of two application strategies: a sequential approach or an average approach. In the sequential approach, the system is subdivided into multiple zones, the number of which will correspond to the number of sets of boundary conditions. In this case, the first set of observations is used to obtain calibration factors for the first zone. These factors are then fixed, another set of factors is determined for the second zone, and so on. In the average approach, final calibration factors are obtained by averaging the calibration factors for each individual calibration application.

14.8.3 Optimization Approaches

The primary alternative to the simulation approach is an optimization approach. When an optimization approach is used, the calibration problem is formulated as a nonlinear optimization problem consisting of a nonlinear objective function subject to both linear and nonlinear equality and inequality constraints. Using standard mathematical notation, the associated optimization problem can be expressed as follows:

Minimize $\quad\quad\quad z = f(\mathbf{X}) \quad\quad\quad (14.13)$

Subject to

$$\mathbf{g}(\mathbf{X}) = \mathbf{0} \quad\quad\quad (14.14)$$

$$L_h \leq h(X) \leq U_h = 0 \tag{14.15}$$

$$L_x \leq X \leq U_x \tag{14.16}$$

where **X** is the vector of decision variables (e.g., pipe–roughness coefficients, nodal demands),

$f(X)$ is the nonlinear objective function,

$g(X)$ is a vector of implicit system constraints,

$h(X)$ is a vector of implicit bound constraints, and

L and **U** are the vectors of lower and upper bounds respectively on the explicit system constraints and the decision variables.

Normally, the objective function will be formulated in a way that minimizes the square of the differences between observed and predicted values of pressures and flows. Mathematically, this can be expressed as:

$$f(X) = a \sum_{j=1}^{J}(OP_j - PP_j)^2 + b \sum_{p=1}^{P}(OQ_p - PQq)^2 \tag{14.17}$$

where OP_j = the observed pressure at junction j, PP_j = the predicted pressure at junction j, OQ_p = the observed flow in pipe p, PQ_p = the predicted flow in pipe p, and a and b are normalization weights.

The implicit bound constraints on the problem may include both pressure-bound constraints and flow rate-bound constraints. These constraints can be used to ensure that the resulting calibration does not produce unrealistic pressures or flows as a result of the model calibration process. For a given vector of junction pressures **P** these constraints can be expressed mathematically as

$$L_P \leq P \leq U_P \tag{14.18}$$

Similarly, for a given vector of pipe flows **Q**, these constraints can be expressed as

$$L_Q \leq Q \leq U_Q \tag{14.19}$$

The explicit bound constraints can be used to set limits on the explicit decision variables of the calibration problem. Normally, these variables will include the roughness coefficient of each pipe and the demands at each node. For a given vector of pipe-roughness coefficients **C**, these constraints can be expressed as

$$L_C \leq C \leq U_C \tag{14.20}$$

Similary, for a given vector of nodal demands **D**, these constraints can be expressed as

$$L_D \leq D \leq U_D \tag{14.21}$$

The implicit system constraints include nodal conservation of mass and conservation of energy. The nodal conservation of mass equation $F_c(Q)$ requires that the sum of flows into or out of any junction node n minus any external demand D_j must be equal to zero. For each junction node j, this may be expressed as

$$F_c(Q) = \sum_{n \in \{j\}}^{N_j} Q_n - D_j = 0 \tag{14.22}$$

where N_j = the number of pipes connected to junction node j and $\{j\}$ is the set of pipes connected to junction node j.

The conservation of energy constraint $F_e(Q)$ requires that the sum of the line loss (HL_n) and the minor losses (HM_n) over any path or loop k, minus any energy added to the liquid by a pump (EP_n), minus the difference in grade between two points of known energy (DE_k) is equal to zero. For any loop or path k, this may be expressed as

$$F_e(Q) = \sum_{n \in \{j\}}^{N_k} (HL_n + HM_n - EP_n) - DE_k = 0 \tag{14.23}$$

where N_k = the number of pipes associated with loop or path k, and $\{k\}$ is the set of pipes associated with loop or path k. It should be emphasized that HL_n, HM_n, and EP_n are all nonlinear functions of the pipe discharge Q.

Although both the implicit and explicit bound constraints have traditionally been incorporated directly into the nonlinear problem formulation, the implicit system constraints have been handled using one of two different approaches. In the first approach, the implicit system constraints are incorporated directly within the set of nonlinear equations and are solved using normal nonlinear programming methods. In the second approach, the equations are removed from the optimization problem and are evaluated externally using mathematical simulation; Lansey and Basnet, 1991; Ormsbee, 1989). Such an approach allows for a much smaller and more tractable optimization problem because both sets of implicit equations (which constitute linear and nonlinear equality constraints to the original problem) can now be satisfied much more efficiently using an external simulation model (Fig. 14.7). The basic idea behind the approach is to use an implicit optimization algorithm to generate a vector of decision variables, which are then passed to a lower-level simulation model for use in evaluating all implicit system constraints. Feedback from the simulation model will include numerical values for use in identifying the status of each constraint as well as numerical results for use in evaluating the associated objective function.

Regardless of which approach is chosen, the resulting mathematical formulation must then be solved using some type of nonlinear optimization method. In general, three different approaches have been proposed and used: (1) gradient-based methods, (2) pattern-search methods, and (3) genetic optimization methods.

Gradient-based methods require either first or second derivative information to produce improvements in the objective function. Traditionally, constraints are handled using either a penalty method or the Lagrange multiplier method (Edgar and Himmelblau, 1988). Pattern search methods employ a nonlinear heuristic that uses objective function values only to determine a sequential path through the region of search (Ormsbee, 1986, Ormsbee and Lingireddy, 1995). In general, when the objective function can be differentiated explicitly with respect to the decision variables, the gradient methods are preferable to search methods. When the objective function is not an explicit function of the decision variables, as normally is the case with the current problem, then the relative advantage is not as great, although the required gradient information can still be determined numerically.

Recently, several researchers have begun to investigate the use of genetic optimization to solve such complex nonlinear optimization problems (Lingireddy and Ormsbee, 1998; Lingireddy et.al., 1995; Savic and Walters, 1995). Genetic optimization offers a significant advantage over more traditional optimization approaches because it attempts to obtain an optimal solution by continuing to evaluate multiple solution vectors simultaneously (Goldberg, 1989). In addition, genetic optimization methods do not require gradient information. Finally, because these methods use probabilistic transition rules as opposed to deterministic rules, they have the advantage of insuring a robust solution methodology.

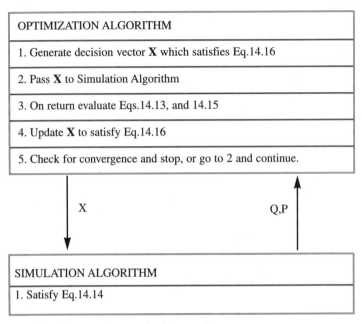

FIGURE 14.8 Bi-level computational framework.

Genetic optimization begins with an initial population of randomly generated decision vectors. For an application to network calibration, each decision vector could consist of a subset of pipe-roughness coefficients, nodal demands, and so on. The final population of decision vectors is then determined through an iterative solution method that uses three sequential steps: evaluation, selection, and reproduction. The evaluation phase involves determination of the value of a fitness function (objective function) for each element (decision vector) in the current population. On the basis of these evaluations, the algorithm then selects a subset of solutions for use in reproduction. The reproduction phase of the algorithm involves the generation of new offspring (additional decision vectors) using the selected pool of parent solutions. Reproduction is accomplished through the process of crossover in which the numerical values of the new decision vector are determined by selecting elements from two parent decision vectors. The viability of the solutions thus generated is maintained by random mutations that occasionally are introduced into the resulting vectors. The resulting algorithm is thus able to generate a whole family of optimal solutions and thereby increase the probability of obtaining a successful calibration of the model.

Although optimization in general and genetic optimization in particular offer very powerful algorithms for use in calibrations a water distribution model, the user should always recognize that the utility of the algorithms are very much dependent upon the accuracy of the input data. Such algorithms can be susceptible to convergence problems when the errors in the data are significant (e.g., headloss is on the same order of magnitude as the error in headloss). In addition, because most network model calibration problems are under-specified (i.e. roughness coefficients, junction demands) can give reasonable pressures if the system is not reasonably stressed when the data are collected.

14.9 FUTURE TRENDS

With the advent and use of nonlinear optimization, it is possible to achieve some measure of success in the area of micro-level calibration. Of course, the level of success will be highly dependent upon the degree that the sources of macro-level calibration errors have first been eliminated or at least significantly reduced. Although these sources of errors may not be identified as readily with conventional optimization techniques, it may be possible to develop prescriptive tools for these problems using expert system technology. In this case, general calibration rules could be developed from an experiential database that could be used by other modelers attempting to identify the most likely source of model error for a given set of system characteristics and operating conditions. Such a system also could be linked with a graphical interface and a network model to provide an interactive environment for use in model calibration.

In recent years, there has been a growing advocacy for the use of both geographic information systems (GIS) technology and Supervisor Control and Data Acquisition (SCADA) system databases in model calibration. GIS technology provides an efficient way to link customer's billing records with network model components for use in assigning initial estimates of nodal demands (Basford and Sevier, 1995). Such technology also provides a graphical environment for examining the network database for errors. Among the more interesting possibilities with regard to network model calibration is the development and implementation of an on-line network model through linkage of the model with an on-line SCADA system. Such a configuration provides the possibility for a continuing calibration effort in which the model is continually updated as additional data are collected through the SCADA system (Schulte and Malm, 1993).

Finally, Bush and Uber (1998) have recently developed three sensitivity-based metrics to rank potential sampling locations for use in model calibration. Although the documented sampling application was small, the approach the authors developed provides a potential basis for selecting improved sampling sites for improved model calibration. This area of research is expected to see additional activity in future years.

14.10 SUMMARY AND CONCLUSION

Network model calibration should always be performed before any network-analysis planning and design study is conducted. A seven-step methodology for network model calibration has been proposed. Historically, one difficult step in the process has been the final adjustment of pipe-roughness values and nodal demands through the process of micro-level calibration. With the advent of recent computer technology it is now possible to achieve good model calibration with a reasonable level of success. As a result, little justification remains for failing to develop good calibrated network models before conducting a network analysis. Future developments and applications of both GIS and SCADA technology as well as optimal sampling algorithms should lead to even more efficient tools.

REFERENCES

Basford, C., and C. Sevier, "Automating the Maintenance of a Hydraulic Network Model Demand Database Utilizing GIS and Customer Billing Records," *Proceedings of the 1995 AWWA Computer Conference*, Norfolk, VA, 1995, pp. 197–206.

Boulos, P., and L. Ormsbee, "Explicit Network Calibration for Multiple Loading Conditions," *Civil Engineering Systems*, 8: 153–160, 1991.

Brion, L. M., and L. W. Mays, "Methodology for Optimal Operation of Pumping Stations in Water Distribution Systems," *ASCE Journal of Hydraulic Engineering*, 117(11), 1991.

Bush, C. A., and J. G. Uber, "Sampling Design Methods for Water Distribution Model Calibration," *ASCE Journal of Water Resources Planning and Management*, 124:334–344, 1998.

Cesario, L., *Modeling, Analysis and Design of Water Distribution Systems*, American Water Works Association, Denver, CO, 1995.

Cesario, L., J. R. Kroon, W. M. Grayman, and G. Wright, "New Perspectives on Calibration of Treated Water Distribution System Models," *Proceedings of the AWWA Annual Conference*, Toronto, Canada, 1996.

Coulbeck, B., "An Application of Hierachial Optimization in Calibration of Large Scale Water Networks," *Optimal Control Applications and Methods*, 6:31–42, 1984.

Cruickshank, J. R, and S. J. Long, "Calibrating Computer Model of Distribution Systems," Proceedings of the 1992 AWWA Computer Conference, Nashville, TN, 1992.

Edgar, T. F., and D. M. Himmelblau, *Optimization of Chemical Processes*, McGraw Hill, New York, pp 334–342, 1998.

Gofman, E., and M. Rodeh, "Loop Equations with Unknown Pipe Characteristics," *ASCE Journal of the Hydraulics Division*, 107:1047–1060, 1981.

Goldberg, D. E., *Genetic Algorithms in Search, Optimization and Machine Learning*, Addison-Wesley, Reading, MA, 1989.

Grayman, W. M., "Use of Trace Studies and Water Quality Models to Calibrate a Network Hydraulic Model," in *Essential Hydraulics and Hydrology*, Haested Press, 1998.

Kennedy, M., S. Sarikelle, and K. Suravallop, "Calibrating Hydraulic Analyses of Distribution Systems Using Fluoride Tracer Studies," *Journal of the American Water Works Association*, 83(7):54–59, 1991.

Lamont, P.A., "Common Pipe Flow Formulas Compared with the Theory of Roughness," *Journal of the AWWA*, 73(5),274, 1981.

Lansey, K, and C Basnet, "Parameter Estimation for Water Distribution Networks," *ASCE Journal of Water Resources Planning and Management*, 117(1), 126-145, 1991

Lingireddy, S., L.E. Ormsbee, and D.J., Wood, User's Manual – KYCAL, Kentucky Network Model Calibration Program, Civil Engineering Software Center, University of Kentucky, Lexington, KY, 1995.

Lingireddy, S., and L. E., Ormsbee, "Neural Networks in Optimal Calibration of Water Distribution Systems," I. Flood and N. Kartam (eds.), *Artificial Neural Networks for Civil Engineers: Advanced Features and Applications*, American Society of Civil Engineers, p. 277, 1998.

McEnroe, B., D., Chase, and W. Sharp, "Field Testing Water Mains to Determine Carrying Capacity," *Technical Paper EL-89*, Environmental Laboratory of the Army Corps of Engineers Waterways Experiment Station, Vicksburg, MS, 1998.

Meredith, D. D., "Use of Optimization in Calibrating Water Distribution Models," *Proceeding of ASCE Spring Convention*, Philadelphia, PA, 1983.

Ormsbee, L.E., "Implicit Pipe Network Calibration," *ASCE Journal of Water Resources Planning and Management*, 115(2):243–257, 1989.

Ormsbee, L. E., "A nonlinear heuristic for applied problems in water resources," *Proceedings of the Seventeenth Annual Modeling and Simulation Conference*, University of Pittsburgh, 1986, pp. 1117–1121.

Ormsbee, L. E., D.V. Chase, and W. Grayman, "Network Modeling for Small Water Distribution Systems," *Proceedings of the AWWA 1992 Computer Conference*, Nashville, TN, 1992, pp. 15–19.

Ormsbee, L., D. V. Chase, and W. Sharp, "Water Distribution Modeling", *Proceedings of the 1991 AWWA Computer Conference*, Houston, TX, April 14–17, 1991, pp. 27–35.

Ormsbee, L. E. and D. V. Chase, "Hydraulic Network Calibration Using Nonlinear Programming," *Proceedings of the International Symposium on Water Distribution Modeling*, Lexington, KY, 1988, pp. 31–44.

Ormsbee, L. E. and S. Lingireddy, "Nonlinear Heuristic for Pump Operations," ASCE *Journal of Water Resources Planning and Management*, 121 (4):302–309., 1995.

Ormsbee, L. E., and D.J. Wood, "Explicit Pipe Network Calibration," *ASCE Journal of Water Resources Planning and Management*, 112(2):166–182, 1986.

Rahal, C. M., M.J.H. Sterling, and B. Coulbeck, "Parameter tuning for Simulation Models of Water Distribution Networks," *Proceedings of the Institution of Civil Engineers*, London, UK, 69(2):751–762, 1980.

Rossman, L., EPANET User's Manual, Drinking Water Research Division, Risk Reduction Engineering Laboratory, Cincinnati, OH, 1994.

Savic, D. A., and G.A. Walters, Genetic Algorithm Techniques for Calibrating Network Models, Report No. 95/12, Center for Systems and Control, University of Exeter, UK, 1995.

Schulte, A. M., and A. P. Malm, "Integrating Hydraulic Modeling and SCADA Systems for System Planning and Control," *Journal of the American Water Works Association*, 85(7):62–66, 1993.

Walski, T. M. "Standards for model calibration," *Proceedings of the 1995 AWWA Computer Conference*, Norfolk, VA, pp. 55–64, 1995.

Walski, T. M., "Sherlock Holmes Meets Hardy Cross, or Model Calibration in Austin, Texas, *Journal of the American Water Works Association,* 82(3):34, 1990a.

Walski, T. M., *Water Distribution Systems: Simulation and Sizing*, Lewis Publishers, Chelsea, MI, 1990b.

Walski, T. M., *Analysis of Water Distribution Systems*, Van Nostrand Reinhold, New York, 1984.

Walski, T. M, "Technique for Calibrating Network Models," *ASCE Journal of Water Resources Planning and Management*, 109(4):360–372, 1983.

Water Authorities Association and WRc, *Network Analysis — A Code of Practice*, WRc, Swindon, UK, 1989.

Wood, D. J., Comprehensive Computer Modeling of Pipe Distribution Networks, Civil Engineering Software Center, College of Engineering, University of Kentucky, Lexington, KY, 1991.

CHAPTER 15
OPERATION OF WATER DISTRIBUTION SYSTEMS

Donald V. Chase
*Department of Civil and Environmental Engineering
and Engineering Mechanics, University of Dayton,
Dayton, OH*

15.1 INTRODUCTION

A water distribution system, like any large complex system, must be operated properly so that it performs at an acceptable level of service. Many water utilities use human operators whose primary function is to monitor the pulse of the water distribution system and provide system control when needed. When the characteristics of the water supply system begin to change—for example, when tank levels increase or pressures fall—the operator initiates an action to ensure that the system operates within reasonable bounds. For example, when tank water levels fall in a particular part of the system, the operator may place a pump into service. When pressures within another part of the system get too high, the operator may turn off a pump that serves the area. For complex systems, operators may even operate valves and regulators within the system so that pressures, flows, and tank levels are kept within acceptable limits.

This chapter details the general nature of water distribution system operations for water utilities across the United States. Of course, each water supply system will have its own unique characteristics that require special consideration from an operational perspective. The role of operations with regard to water quality and emergency response are detailed as well. Most water utilities now use some form of Supervisory Control and Data Acquisition (SCADA) systems in their daily operations. This chapter presents the nuts and bolts of SCADA. The use of SCADA systems in monitoring and controlling a system's behavior also is discussed.

The chapter does not discuss the role of maintenance in distribution system operations. Although preventive and emergency maintenance certainly is crucial to proper operation of any water distribution system. The chapter also does not discuss operations in water treatment plants that actually could require an entire chapter by itself. Instead, this chapter focuses on the actions that take place to ensure that sufficient volumes of water are delivered throughout the distribution system at an acceptable level of service.

As more and more water utilities become comfortable with technology and the use of technology in their daily operations, more and more water utilities may investigate the possibility of unattended operations. Unattended operations are discussed in this chapter, as are the advantages and disadvantages associated with automatic control. New technology also offers the ability to manage energy consumption and reduce the cost of operating the system. In this chapter, the role of optimal control models within the framework of system operations is presented.

Most simulation models capable of predicting the hydraulic, energy consumption, or water-quality characteristics of a distribution system require information about the system itself. Such information typically includes boundary conditions, such as tank levels, pump on/off status, and valve settings. Much of this information can be obtained from a SCADA system. In fact, there has been a significant amount of activity linking SCADA systems with analysis and control models. This chapter also discusses the fundamentals of linking analysis and control models with SCADA systems to assist with and improve operations.

There is a movement within the water works industry away from housing pockets of fragmented data used only by a single department toward using a centralized database shared by all departments within the utility. This so-called data centric approach has the distinct advantage of ensuring that the most up-to-date information is used for a particular application. The use of a centralized database in water distribution operations will be discussed in this chapter. Finally, future trends in water system operations will be presented.

15.2 HOW SYSTEMS ARE OPERATED

Conceptually speaking, operating a water distribution system is not that difficult. All one needs to do is keep an eye on measurements of system performance whether the measurements are in the form of pressure, flow, or tank water levels. If a system operator notices that a quantity, such as pressure, is not within acceptable limits, appropriate action is taken to remedy the situation.

However, consider that the system may have four or five pressure zones and that each pressure zone may have multiple storage tanks or multiple pumping stations. Also consider that the zones may be hydraulically connected so that actions taken in one pressure zone may have an effect on other zones. Finally, consider that water system operations are inherently time-dependent and one may quickly agree that although the operations of a water distribution system appear to be simple, they can require a great deal of skill, especially if a system is large and complex.

Water distribution systems can be operated either manually or automatically. Many small systems in the United States have been operated automatically for years. Typically, operations are based on tank water levels. For example, a liquid-level switch senses the water level in a water storage tank. Field instrumentation sends a signal to a controller that will either turn a pump on or off, depending on the water level in the tank. Larger systems, because of their higher degree of complexity, normally have human operators whose primary function is to monitor the pulse of the system and initiate actions based on system behavior. The criteria that the operator uses to indicate whether the system is operating properly largely depends on what is measured throughout the system. Put another way, for many systems in the United States, the quantities that factor directly into operator decision-making will be measured.

15.2.1 Typical Operating Indexes

Water distribution systems are usually operated on the basis of pressure, flow rate, tank water levels, or combinations of the above. For an operator to know if he or she is operating the system in an acceptable manner, the parameters or indexes that form the basis of operations need to be measured. Accordingly, in many systems, instruments and equipment are used to measure, record, and store systemwide pressures, flows, and tank water levels. Other quantities, such as pump vibration or motor temperature, also can be measured, but they typically do not factor directly into system operations as it relates to an acceptable level of service.

In the United States, flow measurements are generally taken at only a few selected locations, including water treatment plants, pump stations, and boundaries with other systems. Some systems do not even record plant and pump-station discharges. Like flow, pressures also are recorded at a few key locations, usually at pump stations. In addition, pressures may be recorded at the highest and lowest elevations within the system or at sites within a pressure zone to determine the lowest and highest pressures in the system. Almost all water systems in the United States record tank water levels.

Many European systems, on the other hand, take pressure and flow rate measurements throughout the entire system. In Paris, for example, pressures and flows are measured at more than 30 locations, (Gagnon and Bowen, 1996). Not only can comprehensive systemwide measurements assist in system operations, but these data can be used to help calibrate computer models of hydraulics and water quality in a distribution system.

15.2.2 Operating Criteria

For systems whose operations are based on pressure, operators typically operate pumps and possibly valves so that systemwide pressures are maintained within acceptable limits. Although what is considered to be acceptable may vary from system to system, pressures in most cases should be kept above 207 kPa (30 psi) and below 689 kPa (100 psi) during normal operations. Pressures much greater than 689 kpa (100 psi) tend to waste water through leaks and could damage residential and commercial plumbing systems or possibly cause main breaks. A pressure of 207 kPa (30 psi) allows water to be supplied to the top floors of a multistory building.

During emergency conditions, such as a fire, pressures should be maintained above 138 kPa (20 psi) throughout the entire system. The 138 kPa minimum for fire flows is a generally accepted rule of thumb and provides enough pressure to supply the suction side of pumps on a fire pumper truck. More important, pressures of 138 kPa can help prevent contamination of the potable supply from cross-connections. During main breaks, when the pressure can drop below 138 kPa, it is not uncommon for the water utility to issue a "boil water" advisory because of the possibility of system contamination from cross-connections. In an effort to protect public health, many states and communities have adopted minimum pressure requirements.

Depending on the nature of the water supply system, minimum pressures greater than 207 kPa (30 psi) may have to be maintained at certain locations within the system. For example, certain industries or hospitals may require a minimum pressure above 207 kPa so that equipment within the facility will function properly. If the water system sells bulk amounts of water to an adjacent community, that community may require the water to be supplied at a minimum pressure of 345 kPa (50 psi) or higher. Operators must consider such unique circumstances in their operating decisions.

Flow also can be used as a parameter to control a water distribution system. Many systems measure flow at pumping stations and at interconnections with other systems. What constitutes an acceptable range of flows generally is be dictated by the nature of the water distribution system. For example, if the purpose of the system is to sell bulk amounts of water to neighboring communities, operators need to ensure that sufficient volumes of water are delivered to the system's customers. In addition, the water may have to be delivered at or higher than a specified pressure.

Flow and pressure are directly related to one another. When the flows in a pipeline increase, the pressure at the end of the line will decrease. Therefore, although some operators may operate the system according to pressure, one also can think of those operators as operating the system according to flow. In other words, when the pressure in part of the system falls below acceptable limits, it does so because the usage in that part of the system is most likely to be high. Consequently, flows and pressures into that part of the system must be increased by placing a pump into service. Operators also can control valves to direct water to areas where it is needed, but this usually is not done in municipal water distribution systems.

Among the more important parameters that an operator monitors is perhaps the water level in system tanks. Tank levels can provide an indication of the overall pressure throughout a pressure zone or even the entire system. Generally speaking, the higher the tank level, the higher the system pressure. In fact, operations in many systems are based solely on tank levels. For example, over time an operator may have developed an intrinsic feel for systemwide pressures as a function of the level in one or more storage facilities.

Operators must ensure that sufficient volumes of water are stored in tanks at all times in the event of an emergency, such as a fire, power failure, or source outage. In fact, common operating practice in the recent past (and possibly even today in some systems) was to keep storage tanks as full as possible at all times. However, for the most part, operators now recognize the relationship between storage tank water levels and the water quality in the tank. As a result, they usually try to provide some change in tank levels over the course of a day.

15.2.3 Water Quality and Operations

Within the past decade, there has been a much greater awareness of water quality within water distribution systems. This increased awareness has been driven in part by new federal regulations mandating that water-quality standards must be met at the customer's tap. Although hydraulic performance remains the primary basis by which operators make their decisions, more and more attention is being paid to water-quality behavior in the distribution system.

Operations provide a great opportunity to affect water quality in existing distribution systems. For example, through their actions, operators can directly influence tank water levels and pumping operations. Through these actions, they may be able to bring fresh water from a treatment plant and direct it toward a certain part of the service area. Operators in the United States generally do not operate valves within the system to direct water to specific parts of the system. However, in the future such an approach may offer more control than traditional means of operating water distribution systems.

In the near future, more water supply systems may consider using in-line booster disinfection stations or possibly even mini in-line treatment plants in an effort to improve overall water quality in the system. Because these elements are located within the distribution system, their operation will become the system operator's domain. Chemical feed rates would be monitored and controlled by the operator to maximize water quality.

15.2.4 Emergency Operations

Possibly, the real reason that human operators are used in larger systems is to respond to such emergencies as fires, main breaks, source contamination, source outage, or power failures. Operators must be able to respond to any emergency that arises and ensure that system performance remains at an acceptable level of service.

During fires, operators may place more pumps into service to deliver higher rates of flow out into the system. During a power failure one of the operator's tasks may be to place a diesel or natural gas generator into service so that pumps can continue to operate. For many systems, however, backup generators automatically enter into service when a power failure occurs. During contamination of a source, the operator may have to close valves to isolate part of the system. Needless to say, emergency operations are an extremely important component of the operator's duties.

15.3 MONITORING OF SYSTEM PERFORMANCE WITH SCADA SYSTEMS

As discussed above, water utilities typically use human operators to monitor the pulse of the water distribution system. To do their job, operators need information about tank levels, pressures, flows, and so forth. For most utilities, SCADA systems—also called telemetry—provide this information. A SCADA system is a collection of field instrumentation, communications systems, and hardware and software systems that permit a system's behavior to be monitored and controlled, typically from a remote site (ASCE,1991). The following example provides a quick summary of the functionality of a SCADA system.

Suppose that we have been asked to operate a water supply network that utilizes a SCADA system to provide monitoring and control. We have a computer screen in front of us that we use to view the status of the distribution system and its components. For example, we can cause the SCADA system to display current water levels in each elevated or ground-storage tank. Because some systems use touch screens, we might even be able to touch a tank on the screen and the SCADA system would draw a chart showing the water levels in the tank for the past 24 h.

Suppose that during our shift, we notice that the water level in a particular tank falls below half full. We know from experience that whenever water levels in this particular tank fall below half full, pressures in some parts of the pressure zone served by the tank are unacceptably low. So from our control panel, we place a booster station into service by pressing the "On" button for this pump station. In short order, we can see that the tank water level has begun to rise and, as a result, the pressures in the pressure zone are kept within acceptable levels.

Of course, the use of SCADA systems is not limited to the distribution system, nor is it limited to the water works industry. Some water utilities have SCADA systems that provide process monitoring and control for both the water treatment facilities and the water distribution system. In fact, SCADA systems are used wherever there is a need to monitor or control a process. This spans many other industries, including other utilities. The common theme is to monitor the behavior of a process or a system and to feed that information back to a central location where decisions can be made and actions can be taken.

Much of the boundary information for a water distribution system hydraulic model can be obtained from a SCADA system. Such information might include tank water levels, pump on/off status, pump speed, and valve status. If a utility wishes to conduct realtime simulations, in support of an emergency response, for example, then up-to-date boundary information can be obtained from the SCADA system quickly and easily. Another valuable use of the information provided by a SCADA system is model calibration. Since many SCADA systems archive data for some period of time, historical information can be obtained.

15.3.1 Anatomy of a SCADA System

As mentioned above, a SCADA system typically consists of field instrumentation, communications, and hardware and software systems that allow remote monitoring and control. Figure 15.1 presents a general schematic of the individual elements found in a typical SCADA system used in water distribution. The purpose of the field instrumentation is to collect information on the state of the hydraulic system. Such instrumentation may include programmable logic controllers (PLCs), remote terminal units (RTUs), liquid-level switches, or other instruments (AWWA, 1983). These devices are capable of measuring and recording system indexes, such as pressure, flows, or tank water levels. In some cases, these devices are capable of providing localized control in the event of a communications failure.

In the past, RTUs generally were used to collect field data and to send these data to a central computer. The on-site control capabilities of RTUs were limited. PLCs, on the other hand, typically were used to provide some type of localized control, but they had limited data collection and storage features. Today, because the capabilities of RTUs and PLCs are merging, they provide similar functionality. Current RTUs can be programmed to provide some localized control, whereas current PLCs can store data and exchange this information with a central computer. An example of an RTU is shown in Fig. 15.2.

Regardless of whether control is maintained at the local level or from a central location, or no matter whether data are transmitted to a master computer or kept at a local site, quantities must be measured in the field. Quantities that typically are measured in water distribution systems include pressure, flow, and tank water levels. Therefore, some type of measuring device such as a pressure transducer, a flow transducer or a liquid-level

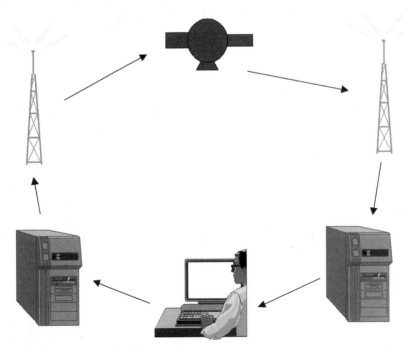

FIGURE 15.1 Elements of a SCADA system.

Operation of Water Distribution Systems 15.7

FIGURE 15.2 Remote terminal unit (RTU) (Courtesy ATSI, Inc.).

FIGURE 15.3 Pressure transducer (Courtesy ATSI, Inc.).

sensor must be installed. Figure 15.3 shows an example of a pressure transducer, and Fig. 15.4 shows an example of a device that collects flow data. Flow data also can be collected using Venturi tubes connected to pressure transducers.

The next link in a SCADA system is transmitting information collected by the field units to a central location. Communications can be accomplished using telephone lines, fiber optics, microwave, radio, or satellite. Each type of communication has features that make it suitable for a particular application. For example, microwaves may be more

FIGURE 15.4 Example of flow measuring device (Courtesy of ATSI, Inc.).

15.8 Chapter Fifteen

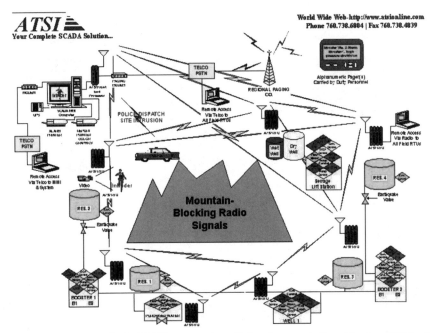

FIGURE 15.5 Schematic of a SCADA system using radio-based communications. (Courtesy ATSI, Inc.)

reliable than telephone lines. However, microwaves may not be suitable for systems having significant elevation differences since the transmitter, receiver, and relay stations must be in visual contact with one another. Figure 15.5 shows some of the components of a radio-based SCADA system.

Information that is transmitted by the communications system is generally sent to a central location where the operations staff reside. For digital SCADA systems, the information is collected by a receiver and is delivered directly to a computer system. For analog systems, the data must somehow be provided to the computer system. Software on the computer system provides the man-machine interface (MMI), which enables operators to monitor the system visually, typically from a central console. The MMI also provides functionality to allow system operators to control field units. In systems using analog monitoring and control, the information received from the field usually is delivered to circular charts or strip charts.

Because the SCADA system is such a vital component of the overall operations of a water distribution system, reliability is extremely important. Many water utilities recognize this and use redundant systems. Frequently, two computer systems are used: One functions as the primary SCADA computer and the other supports some monitoring and control features. If the primary computer fails for whatever reason, the secondary computer takes its place.

The water works industry is certainly moving toward digital monitoring and control; in fact, a large number of utilities control their systems using digital SCADA systems with personal computers. The information is often displayed in a graphical format that makes it easier for operators to visualize what is going on throughout their system.

An equally important component to the monitoring portion of the SCADA system is the ability to control field elements from a central location. If a pump needs to be placed into service or a valve must be closed, the operator initiates the action at a central location sending a signal through the communications link back to the remote site. Field units receive and interpret the signal and implement the requested action.

Another common feature of SCADA systems involves alarm recording. Many elements in the system may fail, but the failure may not be catastrophic. For example, when a storage tank overflows, the system continues to operate. Of course, an overflowing storage tank is undesirable; therefore, the SCADA system sounds an alarm indicating a problem with the tank. An operator can then take corrective action. Some SCADA systems even have the ability to telephone specified individuals, such as the director of operations, and notify them of an alarm condition.

15.3.2 Data Archiving

Many of today's SCADA systems offer data retrieval features that allow historical information describing the performance of the system to be displayed. For example, data describing tank levels, pump status, and system pressures during a main break that occurred several weeks ago might be able to be displayed on the console with the push of a button. Storing historical SCADA information can require a tremendous amount of data storage. Consider a large system that may have as many as 100 individual elements that must be monitored. Now consider that data on each element may be delivered to the central location every 30 s. One can see that a large amount of information can be generated even for a single day.

Data archiving can be a valuable asset when training operators, or hindcasting and possibly for litigation. For instance, the actual response of the operations staff can be cataloged and retrieved at a later date to determine whether the appropriate course of action was taken. Similarly, information about the behavior of the system in response to a particular emergency can be used to train new operators. Finally, data on system performance can be used to calibrate mathematical models that simulate system performance.

15.4 CONTROL OF WATER DISTRIBUTION SYSTEM

Several items in a water distribution system can be controlled by an operator, but by far the most common elements are system pumps. Pumps can be placed on-line or be taken out of service at high-service pump stations or smaller booster stations. High-service pump stations generally deliver water from water treatment plants, or possibly from ground reservoirs, out into the distribution system. They act as a point of entry for water into the distribution system. Booster pumps, on the other hand, are usually in-line pumps whose function is to boost pressures or flows to a particular location in the system. A common application of booster pumps is at the interface between two pressure zones.

Another common element that is controlled are valves—usually those on the discharge side of pumps. Although pumps are commonly started against closed valves, some units are started with the discharge valves open. Check valves usually are used in these cases to prevent backflow through the pumps. Pump discharge valves must be opened slowly to prevent line surges or waterhammer from occurring. Operators also can operate valves to control the flow into or out of storage facilities.

15.4.1 Control Strategies

Several methods of controlling water distribution systems are available, each representing an increasing level of automation. The American Water Works Association Research Foundation (AWWARF) recently supported a study for water treatment facilities that identified three levels of control (Younkin and Huntley, 1996). The three levels of control also can be adapted to water distribution systems, as described in the following sections.

15.4.1.1 Supervisory control. Many water utilities in the United States are operated today by supervisory control. A human operator monitors the behavior of the water distribution system 24 h a day, 7 days a week to make certain the system is operating properly. The operator makes decisions based on his or her knowledge and experience—sometimes gained over a long period. These decisions are then implemented manually by adjusting controls or pressing buttons.

15.4.1.2 Automatic control. This type of control represents the case where instrumentation and control equipment are used to control the distribution system automatically. Such control can be implemented either locally at the facility or throughout the system. Typically, simple operating rules are used to determine which component is operated and how it is operated. An example of automatic control described earlier is the use of liquid-level switches in tanks to control a pump's on/off status.

Coincident with automatic control is the idea of unattended operations. As the name implies, unattended operations have no human operator on duty. Smaller systems have used unattended operations for some time. In these cases, the on/off status of a pump usually is controlled by the water level in a storage tank. Because of their relatively simple nature, unattended operations seemed to be the natural way to operate small systems.

Automatic control is not limited to unattended operations. Human operators may be on duty 24 h a day, 7 days a week even though the system is operating automatically. In these cases, automatic control generally describes the use of computers and control logic to run the system while human operators remain on standby.

15.4.1.3 Advanced control. Systems that rely on advanced control use optimization algorithms, decision support systems, artificial intelligence, or control logic to control the distribution system. Usually, the methodologies used to develop control logic are much more complex and sophisticated than are those used in automatic control. Chapter 16 discusses the fundamentals of control models that can be used in operations. The use of advanced control and automatic control can be combined with one another, with the advanced-control algorithms supplying operating rules and the automatic-control features implementing the rules.

Given the capabilities of today's computer and control technology, a number of water utilities are investigating the possibility of completely automated and advanced control. Although process-monitoring and control technology has become increasingly reliable over the past several years, the primary driving force behind more sophisticated operations is cost reduction.

Personnel costs are the single largest item in the budgets of most water utilities. In fact, it may cost as much as $400,000 annually to staff even a small facility, (Younkin and Huntley, 1996). The next highest budget item after personnel costs are pumping costs. Automatic control can reduce staff requirements, thus reducing costs associated with personnel. Advanced control can reduce operating costs even further through the use of optimized pumping or operating schemes.

A disadvantage associated with automatic control is the perceived *lack of control*. Water utility operators in the United States seem to be extremely cautions. For the most part, they seem unwilling to allow a computer to operate their system. In some measure, this reluctance has hindered the development and subsequent use of advanced control. However, the AWWARF recently funded a study to establish a set of standards for software that will be capable of enabling advanced control.

15.4.2 Centralized Versus Local Control

Water distribution systems can be operated in one of two modes: from a central location or locally with control originating at the facility. Many systems implement a combination of the two methods in which centralized control is in place most of the time. Under emergency conditions such as a power or communications failure, localized control governs the system. Centralized control can be automated, although for many water distribution systems in the United States, humans oversee control of the systems.

In the case of *centralized control*, all decisions are made at a single location. Of course, all system parameters must be delivered to the central location to aid in decision making. Centralized control is straightforward because all control decisions usually are made by human operators. Alternatively, system elements may automatically be placed into or taken out of service according to predefined operating rules.

In the case of *localized control*, all control at a facility, such as a pump station or storage tank, takes place at the facility. Typically, control logic is built into the controllers so that appropriate action occurs. For example, suppose that the pressure at a booster pump station falls below a prescribed value. Furthermore, suppose that all pumps at the station are off. Controller logic at the station might cause the largest pump to be placed into service. If pressures at the pumping station continue to be unacceptable, the next largest pump can be placed into service and so on until pressures are within acceptable limits. Control logic can be specified to address such operational and maintenance constraints as pumps that are unavailable for service or pumps that have recently completed a cycle of operation.

15.5 LINKING OF SCADA SYSTEMS WITH ANALYSIS AND CONTROL MODELS

A recent development in the waterworks industry that will certainly see greater usage in the near future is the linkage of SCADA systems with other analysis and control models or decision support systems. Analysis and control models, such as hydraulic network, optimal control, and water-quality models, can be used by operations staff in a variety of ways, including operator training, emergency response, energy management, and water-quality behavior. (Chapter 16 describes control models in greater detail and discusses how they can be used in operations.)

- Operator Training
- Emergency Response
- Energy Management
- Water Quality Behavior

Hydraulic network, optimal control, and water-quality models require information on the current state of the system, usually in the form of boundary conditions and system loadings. Boundary information, such as tank levels, pump status, and valve settings, can be obtained directly from the SCADA system. Loading conditions describe the demands placed on the system. Although this information may not be available directly from the SCADA system, information provided by SCADA can be used to estimate system loads.

15.5.1 Data Requirements of Analysis and Control Models

The particular data needs of the analysis and control models will, of course, depend on what the model ultimately computes. In the case of hydraulic network models, distribution of pressure and flow throughout the hydraulic network is determined. This includes flows into and out of storage tanks and the operating characteristics of system pumps and valves. Many hydraulic models are capable of performing time simulations indicating that the performance of the system over a specified time horizon, such as 24 h.

Optimal control models can be used to indicate what pumps should be run and when they should be operated so that energy use and operating costs can be minimized. It is important to note that even though energy costs are reduced, the system must be operated at an acceptable level of service. An optimal control model should consider acceptable operating characteristics in its problem formulation.

Water-quality models can predict the concentrations of specified water-quality constituents throughout the distribution system. For example, these models can be used to determine the concentration of chlorine at various locations in the system. Other water-quality parameters of interest that these models can determine include the age and amount of water delivered from individual storage tanks and treatment plants. A hydraulic network model can supply much of the information needed by optimal control or water-quality models. In fact, many control and quality models are integrated with hydraulic network models.

A decision support system used by system operators can include a number of components, including hydraulic network models and optimal control models. Alternatively, a decision support system may consist only of general operating rules developed over many years of operating the distribution system. The difficulty with using general operating rules is that new pump station or tank construction could make the operating rules obsolete.

Information that can be supplied by a SCADA system and used directly in a hydraulic network model include tank and reservoir levels, pump on/off status, pump speed, valve status, and valve setting. Estimates of total system use also can be extracted from a SCADA system using a mass balance approach. If the SCADA system monitors high-service pump station flows and tank water levels, the total usage of system water can be determined from the expression below. Notice that this expression considers that multiple pumps can deliver flow into multiple storage tanks.

$$Q_{Sys} = \Sigma Q(j)_{Pump} + \Sigma \frac{(Level\ (k)_{t + \Delta t} - Level\ (k)_t) * \overline{Area\ (k)}}{\Delta t} \qquad (15.1)$$

where Q_{sys} = average system over time step Δt, $Q(j)_{Pump}$ = average discharge of pump j over time step Δt, level$(k)_{t + t}$ = water level in tank k at time step $t + \Delta t$, level$(k)_t$ = water level in tank k at time step t, and $\overline{area(k)}$ = average area of tank k over time step Δt.

As mentioned above, most available hydraulic network models can perform a time simulation that represents the temporal nature of the distribution system. When performing a time simulation, system demands at various times of the simulation must be supplied. For example, estimates of system demand may need to be supplied every hour during a 24 h simulation. A combination of the mass balance approach and a curve-fitting procedure can provide these estimates.

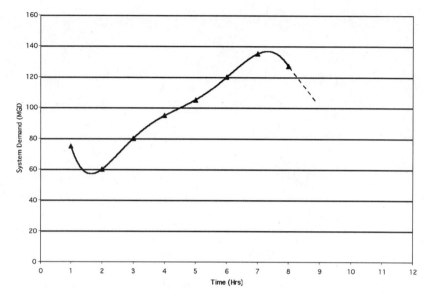

FIGURE 15.6 Projecting future hourly demands from past demands.

If one were to plot system demands over the course of a day, one would find that they are generally smooth and continuous. In other words, there would be no "kinks" in the plot of system demand. At any hour, estimates of system demand could be computed using a mass balance approach, as described above, once pump-flow and tank-level data have become available.

Assuming that demands are indeed smooth and continuous, a curve could be fitted through demand estimates of the past 3 or 4 h. This curve can then be used to extrapolate demands for the next time step. Figure 15.6 graphically shows this approach. Of course, as soon as estimates of actual system demands are available, they should be used to update the curve.

An alternate approach is to use an areawide demand adjustment (Schulte and Malm, 1993). Assuming that meters are placed on key mains within the distribution system and that this information is supplied to the SCADA system, a simple mass balance can be performed. Figure 15.7 illustrates the concept. The measured outflow in pipe P3 is subtracted from the measured inflows through pipes P1 and P2. The flow difference represents the usage in the area, which could be an entire pressure zone.

Another approach would be to use so-called "smart meters" at interconnections with other systems or at points of high water use. Telemetry located in the meter pit would transmit data on real-time water use back to a central location; the information also could be recorded on-site for later retrieval.

15.5.2 Establishment of the Link

Clearly, data that can be used by analysis and control models resides on the SCADA system. The question therefore is, how can data from the SCADA system be transferred

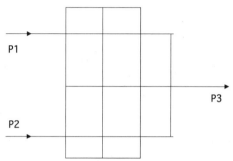

FIGURE 15.7 Mass balance approach for estimating water use.

to the analysis and control models? Similarly, how can results from the analysis and control models be sent back to the SCADA system? This latter step is necessary for comparison purposes or for implementation of automatic control.

Most if not all SCADA systems use databases to store information. A *database* is nothing more that a collection of information that may or may not be related. Possibly, the most common means of exchanging data stored in a database with other applications is to use the Open Database Connectivity (ODBC) interface. Roughly, ODBC consists of a set of software drivers that allow a computer program to exchange data with ORACLE, Microsoft Access, dBASE, Microsoft Excel, Microsoft FoxPro, Borland Paradox, or other software that uses the ODBC interface. Data are not actually exchanged between programs. Instead, programs that use the ODBC interface are capable of accessing the same database generated by the SCADA system.

Applications can manipulate the databases using Structured Query Language (SQL), a set of statements and commands supported by various programming languages, such as Visual Basic, C/C++, and others. A program can be written using SQL to perform a variety of functions on the data stored in the database.

For example, a database containing information on the behavior of the distribution system can be generated by a SCADA system. A hydraulic network model can access the database using ODBC. Using SQL statements built into the hydraulic network model the database can be manipulated to find the date representing maximum water use: that is, the maximum daily condition of the water distribution system. Again, using SQL, the hydraulic network model can extract the boundary conditions on the given date. The network model can then be executed to determine systemwide pressures and flows. This information can in turn be used to train operations staff about the expected behavior of the system during periods of high demand.

An alternate means of transferring data between a SCADA system and other applications is through the use files. Files adhering to a standard format are shared between the SCADA system and other software. Such an approach can be rigid and inflexible, especially if the SCADA system does not allow files to be generated in a user-specified format. A unique format usually is necessary for each piece of software that needs to supply data to the system measuring performance. In addition, if upgrades to new software are made, it is possible that a new format between programs will have to be established.

15.6 USE OF CENTRAL DATABASES IN SYSTEM CONTROL

Within the past decade, there has been a movement within the waterworks industry away from a fragmented data-management structure toward a *data-centric* approach in which all data are housed in a central database or data repository. The advantage of housing all data in a central database is obvious. Different software applications that use similar data can share the information. For example, both a hydraulic network model and a pipeline inventory program might require information about the size and length of water mains in the system. Similarly, both a SCADA system and a hydraulic model may use information describing tank water levels.

Housing information in a central location can ensure that the most recent data describing the water supply system are available to all users who may need it. Consider a design engineer making changes to as-built drawings that reflect modifications made to the distribution system as part of constructing a large industrial park. Because the as-built drawings depict what is in the ground, this information is extremely valuable to the modeling engineer in the planning department. Unless the modeling engineer is aware of the new pipe that was placed in the ground, there is no guarantee that the most recent network topology will be used in the hydraulic model.

A data-centric approach typically involves a client-server computer arrangement, as illustrated in Fig. 15.8. A single computer called a *server* houses all databases and also may contain the applications that use the data. Other computers called *clients* or *host computers* access the data from the server. Thus, multiple users may be able to access and modify the same database at the same time.

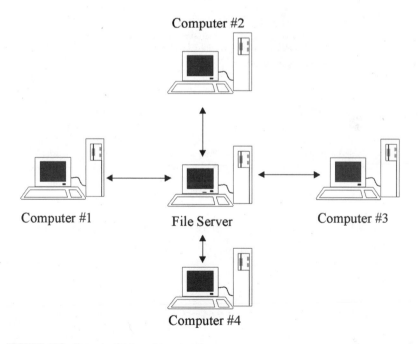

FIGURE 15.8 Example of data-centric approach.

15.7 WHAT THE FUTURE HOLDS

The future of operations in water distribution offers exciting potential, in large measure as a result of the rapid growth of monitoring and control technology as well as the use of advanced control in operations. The following paragraphs describe some of the developments the water distribution system operators may see in the not-to-distant future.

Several water utilities in the United States are using hydraulic network models in their operations, but few systems are actually allowing system operators to use the models. In the future, more and more distribution system operators will use hydraulic network models to assist them in their daily operations and in responding to emergencies. The key to more widespread use of simulation models by system operators is the availability of an easy-to-use graphical interface where operators can specify the "what-if" conditions. More and more systems will link analysis and control models with SCADA systems to permit real-time simulation and control of distribution systems. This linkage can address the need for a graphical interface described above. Much of the data exchange between the SCADA system and the analysis and control models can be transparent to the system operator.

More and more utilities will investigate the benefits associated with optimal control. The increased interest will be driven in part by deregulation of the electrical industry. These control models will allow the development of efficient operating strategies that can reduce energy consumption and associated operating costs. Improvements in control technology will make it easier to implement automatic control, thereby increasing the numbers of unattended operations.

REFERENCES

ASCE, L. E. Ormsbee (ed.), Energy Efficient Operation of Water Distribution Systems, Research Report No. UKCE9104, American Society of Civil Engineers, New York, June 1991.

AWWA, *Automation and Instrumentation*, AWWA Manual M2, American Water Works Association, Denver, CO, 1983.

Cesario, L., *Modeling, Analysis, and Design of Water Distribution Systems*, American Water Works Association, Denver, CO, 1995.

Gagnon, J. L., and P. T. Bowen, "Supply Safety and Quality of Distributed Water—A Contradiction Overcome by the Use of High Performance Models," *Proceedings of the AWWA Computer Conference*, Chicago, IL, 1966.

Schulte, A. M., and A. P. Malm, "Integrating Hydraulic Modeling and SCADA Systems for System Planning and Control," *Journal of the American Water Works Association*, 857, July, 1993.

Younkin, C. S., and G. Huntley, "Unattended Facilities Offer Competitive Advantage," *Proceedings of the AWWA Computer Conference*, Chicago, IL, 1996.

CHAPTER 16
OPTIMIZATION MODELS FOR OPERATIONS

Fred E. Goldman
Goldman, Toy, and Associates, Inc.
Phoenix, AZ

A. Burcu Altan Sakarya
Department of Civil Engineering
Middle East Technical University
Ankara, Turkey

Lindell E. Ormsbee
Department of Civil Engineering
University of Kentucky
Lexington, KY

James G. Uber
Department of Civil and Environmental Engineering
University of Cincinnati
Cincinnati, OH

Larry W. Mays
Department of Civil and Environmental Engineering
Arizona State University
Tempe, AZ

16.1 INTRODUCTION

The operation of a water system involves turning pumps on and off, regulating tank storage, providing disinfection, and delivering good-quality water to customers at a reasonable flow and pressure. The water utility relies on good engineering design and the skill and expertise of its management and operators to meet these goals reliably. Books of procedures and operation manuals are developed to provide standardization and to help staff improve quality control. The performance assistance documents are continually revised and updated as the water system is expanded or modified and when new technologies and methods are adopted. Even with these valuable tools, unpredictable events, such as line breaks, fires, and the dreaded failed coliform test, will occur. Pumps and valves randomly break down and compromise the system's integrity, resulting in interrupted service. The situation may have a remedy available in the performance documents, but, the experience of the operators usually is what saves the day.

Optimization models are being developed that can help the skilled management and personnel to improve the system's operation and respond to emergencies. As computers become more powerful and affordable, the transfer of these methods from research to practice becomes more feasible. Water system operations are extremely complicated. A typical system can have thousands of customers, hundreds of valves, scores of pumps and tanks, and many sources of water of different quality. With the advent of Supervisory Control and Data Acquisition (SCADA) the status of the system can be monitored at remote locations. Pump operation, tank levels, pressures, flows, and water quality can be monitored and reported, using telemetry, to a central location. Many pumping facilities have control systems that monitor pressure and flow and adjust the pump operations automatically to meet the system's requirements using programmed logic controllers. Telemetry can transmit data to a central location, where operations experts can monitor the situation and adjust pumping operations remotely according to the needs of the system. Similarly, chlorine booster stations can be operated remotely to adjust disinfection as necessary. Even with the most sophisticated telemetry, current state of the art for operation of water systems relies on the experience and skill of the operators.

Optimization models offer an effective tool to improve the operation of water systems. Computer models, such as KYPIPE and EPANET, are available to model a water system. The predicted pressures, flows, and water quality can be compared to measured values, and the system's characteristics can be adjusted until the model is calibrated manually or with optimization techniques. Mathematical and combinatorial optimization techniques can be linked with the system model to optimize the operation of a water system for energy efficient operation and water quality.

This chapter describes models that have been developed to optimize water system operations. The descriptions will be general in content but comprehensive enough to provide the reader with insight into the capabilities of each model. Using the references, readers can expand their knowledge of the techniques.

Water system operation is too complicated to expect that computer systems will be able to provide an "automatic pilot" for system operators. What can be expected is the development of computer methods that "learn" the system's characteristics and provide guidance to improve operators' response, resulting in more efficient operation and optimum and cost-effective modifications to the system.

Consider the problem of scheduling pump operation for the next 12 h based on the current status of storage and projected demand. Techniques are available to interpret the data and to offer a menu of pump schedules that will meet system requirements and reduce costs. Consider the situation in which a key pump or well is taken off–line for maintenance or simply fails. The same techniques can be used to provide a set of pump schedules that meet the system requirements efficiently with the available resources.

Finally, the potential uses of optimization techniques are not restricted to water quality or minimization of energy costs. Optimization techniques are available to choose pipes for rehabilitation or replacement, locate valves optimally for efficient operation, and find likely sites of system leakage. Other uses will be identified as computers become more powerful and researchers use their capabilities to improve water system operation.

16.2 FORMULATIONS FOR MINIMIZING ENERGY COST

16.2.1 Energy Management

Water distribution systems are controlled to satisfy various objectives, including hydraulic performance and economic efficiency. Measures of hydraulic performance include pressure levels, fire protection, water quality, and various measures of system reliability. Economic efficiency is influenced by such factors as general operation and maintenance costs and pumping costs. In conventional water supply systems, pumping of treated water represents the major fraction of the total energy budget. In ground-water systems, the pumping costs normally represent the major fraction of the total operating cost. Therefore, most optimal control strategies for water distribution systems have focused on minimizing such operational costs.

Regarding the minimization of pumping costs, the purpose of an optimal control system is to provide the operator with the least-cost operation policy for all pump stations in the water-supply system. The operation policy for a pump station is simply a set of rules or a schedule that indicates when a particular pump or group of pumps should be turned on or off over a specified time period. The optimal policy should result in the lowest total operating cost for a given set of boundary conditions and system constraints.

16.2.2 Management Strategies

When considering the operation of a water distribution system, at least three different levels of automated control can be identified. The lowest level of control is the computer-monitored command structure, or real-time monitoring, shown in Fig. 16.1. At the heart of this arrangement is a computer that collects and logs data, monitors the operational status of the system, and transmits operator's directives to various control devices in the field. The next level of automated control, shown in Fig. 16.2, is the computer-assisted command structure sometimes called real-time simulation. This control structure provides operators with an interactive environment incorporating a SCADA system linked with software capable of predicting the state of the hydraulic system. The computer-assisted

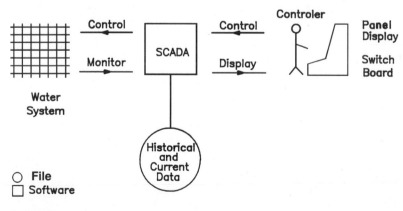

FIGURE 16.1 Computer monitored control.

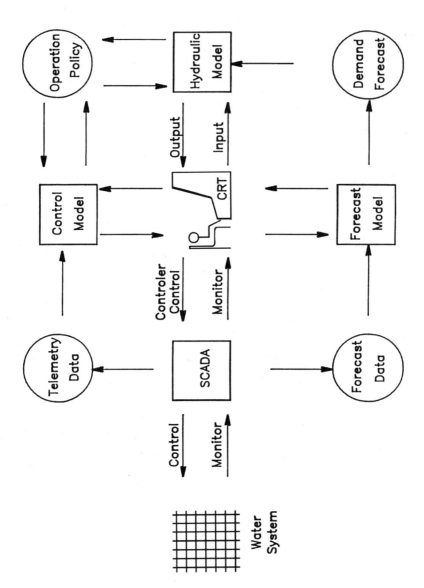

FIGURE 16.2 Computer assistant control.

control configuration enables operators to examine the consequences of their actions before the actions are actually implemented in the field. The final level of control is the computer-directed command structure shown in Fig. 16.3. The computer-directed structure is often called real-time control because the structure is able to provide an operating policy, possibly an optimal one, based on parameter forecasts and to implement the policy automatically via a direct link between the central processing computer and field-sited control devices. Of course, as a safeguard, the operator has complete manual override of the control structure at all times (Caves and Earl, 1979).

Both the computer-assisted and computer-directed systems will contain three major components in addition to the associated SCADA system: a hydraulic network model, a demand forecast model, and an optimal control model. Each of these components is discussed in the following sections.

16.2.3 Management Models

16.2.3.1 Hydraulic network models. To evaluate the cost of a particular pump-operating policy or to assess the associated operational constraints, some type of mathematical model of the distribution system is required. Potential model structures include mass balance, regression, simplified hydraulic, full hydraulic simulation, and the use of artificial neural networks.

1. *Mass-balance models.* In a simple mass-balance model of a single-tank system, the flow into the system equals the demand plus the rate of change in storage in the tank. The pressure-head requirements to achieve the flow into the tank are neglected, and it is assumed that a pump combination is available that achieves the desired change in storage. Nodal pressure requirements are commonly assumed to be satisfied if the tank

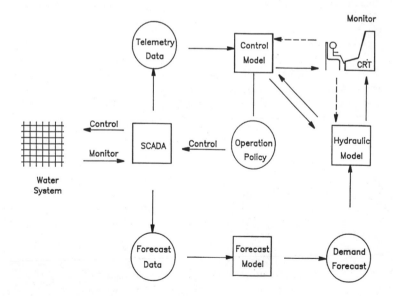

FIGURE 16.3 Computer directed control.

remains within a desired range. When using a mass-balance model, care must be taken when determining the cost to pump a given flow because the operating cost is related to both the discharge and energy added to the flow.

Multidimensional mass-balance models also have been developed. These models consist of weighted functional relationships between tank flow and pump-station discharge. The weights associated with the functional relationships can be determined using linear regression (Sterling and Coulbeck, 1975a) or linearization of the nonlinear network (Fallside and Perry, 1975).

The main advantage of mass-balance models is that the system's response can be determined much faster than it can from simulation models. Thus, these models are well suited for use with optimization strategies that require large numbers of simulation analyses (Joalland and Cohen, 1980). In general, mass-balance models are more appropriate for regional supply systems in which flow is carried primarily by major pipelines rather than by distribution networks in which the hydraulics are commonly dominated by looped piping systems.

2. *Regression Models.* Instead of using a simple mass-balance model, the nonlinear nature of the system hydraulics can be represented more accurately by using a set of nonlinear regression equations. Information required to construct such models can be obtained in a variety of ways. Regression curves can be generated by repeated execution of a calibrated simulation model for different tank levels and loading conditions (Ormsbee et al., 1987) or by the use of information from actual operating conditions to form a database relating pump head, pump discharge, tank levels, and system demands (Tarquin and Dowdy, 1989).

Regression models have the advantage of being able to incorporate some degree of system nonlinearity while providing a time-efficient mechanism for evaluating system response. However, regression curves and databases contain information only for a given network over a given range of demands. If the network changes appreciably or if forecasted demands are outside the range of the database, such an approach provides erroneous results. Moreover, regression curves are approximations of the system's response. Unless the curves are close approximations of the actual response, errors may accumulate over the course of operation that can adversely affect the optimization algorithm and the accuracy and acceptability of its results.

3. *Simplified network hydraulic models.* As an intermediate step between a nonlinear regression model and a complete nonlinear network model, simplified hydraulic models can be used. In such cases, the network hydraulics can be approximated using a macroscopic network model or be analyzed using a system of linearized hydraulic equations. Macroscopic models represent the system by using a highly skeletonized network model. Typically, only a pump, a lumped resistance term (a pipe), and a lumped demand are included. DeMoyer and Horowitz (1975) and Coulbeck (1984) used macroscopic models that had multiple terms relating the effect of various system components but in a single equation.

In certain cases (i.e., where the system boundary conditions are essentially independent of pump-station discharge), it may be possible to represent the system hydraulics using a simple linear model. Jowitt and Germanopoulos (1992) appropriately used an approximate linear model for a system dominated by large pump heads. In this case, small variations in tank levels did not have a significant impact on pump operations. In a similar application, Little and McCrodden (1989) developed a simple linear model for a supply system in which the head in the controlling tank was held constant. The coefficients for both model types may be determined after extensive system analysis. As a result, such models must be evaluated on a system-dependent basis to judge their acceptability.

4. *Full hydraulic simulation models.* Network simulation models provide the capability to model the nonlinear dynamics of a water distribution system by solving the governing set of quasi-steady-state hydraulic equations. For a water distribution system, the governing equations include conservation of mass and conservation of energy. These equations can be solved in terms of adjustment factors for junction grades (Shamir and Howard, 1968), loop flow rates (Epp and Fowler, 1970), and pipe flow rates (Wood and Charles, 1972).

   In contrast to both mass-balance and regression models, simulation models are adaptive to both system changes and variations in spatial demands. For example, if a tank or large main were suddenly taken out of service, a well-calibrated simulation model could still provide the hydraulic response of the modified system. A mass-balance or regression model, on the other hand, would require modification of the database or regression curves to account for the changes in the system's response. Although simulation models are more robust than either mass-balance or regression models, they generally require more data to formulate. They also require a significant amount of work to calibrate properly. Because such models require a greater computational effort than either mass-balance or regression models do, they generally are more useful with optimal control formulations that require a minimum number of individual system evaluations.

5. *Neural network models.* To reduce the computational requirements of a full hydraulic simulation model, the model can be replaced with a neural network representation of the system's response (Ormsbee and Lingireddy, 1995b). In this case, the neural network can be completely trained off-line, then used instead of the network model. The data required to train the neural network can come from multiple applications of a previously calibrated hydraulic simulation model. Alternatively, the neural network can be trained on-line using real time or archived data obtained from a SCADA system

   Neural networks comprise a set of highly interconnected but simple processing units, each responsible for carrying out only a few rudimentary computations. When provided with a sequential set of inputs and outputs for a given system, the network can organize itself internally in a way that allows it to reproduce an expected output for another given input. The internal process of self-organization or developing generalized representation of the system is referred to as the training process and is crucial for the efficient reproduction phase of the neural network. A neural network is said to be well trained if the deviation between the output from the neural network and the specified output is within a tolerable limit. On the basis of the network topology, node characteristics, and learning process, several types of neural networks can be developed.

16.2.3.2 Demand forecast models. To develop an optimal pump-operating policy, network system demands must be known. Because the actual daily demand schedule for a municipality is not known in advance, the optimal operating policy is estimated using forecasted demands from a demand-forecast model. Forecasted demands can be incorporated into the optimal control model using a lumped, proportional, or distributed approach. In a lumped approach, system demands typically are represented by a single lumped value. Such an approach normally is used in conjunction with mass-balance hydraulic models. Proportional demand models are normally used in conjunction with regression-based hydraulic models. In such instances, regression relationships are derived from a single demand pattern that may vary proportionally to the total system demand. A distributed demand approach is applicable when using a full network simulation model. In such an approach, the total system demand may be distributed both temporally and spatially among the various network demand points. Such an approach enables the development of optimal control policies that are adaptable to significant variations in system demand that may occur over the course of the designated operating period.

Distributed demand forecast models typically employ three steps: (1) they predict the daily demand, (2) they distribute the daily demand spatially among the junction nodes, and (3) they distribute the junction demands temporarily over a 24-h operating time horizon. Prediction of the daily demand can be accomplished by considering such factors as daily weather conditions, weather forecasts, seasons of the year, and past trends in water use (Maidment et al., 1985; Moss, 1979; Ormsbee and Jain, 1994; Sastri and Valdes, 1989; Smith, 1988; Steiner, 1989). Distribution of the daily system demand among the junction nodes can be accomplished using past meter records or real-time database information. Disaggregation of daily junction demands into smaller time intervals can be accomplished by considering the day of the week and seasonal patterns of diurnal demand (Bree et al. 1976; Chen, 1988a; Coulbeck et al., 1985; Perry, 1981).

Techniques for estimating demand are generally available but the availability of data (both spatial and temporal data) has limited the development and application of many available tools. As a result, additional work is still needed in this area, including better methods for short-interval prediction and spatial desegregation using historical short-term data. With an increase in the availability of comprehensive SCADA databases, improved model formulations and performance is expected to be attainable.

16.2.3.3 Control models. Proper selection of the optimization algorithm for use in solving the associated control model can often mean the difference between a sluggish or even nonperforming control model and one that functions extremely well. The choice of an appropriate optimization algorithm should be governed by the characteristics of the problem to be solved. Several mathematical programming techniques, such as linear programming (LP), dynamic programming, (DP), and nonlinear programming (NLP), are available to solve the energy cost minimization problem. By far, DP has been the optimization algorithm of choice by past researchers. Typically, DP has been used in an implicit control formulation with tank water level generally serving as the control variable. When DP is used, the control problem is broken down into a series of discrete time steps (stages) that have a prescribed set of potential control variable values (states). The optimal solution to the control problem is found by evaluating all state transitions between adjacent stages as opposed to evaluating all state transitions between all stages (i.e., total enumeration). By evaluating the state transitions between individual stages, a complex problem involving multiple subproblems can be reduced to a series of problems involving a single variable. The main problem associated with the use of DP is the "curse of dimensionality," in which the computational efficiency of the method significantly decreases as the number of control variables increases. Attempts to circumvent this problem have relied on the use of spatial decomposition schemes or the recasting of the problem in terms of alternate decision variables and solving using other mathematical programming techniques.

LP is the branch of mathematical programming that is used to solve problems where the objective function and all constraints are linear functions of nonnegative decision variables. Nonlinear problems are frequently solved via LP by assuming that portions of the object function and constrained solution space are approximately linear within a prescribed interval. LP problems are solved using an approach called the simplex method, which originally was developed by Dantzig in the late 1940s. The simplex method offers an efficient means of finding the optimum solution of a linear optimization problem by repeatedly selecting the decision variable that causes the greatest improvement in the objective function. As a result of the nature of the linear solution space, the optimal solution of a LP problem will always lie at the intersection of two or more constraints. The simplex method uses this feature of convex problems to its advantage by traveling along constraints to the intersection of other constraints. Once an initial feasible solution is determined, the algorithm identifies an adjacent point that will improve the objective function, then moves along a constraint to the new point. By examining the gradients of each constraint passing through the current point, a new point is selected and the process is repeated until the optimal solution is found.

The third type of control model uses NLP. As the name implies, NLP is useful for problems where the objective function or the constraints of an optimization problem, or both are nonlinear. Unlike LP and DP, NLP involves a large number of different techniques that can be used to solve an optimization problem. Such techniques range from elaborate gradient-based techniques to conceptually simple direct search methods. Recently, several researchers have begun to investigate the use of more heuristically based methods, such as simulated annealing and genetic algorithms.

16.2.4 Optimization Models

16.2.4.1 Problem formulation. *Decision variables.* The optimal control problem for a water-supply pumping system can be formulated using either a direct or an indirect approach, depending on the choice of the decision variable. Direct formulation of the optimal control problem divides the operating period into a series of time intervals. For each time interval, a decision variable is assigned for each pump, indicating the fraction of time the pump is operating during the time interval. The objective function for the control algorithm is then composed of the sum of the energy costs associated with the operation of each pump for each time interval. The problem can then be solved using either LP or NLP (Chase and Ormsbee, 1989; Jowitt et al., 1988; Ormsbee and Lingireddy, 1995a,1995b). The pump-control policy that results can be classified as explicit (or discrete) because the policy is composed of the required pump combinations and their associated operating times.

Instead of formulating the control problem directly in terms of pump operating times, the problem can be expressed indirectly as a surrogate control variable, such as tank level or pump-station discharge. Use of such a formulation requires prior development of cost functions expressed as the surrogate control variable. Such cost relationships can be developed from multiple regression analyses of actual cost data or from the results of multiple mathematical simulations of the particular system.

When tank level is used as the surrogate control variable, the objective becomes one of determining the least-cost tank-level trajectory over the specified operating period. When pump-station discharge (or pump head) is used as the control variable, the objective is to determine the least-cost time distribution of flows (or heads) from all the pump stations. The pump-control policies that result from such formulations can be classified as implicit (or continuous) since the individual pump operating times associated with the optimal state variables are not determined explicitly (Fallside and Perry 1975; Sterling and Coulbeck, 1975a; Zessler and Shamir 1989). However, the set of state variables associated with such an implicit solution normally can be converted into an explicit (discrete) policy of pump operating times by subsequent application of a secondary optimization program (Coulbeck et al. 1988b; DeMoyer and Horowitz 1975; Lansey and Awumah 1994).

Objective function—minimization of operating costs. The operating cost for a pumping system typically is composed of an energy consumption charge and a demand charge. The energy consumption charge is the portion of the electric utility bill based on the kilowatt-hours of electric energy consumed during the billing period. The demand charge represents the cost of providing surplus energy and usually is based on the peak consumption of energy that occurs during a specific time interval. The majority of existing control algorithms for water distribution systems only consider energy-consumption charges. This is primarily the result of the wide variability of demand-charge-rate schedules and that the billing period for such charges can vary between 1 week and 1 year. When such charges are not explicitly included in the optimal control objective function, they are either ignored or are addressed via the system constraints.

When the demand charges are excluded from the objective function, the objective function can be expressed solely in terms of the energy-consumption charge. In general, energy-consumption charges can be reduced by decreasing the quantity of water pumped, decreasing the total system head, increasing the overall efficiency of the pump station by proper selection of pumps, or using tanks to maintain uniform, highly efficient pump operations. In most instances, efficiency can be improved by using an optimal control algorithm to select the most efficient combination of pumps to meet a given demand. Additional savings can be achieved by shifting pump operations to off-peak water-demand periods through proper filling and draining of tanks. Off-peak pumping is particularly beneficial for systems operating under a variable electric-rate schedule.

Operational constraints. Constraints associated with the optimal control problem consist of physical system limitations, governing physical laws, and externally defined requirements. Physical system constraints include bounds on the volume of water that can be stored in tanks, the amount of water that can be supplied from a source, and valve or pump settings. The physical laws related to a supply and distribution system are the conservation of flow at nodes (conservation of mass) and conservation of energy around a loop or between two points of known total grade. Also included in this set are relationships between headloss and discharge through a pipe, pump, or valve. Typically, the only external requirements are to meet the defined demands and to maintain acceptable system pressure heads. Pressure-head requirements can have both upper and lower bounds to avoid leakage and ensure satisfying user requirements. Additional constraints can be added to restrict the tank levels to stay within a preset range of values.

When solving the optimization problem, the system's state at the time of analysis is known and an assumed final condition is set as a target. The initial state of the system includes the pump operations and tank levels, whereas the final state defines the end of cycle tank levels. The period of analysis usually is a 1-day cycle, although longer periods can be considered. The cycle for most control schemes typically begins with all tanks either completely full or at a preset lower level and ends 24 h later with the same condition (Shamir, 1985).

Although not normally considered explicitly in most control algorithms, it should be recognized that pump maintenance costs may constitute a significant secondary component of any pump operation budget. Pump wear is directly related to the number of times a pump is turned on and off over a given life cycle. As a result, operators will attempt to minimize the number of pump switches while simultaneously determining least- cost operations. This problem is not as significant for newer pumps, which are better designed and made of more durable materials, but it is a major concern in many older systems. Unfortunately, sufficient data are not currently available to permit the incorporation of such costs directly into the objective function. Instead, limits on pump switches normally are set through the use of the system constraints (Lansey and Awumah 1994) or an approximate cost term (Coulbeck and Sterling, 1978).

16.2.4.2 System classification. Many researchers have developed optimal control formulations to minimize the operating costs associated with water-supply pumping systems. This section, with Table 16.1 as the central reference, cites and classifies the various algorithms that have been developed to solve the associated control problem. Model formulations are classified on the basis of the physical composition of the system (i.e., the number of tanks and pump stations). Following the classification, an overall evaluation of the various algorithms is presented.

The key to classifying the various control algorithms is the type of system addressed by the model. Columns 2 and 3 in Table 16.1 define the number of tanks and sources each model can consider. Sources are defined as the number of alternative pumping locations (either individual pumps or pump stations). Following the description of the type of system to which the model is applicable, the type of hydraulic model and demand model used by the algorithm, the type of control algorithm used, and the resulting control policy (explicit or implicit) are identified. The identified control algorithms include DP, linear quadratic programming, NLP, integer programming, mixed-integer linear programming, nonlinear heuristics, and genetic algorithms.

Single-and multiple pump stations with no tanks. The majority of research related to the optimal control of water supply systems has focused on systems with one or more storage tanks. Two investigators have developed control strategies for systems without effective storage. Chen (1988b) considered a network without tanks and determined the optimal allocation of supply between the pump sources. A continuous nonlinear problem was solved assuming the average pump-station efficiency would be reached for each pump station and a lumped-system relationship could be developed. Dynamic programming was then applied to select the actual pumps given the optimal continuous outflows.

In considering a supply system with a constant head discharge, Little and McCrodden (1989) developed an algorithm to select the optimal pump combinations of a single pump source, including the energy usage and peak demand charge in the objective function. Their algorithm used the pump operating times as the decision variables.

Single tank with single- and multiple-pump stations. Among the earliest published optimization efforts applied to pump operations for a single tank system was completed for a portion of Philadelphia by DeMoyer and Horowitz (1975). Their problem formulation used tank level as the state variable in a dynamic programming model. In a similar application, Sterling and Coulbeck (1975a) also applied DP for a single-reservoir, multiple-source problem in which tank level served as the state variable and tank hydraulics were modeled using a mass-balance relationship. Sabet and Helweg (1985) proposed a similar formulation. Later, Coulbeck (1984) extended his original formulation to include both fixed- and variable-speed pumps.

Multiple-tank and multiple-source systems. In general, dynamic programming has been a high efficient algorithm for use in obtaining optimal control policies for single-tank systems. Extension of the approach to multiple-tank systems is greatly limited because of the increased computational burden that results from multiple decision variables and state variables. One way to avoid this problem is to use of spatial decomposition techniques (Coulbeck, 1988a and 1988b. Joalland and Cohen, 1980; Zessler and Shamir, 1989). In this approach, the system is broken down into subnetworks that contain only one or two tanks. Optimal control policies are then developed for each subsystem and are coordinated at an upper control level by using relationships that link the resulting policies together.

Rather than attempting to overcome the limitations to DP through decomposition schemes, other researchers have formulated the control problem using different decision variables other than tank level (Cembrano et al, 1988; Coulbeck and Sterling, 1978; Fallside and Perry 1975; Lansey and Zhong, 1990; Solanos and Montoliu, 1988; Sterling and Coulbeck, 1975a; Tatejawski 1988). By using a continuous variable-such as pump-station discharge or pump head, a dual-level optimization scheme can be developed that allows a direct consideration of multiple-tank systems. Such methods first determine the optimal discharge or added head associated with each pump station. Then the pump operation schedules associated with the resulting optimal discharges or pump heads are determined by solving a secondary series of discrete optimization problems. Instead of developing a dual-level optimization algorithm in which the pump-operating times are

TABLE 16.1 Summary of Optimization Models

References (1)	Number of Tanks (2)	Number of Sources (3)	Hydraulic Model (4)	Demand Model (5)	Control Algorithm (6)	Control policy E (7)	Comments (8)
DeMoyer and Horowitz (1975)	Single	Single	Simplified hydraulics	Lumped	Dynamic programming (DP)	Explicit	Pump combinations determined in a conventional DP model with a simplified hydraulic representation (macroscopic model).
Sterling and Coulbeck (1975b)	Multiple	Multiple	Mass balance	Lumped	Linear quadratic programming (LQP)	Implicit	Pump-station discharges are determined with a mass-balance-type linear hydraulic model. Extended in Coulbeck and Sterling (1978) and Cembrano et al. (1988).
Sterling and Coulbeck (1975a)	Single	Multiple	Mass balance	Lumped	DP	Implicit	Conventional DP model determines continuous pump-station discharge.
Fallside and Perry (1975)	Multiple	Multiple	Mass balance (linearized hydraulic equation)	Lumped	LQP	Implicit	Spatial decomposition with fixed transfers between subsystems. Lagrangian relaxation, like Sterling and Coulbeck (1975a), is applied. Optimization later dropped for heuristic-based pump priority logic.
Joalland and Cohen (1980); Carpentier and Cohen (1984)	Multiple	Multiple	Mass balance	Lumped	DP	Implicit	Extension of Fallside and Perry (1975). Spatial transfers are optimized with discrete pump operations. Each subsystem has a single-tank and pump source.
Coulbeck (1984)	Single	Single	Simplified hydraulics	Lumped	DP	Explicit	Approach for an in-line hydraulics source resistance-tank-demand supply system. Simplified similar to DeMoyer and Horowitz (1975). Extended to special cases of multisource and multi-tank systems.
Sabet and Helweg (1985)	Single	Multiple	Nonlinear regression	Proportional	DP	Implicit	Uses single pump sources but neglects the interaction between pumps in developing hydraulic relationships.
Whaley and Hume (1986)	Multiple	Multiple	Nonlinear system equation	Proportional	Nonlinear programming (NLP)	Explicit	Gradient-based NLP system algorithm with penalty weights.
Solanos and Montoliu (1988)	Multiple	Multiple	Mass balance	Lumped	DP/LQP	Implicit	Optimal lumped controls are determined in iterative DP-optimal control approach.
Chen (1988)	None	Multiple	Mass balance	Lumped	NLP	Implicit	Determines allocation of demand between sources using Lagrangian function.
Jowitt et al. (1988)	Multiple	Multiple	Mass balance	Lumped	Linear programming (LP)	Explicit	Pump run times are determined.
Tatejewski (1988)	Multiple	Multiple	Mass balance	Lumped	LQP	Implicit	Two-level approach with pump station flows determined in an upper level using continuous cost approximations.
Coulbeck et al. (1988a, b)	Multiple	Multiple	Mass balance	Lumped	NLP/integer programming (IP)	Explicit	Three-level model that fixes tank trajectories by NLP in the upper level. These are passed to two lower levels to select pump combinations in an IP.
Ormsbee et al. (1989)	Single	Multiple	Nonlinear regression	Proportional	DP	Explicit	Optimal tank trajectories are determined in a DP model using regression curves for cost and hydraulics. Enumeration is then used to determine exact pump combinations.
Lannuzel and Ortalano (1989)	Single	Multiple	Nonlinear system equation	Distributed	Heuristic	Explicit	Operation heuristics are captured in an expert system linked with a simulator.

TABLE 16.1 *(Continued)*

References (1)	Number of Tanks (2)	Number of Sources (3)	Hydraulic Model (4)	Demand Model (5)	Control Algorithm (6)	Control policy (7)	Comments (8)
Little and McCrodden (1989)	None	Single	Simplified hydraulics	None	Mixed-integer Linear programming (MILP)	Explicit	Pump operation times for a supply system to a constant elevation tank via a pipeline are determined in an integer-LP model.
Zessler and Shamir (1989)	Multiple	Multiple	Mass balance	Lumped	DP	Implicit	Regional supply system is spatially decomposed with a single tank and equivalent-demand node in each subsystem. Optimal flows are determined by DP and converted by logic to discrete operations.
Chase and Ormsbee (1989, 1991)	Multiple	Multiple	Nonlinear system equation	Distributed	NLP	Explicit	Linked optimization-simulation model using pump run times as decisions for fixed time intervals (Chase and Ormsbee, 1989) and variable time intervals (Chase and Ormsbee, 1991).
Lansey and Zhong (1990)	Multiple	Multiple	Nonlinear system equation	Distributed	NLP	Explicit	Linked optimization-simulation determines optimal pump-station added energy. Converted to discrete operations in a DP model.
Ulanicki and Orr (1991)	Multiple	Multiple	Mass balance	Distributed	NLP/MILP	Explicit	Two-level model selects pump run times to meet a desired upper-level tank flows. Approximate mass-balance model used in the upper level and full system equations in the lower level.
Brion and Mays (1991)	Multiple	Multiple	Nonlinear system equation	Distributed	NLP	Explicit	Extension of Chase and Ormsbee (1989) to consider analytical gradients.
Jowitt and Germanopoulos (1992)	Multiple	Multiple	Simplified	Distributed	DP	Explicit	Pump runtimes are decisions with constant hydraulics pump output due to elevation difference between sources and users.
Awumah and Lansey (1992)	Single	Multiple	Nonlinear regression	Proportional	DP	Explicit	Discrete pump operations with pump switching limits.
Ormsbee and Lingireddy (1995a)	Multiple	Multiple	Nonlinear system equation	Distributed	H	Explicit	Uses a heuristic algorithm, based optimization algorithm
Ormsbee and Lingireddy (1995b)	Multiple	Multiple	Neural network	Distributed	GA	Explicit	Links a genetic optimization algorithm with a neural network model of system response.

expressed in terms of some other implicit decision variable, optimal control algorithms can be developed that explicitly consider pump run times as the decision variables. Such formulations can then be solved using linear programming (Jowitt et al. 1988; Jowitt and Germanopoulos 1992), nonlinear programming (Whaley and Hume 1986; Chase and Ormsbee 1989; Brion and Mays 1991; Ulanicki and Orr 1991; Chase and Ormsbee 1991), or heuristic based methods (Ormsbee and Lingireddy, 1995 a,b).

16.2.5 Summary and Conclusions

As can be seen from the previous citations, numerous methodologies have been proposed to develop optimal control algorithms for water supply pumping systems. The choice of the appropriate algorithm for a particular application will depend largely on the physical characteristics of the system. The most straightforward approach for single-tank systems is a formulation with tank level as the state variable in a DP model. Such an approach is generally efficient when the system demands are lumped at a single node or are assumed to vary proportionally. Attempts to incorporate the impact of the spatial variability of demand or changes in the operational status of various system components normally requires the use of an alternative formulation. For systems that contain a reasonable number of pumps, it may be plausible to use a pump-run-time model (Chase and Ormsbee, 1991; Ormsbee and Lingireddy, 1995a, 1995b). When the total number of pumps is large, the use of an implicit pump-station decision variable may be more appropriate (Lansey and Zhong, 1990).

For multisource-multitank systems that are highly serial or permit a convenient subdivision into distinct hydraulic units, a dynamic programming spatial decomposition approach may be feasible. However, for systems that do not readily permit spatial decomposition, control algorithms normally require lumped-pump-station models or a pump-run-time approach. Both the lumped-pump-station models and the pump-run-time models normally are solved using some form of nonlinear optimization. When significant approximations to the system hydraulics are feasible, it may be possible to solve the formulation using quadratic programming or even LP. However, it is the capability of both the lumped pump-station parameter models and the pump-run-time models to accommodate directly the nonlinear dynamics of most multisource/multitank systems that makes the use of nonlinear optimization an acceptable trade-off. As more tanks and distributed demands are considered, a more detailed simulation model is necessary. The trade-off is then between optimization time requirements, accuracy, and the precision of the associated hydraulic model. Typically, these trade-offs must be evaluated on a network-by-network basis because rules of thumb are difficult to derive.

When using pump-station discharge as a surrogate control variable, the selection of a discharge-cost relationship must be made with extreme care. In most cases, pump-station discharge will vary with both demand and tank level. As a result, the associated cost and hydraulic relationships must have two independent variables (demand and tank level) (Ormsbee et al, 1989), or they must account for the required pressure head in other approximate ways (Coulbeck, 1984). In addition, using pump discharge as the decision variable in a lumped hydraulic model implicitly assumes there is a combination of pumps that will supply the optimal flow under the correct amount of pressure to cause the desired change in tank level. This assumption can be increasingly difficult to satisfy as the network hydraulics become more complex in multiple-source and multiple-tank systems.

In general, as the number of pumps or pump combinations increases, so does the computational advantage of the lumped-pump-station parameter approach over the pump-run-time approach. However, it should be remembered that although the pump run-time approach yields the desired pump operational policy directly, the solution obtained using the lumped-pump-station-parameter approach subsequently must be translated into an appropriate pump policy. Although the computational time associated with this subproblem typically is a small fraction of the time required to solve the implicit control problem, it still can be significant.

In general, the majority of optimal control algorithms have been developed for applications with fixed-speed pumps. Variable-speed pumps can simplify or exacerbate the difficulty of the problem, depending on the decision variable. If pump run time is chosen, each variable speed pump can be represented by a series of fixed-speed pumps. However, such a formulation increases the total number of decision variables and, hence, computation times. On the other hand, the wider continuous-range pump output of variable-speed pumps provides a better mechanism for implementing the continuous solutions associated with formulations of lumped-pump-station parameters. Alternatively, pump speed can be chosen as a continuous decision variable in the lumped-system formulation (Lansey and Zhong, 1990).

Despite the multitude of control algorithms that have been developed for optimal control of water supply pumping systems, several areas of potential research still remain. For example, few researchers have investigated the development of optimal control policies for long-term (weekly) planning horizons. Similarly, little research has been conducted on the impact of final pump operations on pump maintenance requirements. Robustness of operations also has been a neglected area. Finally, the design of water distribution systems is a well-examined area, but little emphasis has been placed on the implications of design on operation and vice versa.

Although the use of expert-system technology or neural-network technology in either developing or implementing optimal control strategies seemingly has great potential, little work has been conducted in this area. Two applications of knowledge-based selection were described by Fallside (1988) and Lannuzel and Ortolano (1989). Fallside and Perry (1975) applied a decomposition approach to an existing system; however, after gaining experience and performing extensive systems analysis, they dropped the scheme in favor of a heuristic described as "pump priority logic" (Fallside, 1988). Lannuzel and Ortolano (1989) also examined a water supply pumping system and developed an operational heuristic from experience. These rules of thumb were then combined with a simulation model in an expert system. Although both studies have limited applicability to other systems, they nevertheless provide some insight into the usefulness of such an approach.

Although several successful applications of optimal pumping control exist in Europe and Israel (Alla and Jarrige, 1989; Orr and Coulbeck, 1989; Orr et al., 1990; Zessler and Shamir, 1989), widespread application of such technology in the United States has been severely limited. With the exception of a control system for Albuquerque, New Mexico. (Jentgen and Hume, 1989), and Detroit, Michigan (Chase et al., 1994), no other large municipalities appear to have the capability to control their pump operations using computer-generated pump policies. Although it is true that several municipalities have implemented some form of computer assisted pump selection or are considering an investigation of computer-control technology for their system (Hutchinson, 1991), actual applications appear to be limited.

Future widespread applications of optimal control technology to domestic water supply systems are likely to depend on increased use of more sophisticated SCADA systems and the availability of more commercially available off-the-shelf control software. Additional problems to overcome include the necessity of well-calibrated network models and the availability of accurate demand-forecast models. Even when such technical problems can be overcome, however, it appears that one great roadblock to the implementation of such technology is not the lack of the necessary tools but the unwillingness of utility staff to use them. Previous attempts at developing energy cost minimization programs have revealed that many pump-station operators have an intrinsic mistrust of computers in general and automated operations in particular. In part, the reason may be the conservative nature of most water utilities and their justifiable concern for the impact of "optimal policies" on consumers. In other cases, system operators may have significant concerns about the impacts of such technology on their job security.

Such concerns highlight the need for systems analysts to work closely with operations personnel to develop and implement a particular control environment. In most cases, experienced operators already possess valuable insights into the operation of their system that may prove to be crucial to the development of a successful control scheme. Ideally, the system analyst should work in concert with the system operator to develop an environment that the operator is not only comfortable with but feels some degree of "authorship" as well. In particular, the system should reflect the operator's existing wants and needs as much as possible while providing a framework for expanded control capabilities. In the final analysis, the real challenge of system analysis may not lie in the development of more sophisticated computer algorithms but in the development of more efficient strategies and programs for their implementation.

16.3 FORMULATIONS TO SATISFY WATER QUALITY

Computer models that simulate the hydraulic behavior of water distribution systems have been available for many years. More recently, these models have been extended to analyze the quality of water as well as the hydraulic behavior. These models are capable of simulating the transport and fate of dissolved substances in water distribution systems and can be used to predict water quality at the point of delivery to the customer. The Safe Drinking Water Act in 1974 and its amendments in 1986 require the measurement of concentrations of residual chlorine, lead, and copper at the point of delivery. Previously, regulatory concerns focused on water as it left the treatment plant before entering the distribution system (Clark, 1994), disregarding the variations in water quality that occurred in the water distribution systems.

Several optimization models have been developed to determine the optimal operation schedule of the pumps in a water distribution system for a predefined time horizon for water-quality purposes and to minimize pumping costs while satisfying the hydraulic constraints, the water-quality constraints, and the bound constraints on pump operation times, pressures, and tank-water storage heights.

Several objective functions can be formulated that address the optimal operation of water distribution systems for water-quality purposes. Three such objective functions are (1) minimization of deviations of the actual concentrations of a constituent from the desired concentration values, (2) minimization of the total pump operation times, and (3) minimization of the total energy cost. The constraints that must be considered for all time periods for all cases are basically the hydraulic, water-quality, and bound constraints.

Consider a water distribution system with M pipes, K junction nodes, S storage nodes (tank or reservoir), and P pumps that are operated for T time periods.

The objective function that minimizes the deviations of concentrations from the desired range of concentrations is

$$\text{Min } Z_I = \text{Minimize} \sum_{t=1}^{T} \sum_{n=1}^{N} \min\left[0, \min\left(C_{nt} - \underline{C}_{nt}, \overline{C}_{nt} - C_{nt}\right)\right]^2 \quad (16.1)$$

where N is the total number of nodes (junction and storage), C_{nt} is the substance concentration at node n at time t (M/L³) and $\underline{C}_{nt}$ and $\bar{C}_{nt}$ are the lower and upper bounds, respectively, on substance concentration at node n at time t (M/L³).

The objective function that considers the minimization of the total pump durations is

$$\text{Min } Z_{II} = \text{Minimize } \sum_{p=1}^{P} \sum_{t=1}^{T} D_{pt} \qquad (16.2)$$

where D_{pt} is the length of time pump p operates during time period t (T). The objective function for the minimization of the total energy cost is

$$\text{Min } Z_{II} = \text{Minimize } \sum_{p=1}^{P} \sum_{t=1}^{T} \frac{UC_t \, 0.746 \, PP_{pt}}{EFF_{pt}} D_{pt} \qquad (16.3)$$

where UC_t is the unit energy or pumping cost during time period t ($/Kwh), PP_{pt} is the power of pump p during time period t (h), D_{pt} is the length of time pump p operates during time period t (h), EFF_{pt} is the efficiency of pump p in time period t. The efficiency of the pump and the unit energy cost are considered to be constant for all time periods.

The distribution of flow throughout the network must satisfy the conservation of mass and the conservation of energy which are defined as the hydraulic constraints. The conservation of mass at each junction node, assuming water is an incompressible fluid, is

$$\sum_i (q_{ik})_t - \sum_j (q_{kj})_t - Q_{kt} = 0 \quad k = 1, ..., K \text{ and } t = 1, ..., T \qquad (16.4)$$

where $(q_{ij})_t$ is the flow in the pipe m connecting nodes i and j at time t (cfs) and Q_{kt} is the flow consumed (+) or supplied (−) at node k at time t(cfs). The conservation of energy for each pipe m connecting nodes i and j, in the set of all pipes, M, is

$$h_{it} - h_{jt} = f(q_{ij})_t \quad \forall \; i, j \in M \text{ and } t = 1, ..., T \qquad (16.5)$$

where h_{it} is the hydraulic grade line elevation at node i (equal to elevation head, E_i plus pressure head, H_{it}) at time t (h) and $f(q_{ij})t$ is the functional relation between headloss and flow in a pipe connecting nodes i and j at time t (h).

The total number of hydraulic constraints is $(K + M)T$, and the total number of unknowns also is also $(K + M)T$, which are the discharges in M pipes and the hydraulic grade-line elevations at K nodes. The pump operation problem is an extended-period simulation problem. The height of water stored at a storage node for the current time period y_{st} is a function of the height of water stored from the previous time period; which can be expressed as

$$y_{st} = y_{s(t-1)} + \frac{q_{s(t-1)}}{A_s} \Delta t \quad s = 1, \ldots, S \text{ and } t = 1, \ldots, T \qquad (16.6)$$

The water-quality constraint which is the conservation of mass of the substance with in each pipe m connecting nodes i and j in the set of all pipes, M, is

$$\frac{\partial (C_{ij})_t}{\partial t} = -\frac{(q_{ij})_t}{A_{ij}} \frac{\partial (C_{ij})_t}{\partial x_{ij}} + \theta(C_{ij})_t \quad \forall \; i, j \in M \text{ and } t = 1, ..., T \qquad (16.7)$$

where $(C_{ij})_t$ is the concentration of substance in pipe m connecting nodes i and j as a function of distance and time (mass/ft³); x_{ij} is the distance along pipe (ft); A_{ij} is the cross-sectional area of pipe connecting nodes i and j (ft²); and $\theta(C_{ij})$t is the rate of reaction of constituent within the pipe connecting nodes i and j at time t (M/L³/day).

All the constraints considered so for are equality constraints, which are handled by the network simulator. In addition, inequality constraints exist that are the bound constraints. The lower and upper bounds on pump operation time are given as

$$\Delta t_{min} \leq D_{pt} \leq \Delta t_{max} \quad p = 1, ..., P \text{ and } t = 1, ..., T \quad (16.8)$$

where D_{pt} is the length of the operation time of pump p at time t, and Δt_{min} and Δt_{max} are the lower and upper bounds on D_{pt}, respectively. Δt_{min} can be zero in order to simulate an idle pump and Δt_{max} is equal to the length of one time period.

The nodal pressure head bounds are

$$\underline{H}_{kt} \leq H_{kt} \leq \overline{H}_{kt} \quad K = 1,... K \text{ and } t = 1, ...T \quad (16.9)$$

where $\underline{H}_{kt}$ and $\overline{H}_{kt}$ are the lower and upper bounds, respectively, on pressure head at node k at time t. No universally accepted values have bean established for the lower and upper bound values. The range of 20-40 psi is acceptable for minimum pressure for average loading conditions, but it may be lowered during emergency situations, such as fires. High pressures are usually 80-100 psi, depending on the distribution system.

The bounds on the height of water storage are

$$\underline{y}_{st} \leq y_{st} \leq \overline{y}_{st} \quad s = 1, ..., S \text{ and } t = 1, ..., T \quad (16.10)$$

where and are the lower and upper bounds, respectively, on the height of water stored at node s at time t, y_{st}. These limits are caused by physical limitations of the storage tank or the reserve required for fire protection.

Cohen (1982) stated that if there is no requirement for the periodicity in a network's operation, the optimization of the operation has no meaning. Hence, to achieve this, final storage bounds traditionally are tightened so that the storage in the tanks will be more or less the same as the initial states. However, it has been found that the system reaches steady state if the daily pump operation schedules repeat themselves for a certain period of time. The tank water levels at the beginning and the end of the day are equal to each other when the steady-state conditions are reached and have been adjusted to the time that the pumps are operating during the simulation. Hence, the constraint that forces the tank water levels at the end of the day to be more or less equal to the initial levels does not need to be met and should not be used in simulations that include water quality because it takes many days for the system to reach steady state.

The minimization of pump operation times and pump operation costs considers the same set of constraints defined by Eqs. (16.4) through (16.10) as the minimization of the deviations of concentrations from the desired values, with an additional constraint for the substance concentration values to be within their desired limits.

The bounds on substance concentrations are

$$\underline{C}_{nt} \leq C_{nt} \leq \overline{C}_{nt} \quad n=1, ... N \text{ and } t = 1,..., T \quad (16.11)$$

where $\underline{C}_{nt}$ and $\overline{C}_{nt}$ are the lower and upper bounds, respectively, on substance concentration at node n at time t.

The above formulation results in a large-scale NLP problem with decision variables, $(q_{ij})_t$, H_{kt}, y_{st}, D_{pt}, and $(C_{ij})_t$. One method of developing an optimization algorithm is to partition the decision variables into two sets: control (independent) and state (dependent) variables. The pump operation times are the control variables. The problem is formulated above as a discrete time optimal control problem. Equations (16.1) (16.2), and (16.3) are examples of objective functions for the minimization of the deviations of concentrations from the desired range of values, the minimization of the pump operation times and the minimization of the energy cost, respectively. Equations (16.9) (16.10), and (16.11) define typical hydraulic equality constraints and water-quality inequality constraints. The inequality constraints are bounds in the state variables: pressure, tank elevation. and nodal contaminant concentrations. Equation (16.8) is the bound constraint for the control variable, the duration of pumping during a time period.

16.4 SOLUTION METHODS AND APPLICATIONS FOR WATER-QUALITY PURPOSES

This section describes two methods of determining the optimal operation of water distribution systems for water-quality purposes. These methodologies are based on describing the operation as a discrete-time optimal-control problem that can be used to determine the optimal operation schedules for the pumps in distribution systems. One method is based a mathematical programming approach, the other method is based on a simulated annealing approach. The following sections describe the two methods and present sample applications with comparisons.

16.4.1 Mathematical Programming Approach

The solution algorithm used in the mathematical programming approach is a reduction technique, similar to the algorithms used for ground-water management (Wanakule et al. 1986), for water distribution system design (Lansey and Mays, 1989), operation of pumping stations in water distribution systems (Brion and Mays, 1991), optimal flood control operation (Unver and Mays, 1990), and optimal determination of fresh water inflows to bays and estuaries (Bao and Mays, 1994a, 1994b). In all these applications, the optimal solution of the problem is obtained by using nonlinear optimization code linked to a hydraulic simulation code.

The mathematical approach reformulates the problem using an optimal control framework that results in linking a simulation code, EPANET (Rossman, 1994), with an optimization code, GRG2 (Lasdon and Waren, 1986), to find the optimal solution. The decision variables are partitioned into control variables and state variables in the formulation of the reduced problem. The control variables are the amount of time that the pump operates during each time period. The series of period operating times results in a pump operation schedule. The control variable values (pump operation schedule) are determined by the optimizer, and they are given as input to the simulator, which solves for the state variables (pressure, water quality, and tank levels). Hence, the state variables are obtained as implicit functions of the control variables. This results in a large reduction in the number of constraints as the equality constraints are solved by the simulator and only the inequality bound constraints of the hydraulics and water quality are left to be solved by the optimizer.

Improvements in the objective function of NLP problems are obtained by changing the control variables of the reduced problem. NLP codes restrict the step size by which the control-variables change so that the control-variable bounds are not violated. The state variables, which are implicit functions of control variables, are not considered in the determination of step size. If the bounds of the state variables are violated, more iterations will be needed to obtain a feasible solution. The penalty function method is used to overcome this problem. The state variable-bound constraints are included in the objective function as penalty terms. The application of this technique is also beneficial because it reduces the size of the problem and the number of the constraints.

Many kinds of penalty functions can be used to incorporate the bound constraints into the objective function. In this application, two different penalty functions, the bracket and the augmented lagrangian, are used.

The bracket penalty function penalty method (Li and Mays, 1995; Reklaitis, et al. 1983) uses a simple penalty function which has the following form:

$$PB_j(V_{j,i}, R_j) = R_j \sum_i \left[\min(0, V_{j,i})\right]^2 \quad (16.12)$$

The augmented Lagrangian method (Fletcher, 1975) uses the following penalty function:

$$PA_j(V_{j,i}, \mu_{j,i}, \sigma_{j,i}) = \frac{1}{2} \sum_i \sigma_{j,i} \min\left[0, \left(V_{j,i} - \frac{\mu_{j,i}}{\sigma_{j,i}}\right)\right]^2 - \frac{1}{2} \sum_i \frac{\mu_{j,i}^2}{\sigma_{j,i}} \quad (16.13)$$

where the index j is the representation of H for pressure head, C = concentration, and y = water-storage-height bound constraints. The index i is a one-dimensional representation of the double index (k, t) for the pressure-head penalty term, (n, t) is for the concentration penalty term, and (s, t) is for the storage bound penalty term. PB_j and PA_j define the bracket and the augmented Lagrangian penalty functions for bound constraint j, respectively. V_j, i is the violation of the bound constraint j. R_j is a penalty parameter used in the bracket penalty method, and $\sigma_{j,i}$ and $\mu_{j,i}$ are the penalty weights and Lagrangian multipliers used in the augmented Lagrangian method, respectively. (See Sakarya, 1998, for a detailed description of the bracket and the augmented Lagrangian penalty-function methods and the methods to update the penalty function parameters).

The violation of the pressure head constraint is defined as

$$V_{H,kt} = \min\left[(H_{kt} - \underline{H}_{kt}), (\overline{H}_{kt} - H_{kt})\right] \quad (16.14)$$

Similarly, the violations of the substance concentration and the water-storage-height bound constraints can be defined.

The reduced problem for minimizing the deviations of the concentrations from the upper and lower bounds is

$$\text{Min } L_I = P_C(V_{C,nt}, F_C) + P_H(V_{H,kt}, F_H) + P_y(V_{y,st}, F_y) \quad (16.15)$$

subject to

$$0 \leq D_{pt} \leq \Delta t \quad p = 1,..., P \text{ and } t = 1,..., T \quad (16.16)$$

where P_C, P_H, and P_y define the bracket or augmented penalty terms associated with the concentration, pressure head, and storage bound constraints, respectively, depending on the penalty function method used. Similarly, F_C, F_H, and F_y define the penalty function parameters which are the penalty parameters for the bracket penalty method or the penalty weights and the Lagrangian multipliers for the augmented Lagrangian penalty method, associated with the concentration, pressure head, and storage bound constraints, respectively.

For minimizing pump operation time, the reduced objective function is subjected to the same constraint defined by Eq. (16.16), and has the following form.

$$\text{Min } L_{II} = \sum_{p=1}^{P} \sum_{t=1}^{T} D_{pt} + P_C(V_{C,nt}, F_C) + P_H(V_{H,kt}, F_H) + P_y(V_{y,st}, F_y) \quad (16.17)$$

The reduced objective function for the minimization of pump energy is defined as

$$\text{Min } L_{III} = \sum_{p=1}^{P} \sum_{t=1}^{T} \frac{UC_t \, 0.746 PP_{pt}}{EFF_{pt}} D_{pt} + P_C(V_{C,nt}, F_C) + P_H(V_{H,kt}, F_H) + P_y(V_{y,st}, F_y) \quad (16.18)$$

subject to the constraint defined by Eq. (16.16).

The solution of the final form of the problem is obtained by the two-step optimization procedure described in Brion and Mays (1991), Lansey and Mays (1989), Mays (1997) and Wanakule et al. (1986). Finite difference approximations were used to calculate the derivatives of the objective function with respect to the control variables, the pump operation times. These derivatives are the reduced gradients that the optimization code needs to find the optimal solution. Figure 16.4 shows the flowchart of the optimization

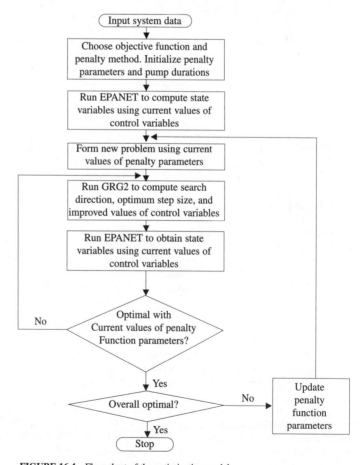

FIGURE 16.4 Flow chart of the optimization model.

model. Initially, the original objective function and the penalty method are chosen depending on the case being considered. The optimization procedure is divided into inner and outer levels. The outer level-code selects values for the penalty function parameters, which are Lagrangian multipliers and penalty weights for the augmented Lagrangian penalty functions, and penalty parameters for the bracket penalty function associated with three terms in the objective function. The fixed Lagrangian multipliers and fixed penalty parameters are programmed into GRG2, and the inner-level optimization is made linking GRG2 and EPANET. GRG2 finds a set of the control variables D_{pt}, which is the trial pump-operation schedule. Using the trial pump operation schedule, the objective function is evaluated. If there is little or no improvement from the previous trial schedule, the pump schedule is adopted as optimum. If there is significant change, the outer-level loop is carried out to update the penalty function parameters. The GRG2 Lagrangian multipliers and fixed-penalty parameters are updated, and a new GRG2-EPANET optimization is carried out again to find a new pump operation schedule. The procedure is repeated until the Lagrangian and penalty function parameters are not updated (i.e. the overall optimum is found), the iteration limit of the outer loop is reached, or no improvements are achieved for some predefined consecutive iterations.

During the solution procedure, GRG2 needs the values of reduced objective function and the reduced gradients while searching for the optimum solution. For each iteration step in the inner loop, GRG2 changes the control variables and provides a new pump operation schedule. The gradients are found using finite elements.

A simplified method is used to reduce the number of EPANET calls. The main idea of this simplified method is that if the maximum change in the control variables between consecutive iterations is small, the change that occurs in the state variables also is small. Thus, if the maximum change in the control variables between consecutive iterations is smaller than a specified limit, the change in the values of the state variables also will be small. EPANET will not be called at that iteration to calculate the state variables, and the previous values will be used.

16.4.2 Simulated Annealing Approach

Simulated annealing is a combinatorial optimization method that uses the Metropolis algorithm to evaluate the acceptability of alternate arrangements and slowly converge to an optimum solution. The method does not require derivatives and is flexible enough to consider many different objective functions and constraints. Simulated annealing uses concepts from statistical thermodynamics and applies them to combinatorial optimization problems. Kirkpatrick et al. (1983) explained the simulated annealing methodology and applied the method to the "traveling salesman" and computer design problems. Kirkpatrick (1984) provided additional insights and applications, including graphical partitioning, which is useful in the electronics industry for the design of circuits. Dougherty and Marryott (1991) applied simulated annealing to groundwater remediation.

Combinatorial optimization requires that the decision variables be restricted to a set of discrete values. The set of all possible combinations is called the configuration space. For example, consider a pump that operates for 6-h divided into 1-h periods, where the pump can either be on or off for any period. The number of pump-operation combinations is 2^6 = 64. A pump that operates for 24-h divided into 1-h periods has 2^{24} = 16,777,216 combinations. The 6-h example can be solved by trial and error, but the 24-h example is too large to solve by trial and error. For large combinatorial optimizations problems, simulated annealing provides a manageable solution strategy.

Kirkpatrick et al. (1983) explained how the Metropolis algorithm was developed to provide a statistically based mathematical simulation of a system of atoms at a high temperature that cools slowly to its ground energy state. If an atom is given a small random change, there will be a change in system energy ΔE. If $\Delta E < 0$, the new configuration is at a lower energy state and is accepted. If $\Delta E > 0$, the decision to change the system configuration to a higher energy state is treated probabilistically and is calculated by

$$P(\Delta E) = \exp(-\Delta E/k_B T) \quad (16.19)$$

where T = temperature and k_B = Boltzmann constant. A random number that is distributed evenly between 0 and 1 is chosen. If the number is smaller than $P(\Delta E)$, the new higher energy configuration is accepted; otherwise, it is discarded and the old configuration is used to generate the next arrangement. Reviewing Eq. (16.19), it can be seen that for extremely high temperatures, $P(\Delta E)$ approaches 1 and nearly all higher-energy system configurations will be accepted. As the system cools and the temperature decreases, $P(\Delta E)$ will become a smaller number and fewer higher-energy configurations are accepted. When the system nears its bottom temperature, $P(\Delta E)$ will be so low that the probability of accepting a higher-energy configuration will be small. The Metropolis algorithm simulates the random movement of atoms in a water bath at temperature T. By using Eq. (16.19), the system becomes a Boltzmann distribution. Figure 16.5 shows the flowchart of the Metropolis algorithm.

Kirkpatrick et al. (1983) again used a physical analogy to describe the annealing process applied to engineering problems. They replaced $k_B T$ by an "effective temperature" T. The "effective temperature" for a pump optimization problem, is a scalar number that consists of the pump energy cost and penalties for violation of system constraints (such as high or low pressure or extreme chemical concentrations). First, the system being optimized is "melted" by choosing a high "effective temperature" then the temperature is lowered slowly until the system "freezes." The process is carried out at each temperature until the system reaches a "steady state." The "annealing schedule" is the number of trial system rearrangements tested at each temperature and the sequences of temperatures.

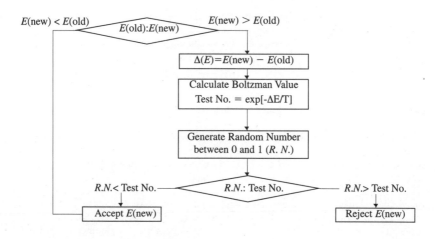

FIGURE 16.5 Metropolis algorithm.

By using the Metropolis algorithm, simulated annealing differs from iterative improvement because transitions out of local optimum are always possible even at low temperatures, whereas iterative improvement is likely to get stuck in a local optimum. This prevents the solution from becoming anchored into a local minimum. After a sufficient number of trials at a given temperature, the system is considered to be in equilibrium. The temperature is then lowered gradually, which slowly reduces the likelihood that a new a configuration with a higher cost is accepted. The process continue until the process has cooled sufficiently, which can be determined by the number of trials without accepting a "better configuration" or, as is often the case with real engineering problems, until a reasonable amount of computer time is used.

The analogy of annealing metals infers that the atoms are the same and there is a single basement energy level. Real engineering systems have constraints that interfere with each other, and is no configuration is likely to meet every constraint. Kirkpatrick et al. (1983) used the example of a system of bipolar magnets that could attract or repel each other. Such systems are a class of Hamiltonians, and the interference between constraints is defined as *frustration*. No single configuration can satisfy all the interactions simultaneously. This implies that there are many ground-states of nearly equal energy. Usually, the ground state energies are extremely low compared to the original random state and, when the process ends, transferring from one state into another requires considerable rearrangement.

Annealing-type optimization has several implications. Even in the presence of frustration, significant improvements can be expected over the random starting arrangement. Because many good, near-optimal solutions, should be available, a stochastic search procedure, such as annealing, should have a good chance of finding some ground states. No one ground state is expected to be significantly better than the others; therefore, searching for the absolute optimum solution is not useful.

Entropy is a measure of the variation of energy at a given temperature. At higher temperatures, the variation in energy is significant. The variation declines dramatically as the temperatures are lowered. This implies that the annealing process should not get "stuck" because transitions out of an energy state are always possible. Also, the process is a form of "adaptive divide-and-conquer." Gross features of the eventual solution appear at the higher temperatures, with fine details developing at low temperatures, (Kirkpatrick et al.,1983).

The requirements for applying simulated annealing to an engineering problem are (1) a concise representation of the configuration of the decision variables, (2) a scalar cost function, (3) a procedure for generating rearrangements of the system, (4) a control parameter (T) an annealing schedule, and (5) a criterion for termination (Dougherty and Marryott, 1991).

16.4.3 Development of Cost Function

The cost function for a given configuration is used in place of energy of a system of atoms. The cost function should include the cost of pumping and penalties for violations of the storage tank level, pressure, and water-quality bounds. The pumping cost is calculated for each period for each pump running and is a function of the pump flow rate, head, pump efficiency, and electricity rate during that period. The cost of violations of the pressure and water quality bounds are calculated using penalty functions. By adjusting the penalty functions, the optimization problem can be adjusted to bias one constraint over another constraint. The temperature T can be considered to be a control parameter that has the same units as the cost function.

The pump cost function can be described as

$$\text{COST}_{\text{pumps}} = K \sum_{pt} \frac{Q_{pt} TDH_{pt} P_t D_{pt}}{EFF_{pt}} \qquad (16.20)$$

where K = unit conversion factor, Q_{pt} = flow from pump p during period t, TDH_{pt} = operating-point total dynamic head for pump p during period t, P_t = the power rate per K wh during period t, D_{pt} = length of time that pump operates during period t (either 0 or 1 times the length of the period), and EFF_{pt} is the wire to water efficiency of pump p.

Storage tanks typically have a minimum level to provide emergency fire flow-storage. If the tanks are depleted below this level, fire protection is compromised. The following penalty term based on the constraint penalty terms developed by Brion (1990) has been developed to account for this constraint.

$$P_1 = \Sigma_t \beta_s (\min[0, y_{\min,s} - y_{st}])^2 \qquad (16.21)$$

where y_{st} = water level for tank s during period t, $y_{\min,s}$ = lower bound or the minimum level in tank s, and β_s = penalty term for tank low-level constraint for tanks.

Cohen (1982) stated that "optimization of a network over a limited horizon of, say, 24 hours has no meaning without the requirement of some periodicity in operation, a simple way to do that is to constrain the final states to be the same as the initial ones." A constraint is developed to generate a cost if the tank levels do not return to their starting elevation.

$$P_2 = \Sigma_s \beta_{s2} (\min[0, 1 - |y_{s,1} - y_{s,T}|])^2 \qquad (16.22)$$

where $y_{s,1}$ = water level for tank s at beginning of the simulation during period 1, $y_{s,T}$ = water level for tank s at the end of the simulation during the final period T, and β_{s2} = penalty term for beginning and ending tank level constraint for tank s.

For a 24-h simulation, the concept of returning tanks to their starting level is somewhat addressed by using a starting configuration in which the pumps supply a volume of water equal to the sum of the nodal demands. Providing a total volume of pumped water equal to the total volume of the system demands will return the tanks to their original level exactly if there is only one tank. Even if pumping equals exactly demand, the tanks may not return to their original level if there are several tanks unless all the tanks are full at the beginning. Because one tank may "supply" water to another tank during the simulation, the total volume stored will be the same but shifted from one tank to another.

The Cohen condition may be unnecessarily restrictive. The example used to compare the method involved modifying pump operations for a 24 h period that was repeated for 12 days to allow the variations in water quality to overcome the initial conditions. It was observed that the tanks exhibited periodic behavior and adjusted themselves until the pumps were supplying a quantity equal to the demands. If the pumps operated for longer periods, the tanks remained closer to full and the pump heads moved to the left of the system curve, reducing the flow. If the pumps operated for shorter periods, the tank levels lowered until the pump operating point moved along the pump operating curve until pumped flows increased to meet the demands. By running the simulations over several days, the pump operation can be scheduled to optimize efficiency and perhaps reduce cost.

A water distribution system needs to deliver water at sufficient pressure to service the system's customers but at a pressure that will not damage water systems or customer's facilities. The Uniform Plumbing Code sets the normal range of pressure as 15–80 psi (IAPMO, 1994). A city may have a range in a pressure zone from 40 to 80 psi with a 20-psi residual during a fire flow (Malcolm Pirnie, 1996). The water system needs to operate between two extreme pressures, $p_{\min}$ and $p_{\max}$.

The constraints and penalty functions for pressure bounds at each node are

$$P_3 = \Sigma_{k,t}\, \gamma_1\, (\min[0,\ p_{k,t} - p_{\min}])^2 \qquad (16.23)$$

and

$$P_4 = \Sigma_{k,t}\, \gamma_2 (\min[0,\ p_{\max} - p_{k,t}])^2 \qquad (16.24)$$

where $p_{\min}$ = minimum system pressure bound, $p_{\max}$ = maximum system pressure bound, p_{kt} = pressure at node k during time period t, and γ_1 and γ_2 = penalty terms for minimum and maximum pressure violations, respectively.

The EPANET program (Rossman, 1994) solves distribution systems hydraulically, then routes chemicals or contaminants through the water system during the time period. EPANET also can consider chemical reactions, such as the decay of chlorine, with time. The combination of hydraulic solver and water-quality calculations allows the program to predict the chlorine residual at any node in the system at any time period. A penalty term is used to consider violations of upper-and lower-limit free chlorine bounds.

The penalty terms for minimum and maximum free chlorine concentration are:

$$P_5 = \Sigma_{kt}\, \gamma_3 \left[\min(0,\ C_{kt} - C_{\min}) \right]^n \qquad (16.25)$$

$$P_6 = \Sigma_{kt}\, \gamma_4 \left[\min(0,\ C_{\max} - C_{kt}) \right]^n \qquad (16.26)$$

where $C_{\min}$ = minimum free chlorine concentration, $C_{\max}$ = maximum free chlorine concentration, C_{kt} = chlorine residual concentration at node k during time period t, and γ_3 and γ_4 = penalty terms for minimum and maximum chlorine residual pressure bound violations, respectively.

The value of n will usually be 2. In cases where the lower concentration bound is more important, a value of $n < 1$ will place a higher penalty on violations of minimum free chlorine. There also will be a cost term for the amount of chlorine used. The goal is to meet the chlorine residual bounds while using the smallest amount of chlorine. Not only will the operational cost be decreased, but the creation of total trihalomethanes will be reduced as well.

16.4.4 Sample Application

To illustrate the two approaches to solution, the primary zone of the North Marin Water Distribution System shown in Fig. 16.6 was used (Rossman, 1994, Vasconcelos et al. 1996). The system contains 115 pipes, 91 junction nodes, 2 pumps, 3 storage tanks, and 2 reservoirs. The minimum and maximum pressures at demand nodes were set at 20 and 100 psi, respectively. The desired minimum water storage height in all tanks was 5 ft. The minimum and maximum allowable concentration limits at all demand nodes were set at 50 µg/L and 500 µg/L respectively. The bulk and the wall-rate coefficients used in the simulation were -0.1 day^{-1} and -1 ft/day, respectively. The simulation was conducted for a total of 12 days, and the values at the last day were used to evaluate the objective function and constraints. The unit cost of energy was assumed to be constant at 0.07 $/kWh for all time periods, and the efficiency of both pumps was a constant 0.75.

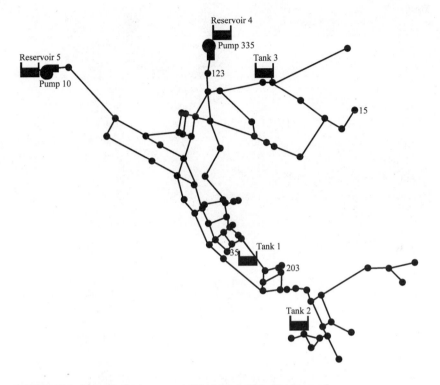

FIGURE 16.6 Water distribution system of North Marin Water District zone I (*Source:* Rossman, 1994).

Table 16.2 lists the optimized solutions obtained by using the mathematical programming and the simulated annealing approaches with a 500 μg/L concentration at both reservoirs, held constant throughout the simulation to minimize pumping time and pumping cost applications. The final results obtained by both approaches had no violations of any penalty term. Because pump 335 is larger than pump 10, closing pump 335 for a certain period of time has more effect than does closing pump 10. Since the strongest gradient resulted from closing pump 335, it was always closed first for the mathematical programming approach, which solves for the optimum solution by using reduced gradients. The simulated annealing approach had no bias because it randomly chose a pump and a time period to make a change (off to on or on to off). The total pump-operation times obtained from the mathematical programming approach were greater than the ones obtained from the simulated annealing approach. However, the total pump-operation times of pump 335 and the total 24-h cost of energy obtained from the mathematical programming approach were lower than the ones obtained from the simulated annealing approach because the mathematical programming approach preferred to close the larger pump (335) first. The total number of EPANET calls were 425 and 356 for minimizing the pump operation time and pump operation cost, respectively, for the mathematical programming approach, whereas 1500 calls were made for both cases when the simulated annealing approach was used.

TABLE 16.2 Optimized Solutions Obtained by Using Different Solution Approaches for 500 ug/L of Concentration at Both Reservoirs

	Minimize Pump Operation Time		Minimize Pump Operation Cost	
	Mathematical Programming	Simulated Annealing	Mathematical Programming	Simulated Annealing
Original objective function	34.47	24.00	399.95	429.53
Concentration violation	0.00	0.00	0.00	0.00
Pressure violation	0.00	0.00	0.00	0.00
Tank water level violation	0.00	0.00	0.00	0.00
Operation time of pump 10 (h)	19.97	5.00	24.00	14.00
Operation time of pump 335 (h)	14.50	19.00	13.62	17.00
Total pump operation time (h)	34.47	24.00	37.62	31.00
Total 24 h energy cost ($)	401.06	433.63	399.95	429.53

Figures 16.7 and 16.8 show the optimal pump operation times for pumps 10 and pump 335 obtained by using the mathematical programming approach and the simulated annealing approach for minimizing pump operation time, respectively. Similarly, the optimum pump operation times of pumps 10 and 335 obtained by using both approaches for minimizing pump operation cost, are shown in Figs. 16.9 and 16.10, respectively.

16.4.5 Advantages and Disadvantages of the Two Methods

Both solution approaches are capable of finding the optimum solution for the pump operation problem for water-quality purposes. The mathematical programming approach requires the calculation of the derivatives of the objective function with respect to pump operation times (reduced gradients), which makes the simulated annealing approach more flexible and adaptable.

Because the mathematical programming approach tries to reduce the most costly (larger) pump operation times the result is higher total pump operation times compared to the simulated annealing approach. There were distinct differences in the total pump operation times, whereas similar results were obtained for the total 24-h energy costs.

The mathematical programming approach tries to find one "global optimum" solution, whereas the simulated annealing approach finds many solutions that have total penalties close to each another. The solutions obtained using the mathematical programming approach depend on the initial values of the penalty function parameters, which makes finding the global optimum more difficult. The global optimum solution cannot be guaranteed because convexity of the objective function cannot be proved.

16.5 OPTIMAL SCHEDULING OF BOOSTER DISINFECTION

Water utilities and regulatory agencies often want a detectable disinfectant residual at all points of water consumption in a water distribution network. Such a residual can be difficult to maintain when disinfectant is added at a single location, such as a treatment plant, because the spatially distributed nature of water storage facilities and consumer

Optimization Models for Operations **16.29**

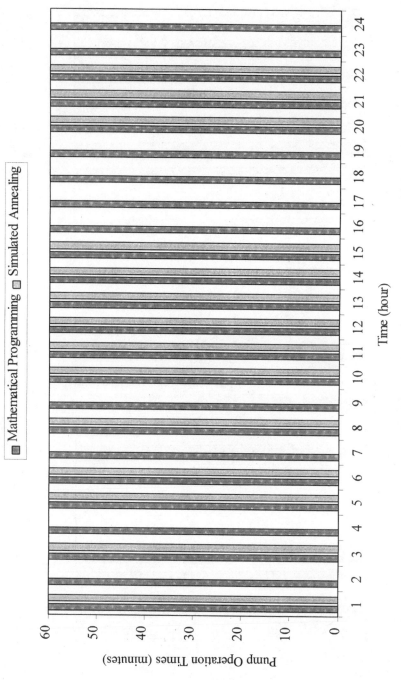

FIGURE 16.7 Optimal operation times of pump 10 used in the North Marin Water District to minimize the pump's operation costs.

16.30 Chapter Sixteen

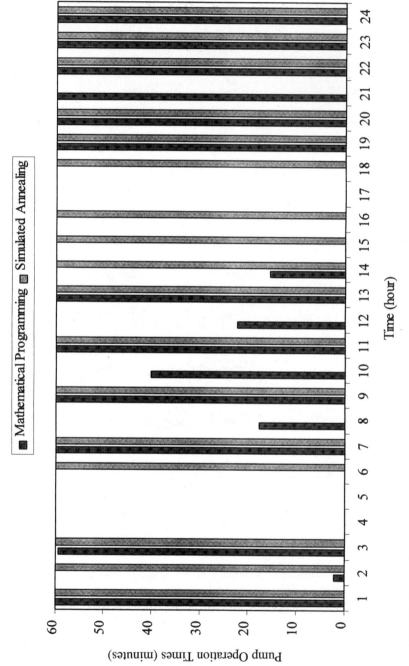

FIGURE 16.8 Optimal pump operation times of pump 335 used NMWD for minimizing pump operation time.

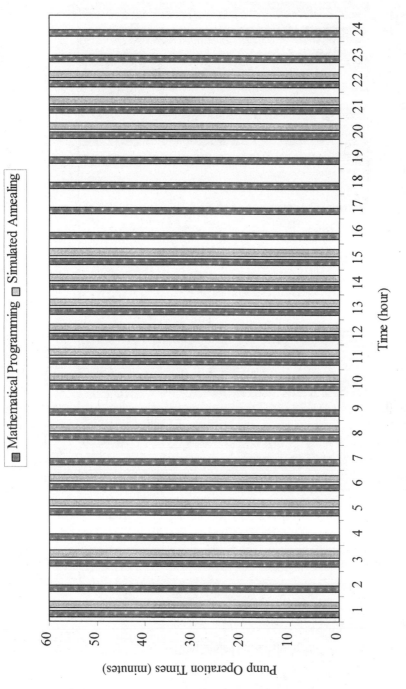

FIGURE 16.9 Optimal operation times of pump 335 used in the North Marin Water District to minimized the pump's operation costs.

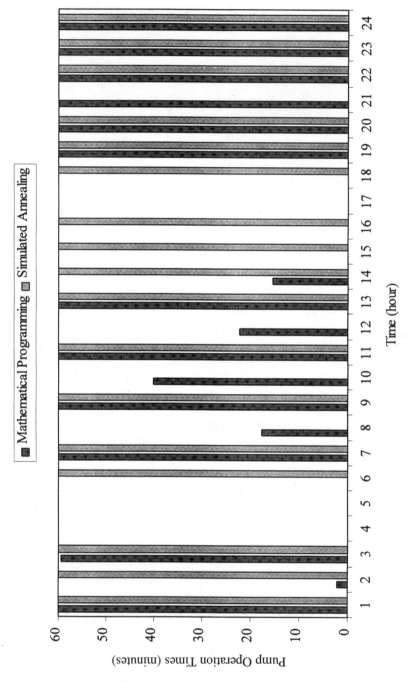

FIGURE 16.10 Optimal pump operation times of pump 335 used NMWD for minimizing pump operation costs.

demands leads naturally to a wide distribution in water travel times. Because all disinfectants that can maintain a residual will react with a variety of inorganic and organic compounds, they decay over time; thus, the addition of any single disinfectant must be sufficient to maintain a detectable residual for the longest travel time. This requirement can lead to unacceptably high residuals close to the point of addition and to an uneven spatiotemporal residual distribution. Both situations may by undesirable from the perspective of unpleasant tastes and odors or public health (e.g., unpleasant tastes of excessive chlorine or chlorinated organic compounds or the health impact of chlorinated by-products (Bull and Kopfler, 1991).

To address the difficulties of residual maintenance in distribution systems, one may decide to distribute injections of disinfectant to strategic locations in the distribution network. This practice of "booster disinfection" has been practiced by utilities for some time. However, it can be difficult to determine: (1) the best number and locations of disinfectant additions, and (2) the best way to schedule or control the dose at each location over time. These difficulties arise principally from complex dynamic hydraulics that affect transport of disinfectant, coupled with decay of disinfectant caused by bulk and wall reactions.

This section presents mathematical programming formulations (optimization models) and an associated design approach for locating and scheduling additions of disinfectant in distribution networks. Three formulations are presented. The first formulation attempts to identify optimal dosage rates of disinfectant at booster stations—in the sense that the total disinfectant dosage is a minimum—by assuming that the number of booster stations and their locations are known. The resulting LP problem can be solved by the standard simplex algorithm. The second formulation combines the function of the first formulation with a determination of the best locations of the booster stations from among a set of potential locations. This formulation is a mixed-integer linear programming (MILP) problem with no particular structure and is much more difficult to solve using standard techniques. The third formulation shows how a pure optimal location problem—which minimizes the number of booster stations so that the ratio of the minimum to maximum residual concentrations is controlled—can be expressed as a particular type of integer linear programming (ILP) problem called the maximum set-covering problem. Doing so is a great advantage because such ILP problems can usually be solved easily and efficiently using the simplex method or heuristics.

All the above formulations share the following characteristics: (1) they assume known time-varying network hydraulics (i.e., a known and repeating demand pattern) and first-order disinfectant decay kinetics, (2) they consider the dynamic hydraulic behavior in the network explicitly and logically to achieve a goal of adequate long-term operation, (3) they effectively embed the information and assumptions contained within standard extended-period network water-quality models within linear optimal design formulations, and (4) they can be solved in reasonable computer time for practical networks, with the possible exception of the second formulation above.

Following presentation of the model formulations, solution approaches are discussed along with a suggested design approach that uses the optimization model formulations in concert so that the maximum set-covering model is used as a screening model to select locations for use by either the optimal location and scheduling (MILP) or optimal scheduling (LP) models. Available software to facilitate development and solution of these models also is discussed.

16.5.1 Background 1: Linear Superposition

In a water distribution system, the monitoring and booster locations coincide with nodes of a dynamic hydraulic and water-quality network model (Boulos et al., 1995; Grayman et. al., 1988; Liou and Kroon, 1987; Rossman and Boulos, 1996; Rossman et al., 1993).

The concentration of disinfectant at monitoring node j and time period m, denoted $c_j^m(u)$ [M/L^3], generally will depends on the mass injection rate of disinfectant at booster node i and time period k, denoted u_i^k [M/T], for all booster locations i and time periods $k \leq m$ (u is the vector of all such mass injection rates). This dependence can be complicated owing to significant time delays for transporting water (and thus disinfectant) between booster and monitoring nodes and to multiple transport paths between any pair of booster and monitoring locations. Under certain assumptions that are common to network water-quality simulation models. The principle of linear superposition can be applied to express the concentration $c_j^m(u)$ as a *linear function* of the mass injection rates u. This linear relationship between injection rates and concentration is a powerful tool for developing practical optimal design formulations, and it underlies each optimization model formulation discussed in this section; thus, it is worthwhile to discuss briefly how linear superposition can be applied to modeling the dynamics of chlorine concentrations resulting from dosages at multiple booster stations.

We assume that (1) the water distribution system hydraulics are known and time-varying, (2) disinfectant decay kinetics are first-order with respect to disinfectant concentrations, and (3) reaction rate coefficients are independent of the booster injections u. Boccelli et al. (1998) showed that, under these conditions, transport of disinfectant in water distribution networks with time-varying hydraulics and disinfectant doses can be described mathematically as a linear dynamic system; thus, the principle of linear superposition is applicable (e.g. Luenberger, 1979). They also give a more intuitive explanation of linear superposition by way of a simple network example. Using a different approach, Zierolf et. al., (1997) derived a recursive expression for $c_j^m(u)$ by backtracking through the distribution system over all transport paths to find the superimposed impact of all sources of disinfectant. These works show that, for any arbitrarily complicated distribution system, the monitoring concentration $c_j^m(u)$ can be defined as a linear summation of individual booster-injection influences:

$$c_j^m(u) = \sum_{i=1}^{n_b} \sum_{k=1}^{m} a_{ij}^{km} \, u_i^k \qquad (16.27)$$

Where $a_{ij}^{km} = \partial c_j^m / \partial u_i^k$ ((M/L^3)/(M/T)) corresponds to the coefficients of the discretized impulse response function (Chow et al., 1988; Oppenheim and Willsky, 1997) describing the effect of dose u_i^k on the concentration at monitoring node j and time m. The impulse response coefficients can be calculated via network water-quality simulation software (see below); once computed, Eq. (16.27) is an accurate mathematical statement of the effect of changes in u on residual concentrations and, because of its simple form, allows development of efficient mathematical models to optimize dose magnitudes and their locations.

16.5.2 Background 2: Dynamic Network Water-Quality Models in a Planning Context

To satisfy engineering objectives, the booster disinfection dosages should maintain adequate disinfectant residuals at all times, much as the hydraulic infrastructure should be designed always to maintain minimum pressures. Nevertheless, in a planning context, assumptions about the *future variation* in dose injections and network hydraulics are needed to construct practical models for locating and scheduling booster dosages. Such assumptions are similar in concept to adopting peak demand plus fire flow as a basis for the design of hydraulic networks. One assumes that networks designed on that basis will operate in an acceptable manner despite unknown future disturbances (i.e., the actual demands will, in general, never equal the demand scenario assumed for design. However, the assumed scenario is nevertheless useful because it leads to acceptable designs). The essential problem here is that when using dynamic models as a basis for design (as opposed to steady-state models), one must treat the initial conditions carefully because, although the dynamic water quality response to booster dosing must be reflected in the design, these dynamics cannot be based on any arbitrary initial conditions. The idea is to make certain that the water-quality dynamics are periodic, then base the design on a representative "snapshot" of those periodic dynamics. These conceptual design issues are discussed more fully below, after presentation of the design assumptions and conditions.

We assume that discrete mass injection rates are periodic so that the mass dosage variation at each booster station has a cycle time $T_s = n_s \Delta t$, where n_s is the number of injection rates contained in one scheduling cycle and Δt is the duration of one discrete mass injection period (cycle time is analogous to wavelength). Such an assumption defines periodic mass dose rates v_i^k such that $v_i^k = u_i^{k+qn_s}$, $\forall q, k = 1,\ldots, n_s$. Thus Eq. (16.27)

$$c_j^m(v) = \sum_{i=1}^{n_b} \sum_{k=1}^{m} \alpha_{ij}^{km} v_i^k \qquad (16.28)$$

where the composite impulse response coefficients α_{ij}^{km} quantify the response of concentration at a monitoring location to a unit periodic dose at a booster station.

If, in addition to injection rates, the α_{ij}^{km}, $\forall i,k$, are periodic with an impulse response cycle time $T_a = n_a \Delta t_m$ where n_a is the number of monitoring times contained in one impulse-response cycle and Δt_m is the disinfectant residual-monitoring time step, the monitoring concentration c_j^m are themselves periodic with a cycle time T_a (Eq. 16.28). Accordingly, the composite impulse response coefficients are assumed to be periodic with cycle time T_a, in which case it is sufficient for design purposes to consider only one cycle of these dynamics (this assumption, as well as the conditions under which a_{ij}^{km} is periodic, is discussed below). Such a periodicity assumption is one practical solution to the indeterminacy inherent in many long-range planning problems, in which the underlying processes are dynamic and suggest a strong cyclical component.

If the optimal disinfectant concentrations are to be periodic, other characteristics of the system must be consistent with such behavior (i.e., merely asserting that residual concentrations must be periodic does not make them so). First, we have already discussed that optimal disinfectant doses are assumed to be periodic with a cycle time T_s. Intuitively, the monitoring node concentrations could not be periodic if a key driving force—the dose schedules—were not. Second, the hydraulic dynamics obviously play an important role in the resultant concentration dynamics. Specifically, we assume periodicity of the network's hydraulic dynamics with cycle time T_h (i.e., the flow rates, tank elevations, and pressures are periodic with cycle time T_h). Periodic network hydraulics would, on a practical level, require periodic water demands at the network nodes. Of course, the actual demands will differ from those assumed for design; thus, the design, if implemented, will perform differently from model predictions. Again, there is no way completely around this difficulty; assumptions about the future are a routine and necessary part of every long-term design or planning problem.

Boccelli et al. (1998) have shown that the monitoring node concentrations and the composite impulse response coefficients have a cycle time $T_a = \eta T_s = \mu T_h$ for some integers $\eta, \mu > 0$. Thus, long-term periodicity of monitoring node concentrations requires long-term periodicity of both dose schedules and assumed hydraulic dynamics, and the relationship between the independent cycle times T_s and T_h is sufficient to determine the cycle time T_a. To take a specific example, if hydraulic dynamics are periodic on a 24-h cycle and booster dosages are periodic on a 12-h cycle, then the residual concentrations will become periodic on a 24-h cycle ($n = 2, h = 1$).

By achieving periodic concentrations of disinfectant assuming periodic dose schedules and network hydraulics, important dynamic processes are allowed to influence the optimal schedule while maintaining reasonable data requirements. This point is emphasized by considering other plausible assumptions about the disinfectant concentration dynamics, and the hydraulic forcing that drives those dynamics. For example, one might assume steady-state concentrations forced by steady-state hydraulics. Arguably, the assumptions of dynamic concentrations forced by dynamic hydraulics—even with the pragmatic assumption of periodicity—leads to a more realistic representation of transport and decay of disinfectant in water distribution systems, a point emphasized by some of the results Boccelli et al., (1998) discussed.

16.5.3 Optimal Scheduling of Booster-Station Dosages as a Linear Programming Problem

This section presents a formulation for optimizing booster dose schedules defined by the periodic dose rates v_i^k, $i = 1,..., n_b$, $k = 1,..., n_s$. The locations of n_b booster stations are assumed to be known. The design objective is to minimize the total disinfectant mass rate applied over one period of the concentration dynamics. This objective is intended as a surrogate for minimizing the formation of disinfection by-products, chemical costs, and objectionable tastes and odors associated with chlorination of natural waters. However, actual reductions in the formation of disinfection by-products or objectionable tastes and odors cannot be quantified at this time. The mass injections are required to satisfy lower and upper bound constraints on disinfectant residual at n_m monitoring locations and over all time. These constraints are consistent with environmental regulations calling for a detectable residual at all points of consumption to serve as a barrier against microbiological contamination of the distribution system.

The model formulation is stated as the following LP problem:

$$z = \min \sum_{i=1}^{n_b} \sum_{k=1}^{n_s} v_i^k \tag{16.29}$$

subject to

$$c^{\min} \leq c_j^m(v) = \sum_{i=1}^{n_b} \sum_{k=1}^{n_s} \alpha_{ij}^{km} \cdot v_i^k \leq c^{\max} \tag{16.30}$$

$$j = 1,..., n_m, m = 1,..., n_a$$

$$v_i^k \geq 0, i = 1,..., n_b, k = 1,..., n_s \tag{16.31}$$

where $c^{\min}$ and $c^{\max}$ = minimum and maximum concentration limits within the distribution system and n_a = number of monitoring time periods contained in one period of the concentration dynamics.

16.5.4 Optimal Location and Scheduling of Booster-Station Dosages as a Mixed-Integer Linear Programming Problem

This formulation extends the above optimal scheduling model to consider the optimal booster-station locations. The extension of Eqs. (16.29)–(16.31) yields MILP problem

(Hillier and Lieberman, 1980). Once again, the objective is to minimize the total mass dose rate during one period of the concentration dynamics:

$$z = \min \sum_{i=1}^{n_b} \sum_{k=1}^{n_c} v_i^k \qquad (16.32)$$

subject again to restrictions on the concentrations at monitoring nodes, over all time:

$$c^{\min} \leq c_j^m(v) = \sum_{i=1}^{n_b} \sum_{k=1}^{n_s} \alpha_{ij}^{km} \cdot v_i^k \leq c^{\max} \qquad (16.33)$$

$$j = 1,...n_m, \ m = 1,..., n_a$$

Binary variables δ_i are introduced to determine whether a new booster station is ($\delta_i = 1$) or is not ($\delta_i = 0$) to be built at location i. If there is no booster station at location i, then additional constraints must ensure that the total mass dose is zero at that location:

$$\sum_{k=1}^{n_s} v_i^k \leq M_i \cdot \delta_i, \ i = 1,..., n_b \qquad (16.34)$$

where M_i is a positive constant equal to an upper bound on the total mass dose at booster location i (assumed to equal the dose for which an entering concentration of $c^{\min}$ would exit location i at concentration $c^{\max}$, which can be derived from knowledge of the total flow rate exiting location i over time); thus, when $\delta_i = 1$, the total dose at location i is unrestricted, whereas when $\delta_i = 0$, the total dose is constrained to equal zero. This "big M" formulation is common in problems involving a facility's location or in "fixed-charge" problems in general, and it is used because it preserves the linearity of the formulation (it does not involve any products of δ_i and v_i^k, which would result, for example, if instead of Eq. (16.34), one replaced the v_i^k in Eq. (16.33) with the product $v_i^k \delta_i$).

A restriction is placed on the total number of booster stations to be built, $\bar{n}_b < n_b$, which is a surrogate for the cost of installing booster stations:

$$\sum_{i=1}^{n_b} \delta_i \leq \bar{n}_b \qquad (16.35)$$

Finally, the variables δ_i are restricted in value to zero or one, and again the dose rates must be positive:

$$\delta_i = \{0, 1\}, \ i = 1, ..., n_b \qquad (16.36)$$

$$v_i^k \geq 0, \ i = 1, ..., \bar{n}_b, \ k = 1, ..., n_s \qquad (16.37)$$

The above MILP problem [(Eqs. (16.32)–(16.37)] can be solved by the branch-and-bound technique with the simplex method (e.g., Hillier and Lieberman, 1980). A minor modification of the formulation allows consideration of booster stations that already exist, combined with new potential locations, so that the operation of preexisting facilities can be optimized simultaneously with the location of new ones. However, this MILP model is significantly more difficult to solve than the LP scheduling model (solution of the scheduling model is essentially incorporated as a subtask of the branch-and-bound procedure for solving the MILP location model). In practice, it may be that the MILP model can be applied only for analysis of large networks if a relatively small set of potential locations has been determined by experience or other means (i.e., if n_b is small—most likely much smaller than the total number of network nodes). This computational concern leads us to the following formulation of the location model as a maximum set-covering model, which might be used as a method of locating booster stations in its own right, or as a screening tool to select good potential locations for the MILP formulation.

16.5.5 Optimal Location of Booster Stations as a Maximum Set-Covering Problem

This section shows how the booster-station location problem can be formulated as a classical integer linear programming (ILP) model called the maximum set-covering (MSC) model. Such a formulation brings with it significant computational advantages, to the point where medium to large networks can be treated. All network nodes can be considered to be potential booster locations, if that is desired, while maintaining a computationally tractable model. To gain these computational advantages, we give up the ability to optimize the dose schedules, preferring instead to treat all booster stations as if they would be operated in a rather typical fashion. So, unlike the MILP formulation presented above, the MSC formulation does not build on the optimal scheduling LP model directly and thus does not combine the two functions of optimal scheduling and optimal location. For this reason, we suggest that the MSC model be used in concert with either the optimal scheduling or optimal scheduling\location model formulations presented above. This work draws on earlier results in facility location models and, in particular, the maximal covering location model (Church and Revelle, 1974), which also has provided inspiration for models that locate wells in groundwater monitoring networks (Meyer and Brill, 1988).

The basic MSC booster-station location model is relatively simple to state mathematically; we do this first, leaving some of the details and motivation for later. Define a set of indexes of potential booster locations N_j^m so that booster location i is included in N_j^m if, and only if, disinfectant dosing at i elicits a "significant response" at monitoring location j and time period m. Thus N_j^m is the set of indexes of all potential booster locations that are assumed to have an effect on residual at monitoring location j and time m. We discuss the precise definition of a "significant response" and thus of the sets N_j^m below. For now, one can simply assumed that if a booster location elicits a "significant response" at a monitoring location and time, then under reasonable conditions for operation of that booster station, an acceptable disinfectant residual can be maintained at that location and time by operation of only that one booster station.

Again, define the binary variables δ_i to indicate whether a booster station is ($\delta_i = 1$) or is not ($\delta_i = 0$) to be constructed at location i. We seek to minimize the number of booster stations (a surrogate objective for minimizing the cost of booster station installation and operation) using

$$\min \sum_{i=1}^{n_b} \delta_i \qquad (16.38)$$

where n_b, again, is the number of potential booster-station locations but is likely to be a much larger set than would be used in the MILP formulation (e.g., the set of all network nodes might be used as the potential booster locations). In minimizing the number of booster stations, we require each monitoring location and time to include a significant response from at least one selected booster location. Otherwise, it will be impossible to maintain the residual concentration at that monitoring point. Mathematically, we have the following constraints:

$$\sum_{i \in N_j^m} \delta_i \geq 1, j = 1,\ldots, n_m, m = 1,\ldots, n_a \tag{16.39}$$

Whereas with the above LP and MILP models, the constraints (Eq. 16.39) are written for one cycle of the periodic residual-concentration dynamics. Solution of Eq. (16.38) subject to Eq. (16.39) will yield the minimum number of booster locations that provide spatial and temporal "coverage" for the network because the residual concentration at each monitoring location and time is influenced by at least one selected location.

The coverage set N_j^m is defined, more precisely, to include booster location i if, and only if, a unit set point concentration at location i elicits a residual concentration response exceeding a threshold $0 < \eta \leq 1$ at monitoring location j and time period m. The reason for specifying the response in terms of a unit set point concentration is to decouple the decision about the optimal dose schedules from that about the optimal dose locations. If this decoupling is to be accomplished, one must assume a logical scenario for operation of the booster stations. Here, we define such a scenario as that where each potential booster station is operating in an identical manner so that the mass dose schedule applied leads to a uniform unit exit concentration from the booster station. Such a mode of operation would be consistent with a local feedback controller that adjusts the dose rate to maintain a unit set point.

The coverage sets can now be defined in terms of the impulse response coefficients α_{ij}^m:

$$N_j^m = \{i | \alpha_{ij}^m \geq \eta\} \tag{16.40}$$

Where the impulse response coefficients α_{ij}^m are defined in a similar fashion as in Sec. 16.5.3 except they indicate the periodic concentration response at monitoring location j and time m as a result of a particular periodic mass dosage at booster location i so that a unit set point concentration is maintained at location i at all times.

To attach some meaning to the residual coverage parameter η, we proceed as follows. First, we could just as well define the coverage sets as follows:

$$N_j^m = \{i | c_j^m \geq c^{\min}\} \tag{16.41}$$

where c_j^m is the concentration that results from maintaining a set point concentration at booster station location i and $c^{\min}$, again, is the minimum disinfectant residual allowed or desired. Furthermore, the maximum residual concentration in the network is the set point concentration maintained at the booster locations because of disinfectant decay. Since all booster stations are assumed to be operating identically (at the same exit set-point concentration), let us call this set-point concentration $c^{\max}$ —the maximum anywhere in the network. Linear superposition then allows us to define the concentration c_j^m in Eq. (16.41) in terms of the impulse response coefficient a_{ij}^m: $c_j^m = a_{ij}^m c^{\max}$; substitution in Eq. (16.41) and rearranging yields

$$N_j^m = \{i | \alpha_{ij}^m \geq c^{\min}/c^{\max}\} \tag{16.42}$$

Thus, comparing Eqs. (16.40) and (16.42), the residual coverage parameter $\eta = c^{\min}/c^{\max}$ has a meaningful interpretation as the ratio of the minimum to maximum residual concentration in the network—arguably a useful measure of residual variability. In fact, η is interpreted more accurately as a lower bound on $c^{\min}/c^{\max}$ for booster stations operated to maintain an identical set point since any set of booster stations satisfying Eq. (16.39) will include at least one booster station for every monitoring location and time so that $a_{ij}^m \geq \eta$.

The residual coverage parameter will be a key determinant of solutions to the MSC booster location model. As a mental exercise, one can consider a network where every node (and thus every monitoring location) is a potential booster location. The optimal MSC solutions for the extreme values of $\eta = \epsilon$ (where ϵ is an arbitrarily small strictly positive constant) and $\eta = 1$ are known immediately. When $\eta = \epsilon$ the solution is a minimal set of booster locations restricted to those associated with network water sources (this is the conventional case where disinfectant is added only at treatment nodes). When $\eta = 1$, the solution includes every node in the network, which is the only way to have a perfectly uniform concentration at each monitoring node. Thus, η will be related to the number of booster stations required to achieve "coverage" of the monitoring locations and times, and a plot of the optimal number of booster stations (from solution of the MSC problem) versus the residual coverage parameter η indicates the optimal trade off between the variability of disinfectant residual and a measure of residual maintenance costs.

16.5.6 Solution of the Optimization Models

A critical step in solving each optimization model described above is calculation of the appropriate impulse response coefficients. Furthermore, these impulse response coefficients must convey the influence of a periodic dosage on the resulting periodic concentration dynamics. We stress that, as was discussed above, hydraulic and disinfectant-dose periodicity is a necessary condition for long-term residual concentration periodicity, yet little can be said about the time required for the residuals to become periodic (say, in a network simulation). Briefly, the dynamic concentrations can be said to consist of two components: nonperiodic dynamics resulting from the initial conditions (initial concentrations in pipes and reservoirs, as specified in a simulation) and periodic dynamics from the periodic dosages and hydraulics. According to the design approach advocated here, the initial conditions are considered to be arbitrary and thus their effect must be ignored when computing the impulse response coefficients. Software is available which accomplishes this when the response coefficients are calculated via perturbation methods (see Sec. 16.5.7).

The LP scheduling model is relatively straightforward to solve using existing methods for general LP problems and, specifically, the simplex method. The size of the problem for practical networks is within the capability of commercial LP software because the number of decision variables depends on the selected number of scheduling periods times the number of booster stations $n_s \times n_b$ and is otherwise independent of the size of the pipe network and system dynamics. However, the size of the constraint set will depend on the pipe network through the number of monitoring locations n_m and the impact cycle time T_α. In any case, at least for the most straightforward method of calculating the impulse response coefficients, the total efficiency of solution seems to hinge more on the problem setup than on the actual LP solution. The calculation of composite impulse response coefficients α_{ij}^{km} will at least be a significant component of the total computational burden.

Much of what has been said above also applies to solving the MILP scheduling/location model. However, solution by branch-and-bound with the simplex method is far more intensive computationally, and no reliable methods exist for estimating the solution time a priori. It also is likely that algorithmic options or problem-specific details will greatly influence the computation time. In short, solution is not straightforward, but it is possible for practical networks, provided that the number of potential booster locations is not too large and that sufficient time is devoted to understanding the effect of algorithmic options and tolerances on branch-and-bound performance.

The MSC model formulation is a special category of ILP optimization models. Models of this type are usually efficient to solve compared with IP models of the same dimension that lack any particular structure; it is not unusual that the relaxed solution of an MSC model (where the binary variables are allowed to vary continuosly between zero and one, yielding a pure LP problem) is a natural integer solution, in which case the solution is efficient to obtain via the simplex algorithm. Heuristics also have been invented on the basis of the special form of the constraints. Similar models (in other application contexts) have been solved using LP plus heuristics, LP plus branch-and-bound implicit enumeration, and simulated annealing (Church and Revelle, 1974, Meyer and Brill, 1988; Meyer et al., 1994).

16.5.7 Available Software

A computer code called Booster Disinfection Design Algorithms (BDDA) (Uber et al., 1999) has been developed to interface with EPANET (Rossman, 1994), a distribution system water-quality model, to produce a specification of the above optimal booster scheduling, location problems, or both automatically, given a calibrated dynamic network water-quality model. The code can produce a specification of the optimization model formulations in standard MPS format (Murtagh, 1981). This specification of the LP/MILP/MSC problem can then be solved using any available implementation of the simplex algorithm (LP) or branch-and-bound with the simplex method (MILP/MSC). Alternatively, the code interfaces directly with the CPLEX LP and MILP solution algorithms (CPLEX, 1988) for users of that commercial software. The standard EPANET data file contains information about the physical and chemical characteristics of the distribution system. The BDDA application requires additional information that is relevant to the optimization model formulation, including potential and existing booster locations, booster schedule cycle time (T_s), number (n_s) and length of mass injection intervals for each booster location, monitoring of node locations, and lower and upper residual bounds (c^{min} and c^{max}, respectively).

Figure 16.11 illustrates the steps required to set up the LP optimal scheduling problem for solution; the MILP and MSC problem setup is similar. First, the network hydraulics are simulated for use during subsequent water-quality simulations; the cycle time T_h is determined from this hydraulic solution. T_a can be determined by values of $\eta, \mu > 0$ so that $T_a = \eta T_s = \mu T_s = \mu T_h$ is a minimum. The matrix generator then computes the composite impulse-response coefficients by a straightforward perturbation method. Each booster location i and periodic dose interval k is selected in turn and is modeled as a periodic source of disinfectant (so that $u_i^{k+pm_s} = \hat{v}, \forall p \geq n$). Because the quantity $\hat{v}$ is a sufficiently large dose rate, roundoff error is avoided in the calculation of the impulse response coefficient. The matrix generator uses simulated concentrations for the n_m monitoring nodes (c_j^m) to calculate the composite impulse-response coefficients corresponding to the periodic mass injections ($a_{ij}^{km} = c_j^m/\hat{v}, m = M, ..., M + n_a - 1$). The value of the time period M is estimated as the time when differences in impulse-response coefficient values over successive impact cycles (of length T_a) are considered to be insignificant ($|\alpha_{ij}^{km} - \alpha_{ij}^{k,m-n_\alpha}| \leq \epsilon$). Once M has been determined (for a particular booster node i and injection period k), the BDDA records the column of the LP coefficient matrix corresponding to v_i^k. This iterative procedure continues until all booster locations and injection periods have been analyzed. The BDDA then adds the neccesary constraint information to the MPS description of the LP problem for solution by a commercial optimization algorithm, or it interfaces directly with the appropriate CPLEX modules.

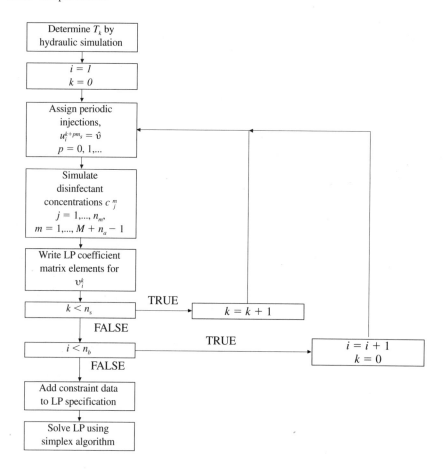

FIGURE 16.11 Setup of optimal scheduling of booster disinfection dosages as a linear programming problem.

16.5.8 Summary

Optimization model formulations were described for optimal location and scheduling of booster disinfection dosages throughout a distribution network and throughout time. Objectives include minimizing the total mass of disinfectant applied at multiple points in a distribution system and minimizing the total number of booster stations constructed. Constraints ensure the maintenance of adequate disinfection residuals at all monitoring nodes and at all times. The key to these formulations is to notice that linear superposition is applicable to transport of disinfectant in distribution networks, assuming that the future dynamic hydraulics are known. By assuming periodicity of the hydraulics and requiring the optimal dose schedules to be periodic, formulations were derived that logically incorporate the important influence of disinfectant concentration dynamics into the booster scheduling and location designs.

REFERENCES

Alla, P., and P. A. Jarrige, "Optimal Control of the West Parisian Area Water Supply Network," *Proceedings of the 16th Annual Conference of ASCE Water Resources Planning and Management Division,* pp. 661–664. 1989,

Awumah, K., and K. Lansey, "Energy Efficient Pump Station Operation with a Pump Switching Constraint," *Proceedings of the Water Resource Sessions at Water Forum 92,* ASCE, pp. 604–609. 1992,

Bao, Y. X., and L. W. Mays, "New Methodology for Optimization of Freshwater Inflows to Estuaries," *ASCE Journal of Water Resources Planning and Management,* 120:199–217, 1994a.

Bao, Y. X., and L. W. Mays, "Optimization of Freshwater Inflows to the Lavaca-Tres Palacios, Texas, Estuary," *ASCE Journal of Water Resources Planning and Management,* 120:218–236, 1994b.

Boccelli, D., M. Tryby, J. Uber, L. Rossman, M. Zierolf, and M. Polycarpou, "Optimal Scheduling of Booster Disinfection in Water Distribution Systems," *ASCE Journal of Water Resources Planning and Management,* 124: 1998.

Boulos, P. F., T. Altman, P. Jarrige, and F. Collevati, "Discrete Simulation Approach for Network-Water-Quality Models," *ASCE Journal of Water Resources Planning and Management,* 121:49–60, 1995.

Bree, D. W., D. H. Budenaers, D. N. Horgan, and L. C. Markel, "Improved Methodology for Design and Operation of Water Distribution Systems," Final Technical Report, OWRT Project No. C-6198, Systems Control, Palo Alto, CA, June 1976.

Brion, L. M., "Methodology for Optimal Operation of Pumping Stations in Water Distribution Systems," Doctoral dissertation, University of Texas, Austin, Texas, 1990.

Brion, L. M., and L. W. Mays, "Methodology for Optimal Operation of Pumping Stations in Water Distribution Systems," *ASCE Journal of Hydraulic Engineering,* 117(11):1551–1569, 1991.

Bull, R. J., and R. C. Kopfler, *Health Effects of Disinfectants and Disinfection By-products,* American Water Works Association Research Foundation, Denver, CO, 1991.

Carpentier, P., and G. Cohen, "Decomposition, Coordination and Aggregation in the Optimal Control of a Large Water Supply Network," *Proceedings of the 9th Triennial IFAC World Congress,* Budapest, Hungary, pp. 3207–3212, 1984.

Caves, J. L., and T. C. Earl, "Computer Applications: A Tool for Water Distribution Engineering," *Journal of the American Water Works Association,* 75:230-235, 1979.

Cembrano, G., M. Bryds, J. Quevedo, B. Coulbeck, and C. Orr, "Optimization of a Multi-Reservoir Water Network Using a Conjugate Gradient Technique: A Case Study," *Proceedings of the 8th INRIA International Conference on Analysis and Optimization of Systems,* Antibes, France, pp. 987–999, 1988.

Chase, D., and L. Ormsbee, Optimal Pump Operation of Water Distribution Systems with Multiple Storage Tanks, *Proceedings of the American Water Works Association Computer Specialty Conference,* American Water Works Association, pp. 205–214, 1989.

Chase, D., and L. Ormsbee, "An Alternate Formulation of Time as a Decision Variable to Facilitate Real-Time Operation of Water Supply Stems," *Proceedings of the 18th Annual Conference of the ASCE Water Resources Planning and Management Division,* pp. 923–927, 1991.

Chase, D., D. Guastella, K. Lai, and L. Ormsbee, "Energy Management for Large-Scale Water Supply Systems," *Proceedings of the AWWA Computer Conference,* Los Angeles, CA, April 1994.

Chen, Y., "Application of Time Series Analysis to Water Demand Prediction," in B. Coulbeck and C. Orr, eds., *Computer Applications in Water Supply,* Vol. 2, Research Studies Press, Somerset, UK, pp. 289–295, 1988a.

Chen, Y., "Simplification of Water Distribution Systems for Optimal Operations," *Computer Applications in Water Supply,* Vol. 2, Research Studies Press, Somerset, UK, pp. 208–224, 1988b.

Chow, V. T., D. R. Maidment, and L. W. Mays, *Applied Hydrology,* McGraw-Hill, New York, pp. 204–213, 1988.

Church, R., and C. Revelle, "The Maximal Covering Location Problem," *Pap. Reg. Sci. Assoc.,* 32:101–118, 1974.

Clark, R. M., "Applying Water Quality Models," in M. H. Chaudry and L. W. Mays, eds., *Computer Modeling of Free-Surface and Pressurized Flows*, Kluwer Academic Publishers, Netherlands, pp. 581–612, 1994.

Clark, R. M., W. M. Grayman, R. M. Males, and A. Hess, "Modeling Contaminant Propagation in Drinking Water Distribution Systems," *ASCE Journal of Environmental Engineering*, 119(2):349–364, 1993.

Cohen, G., "Optimal Control of Water Supply Networks," in S. G. Tzafestas, ed., *Optimization and Control of Dynamic Operational Research Models*, Vol. 4, North-Holland Publishing Company, Amsterdam, 1982.

Coulbeck, B., "Optimization of Water Networks," *Transactions of the Institute of Measurement and Control*, 6:271–279, 1984.

Coulbeck, B., and M. Sterling, "Optimal Control of Water Distribution Systems," *Proceedings of the Institute of Electrical Engineers*, 125:1039–1044, 1978.

Coulbeck B., S. Tennant, and C. Orr, "Development of a Demand Prediction Program for Use in Optimal Control of Water Supply," *Systems Science*, 11(1):59–66, 1985.

Coulbeck, B., M. Bryds, C. Orr, and J. Rance, "A Hierarchal Approach to Optimized Control of Water Distribution Systems: Part I. Decomposition," *Journal of Optimal Applications and Methods*, 9(1):51–61, 1988a.

Coulbeck, B., M. Bryds, C. Orr, and J. Rance, "A Hierarchal Approach to Optimized Control of Water Distribution Systems: Part II. Lower Level Algorithm," *Journal of Optimal Control Applications and Methods*, 9(2):109–126, 1988b.

DeMoyer, R., and L. Horowitz, *A Systems Approach to Water Distribution Modeling and Control*, Lexington Books, Lexington, MA, 1975.

Dougherty, D. C., and R. A. Marryott, "Optimal Groundwater Management - 1. Simulated Annealing," *Water Resources Research*, 27:2493–2508, 1991.

Epp, R., and Fowler, A. G., "Efficient Code for Steady-State Flows in Networks," *ASCE Journal of the Hydraulics Division*, 96(1):43–56, 1970.

Fallside, F., "Computer Techniques for On-line Control of Water Supply Networks," in B. Coulbeck and C. Orr, eds., *Computer Applications in Water Supply*, Vol. 2, Research Studies Press, Somerset, UK, pp. 313–328 1988.

Fallside, F., and P. Perry, "Hierarchical Organization of a Water-Supply Network," *Proceedings of the Institute of Electrical Engineers*, 122:202–208, 1975.

Fletcher, R., "An Ideal Penalty Function for Constrained Optimization," *Journal of the Institute of Mathematics and Its Applications*, 15:319–342, 1975.

Goldman, F. E., "The Application of Simulated Annealing for Optimal Operation of Water Distribution Systems," Doctoral dissertation, Department of Civil and Environmental Engineering, Arizona State University, Tempe, 1998.

Grayman, W. M., R. M. Clark, and R. M. Males, "Modeling Distribution-System Water Quality: Dynamic Approach," *ASCE Journal of Water Resources Planning & Management*, 114:295–312, 1988.

Hillier, F. S., and G. J. Lieberman, *Introduction to Operations Research*, Holden-Day, Oakland, CA, pp. 68–91 1980.

Hutchinson, B., "Operational Control of Water Distribution Systems," *Proceedings of the 1991 American Water Works Association Conference on Computers in the Water Industry*, pp. 53–59 1991.

IAPMO (International Association of Plumbing and Mechanical Officials), Uniform Plumbing, 1994.

Jentgen, L. A., and R. Hume, "Applications of a SCADA and Optimum Pump Control System for the Enhancement of Water Distribution System Operations and Maintenance", *Proceedings of the 1989 American Water Works Association Conference on Computers in the Water Industry*, 459-480, 1989.

Joalland, G., and G. Cohen, "Optimal Control of Water Distributions by Two Multilevel Methods," *Automatica,* 16(2):83–88, 1980.

Jowitt, P., R. Garrett, S. Cook, and G. Germanopoulos, "Real-Time Forecasting and Control for Water Distribution," in B. Coulbeck and C. Orr, eds., *Computer Applications in Water Supply,* Vol. 2, Research Studies Press, Somerset, UK, pp. 329–355 1988.

Jowitt, P., and Germanopoulos, G., "Optimal Schaduling in water. Supply Networks" *ASCE Journal of Water Resources Planning and Management* 118: 406–422, 1992.

Kirkpatrick, S., "Optimization by Simulated Annealing: Quantitative Studies," *Journal of Statistical Physics*, 34:975–986, 1984.

Kirkpatrick, S., C. D. Gelatt, and M. P. Vecchi, "Optimization by Simulated Annealing," *Science*, 220(4598):671–680, 1983.

Lannuzel, P., and L. Ortolano, "Evaluation of Heuristic Program for Scheduling Treatment Plan Pumps", *Journal of Water Resources Planning and Management*, 115:457–471, 1989.

Lansey,K. E., and K. Awumah, "Optimal Pump Operations Considering Pump Switches," *ASCE Journal of Water Resources Planning and Management*, 120:17–35, 1994.

Lansey, K. E., and L. W. Mays, "Optimization Model for Water Distribution System Design," *ASCE Journal of Hydraulic Engineering*, 115:1401–1418, 1989.

Lansey, K., and Q. Zhong, "A Methodology for Optimal Control of Pump Stations: Water Resources Infrastructure," *Proceedings of the 1990 Water Resources Planning and Management Specialty Conference*, 58–61, 1990.

Lasdon, L. S., and A. D. Waren, GRG2 User's Guide, Department of General Business, University of Texas at Austin, 1986.

Li, G., and L. W. Mays, "Differential Dynamic Programming for Estuarine Management," ASCE *Journal of Water Resources Planning and Management*, 121:455–462, 1995.

Liou, C. P., and J. R. Kroon, "Modeling the Propagation of Waterborne Substances in Distribution Networks, *Journal of the American Water Works Association*, 79(11):54–58, 1987.

Little, K., and B. McCrodden, "Minimization of Raw Water Pumping Costs Using MILP," *ASCE Journal of Water Resources Planning and Management*, 115:511–522, 1989.

Luenberger, D. G, *Introduction to Dynamic Systems: Theory, Models, and Applications,* John Wiley and Sons, New York, NY, pp. 108-112, 1979.

Maidment, D. R., S. Miaou, and M. Crawford, "Transfer Function Models of Daily Urban Water Use," *Water Resources Research,* 21:425–432, 1985.

Mays, L. W., *Optimal Control of Hydrosystems*, Marcel Dekker, New York, 1997.

Meyer, P., and E. C. Brill, Jr., "A Method for Locating Wells in a Groundwater Monitoring Network Under Conditions of Uncertainty," *Water Resources Research*, 24(8), 1988.

Meyer, P., A. Valocchi, and J. W. Eheart, "Monitoring Network Design to Provide Initial Detection of Groundwater Contamination," *Water Resources Research*, 30: 1994.

Moss, S. M. On Line Optimal Control of a Water Supply Network, Doctoral dissertation, Cambridge University, Cambridge, UK, 1979.

Murtagh, B. *Advanced Linear Programming: Computation and Practice*, McGraw-Hill, New York, 1981.

Oppenheim, A. V., and A. S. Willsky, *Signals and Systems*, 2nd ed., Prentice-Hall, Englewood Cliffs, NJ, pp. 77–90 1997.

Ormsbee, L., and A. Jain, "A Comparison of Daily Demand Forecast Models for Municipal Water Use," *Proceedings of the 1994 AWWA Computer Conference*, Los Angeles, CA, April 1994, pp.

Ormsbee, L., and S. Lingireddy, "Nonlinear Heuristic for Pump Operation," *ASCE Journal of Water Resources Planning and Management*, 121:302–309, 1995a.

Ormsbee, L., and S. Lingireddy, "Pumping System Control Using Genetic Optimization and Neural Networks," *Proceedings of the 7th IFAC/IFORS/IMACS Symposium on Large Scale Systems: Theory and Applications*, London, UK, July 11–13, 1995b.

Ormsbee, L., T. Walski, D. Chase, and W. Sharp, "Techniques for Improving Energy Efficiency at Water Supply Pumping Stations," ETL 1110-1-86, U.S. Army Corps of Engineers, Vicksburg, MS, 1987.

Ormsbee, L., T. Walski, D. Chase, and W. Sharp, "Methodology for Improving Pump Operation Efficiency," *ASCE Journal of Water Resources Planning and Management*, 115:148–164, 1989.

Orr, C.H., and B. Coulbeck, "Computer Applications in Water Systems in the U.K.," Proceedings of the 16th Annual Conference of the ASCE Water Resource Planning and Management Division, pp. 653–656, 1989.

Orr, C.H., M. Parkar, and S. Tenant, "Implementation of an On-line Control Scheme for a City Water System," *ASCE Journal of Water Resources Planning and Management*, 116:709–726, 1990.

Perry, P. F., "Demand Forecasting in Water Supply Networks," *ASCE Journal of the Hydraulics Division*, 107:1077–1087, 1981.

Pirnie, Malcolm, Inc., "Peoria Water Master Plan-Technical Memorandum No. 3-Distribution and Flow Model (Draft)," 1996.

Rao, H. S., and D. Bree, Jr., "Extended Period Simulation of Water Systems-Part A," *ASCE Journal of the Hydraulics Division*, 103(HY2):97–108, 1977.

Reklaitis, G. V., A. Ravidran, and K. M. Ragsdell, *Engineering Optimization: Methods and Applications*, Wiley Interscience, New York, 1983.

Revelle, C. S., D. P. Loucks, and W. R. Lynn, "A Management Model for Water Quality Control," *Journal of the Water Pollution Control Federation*, 39(6): 1967.

Revelle, C. S., D. P. Loucks, and W. R. Lynn, "Linear Programming Applied to Water Quality Management," *Water Resources Management*, 4(1):1–9, 1968.

Rossman, L. A., EPANET Users Guide, Drinking Water Research Division, Risk Reduction Engineering Laboratory, Office of Research and Development, U.S. Environmental Protection Agency, Cincinnati, OH, 1994.

Rossman, L. A., and P. F. Boulos, "Numerical Methods for Modeling Water Quality in Distribution Systems: A Comparison," *ASCE Journal of Water Resources Planning and Management*, 122:137–146, 1996.

Rossman, L. A., P. F. Boulos, and T. Altman, "Discrete Volume Element Method for Network Water-Quality Models," *ASCE Journal of Water Resources Planning and Management*, 119:505–517, 1993.

Rossman, L. A., R. M. Clark, and W. M. Grayman, "Modeling Chlorine Residuals in Drinking-Water Distribution Systems," *ASCE Journal of Environmental Engineering*, 120:803–820, 1994.

Sabet, M., and O. Helweg, "Cost Effective Operation of Urban Water Supply System Using Dynamic Programming," *Water Resource Bulletin*, 21(1):75–81, 1985.

Sakarya, A. B., "Optimal Operation of Water Distribution Systems for Water Quality Purposes," Doctoral dissertation, Department of Civil and Environmental Engineering, Arizona State University, Tempe, 1998.

Sastri, T., and J. B. Valdes, "Rainfall Intervention for On-line Applications," *ASCE Journal of Water Resources Planning and Management*, 115:397–415, 1989.

Seborg, D. E., T. F. Edgar, and D. A. Mellichamp, *Process Dynamics and Control*, John Wiley & Sons, New York, pp. 224–2451989.

Shamir, U., "Computer Applications for Real-Time Operation of Water Distribution Systems," *Proceedings of the ASCE Specialty Conference on Computer Applications in Water Resources*, pp. 379–390 1985.

Shamir, U., and C. D. D. Howard, "Water Distribution Systems Analysis," *ASCE Journal of the Hydraulics Division*, 94(1):219–234, 1968.

Smith, J. A., "A Model of Daily Municipal Water Use for Short Term Forecasting," *Water Resources Research*, 24:201–206, 1988.

Solanos, J., and J. Montoliu, "The Optimal Operation of Water Systems," in B. Coulbeck and C. Orr, eds, *Computer Applications in Water Supply*, Vol. 2, Research Studies Press, Somerset, UK, pp. 356–375 1988.

Steiner, R. C., "Operational Demand Forecasting for Municipal Water Supply," *Proceedings of the 16th Annual Conference of the ASCE Water Resources Planning and Management Division*, 445–448 1989.

Sterling, M., and B. Coulbeck, "A Dynamic Programming Solution to Optimization of Pumping Costs," *Proceedings of the Institute of Civil Engineers*, 59(2):813–818, 1975a.

Sterling, M., and B. Coulbeck, "Optimization of Water Pumping Costs by Hierarchical Methods," *Proceedings of the Institute of Civil Engineers*, 59(2):787–797, 1975b.

Tarquin, A., and J. Dowdy, "Optimal Pump Operation in Water Distribution, *ASCE Journal of Hydraulic Engineering*, 115:–168, 1989.

Tatejewski, P., "A Suboptimal Approach to Scheduling of Reservoir Levels for a Multi-Reservoir Water Distribution Network," in B. Coulbeck and C. Orr, eds., *Computer Applications in Water Supply*, Vol 2, Research Studies Press, Somerset, UK, pp. 225–239, 1988.

Tryby, M. E., D. C. Boccelli, M. T. Koechling, J. G. Uber, R. S. Summers, and L. A. Rossman, "Booster Disinfection for Managing Disinfectant Residuals," *Journal of the American Water Works Association*, 91(1): 1997.

Ulanicki, B., and C. H. Orr, "Unified Approach for the Optimization of Nonlinear Hydraulic Systems," *Journal of Optimization Theory and Applications*, 68(1):161–171, 1991.

Unver, O. L., and L. W. Mays, "Model for Real-Time Optimal Flood control Operation of a Reservoir System," *Water Resources Management*, 4:21–46, Kluwer Academic, Netherlands, 1990.

Vasconcelos, J. J., P. F. Boulos, W. M. Grayman, L. Kiene, O. Wable, P. Biswas, A. Bhari, L. A. Rossman, R. M. Clark, and J. A. Goodrich, *Characterization and Modeling of Chlorine Decay in Distribution Systems*, AWWA Research Foundation and American Water Works Association, Denver, CO, pp. 258–271 1996.

Wagner, H. M., *Principles of Operations Research with Applications to Managerial Decisions*, Prentice-Hall, Englewood Cliffs, NJ, pp. 361–365, 1969.

Wanakule, N., L. W. Mays, and L. S. Lasdon, "Optimal Management of Large-Scale Aquifers: Methodology and Applications," *Water Resources Research*,22:447–465, 1986.

Whaley, R. S., and R. Hume, An Optimization Algorithm for Looped Water Networks," Proceedings of the PSIG (Pipeline Simulation Interest Group) Annual Meeting", pp. 361–365, 1986.

Wood, D. J., and C. Charles, "Hydraulic Network Analysis Using Linear Theory," *ASCE Journal of the Hydraulics Division*, 98;1157–1170, 1972.

Zessler, U., and U. Shamir, "Optimal Operation of Water Distribution Systems," *ASCE Journal of Water Resources Planning and Management*, 115:735–752, 1989.

Zierolf, M. L., M. M. Polycarpou, and J. G. Uber, "Development and Auto-Calibration of an Input-Output Model of Chlorine Transport in Drinking Water Distribution Systems," *IEEE Transactions on Control Systems Technology*, 1997.

CHAPTER 17
MAINTENANCE AND REHABILITATION/ REPLACEMENT

Thomas M. Walski
Pennsylvania American Water Co
Wilkes-Barre, PA

James W. Male
Department of Civil Engineering
University of Portland
Portland, OR

17.1 INTRODUCTION

When compared with other infrastructure networks, water distribution systems are extremely reliable. It is not uncommon for pipes to perform for more than 100 years without a break or a leak. Numerous pumps have run for over 40 years with only minimal routine maintenance. In fact, in some older cities, wooden mains still convey water. As is the case with systems that require little attention, especially one that is not visible, water distribution systems are usually taken for granted. However, good maintenance practices can extend the life of distribution system components and rehabilitation can further prolong their life.

17.1.1 Maintenance and Rehabilitation Problems

Many water systems are old yet perform adequately. However, a number of factors at work might lead to failure if left unchecked. Some of the primary factors are described below.

17.1.1.1 Normal wear. Most moving components will wear out over time. Distribution systems have few moving parts and thus are not extremely susceptible to wear. Pumps are the components most likely to face wear problems. Depending on the type of pump, routine maintenance can maximize the life and maintain the efficiency of pumps. The moving parts in control valves also need frequent routine maintenance to continue functioning adequately.

17.1.1.2 Corrosion. Many distribution system components are made of metal, and, on contact with an electrolyte, metal tends to corrode. Metal pipes and tanks are the most obvious example of this type of problem. Corrosion also can attack reinforcing wire in concrete structures, and corrosion on valve stems and bolt threads can render these items inoperable. Corrosion is often the root cause of other problems as well: It leads to loss of metal, weakening of the component, and ultimately to failure. The by-products of the corrosion process accumulate on pipe walls and cause reduced carrying capacity and lower pressure. Corrosion also can contribute to poor water quality.

17.1.1.3 Unforeseen loads. Some distribution facilities fail because the load placed on them exceeds the load for which they were designed. Pipes broken by excessive surge pressure or external loads may have been designed correctly at the time they were installed; however, any component (e.g. ,pipes, pumps) can simply become too small when water demands increase over time.

17.1.1.4 Poor manufacture and installation. Even if an item is designed properly, it may be manufactured or installed incorrectly. A common cause of pipe breaks is poor bedding and backfill. Tank corrosion is accelerated by poor coating. A misaligned pump will fail quickly. Utilities that work with reputable suppliers and contractors have well-prepared contract documents, and quality inspections reduce the risk of poor installation.

17.1.2 Preview of the Chapter

With the above list of some of the primary causes of distribution system problems, it is a credit to the water industry that clean water gets to customers with adequate pressure virtually all the time. The subsequent sections of this chapter will discuss specific types of problems and give suggestions for their solution. In the major sections that follow, problems and solutions are grouped according to unaccounted-for water, pipe breakage, hydraulic carrying capacity, and information management.

The chapter is merely an overview of issues associated with maintenance and rehabilitation. Further information can be found in the *AWWA Series on water supply operations* (1996), Deb et al. (1995), Male and Walski (1990), O'Day et al. (1986), O'Day et al. (1987), Utility Infrastructure Manual (1984) and Walski (1987a).

17.2 UNACCOUNTED-FOR WATER

Water produced and delivered to the distribution system is intended to be sold to customers and not lost or taken from the distribution system without authorization. At one time, a large fraction of water companies sold water at a flat rate without metering. As water has become more valuable and metering technology has improved, almost all water systems in the United States now meter all water customers. Although all customers may be metered in a given utility, a fairly sizable portion of the water produced by most utilities does not pass through customer meters. Some is taken for authorized purposes, such as fire fighting and flushing and blowoffs for water-quality reasons. These quantities are usually fairly small, and the primary cause of excessive unaccounted-for water is often leakage.

17.2.1 Indicators for Unaccounted-for Water

Unaccounted-for water is the difference between water produced by the utility (and usually measured at the treatment facility) and metered use (i.e., sales plus non-revenue producing metered water). Unaccounted-for water can be expressed in flow units [e.g., millions gallons per day (mgd)] but is usually discussed as a percentage of water production:

$$\text{Unaccounted-for water (\%)} = \frac{(\text{Production} - \text{metered use} \times 100\%)}{(\text{Production})} \qquad (17.1)$$

Although various groups have (e.g., AWWA M-36, 1996) proposed a consistent definition for unaccounted-for water, there are almost as many ways to calculate unaccounted-for water as there are individuals performing the calculation. For example, some utilities may subtract an estimate of water used for blowoffs and flushing from water production, whereas others may add blowoffs and flushing to metered use or include it as unaccounted-for. Some utilities include estimates of known, unrepaired leaks as accounted-for use. Because unaccounted-for water is used as an indicator of system performance, an universally agreed-on definition is not so important as long as comparisons made between utilities and, in a given utility over time, use the same definition.

Elimination of unaccounted-for water is a goal of all utilities, but it is impossible for utilities to reach this goal. A commonly accepted rule-of-thumb for acceptable levels of unaccounted-for water is 15 percent, although this value is highly site specific. The real rule for deciding whether unaccounted-for water exists at an acceptable level is an economic one; the economic savings in water production at least offsets the cost of reducing unaccounted-for water. For example, on a present-worth basis, the cost of a leak detection and repair program should be less than the value of water no longer leaked plus any damages associated with leaking water. In an area with costly treatment requirements and limited source capacity, it may be worthwhile to reduce unaccounted-for water to less than 10 percent. In a utility with excess capacity, little growth, and inexpensive treatment and pumping, unaccounted-for water exceeding 20 percent may be acceptable.

An unexpected increase in production at a well or water treatment plant can indicate that a new leak has occurred but has not yet been detected. If daily water production can be monitored for smaller zones within the distribution system, even better insights into the location of new leaks can be obtained.

17.2.2 Understanding the Causes of Unaccounted-for Water

To reduce unaccounted-for water, it is necessary to understand where this water is going (Hudson, 1978; Siedler, 1985; Wallace, 1987). It could be lost to leaks, theft, meter underregistration, authorized unmetered use, and flat-rate users. There are some rules of thumb, however, to help identify the primary cause (Siedler, 1982). Once the primary cause of large amounts of unaccounted-for water has been isolated, an effective program to reduce unaccounted-for water can be implemented.

Per capita water usage can be calculated using several formulas to shed some insight on the source of unaccounted-for water. The simplest definition of per capita use is

$$\text{Systemwide per capita use} = \frac{(\text{Water production})}{(\text{Population served})} \qquad (17.2)$$

Usage on the order of 550 Lpcd (liters per capita per day) or 150 gpcd (gallons per capita per day) is considered the norm in cases where exterior irrigation use and industrial demand are typical. This value is easy to obtain because all utilities should keep track of production. If this number is calculated monthly, it can provide insights into seasonal patterns. For example, in areas where summer irrigation demands are high, winter per capita usage should still be near 550 Lpcd (150 gpcd). This parameter can be calculated even for systems where customer metering is not practiced.

Other indicators can go further in predicting the cause of high unaccounted-for usage. For example, nonindustrial per capita use can be calculated as

$$\text{Nonindustrial use} = \frac{(\text{Production} - \text{Commercial and industrial use})}{(\text{Population served})} \quad (17.3)$$

This quantity eliminates the effect of industrial usage on water demand and should reflect use by domestic customers. If this value is greater the about 260 Lpcd (70 gpcd) in the nonirrigation season, excessive leakage is likely.

Another indicator involves looking at *domestic usage* directly, rather than as production minus industrial use, and is defined as

$$\text{Domestic use} = \frac{(\text{Domestic metered consumption})}{(\text{Population served})} \quad (17.4)$$

Since it is based on actual meter readings of domestic customers, this value is a more direct measure of domestic usage. If this parameter is less than 260 Lpcd (70 gpcd), underregistration by domestic meters is a likely contributor to unaccounted for water. By combining the three indicators in Eqs. (17.2), (17.3), and (17. 4), it is possible to get a feel for the causes of unaccounted-for water in a typical system. The indicators may need to be modified for use in atypical water systems. Once the utility understands the cause of unacceptable unaccounted-for water, it can develop a program to reduce the levels. Corrective measures to address each cause are detailed below.

Although most water use follows a typical diurnal pattern, leakage usually occurs at a steady rate for all 24 h of a day. This knowledge can be used to develop another indicator of the magnitude and probable cause of unaccounted-for water. It is the minimum nighttime ratio, defined as

$$(\text{Minimum nighttime ratio (\%)}) = \frac{(\text{Use during minimum hour}) \times (100\%)}{(\text{Average hourly use})} \quad (17.5)$$

Determining the hourly use involves calculating the hourly source production and change in storage tank levels on an hourly basis throughout the day. The minimum hour value is then divided by the average use. In a typical system, the value should be less than 40 percent. A higher value indicates either leakage or a customer(s) with high continuous use (e.g., a 24-hr manufacturer). If there is a single industry with a high use on a 24-hr basis, it may be possible to use data logging to determine that customer's hourly use and subtract it from each hourly use to find the minimum nighttime ratio without that customer's influence.

The minimum nighttime ratio can be calculated on a systemwide basis, but it is most valuable if it can be calculated on some smaller portion of the system (e.g., a single pressure zone). This approach not only will indicate whether leakage is a problem but also will point out which portion of the system is likely to have the greatest leakage.

17.2.3 Components of Unaccounted-for Water

The various components that contribute to unaccounted-for water can differ a good deal; some are actual losses of water from the system, whereas others can be attributed to lack of knowledge of water use. Components of unaccounted-for water are described below. Further discussion can be found in AWWA M-36 (1996), Jeffcoat and Saravanapavan (1987); and Male, (1985).

17.2.3.1 Water main leakage. In most cases, the largest portion of unaccounted-for water is lost through main leaks. Water utility personnel use the terms "leak" and "break" in different ways; there is no standardized definition. For our purposes, a break in a main requires emergency repair, whereas a leak does not. In addition, evidence of a break is obvious, whereas detection of a leak may require special equipment. Although there are no hard and fast rules, Table 17.1 lists the characteristics of leaks as opposed to breaks.

Failed joints often contribute to leaking. Their failure can be attributed to a number of factors, including poor design, improper installation, and poor joint material. Joints near bends can fail because of poor thrust restraint, and joints with gaskets can fail because of poor installation. Old lead or leadite joints often become brittle with age and crack. Because repouring these old joints is becoming a lost art, replacement of a pipe section may be required. For large pipes, it is possible to seal poor joints from the inside. In addition to problems with the joint sealer, bells on older pipes can crack. Although bell repair clamps are available, pipe replacement may be more appropriate, depending on conditions.

Leak detection usually involves the use of sonic leak-detection equipment, which listens for the sound of water escaping a pipe. These devices can include pinpoint listening devices that make contact with valves (Fig. 17.1) and hydrants and geophones that listen directly on the ground (Fig. 17.2). In addition, correlator devices (Fig. 17.3) can listen at two points simultaneously to pinpoint the exact location of a leak. In addition, experienced personnel can estimate the size of a leak accurately (Male, et al. 1985).

Large leaks do not necessarily contribute to a greater volume of lost water, particularly if water reaches the surface; they are usually found quickly, isolated, and repaired. Undetected leaks, even small ones, can lead to large quantities of lost water since they might exist for long periods of time. Ironically, small leaks are easier to detect because they are noisier and more easily heard using hydrophones. The most difficult leaks to detect and repair are usually those under stream crossings.

Leak detection efforts should focus on the portion of the distribution system with the greatest expected leakage. Indicators, such as high minimum nighttime flows, or high

TABLE 17.1 General Characteristics of Leaks and Breaks

Leak	Break
Scheduled repairs are possible	Requires emergency repair
Specific means of detection are necessary	Detection is obvious (e.g., surfacing water, low pressure)
Repair does not usually interrupt service	Repair requires service shut down
More frequently occurs at pipe joints and service lines	Often occurs along the pipe barrel

FIGURE 17.1 Contact point sonic leak detection. (Photograph by T. M. Walski).

FIGURE 17.2 Ground microphone sonic leak detection. (Photograph by T. M. Walski).

FIGURE 17.3 Leak correlator equipment used for pinpointing leaks. (Photograph by T. M. Walski).

unaccounted-for water in specific portions of the distribution system can point to areas with a high incidence of leakage. To access specific sections of the system, the utility must be able to assure the flow of water into that region. This effort usually involves closing valves to isolate the section and measuring flow at a few key points. Installation of additional metering at key locations is advisable in systems where pressure-zone boundaries, valves, or both can make subdivision of the system feasible.

Of course, detecting leaks is only the first step in eliminating leakage. Leak repair is the more costly step in the process. Repair clamps, or collars, are the preferred method for repairing small leaks, whereas for larger leaks, replacement of one or more sections of pipe may be required. Methods of pipe repair are discussed in a later section.

On the average, the savings in water no longer lost to leakage outweigh the cost of leak detection and repair (Moyer, 1985, et al. 1983). In most systems, assuming detection is followed by repair, it is economical to survey the system completely every 1 to 3 years. Utilities also find it advantageous to survey sections of the system more frequently that historically have shown higher leakage rates.

Instead of repairing leaking mains, some argue that replacing more leak-prone (generally older) pipes is preferable. The selection of a strategy depends on the frequency of leaks in a given pipe and the relative costs of replacement and repair. The decision whether to emphasize detection and repair over a program of replacement depends on site-specific leakage rates and costs. In general, detection and repair result in an immediate reduction in lost water, whereas replacement will have a longer-lasting impact to the extent that it eliminates the root cause of leakage. Replacement also may be advantageous before taking other actions, such as street repaving or other underground utility work. Decisions regarding repair are described in more detail in section 17.3.

17.2.3.2 Service pipe leakage. Methods similar to those used for main leaks are applied to detect service line leaks. Leaking services are usually replaced rather than repaired unless they are in a location where access is difficult. In many instances, the water utility only owns the service line to the curb stop, not to the meter, which may be located in the customer's basement or crawl space. Although customers may own piping between the curb stop and the meter, they have little motivation to repair the service line because they are not paying for the leaked water. Some utilities have provisions in the service agreements or tariffs that give them the authority to terminate service after proper notification to customers with service-line leaks.

17.2.3.3 System pressure. Water pressure in the distribution system can affect unaccounted-for water in several ways. First, high pressure can lead to higher break rates, and waterhammer can lead to breaks and separation at bends. These problems also will be discussed more in later sections. In terms of leakage, high pressures can lead to higher leakage rates once a leak has occurred. A set of studies showed a 6 percent reduction in water use following to a 30 psi drop in pressure (Hoag, 1984a). Utilities attempt to keep pressures below 100 psi in typical distribution mains. Where higher pressures occur (e.g., crossing a valley), utilities often ensure that a pipe with an adequate pressure rating is installed.

Once the system has been laid out and pressure zones have been established, it is difficult to reduce pressure in an area. Some utilities, especially those outside the United States, have been known to reduce pressure at night using valving to reduce leakage (Jowitt and Xu, 1989; National Water Council, 1980). This is not feasible in zones served directly by tanks that float on the system, but it is possible in pressure zones served by pressure–reducing valves (PRVs) and pump stations discharging into closed systems, in particular those with variable-speed drives.

17.2.3.4 Fire fighting. Water used for fire fighting is unaccounted-for and, over short periods of time and during catastrophic fires, can be a significant portion of production. However, over long periods of time, fire fighting accounts for less than 1 or 2 percent of total production. It is unlikely to be a major cause for high unaccounted-for water. Similarly, water used for occasional fire-hydrant flow testing is not a significant portion of unaccounted-for water.

17.2.3.5 Main flushing. Although main flushing through hydrants does not result in as high a flow rate as fire fighting does, virtually all mains are operated during a flushing program; hence, the total volume of water used during flushing can be significant. Flushing crews should log flow rates and duration of flushing so that estimates can be made of water used during the process. It is desirable to measure the flow rate with a Pitot gauge to get a more accurate estimate of the flushing rate. For directional flushing, valves are operated to maximize flushing velocity and to maximize the amount of fresh water from the system source delivered to the flowed hydrant (Oberoi, 1994). Directional flushing can reduce the amount of water used.

Maintaining a data-base of hydrant flow rates during flushing also can point to loss of carrying capacity in mains. In particular, a sudden drop in flow from one year to the next may indicate a closed valve in the system near the hydrant.

17.2.3.6 Blowoffs. The term "blowoffs" in this chapter refer, to valves in the distribution system (usually at the end of dead-end lines) that are opened to draw fresh water into the dead–end line to improve water quality or to prevent freezing by maintaining velocity in a main. Blowoffs should be run only as long as necessary and only at a rate needed to maintain water quality or prevent freezing. Recording the flow rate and run duration for each blowoff allows the utility to estimate its contribution to unaccounted-for water. Ideally, when blowoff usage becomes a significant part of production, steps to reduce these flows are implemented. Options are usually long-term efforts, such as eliminating dead-ends (especially in old unlined metal pipe) and replacing older pipes with new pipes or with pipes buried at depths where freezing will not be a problem.

17.2.3.7 Flat rate customers. Flat-rate customers are usually those without meters. Therefore, they are not charged for water used but for a set amount for each billing period. The charge is often based on the size of the house or, even more often, on the size of the service connection. This type of rate schedule does not discourage waste of water. Nor does it provide an indication of whether the water is being used productively at a customer's location or is being lost through leakage or theft. Even when water is sold on a flat rate basis, there is value in metering as a minimum to provide some accountability. In one study, Hoag (1984b) showed a decrease of 20 percent in water use with metered homes compared with homes without meters

17.2.3.8 Authorized unmetered uses. Some water utilities provide free service for some uses, such as municipally owned buildings, park irrigation, street sweeping, and municipal pools and fountains. Provision of free water service is the prerogative of the water utility; however, these uses often are metered to provide accountability.

17.2.3.9 Meter under registration. Most consumer water meters tend to run slowly as they become older and register less water than is actually used. The cost of replacing meters prohibits utilities from doing so often enough to maintain 100 percent registration. It is best to replace water meters when the cost of lost water caused by underregistration exceeds the cost of meter replacement.

The cost of replacing a meter depends on the size and type of meter being replaced, ease of access to the property, and whether the replacement is done as part of a systematic meter replacement program or is done on a case-by-case basis. The cost of lost water depends on the estimated average flow rate, the accuracy of the meter, and the value of water (Male, et al., 1985). Unlike the case of repairing leaks when the value of lost water is the variable O&M cost for water production, the value of lost water in this instance is the actual cost of water to the customer.

The accuracy of water meters can be determined by testing them. For small meters (e.g., residential meters), a representative sample of meters is removed and taken to a meter testing shop to determine the accuracy of that category of meter (e.g., grouping by size, type, age). For larger meters, the analysis is performed on a case-by-case basis using in-place meter testing whenever possible. Methods exist to determine the best frequency of meter testing (California Section Committee Report, 1966; Kuranz, 1942; Noss, et al., 1987; Sisco, 1967).

A meter may be accurate over the range of flows for which it was designed but be significantly inaccurate at lower flow rates. It is important to assess the overall accuracy of the meter at the rate of flow that the meter usually encounters rather than at the customer's peak rate of flow. For example, a meter may be 100 percent accurate from 20 to 150 gpm, but the actual flow rate through the meter may be 10 gpm most of the time. Thus, it is important to test the meter at that flow rate. If a utility does not have a good idea about the actual rate at which a customer uses water, it is possible to attach a data–logging device to most meters that is capable of determining the actual rates of flow.

In some cases a meter may have been sized for a flow larger than current use. In this case, the utility needs to evaluate the current range of water use (including fire flow requirements when industrial and commercial customers have private fire-fighting systems or sprinkler systems). The rate of use can be determined by profiling the customer's consumption using a data logger to record flow rates. Sullivan and Speranza (1992) and Walski et al (1995) describe how data logging can be used to assess meter sizing for existing customers.

If water use is substantially lower than the optimal range for the meter, then it should be replaced by a smaller meter. If the upper end of the meter's range still requires a large meter but most of the flow is below its minimum accurate range, then a compound meter with a wider range may be called for. Sizing and selection of meters is discussed in AWWA M-6 (1986b), AWWA M-22 (1975) and AWWA M-33 (1989).

17.2.3.10 Theft of water. Water can be stolen easily from hydrants, in particular near construction sites when contractors need to fill water trucks. Utilities can provide contractors with portable meters to attach to hydrants. Some utilities have even established stations where contractors can purchase water. Providing authorized locations where water can be purchased ensures that the rate of withdrawal does not adversely affect pressure and that adequate backflow protection will be used.

Theft also can take place at services. Some customers try to steal water either by never installing a meter when a building is constructed or by removing meters after construction. Meters are usually sealed when installed to prevent tampering. If a meter is located in a basement or crawl space, it should be installed where it is virtually impossible to tap the service line upstream of the meter. In addition, meter readers are encouraged to report any structures that they do not read on their route

Many utilities provide water through unmetered fire services or through services-only using a detector meter. A detector meter is not on the main line and measures only a fraction of the water used by the customer. Utilities routinely read detector meters, and when one shows usage each month (in excess of what can be expected from an occasional flow test), there is reason to believe that water is being consumed illegally from the fire service. In such cases, the usage should be removed from the line or the detector meter should be replaced by a fire-line compound meter.

17.2.4 Summary

Unaccounted-for water consists of two basic components: water lost from the system and water used beneficially but not sufficiently documented and often not paid for. This section has detailed how various types of unaccounted-for water can be determined and how they can be documented better. By doing so, the extent of leakage can be assessed more effectively.

Most utilities practice some form of leak detection and repair. The level of effort depends on the condition of the system, excess capacity in the system, management philosophy on maintenance, personnel usage, and the value of the water. Many pipe leakage problems can be traced to poor installation of pipes and thrust blocking. Proper installation and inspection can go a long way for a long time in reducing leakage (Arrequin-Cortes and Ochoa-Alejo, 1997; Boyle Engineering, 1982; Hanke, 1981; Heim, 1979; Kingston, 1979; Lai, 1991; Laverty, 1979; Male et al., 1985; Male and Franz, 1986; Moyer, et al., 1983, National Water Council, 1980; Walski, 1984).

17.3 PIPE BREAKS

This section addresses the structural integrity of water mains and the relationship of that integrity to breaks in pipe segments. Therefore, it differs from the earlier section on leakage because that section focused on the loss of water from the system, whereas this section addresses the causes and impacts of pipe breaks from the viewpoint of system integrity. Each major cause of breaks and their associated remediation measures is discussed below. Although each possible cause is discussed separately, failure of a main often is the result of interacting forces. For example, corrosion may have weakened a main to the point where excessive pressure (internal or external) will cause a break.

The effects of pipe breaks on distribution hydraulics are described in the earlier chapter on reliability and in sources such as Jowitt and Xu (1993). Reducing the impact of a break depends on the ability to isolate the break with valves (Bouchart and Goulter, 1991; Walski, 1993).

17.3.1 Corrosion

Corrosion of pipes is a major contributor to poor system integrity and often leads to other problems with water mains, including, breaks, reduced hydraulic carrying capacity, and poor water quality. In many cases, older mains were installed without considering the impact of corrosion; internal pipe walls were not lined, and external walls were not protected. Inside walls of newly installed pipes are now lined, and the pipes are wrapped or coated to reduce external corrosion.

When pipe segments are replaced, they should be examined for the effects of corrosion. In steel pipes, corrosion shows up as pitting. In cast iron pipes, corrosion takes the form of graphitic corrosion (graphitization) in which the graphite remains behind when the iron is oxidized. The pipe must be scraped or abrasive cleaned to reveal the pits in the cast iron. Dempsey and Manook (1986) described a procedure for evaluating the condition of cast iron pipes. Ideally, the water industry should have nondestructive methods for assessing piping with regard to corrosion and structural integrity. Such methods as ultrasonic, eddy current, and magnetic flux are used on high-value gas and oil pipelines made of steel. The methods are practicable for those industries because of the high value of the product, the nature of steel, and the long uninterrupted runs of piping. These methods have not yet found widespread application in water mains. (Battelle Memorial Institute, 1998; Jackson et al., 1992).

17.3.1.1 External soil corrosion. Although pipe wall thickness in new pipes usually has a considerable safety allowance, several factors contribute to reducing the life of pipe.

External corrosion and aggressive soils, which weaken metal and concrete pipes, are a contributing factor to most breaks even though they may not be the final cause of failure (Fitzgerald, 1968; Gerhold, 1976; Romanoff, 1964, 1968). Loose wrapping or bonded coatings provide substantial protection provided their integrity is not compromised. Cathodic protection may be warranted in larger pipelines.

17.3.1.2 Internal corrosion. Corrosion both weakens the pipe wall by graphitization and reduces the hydraulic carrying capacity of the mains because of tuberculation. The extent of corrosion and tuberculation depends on the water quality in terms of pH, conductivity, dissolved oxygen, use of corrosive inhibitors, and the tendency of the water to precipitate coatings, such as calcium carbonate. Usually, maintenance of a slightly positive Langlier index is regarded as desirable for systems with reasonably high alkalinity (Langlier, 1936; Loewenthal and Marais, 1976; Merrill and Sanks, 1978). Addition of polyphosphates to treated water to form a film on the inside of water mains also can reduce corrosion. Additional material can be found in AWWA and DVGW Forshcungsstelle (1985), Kirmeyer and Logsdon (1983), and Singley, et al. (1984).

17.3.1.3 Stray current corrosion. Stray direct current can cause corrosion holes in pipes. Sources of stray current include electrical equipment, such as subways, trolleys, and electric trains; ground beds for impressed current cathodic protection; and grounding of electrical devices to water piping. Holes produced by stray current are usually small and circular, as opposed to large areas of pitting resulting from soil corrosion. Stray current also can lead to shocking of water distribution employees who come into contact with such piping (Duranceau, et al. 1998).

Identifying the cause of the stray current is helpful in remedying the problem. Elimination of the current's source is desirable but is only possible in some instances (e.g., illegal grounding). In other cases, cathodic protection of the water piping may be needed for protection. Cathodic protection can be provided by either buried sacrificial anodes or impressed current.

17.3.1.4 Bimetallic connections. Contact between two metals in the presence of an electrolyte can lead to corrosion of the less noble metal. In water distribution systems, this can occur at taps and repair clamps, especially in moist clay soil. If a utility notices this problem, it usually conducts an investigation of tapping and repair procedures to eliminate the problem.

17.3.2 External Loads

Excessive external loads are the greatest single cause of water main breaks. They can be grouped into three categories: (1) contact during excavation, (2) poorinstallation practices or poor-quality materials, and (3) changes in surface load or bedding condition. Breaks caused by external loading usually occur as circumferential breaks that look as if someone had cut the pipe with a saw. They are the result of beam loading on the pipe. Smaller pipes are more susceptible to circumferential breaks because they have less beam strength.

Contact during excavation usually occurs when an individual operating excavating equipment breaks the pipe. This generally can be prevented by following the procedures set out in the local "one-call," "dial before you dig," or similar program. Some excavators think it is easier to fix a water main break than to dig carefully in the first place. Fines for failure to comply with a preventive program exist in some utilities.

17.12 Chapter Seventeen

Poor installation practices and use of poor materials need to be addressed before mains are put in the ground. Faulty bedding and backfill or poor joint assembly are usually the cause of subsequent breaks in pipe barrels. Installation of pipe by contractors without full-time inspection can encourage poor workmanship. Pipe manufacturers are clear about bedding and backfill requirements for their pipe, and there is little reason to deviate from those directions without detailed investigations. Because inadequate thrust blocking can lead to problems at bends and joints, backing the thrust block with undisturbed soil is important.

Once a pipe is installed properly, it must be protected from additional loads. In particular, pipes that are not located under paved areas are subject to structures being placed above the pipe even if the utility has easements prohibiting such activity. The pipe may not be designed for these additional loads. Installation of pipes and conduits by other utilities also can undermine the proper bedding of water mains. Small leaks in pipes, if left uncorrected, can undermine the bedding of pipes, eventually leading to a break. Well-run utilities patrol their rights-of-way to prevent encroachment.

Joint leaks are usually caused by an external load causing settlement or internal pressure caused by waterhammer moving a thrust block. Joint deflection is most noticeable on above-ground piping, such as stream-crossing pipes located on piers (Fig. 17.4). Pipes with nonrigid connections (e.g., rubber gaskets) can absorb a limited amount of deflection before they begin to break.

A source of external loads that utilities cannot control is the added load posed by frost penetration. Frost can significantly increase the external load on pipes because it "stiffens" the overlying soil. This impact usually shows up as an increase in pipe breakage in the winter, in particular during the first cold weather of the season. (O'Day, 1986; Walski, 1982) Little can be done to remedy this situation other than by preparing for this increase in breakage or by replacing main segments at a deeper depth (Cohen and Fielding, 1979; Gros, 1976; Monie and Clark, 1974; Smith, 1976).

FIGURE 17.4 Deflection of joints at a stream crossing. (Photograph by T.M. Walski).

17.3.3 Poor Tapping

Pipe tapping can weaken the pipe and lead to breaks. It is important to follow the pipe manufacturer's recommendation for tapping practice. Some pipes can only be tapped with tapping saddles, whereas ductile iron and cast iron can usually be tapped directly for small services. To reduce the possibility of future problems, utilities repair external coatings and wraps after tapping and backfilling have been done correctly.

17.3.4 Pressure-Related Breaks

Breaks related to internal pressure usually show up as a longitudinal rupture of the pipe (Fig. 17.5). Pipes are usually designed to withstand pressures significantly in excess of the normal working pressure. If a pipe ruptures during normal operation, the pipe was likely to be defective initially, be damaged during installation, or be weakened by corrosion. Pipes can be inspected for thin wall sections or other signs of defects.

Most pressure-related breaks are caused by waterhammer surges associated with abrupt changes in the velocity of water in a pipe. When an apparently good pipe ruptures suddenly, a utility usually looks for evidence of waterhammer. Pressure recording charts and Supervisory Control and Data Acquisition (SCADA) system signals near the site of the break are checked for short-term high pressures, although care must be exercised to distinguish between surges that caused the break and surges that resulted from the break.

Pump startup or shutdown, whether the result of normal operation or a power outage, is a likely cause of a surge. Sudden shutdown of a fire hydrant, whether caused by hydrant testing or use during a fire also can cause surges. Experienced operators check pressure signals and pressure charts to look for indications that a surge problem exists. When a waterhammer problem is identified, installation of appropriate valving and careful operation can minimize its effects.

FIGURE 17.5 Longitudinal split in cast iron pipe. (Photograph by T. M. Walski).

17.3.5 Repair Versus Replacement

The options available to utilities when trying to correct for excessive pipe breakage are limited; they can continue to repair the breaks as they occur or replace the offending main segment. Repair is usually done on an emergency basis, requires service shut down, and may involve replacing one or two pipe segments if a collar is not adequate. Fig. 17.6 shows a repair clamp at the discharge of a check valve. Figs. 17.7 and 17.8 show ductile iron repair pipes and the couplings used to connect them. Replacement is usually done for one or more main sections, where each section often corresponds to a city block.

If pipes break at a high rate, the utility may chose to implement a program of systematic replacement rather than continue to repair breaks. The decision to implement a program of replacement should be based on an economic analysis, comparing the present worth of the cost of continued repair with the cost of a replacement program for that main segment or group of segments. Fig. 17.9 illustrates the effect on total cost as the extent of replacement increases from low to high. Cost analysis should compare the costs of both alternatives and include the value of indirect costs that cannot be measured directly, such as interruption of service.

FIGURE 17.6 Repair clamp on a pipe (Photograph by T.M. Walski).

FIGURE 17.7 Ductile iron pipe in storage. (Photograph by T. M. Walski).

FIGURE 17.8 Couplings for repair of a pipe (Photograph by T. M. Walski).

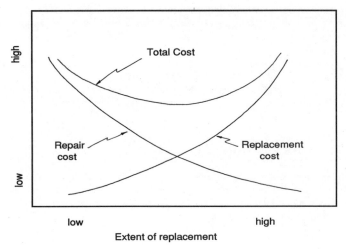

FIGURE 17.9 Relationship between replacement and repair costs.

One approach to quantifying the need for replacement is use of a critical pipe break rate called J^*. If a pipe breaks at a higher rate than J^*, it is a candidate for replacement. If it breaks at a lower rate, it is less expensive to continue to repair breaks in the pipe:

$$J^* = \frac{[(KrC_r - C_d)e^{-bT} - C_w Q_0]}{C_b} \tag{17.6}$$

where J^* = critical break rate [breaks/yr/km (breaks/yr/mi)], K = 1000 if metric units, 5280 if English units, r = discount rate [fraction (i.e,. 0.04 not 4 percent)], C_r = cost to replace pipe [$/m ($/ft)], b = rate of increase of breakage (fraction), T = planning time horizon (yr), C_w = cost of leakage [$/ML ($/MG)], Q_0 = leakage rate, [ML/ km/yr (MG/mi/yr)] C_b = cost of a break ($/break), and C_d = cost of leak detection and repair [$/Km ($/mi)].

The parameter b is an indicator of the rate of deterioration of pipes. In several studies (Walski and Pelliccia, 1982 Walski and Wade, 1987), the annual increase in break rate was found to vary from 2 to 4 percent (i.e., b varies from 0.02 to 0.04). A more detailed explanation of how to determine the parameters in Eq. (17.6) can be found in Walski (1987). Values for C_w are based on estimated leakage rates and the anticipated duration of the leak. Values for C_r and C_b should include, to the extent possible, estimates of any direct costs.

The utility can apply this formula as a rough screening tool by using typical values for leakage rate, cost of pipe installation, and cost of a break based on categories of mains. These categories can be composed of main segments with similar characteristics, such as diameter, material, location, and so forth. The critical break rate for each category is then compared to the break rate of individual pipes (J). A pipe should be considered for replacement if $J > J^*$. Once a set of possible pipes needing replacement has been identified, pipe-specific parameters can be determined, taking into account site-specific aspects.

The above approach accounts for the cost of breaks, repairs, replacement, and lost water. However, replacing pipes also has the benefits of improving water quality and flows (in particular, fire flows, especially if the mains under consideration are old and tuberculated). If either of these factors are significant, they also should be factored into the analysis. However, it is difficult to provide a defensible dollar value on an improvement in fire flow or water quality.

Additional guidance for water main replacement can be found in AWWA Task Group 2850-D (1969), Clark, et al. (1985), Heitzman and Willis (1995), Kutsal (1995); Li and Haimes (1992a, 1992b), Male, et al. (1990a, 1990b), O'Day et al. (1986), Quimpo (1991), Ramos (1995), Shamir and Howard (1979), and Weiss et al. (1985).

17.4 HYDRAULIC CARRYING CAPACITY

In a well-designed distribution system, components should have been sized to provide adequate capacity. Issues regarding system hydraulic design are addressed in earlier chapters. However, over the course of years, demand patterns may change and the physical components of the system may have changed. For example, the service population may have increased dramatically because of annexation, or pipe segments may have lost carrying capacity because of tuberculation. Assuming the design and installation of the system were performed correctly, the utility must address what has occurred since that time. Some common causes of these symptoms and their solutions are addressed in the following sections. Hydraulic problems in a water distribution system usually show up as complaints of low pressure or poor fire-fighting capacity.

17.4.1 Diagnosis of Pressure Problems

Determining the cause, or causes, of pressure problems often is not straightforward. Distribution systems are extremely complex, being composed of a variety of components that were often installed over the course of many years and are subject to demands that vary with time and location.

17.4.1.1 Pressure gauges. Water systems are sized to meet some estimate of future demands. In some growth areas, new demands can exceed the values used in system planning. The problem shows up as a drop in pressure during periods of high demand and can be diagnosed by setting out recording pressure gauges in areas of interest. A drop in pressure can indicate that demands exceed the system's capacity. The utility will then diagnose the cause of the problem, which might be associated with one or two individual components, such as a pipe that is no longer sized adequately or a pump that is unable to meet increased demand. Once identified, carrying capacity problems of main segments can be determined with fire hydrant flow tests. Procedures for carrying out test tests are presented in AWWA M-17 (1989) and NFPA 291 (1988).

17.4.1.2 Hydraulic modeling. Hydraulic models, described elsewhere in this book, are useful tools to diagnose system problems and narrow the focus to individual components. The model should be calibrated using the results of hydrant-flow tests and pipe-roughness tests conducted in the area of reported pressure problems (Fig. 17.10). If the model cannot be made to calibrate in that part of the system, the system may not be behaving as designed. A likely cause for this situation is a valve that was mistakenly left closed or partly closed (see the discussion below for more on this problem). Once the model is calibrated, the modeler can try to replicate the problem of low pressure in the model. This process usually indicates whether the problem is the result of inadequate pumping

FIGURE 17.10 Hydrant flow test. (Photograph by T. M. Walski).

capacity, inadequate main capacity, loss of carrying capacity because of tuberculation and scale, or demands that exceed those for which the system was designed. Once the model can simulate the problem, it can be used to simulate alternative solutions to arrive at the most cost-effective alternative. Continuous pressure problems usually result from trying to serve customers at too high an elevation. Once again, the model can be used to assess potential solutions, which can include adjusting pressure-zone boundaries and installing pumps to adjust PRV settings.

17.4.2 Correction of Pressure Problems

Problems that often occur in distribution systems are highlighted in the following sections, along with potential solutions. Isolating specific causes is often difficult and requires extensive knowledge of the system, engineering judgment, and use of specific procedures outlined in the previous section. In addition, problems associated with low pressure and reduced carrying capacity may have multiple causes.

17.4.2.1 Closed isolating valves. As their name implies, isolating valves are used to isolate portions of the system for maintenance and repairs or to improve flushing. They should only be closed during these times. Isolating valves (usually gate valves although butterfly valves can be used in some instances, especially in larger pipes) should be fully closed along pressure zone boundaries but be fully open in the remainder of the system. It is difficult in a looped system to identify when a valve is incorrectly left closed or partly closed. This difficulty arises because velocities in most systems are so low that water can find alternative paths to all customers without serious loss in pressure. However, during high–flow periods, especially during fires, closed or partly closed valves can seriously limit the system's capacity.

On occasion, model calibration can indicate a closed valve; however, the best method of ensuring that valves are set correctly is to have a valve exercising program. In such a program, every valve in the system is operated periodically (usually every year). This approach accomplishes several purposes: ensuring that the valve is in the correct position, preventing the valve from rusting in place, locating valves that have been paved over since they were last operated, and providing training for new employees on the location of valves. In addition, if a valve should break while being exercised, repair can be accomplished with scheduled maintenance rather than during an emergency.

17.4.2.2 Elevation and pressure zone issues. Pressure zones initially are planned so that each one includes all customers between a planned high and low elevation. Occasionally, pressure zones have a way of growing haphazardly such that customers may be at too high an elevation to receive adequate pressure or at too low an elevation, resulting in excessive pressure. Some utilities review pressure-zone boundaries periodically (usually at the time of master planning studies) to make certain that customers are served correctly. Customers with low pressures usually are quick to notify the water utility, whereas those with excessive pressure become accustomed to it. Surprisingly, reducing a customer's pressure from a high pressure, say 110 psi, to an acceptable pressure, say 80 psi, can result in complaints.

Pressure-zone boundary adjustments must be made carefully so that fire flows are not adversely affected and dead-ends are not created, adding to water-quality problems. Some sprinkler systems are designed to operated on the basis of a certain main pressure and pressure-zone adjustments can affect sprinkler performance. With all the problems in adjusting pressure-zone boundaries, it is important to establish pressure zone boundaries when the system is being designed.

17.4.2.3 Carrying capacity. In older systems, especially those with unlined cast iron pipes, carrying capacity may be lost because of tuberculation in mains, as shown in Fig. 17.11. The tuberculation appears as deposits of oxidized iron on pipe walls (California Section Committee Report, 1962; Hudson, 1966) and can significantly reduce the cross-sectional area of a pipe. In other systems, carrying capacity can be lost as a result of postprecipitation of floc, oxidized manganese, or scale because of calcium carbonate precipita-

FIGURE 17.11 Tuberculation in a water main. (Photograph by T. M. Walski).

tion. The problem can be indicated using a distribution model if low Hazen-Williams C-factors (or high pipe roughness) are needed to calibrate the model. This problem needs to be verified by conducting headloss tests for a section of pipe, in which it is necessary to measure headloss, length of test section, flow rate, and diameter (Walski, 1984). The pipe's C-factor can then be calculated based on the Hazen-Williams equation, rearranged below to solve for C:

$$C = KVD^{-0.63}\left(\frac{h}{L}\right)^{-0.54} \tag{17.7}$$

where: C = Hazen-Williams C-factor, K = 2.85 (metric units) or 1.82 (English units), V = velocity [m/s (ft/s)], D = diameter [m (ft)], h = headloss in test section [m (ft)], and L = length of test section [m (ft)].

C-factor values of less than 100 indicate a significant loss of carrying capacity. Once the low C-factor values have been identified, the cause of the low values needs to be confirmed. Usually the existence of scale or tuberculation has been observed by maintenance personnel during pipe repairs and pipe tapping. When such observations are not available, the utility often cuts into the pipe to remove a sample of the pipe wall for observation

The C-factor that justifies use of cleaning is site specific. Some oversized pipes can still perform adequately with C-factors below 50, whereas others can fail to deliver adequate capacity even at C-factors greater than 100. In general, in old tuberculated cast iron mains, C-factors can be restored to values on the order of 100–120 with cleaning and cement mortar lining. The final C-factor will depend on the pipe's smoothness after rehabilitation and its size.

Hydraulic models can calculate post rehabilitation pressures and flows to predict whether a cleaning project will restore adequate capacity. If some form of rehabilitation can meet requirements, the cost of rehabilitation (i.e., cleaning with or without lining) needs to be compared with the cost of other alternatives (replacement, installation of parallel lines). In general, if cleaning is used solely to restore lost carrying capacity, it will be more economical than will installing new pipe. However, cleaning is limited in the extent to which it can restore capacity. Therefore, before pipe rehabilitation is considered, the leak-and-break history of the pipe should be evaluated. If the pipe has had numerous breaks, leaks, or both then installing new pipe may be the best alternative. If the pipe is structurally sound, it is a candidate for rehabilitation.

17.4.2.4 Inadequate capacity. Problems with a distribution system's capacity are often thought of as problems with individual pipes or groups of pipes that make up the network. However, the system is made up of more than pipes, and these other components, such as pumps and PRVs, can contribute to pressure and capacity problems. In addition, evaluation of pipeline capacity alone may not lead to the most economical solution.

In some cases where the low-pressure problem is the result of small pipes and the length of pipe required to correct the problem is large, it may be economical to install elevated storage (or a ground tank on a hill) near the perimeter of the system to improve pressure during times of peak use. Then, the piping only needs to be large enough to meet peak day demands since the storage tank will provide flow to meet instantaneous peak demand.

If a low-pressure problem exists near the high end of a pressure zone adjacent to a higher zone, it may be economical to install a PRV between the two zones that would open only when the pressure in the lower zone drops significantly. Although this type of connection often is economical, it is important to remember that water pumped to a high-pressure zone, only to flow back down into a lower zone, represents a waste of energy.

If pressure problems are the result of inadequate pumping and the pump does not have a full sized impeller, it may be possible to replace the impeller with a larger one, provided the pump motor can handle the load. If a pump's capacity is simply insufficient, a capital project should be developed to increase capacity to meet the needs.

17.4.3 Pipe Rehabilitation Technology

Restoring lost carrying capacity in water mains involves removing deposits from pipe walls, then lining the interior of the pipe if necessary (AWWA, 1986, Deb et al. 1990). Pipe cleaning is accomplished by hydraulically driven pigs or cable-pulled scrappers. Pigs are made of polyethylene with an abrasive skin (Fig. 17.12) They are forced through a main by water pressure from a point of insertion to a point where the pig exits the pipe. Generally, excavation at each point is necessary; however, the distance between points can be fairly long. The use of pigs usually results in a great deal of dirty water that must be disposed of at the downstream end. In some cases, utilities may have trouble getting discharge permits for the water. If the pipe is going to be relined afterward, there is little advantage to using pigs. Scrapers must be pulled through a section of pipe, requiring the cable to be run through the pipe first. Because of this fact, shorter lengths must be scraped at each step if lining is to follow.

If cleaning simply removes scale or some other soft deposits, it may not be necessary to reline the pipe. However, if the cleaning exposes base metal pipe, the tuberculation process will resume and red-water problems may occur. It is necessary to prevent these reactions either by lining the pipe or by altering water quality (i.e., pH adjustment, corrosion inhibitor) to minimize corrosion.

There are three basic types of linings: sprayed-on linings, slip linings and inversion linings. With sprayed-on linings, cement mortar or epoxy is sprayed onto the inside of the

FIGURE 17.12 Pig for cleaning a water main. (Photo courtesy of Flowmort Services, Inc.).

pipe wall and is allowed to cure before the pipe is placed back in service. Like scrapers, the mortar sprayer is pulled through the main, often followed by a smoothing device, such as a cone (for smaller pipes) or a rotating trowel (for larger pipes). Service connections can be restored by blowing compressed air back into the main to prevent the service from becoming blocked. With slip linings, a new thin-walled pipe is pulled into the newly cleaned main. Services need to be restored manually, although a new approach to restore services from inside the pipe using robotic technology is being developed. Inversion linings are soft linings that are pulled into the main much like an inside-out sock and are cured in place. This technology works well in gravity sewers but not as well in pressure pipe because the lining must be fairly thick. Pipe rehabilitation technology is described in more detail in AWWA M-28 (1987a) and Warren et al., (1990).

Another technology is pipe insertion, in which a mole is pulled through the pipe. The mole, which is larger than the inside diameter of the pipe, breaks the old pipe in place and enlarges the hole. A new pipe is then pulled into the hole left behind. With this technology, it is usually possible to install a slightly larger pipe. This method only works when the old pipe is made of a brittle material, such as cast iron or asbestos cement.

17.4.4 Evaluation of Pipe Rehabilitation

The decision whether to rehabilitate a pipe involves comparing the cost of rehabilitation against the cost of replacing the pipe or installing a parallel main. The pipe should be replaced if it has had a history of leak-and-break problems or is a likely candidate for such problems. If the existing pipe is still an asset for the utility, installation of a parallel main is the usual alternative to replacement. If growth in demand is expected, cleaning is probably not justified because it cannot increase the size of pipes. If growth is expected to be modest, however, cleaning the existing pipe can be economical. Approaches to evaluation involved in deciding on pipeline rehabilitation are given in Evins et al.(1989), Halhal, et al. (1997), Kettler and Goulter (1985), Kim and Mays (1994), Lansey et al. (1992), Walski (1985, 1986) and Woodburn et al, (1987). A hydraulic model can be used to simulate the various alternatives as is described in earlier chapters and in AWWA M-32 (1989), Cesario (1995), and Walski et al. (1990).

Cleaning of pipes also can be based on savings in energy costs, a special situation usually applicable at pumping stations. The evaluation, given in Walski (1982), is based on the a comparison of the cost of cleaning and the present worth of potential savings in energy costs. For the simple case where the pumping equipment is not being changed, cleaning is justified if the cost of energy saved exceeds the cost of cleaning, as given by

$$c_r = \frac{10.43 \ C^* \ (Q_a^{1.85} P \ (swpf) \ (0.0164))}{D^{4.87}} \quad (17.8)$$

where c_r = cost of rehabilitation ($/ft), Q_a = average flow through pump station (gpm), P = price of energy (cents/kWh), $spwf$ = series present worth factor, D = Diameter (in), $C^* = (1/C^{1.85}) - (1/C'^{1.85})$, C = C –factor before cleaning, and C' = C–factor after cleaning.

17.5 MAINTENANCE INFORMATION SYSTEMS

Making good maintenance and rehabilitation decisions depends heavily on having good information. Unfortunately, many such decisions are based on intuition when better information could have lead to better decisions. Some tools for information management are described below.

17.5.1 System Mapping

Having good maps of the water distribution system is a starting point for many kinds of maintenance work and for the modeling needed for good hydraulic analysis. At the minimum, maps should contain the location of and information (such as size) on all pipes, valves, hydrants, regulating valves, tanks, and pumping stations. A large amount of other information is often placed on maps, including pressure-zone boundaries, installation date, rehabilitation date, service connections, hydrant identifier, valve identifier, and pipe material. However, too much information can make a map too cluttered to read.

Utilities have moved away from paper-base maps to maps based on computer-aided drafting (CAD) systems (Martinez, 1995). Computer-based maps provide the capability to produce color-coded details, special maps at any scale, maps with only some of the data displayed (e.g., not showing double-line streets). Computerized maps are relatively easy to update. Although paper maps must still be produced for operation and maintenance crews, the day may come when maintenance crews have their own portable computers with mapping capabilities.

The general distribution system maps are just one type of map that a utility needs. For work in complicated intersections, most utilities use more detailed maps, which may be referred to as intersection details, valve books, gate sheets, or by other names. These sketches are usually drawn in a smaller size, (e.g., 8 1/2 x 11 in paper), and contain information, such as setbacks from curbs and lot lines, and swing ties to permanent facilities, such as utility poles, corners of buildings, and manholes.

Another type of map that is useful for hydraulic analysis is an elevation map or hydraulic profile. This type of map shows major facilities (i.e., tanks, pumps, control valves) to scale in the vertical direction and provides important insights to the operation of the system. Mapping needs to be updated routinely so that all upgrades to the system are added to the maps shortly after the improvement has been completed.

17.5.2 System Database

Information can be maintained on maps or in text databases. Information that is text in nature can be stored just as well in a database as on the map. Such information as date laid, pipe material, installation work order number or contract name, pipe class, valve manufacturer, hydrant manufacturer, and depth of cover are permanent properties of the pipes and can be shown on the map or kept in a database. Information such as the results of hydrant flow tests, pressure recorded, break history, and results of headloss tests are values that can change with time and thus should be kept in databases, not on the record maps. An example of the use of such data for rehabilitation decisions is given in Habibian (1992) and O'Day (1983).

17.5.3 Geographic Information Systems

In the previous sections, the distinction was drawn between graphical data (e.g., drawing of pipes on a map) and numerical values associated with the graphical information (e.g., year installed). Ideally both kinds of information can be linked, and this is the value of geographic information systems (GIS), which merge the function of maps and databases by associating attributes with any entity (lines, points). The cost of setting up a GIS is high, particularly when including the utility's cost of entering data. However, because the costs are dropping steadily, most utilities may be able to afford GIS. Costs can be reduced by sharing base mapping costs with other utilities and government agencies.

Some fundamental questions that must be answered when setting up a GIS, such as what is the basic size entity for which data must be stored. Is a pipe segment a city block or the distance between valves and intersections? Is the pump and motor one entity, or are the pump and motor two separate entities? Or is the pump station an entity? Should pipes be named by their x-y coordinates, the street block face, or some other system? The most difficult job in setting up a GIS is the huge effort involved in entering the large volume of data and ensuring its accuracy and subsequently ensuring that the GIS will be properly maintained. Once a GIS is established, it is a powerful tool for decision-making. An example of the widespread use of GIS is the fact that roughly 25 percent of the annual AWWA Information Technology conference is devoted to GIS. Some examples include Buyens et al.(1996), Cannistra et al. (1992), Dacier et al. (1995), and Holliman and Lin (1985).

GIS allows utilities to maintain an extensive amount of information that can be easily isolated and extracted easily. Larger municipalities often maintain GIS information for multiple underground utilities (e.g., water, sanitary sewer, storm sewer, electric lines) as well as street curbs, property lines, and so on.

17.5.4 Maintenance Management Systems

Utilities usually keep some sort of record of repairs and rehabilitation work. Maintenance management tools are available to record these events systematically and perform some analysis on the records. Some can even track work orders from creation to completion and remind managers of the need for programmed routine maintenance. Information, such as pipe break history, is helpful in performing analyses described earlier to determine whether pipes should be replaced systematically. These maintenance management systems are being interfaced with GIS to realize their full potential.

17.5.5 SCADA Systems

Utilities collect large amounts of operational data to assist in day-to-day operations decisions. The term "Supervisory Control and Data Acquisition" (SCADA) is used to describe an integrated system of remote telemetry and automated control. Operations are described in other chapters of this book, but a good deal of the information from the SCADA system is useful when making maintenance and rehabilitation decisions. Information from pressure sensors usually is most useful for decision-making associated with problems, such as low pressure and waterhammer. SCADA information also is important for calculating quantities, such as minimum nighttime flow, to assist in evaluating unaccounted-for water.

Although SCADA systems are helpful for operations and maintenance, they also require a good deal of maintenance themselves. It is fairly easy to find parts for a 20-year-old pump or valve, but SCADA systems involve electronic components. A 5-year–old SCADA system can be virtually obsolete, and it may be almost impossible to find replacement parts for a 10-year-old SCADA system. Unless utilities purchase a large inventory of spare components, they must be prepared to upgrade the entire system every few years.

REFERENCES

Arrequin-Cortes, F. I., and L. H. Ochoa-Alejo, "Evaluation of Water Losses in Distribution Networks," *Journal of Water Resources Planning and Management,* 123(5):284, 1997.

AWWA, *Cleaning and Lining of Distribution System Pipe,* American Water Works Association, Denver, CO, 1986.

AWWA, *Distribution System Maintenance Techniques,* American Water Works Association, Denver, CO, 1987.

AWWA, *Water Supply Operations Series: Part III. Transmission and Distribution,* American Water Works Association, Denver, CO, 1996.

AWWA M-6, *Water Meters—Selection, Installation, Testing and Maintenance,* American Water Works Association, Denver, CO, 1986.

AWWA M-17, *Installation, Field Testing and Maintenance of Fire Hydrants,* American Water Works Association, Denver, CO, 1987.

AWWA M-22, *Sizing Water Service Lines and Meters,* American Water Works Association, Denver, CO, 1975.

AWWA M-28, *Cleaning and Lining Water Mains,* American Water Works Association, Denver, CO, 1987.

AWWA M-32, *Distribution Network Analysis for Water Utilities,* American Water Works Association, Denver, CO, 1989.

AWWA M-33, *Flowmeters in Water Supply,* American Water Works Association, Denver, CO, 1989.

AWWA M-36, *Water Audits and Leak Detection,* American Water Works Association, Denver, CO, 1996.

AWWA and DVGW Forschungsstelle, *Internal Corrosion of Water Distribution Systems,* American Water Works Association Research Foundation, Denver, CO, 1985.

AWWA Task Group 2850-D, "Replacement of Water Distribution Mains," *Journal of the American Water Works Association,* 61(9):417, 1969.

Battelle Memorial *Institute, Innovative Approaches for New and Existing Water Distribution Systems,* USEPA, National Risk Management Research Laboratory, Edison, NJ, 1998.

Bouchart, F., and I. C. Goulter, "Improvements in Design of Water Distribution Networks Recognizing Valve Location," *Water Resources Research,* 27:3029, 1991.

Boyle Engineering, *Municipal Leak Detection Program Loss Reduction—Research and Analysis,* California Department of Water Resources, 1982.

Buyens, d. J., P. A. Bizier, and C. W. Combee, "Using a Geographical Information System to Determine Water Distribution Model Demands," *Proceedings of the Annual AWWA Conference,* E 131, 1996.

California Section Committee Reports, "Loss of Carrying Capacity of Water Mains," *Journal of the American Water Works Association,* 54:1293, 1962.

California Section Committee Report, "Determination of Economic Period for Water Meter Replacement," *Journal of the American Water Works Association,* 58(6):642, 1966.

Cannistra, J. R., R. Leadbeater, and R. Humphries, "Washington Suburban Sanitary Commission Implements GIS," *Journal of the American Water Works Association,* 84(7):72, 1992.

Cesario, a. L., *Modeling, Analysis and Design of Water Distribution Systems,* American Water Works Association, Denver, CO, 1995.

Clark, R. M., B. W., Lykins, and J. A. Goodrich, *Infrastructure and Maintenance of Water Quality,* USEPA 600/D-85/215, U.S. Environmental Protection Agency, Cincinnati, OH, 1985.

Cohen, A., and M. B. Fielding, "Predicting Frost Depth: Protecting Underground Pipelines," *Journal of the American Water Works Association,* 71(2):113, 1979.

Dacier, N. M., P. F. Boulos, J. W. Clapp, A. K. Dhingra, and R. W. Bowcock, "Taking Small Steps Toward Fully-Integrated Computer Based Environment for Distribution System Operation, Maintenance and Management, in *Computers in the Water Industry,* American Water Works Association, Denver, CO, 1995.

Deb, A. K., J. K. Snyder, J. J. Chelius, J. Urie, and D. K. O'Day, *Assessment of Existing and Developing Water Main Rehabilitation Practices*, American Water Works Association, Denver, CO, 1990.

Deb, A. K., Y. J. Hasit, and F. M. Grablutz, 1995, *Distribution System Performance*, American Water Works Association Research Foundation, Denver, CO, 1995.

Dempsey, P., and B. A. Manook, Assessing the *Condition of Cast Iron Pipes,* Water Research Centre, Swindon, UK, 1986.

Duranceau, S. J., M. L. Schiff, and G. E. C. Bell, "Electrical Grounding, Pipe Integrity and Shock Hazard," *Journal of the American Water Works Association,* 90(7):40, 1998.

Evins, C., G. Stephenson, I. C. Warren, and S. M. Williams, *Planning the Rehabilitation of Water Distribution Systems,* Water Research Centre, Swindon, UK, 1989.

Fitzgerald, J. H., "Corrosion as Primary Cause of Cast Iron Main Breaks," *Journal of the American Water Works Association,* 60(7):882, 1968.

Gerhold, W. F., "Corrosion Behavior of Ductile Cast Iron Pipe in Soil Environments", *Journal of the American Water Works Association,* 68(12):674, 1976.

Gros, W. F. H., "Emergency Distribution System Operations in Cold Weather," *Journal of the American Water Works Association*, 68(1):8, 1976.

Habibian, A., "Developing and Utilizing Data Bases for Water Main Rehabilitation, *Journal of the American Water Works Association,* 84(7), 1992.

Halhal, D., G. A. Walters, D. Ouazar, and D. A. Savic, "Water Network Rehabilitation with Structured Messy Genetic Algorithm, *Journal of Water Resources Planning and Management,* 123(3):127, 1997.

Hanke, S. H., "Distribution System Leak Detection and Control," *Water Engineering and Management,* R-108, 1981.

Heim, P. M., "Conducting a Leak Detection Search," *Journal of the American Water Works Association,* 71(2):66, 1979.

Heitzman, G. C., and K. A. Willis, "Considerations in Implementing a Comprehensive Infrastructure Renewal Program," *Proceedings of the AWWA Annual Conference,* 1995.

Hoag, L. N., *Effect of Water Pressure on Water* Use, U.S. Department of Housing and Urban Development, Washington, DC, 1984a.

Hoag, L. N., *Effect of Water Meters on Water Use,* U.S. Department of Housing and Urban Development, 1984b.

Holliman, T. R., and A. Lin, "GIS Database Design for Municipal Water and Sewer Systems," *Computers in the Water Industry,* American Water Works Association, Denver, CO, 1995.

Hudson, W. D., "Studies of Distribution System Capacity in Seven Cities," *Journal of the American Water Works Association,* 58(2):157, 1966.

Hudson, W. D., "Increasing Water System Efficiency Through Control of Unaccounted for Water," *Journal of the American Water Works Association*, 70(7):362, 1978.

Jackson, R. A., C. Pitt, and R. Skabo, *Nondestructive Testing of Water Mains for Physical Integrity,* American Water Works Association Research Foundation, Denver, CO, 1992.

Jeffcoat, P., and A. Saravanapavan, *The Reduction and Control of Unaccounted-for Water: Working Guidelines,* World Bank Technical Paper No. 72, World Bank, Washington, DC, 1987.

Jowitt, P. W., and C. Xu, "Optimal Valve Control in Water Distribution Networks," *Journal of Water Resources Planning and Management,* 116:455, 1989.

Jowitt, P. S., and C. Xu, "Predicting Pipe Failure Effects in Water Distribution Networks," *Journal of Water Resources Planning and Management,* 119(1):18, 1993.

Kettler, A. J., and I. C. Goulter, "An Analysis of Pipe Breakage in Urban Water Distribution Networks," *Canadian Journal of Civil Engineering,* 12(286): 1985.

Kim, J. H., and L. W. Mays, "Optimal Rehabilitation Model for Water Distribution Systems," *Journal of Water Resources Planning and Management,* 120:674, 1994.

Kingston, W. L., " A Do-It-Yourself Leak Survey Benefit Cost Study, *Journal of the American Water Works Association*, 71(2):70, 1979.

Kirmeyer, G. J., and G. S. Logsdon, "Principles of Internal Corrosion and Corrosion and Corrosion Monitoring," *Journal of the American Water Works Association,* 75(2):57, 1983.

Kuranz, A. P., "Meter Maintenance Practice," *Journal of the American Water Works Association,* 34(1):117–120, 1942.

Kutsal, M., "Criteria and Methods for Evaluating Distribution System Mains," Proceedings of the AWWA Annual Conference, 1995.

Lai, C. C., "Unaccounted-for Water and the Economics of Leak Detection," *Water Supply,* 9(3):IRI-1, 1991.

Langlier, W. F., "The Analytical Control of Anti-Corrosion Water Treatment," *Journal of the American Water Works,* 28(10):1500, 1936.

Lansey, K., "Optimal Maintenance Scheduling for Water Distribution Systems," in Civil Engineering Systems, Gordon and Breach, UK, 1992.

Laverty, G. L., "Leak Detection: Modern Methods, Costs and Benefits," *Journal of the American Water Works Association,* 71(2):61, 1979.

Li, D., and Y. Y. Haimes, "Optimal Maintenance Related Decision Making for Deteriorating Water Distribution Systems: 1. Semi-Markovian Model for a Water Main," *Water Resources Research,* 28:1053, 1992a.

Li, D., and Y. Y. Haimes, "Optimal Maintenance Related Decision Making for Deteriorating Water Distribution Systems: 2. Multilevel Decomposition Approach," *Water Resources Research,* 28:1063, 1992b.

Loewenthal, R. E., and G. V. R. Marais, *Carbonate Chemistry of Aquatic Systems:* Theory and Applications, Ann Arbor Science, Ann Arbor, MI, 1976.

Male, J. W., and S. L. Franz, "Allocating Funds for Repair of Leaky Water Distribution Systems," *TIMS Studies in the Management Sciences, Elsevier Science,* 22:183, 1986.

Male, J. W., and T. M. Walski, *Water Distribution Systems: A Troubleshooting Manual,* Lewis Publishers, 1990.

Male, J. W., T. M. Walski, and A. H. Slutsky, "Analyzing Water Main Replacement Policies," *Journal of Water Resources Planning and Management,* 116:362, 1990a.

Male, J. W., T. M. Walski, and A. H. Slutsky, *New York Water Supply Infrastructure Study—Vol. V: Analysis of Replacement Policy,* Technical Report EL-87-9, Waterways Experiment Station, U.S. Army Corps of Engineers, Vicksburg, MS, 1990b.

Martinez, B., "Water System Maps to CAD in Record Time," in *Computers in the Water Industry,* American Water Works Association, Denver, CO, 1995.

Merrill, D. T., and R. L. Sanks, "Corrosion Control by Deposition of $CaCO_3$ Films, American Water Works Association, Denver, CO, 1978.

Monie, W. D., and C. M. Clark, "Loads on Underground Pipe Due to Frost Penetration," *Journal of the American Water Works Association,* 68(6):353, 1974.

Moyer, E. E., *Economics of Leak Detection*: A Case Study Approach, American Water Works Association, Denver, CO, 1985.

Moyer, E. E., J. W. Male, I. C. Moore, and J. G. Hock, "The Economics of Leak Detection and Repair," *Journal of the American Water Works Association,* 75(1):28, 1983.

National Water Council, *Leakage Control Policies and Practices,* National Water Council, London, UK, 1980.

NFPA, *Fire Flow Testing and Marking of Hydrants,* National Fire Prevention Association, Batterymark, MA, 1988.

Noss, R. R., G. Newman, and J. W. Male, "Optimal Testing Frequency for Domestic Water Meters," *Journal of Water Resources Planning and Management,* 113(1):1-14, 1987.

Oberoi, K., "Distribution Flushing Program: The Benefits and Results", *Proceedings of the AWWA Annual Conference,* 1994. pp.

O'Day, D. K., "Geoprocessing—A Water Distribution Management Tool, *Journal of the American Water Works Association,* 75(1):41, 1983.

O'Day, D. K., R. Weiss, S. Chiavari, and D. Blair, *Water Main Evaluation for Rehabilitation/ Replacement,* American Water Works Association Research Foundation, Denver, CO, 1986.

O'Day, D. K., R. Weiss, S. Chiavari, and D. Blair, *Water Main Evaluation for Rehabilitation/Replacement,* EPA/600/2-87/024, U.S. Environmental Protection Agency, Cincinnati, OH, 1987.

Quimpo, R. G., and U. M. Shamsi, "Reliability Based Distribution System Maintenance," Journal of Water Resources Planning and Management, 117:321, 1991.

Ramos, W. L., "Benefits/Cost Analysis for Determining Water Main Replacement," *Proceedings of the AWWA Annual Conference,* 1985, pp.

Romanoff, M., "Exterior Corrosion of Cast-Iron Pipes," *Journal of the American Water Works Association,* 56:1129, 1964.

Romanoff, M., "Performance of Ductile Iron Pipe in Soils," *Journal of the American Water Works Association,* 60:645. 1968.

Shamir, U., and C. D. D. Howard, "An Analytical Approach to Scheduling Pipe Replacement," *Journal of the American Water Works Association,* 71(5): 248, 1979.

Siedler, M., "Obtaining an Analytical Grasp on Water Distribution," *Journal of the American Water Works Association,* 74(12): 1982.

Siedler, M., "Winning the Water Against Unaccounted-for Water," *Journal of the New England Water Works Association,* 99(20):119, 1985.

Singley, J. E., B. A. Beaudet, and P. H. Markey, *Corrosion Manual for Internal Corrosion of Water Distribution Systems,* USEPA 570/9-84-001, U.S. Environmental Protection Agency, Washington, DC, 1984.

Sisco, R. J., "The Case for Meter Replacement Programs," *Journal of the American Water Works Association,* 59(11):1449–1455, 1967.

Smith, H. E., "Frost Loadings on Underground Pipe," *Journal of the American Water Works Association,* 68(12):673, 1976.

Sullivan, J. P., and E. M. Speranza, "Proper Meter Sizing for Increased Accountability and Revenue," *Journal of the American Water Works Association,* 84(7):53, 1992.

Utility Infrastructure Manual, U.S. Department of Housing and Urban Development, Washington, DC, 1984.

Wallace, L. P., *Water and Revenue Losses: Unaccounted-for Water,* American Water Works Association Research Foundation, Denver, CO, 1987.

Walski, T. M., "Economics Analysis of Rehabilitation of Water Mains," *Journal of Water Resources Planning and Management,* 108(WR3):296, 1982.

Walski, T. M., *Analysis of Water Distribution Systems,* Van Nostrand Reinhold, New York, 1984.

Walski, T. M., Cost of *Water Distribution System Infrastructure Rehabilitation, Repair and Replacement,* TR EL-85-5, U.S. Army Corps of Engineers Waterways Experiment Station, Vicksburg, MS, 1985.

Walski, T. M., "Making Water System Rehabilitation Decisions," *ASCE Water Forum,* 1986.

Walski, T. M., "Replacement Rules for Water Mains," *Journal of the American Water Works Association,* 79(11):33, 1987a.

Walski, T. M., Water Supply System Rehabilitation, American Society of Civil Engineers, New York, 1987b.

Walski, T. M., *"Water Distribution Valve Topology for Reliability Analysis,"* Reliability Engineering and System Safety, 42(1):13, 1993.

Walski, T. M., "Cleaning and Lining vs. Parallel Main," Journal of Water Resources Planning and Management, 111(1):43, 1995b.

Walski, T. M., and A. Pelliccia, "An Economic Analysis of Water Main Breaks," *Journal of the American Water Works Association,* 74(3):140, 1982.

Walski, T. M., and R. Wade, *New York Water Supply Infrastructure Study, Vol, III: The Bronx and Queens,* U.S. Army Corps of Engineers Waterways Experiment Station, Vicksburg, MS, 1987.

Walski, T. M., J. Gessler, and J. W. Sjostrom, *Water Distribution Systems; Simulation and Sizing*, Lewis Publishers, 1990.

Walski, T. M., J. Komensky, S. Rhoades, A. Wood, and E. Rothstein, "Using Flow Rate Date to Assess Meter Selection for Large Water Users," *Proceedings of the AWWA National Convention*, 1995.

Warren, I. C., S. D. Mason, and P. J. Conroy, *In-situ Cement Mortar Lining—Operation Guidelines and Code of Practice*, Water Research Center, Swindon, UK, 1990.

Weiss, R. A., T. M. King, D. K. O'Day, S. K. Lior, F. J. Hood, and N. E. Johnson, *Philadelphia Water Supply Infrastructure Study,* U. S. Army Corps of Engineers and Philadelphia Water Department, Philadelphia, PA, 1985.

Woodburn, J., K. E. Lansey, and L. W. Mays, "Model for Optimal Rehabilitation and Replacement of Water Distribution System Components," *Proceedings of the ASCE National Conference on Hydraulic Engineering*, p. 606, 1987.

CHAPTER 18
RELIABILITY ANALYSIS FOR DESIGN

Ian Goulter
Swinburne University of Technology
Hawthorn, Victoria, Australia.

Thomas M. Walski
Pennsylvania American Water Co.
E. Wilkes-Barre, PA

Larry W. Mays
Department of Civil and Environmental Engineering
Arizona State University
Tempe, AZ

A. Burcu Altan Sakarya
Department of Civil Engineering
Middle East Technical University
Ankara, Turkey

Francious Bouchart
Department of Civil and Offshore Engineering
Heriot-Watt University
Edinburgh, Scotland

Y. K. Tung
Department of Civil Engineering
Hong Kong University of Science and Technology
Clear Water Bay, Kowloon, Hong Kong

18.1 FAILURE MODES FOR WATER DISTRIBUTION SYSTEMS

18.1.1 Need and Justification

Water utilities are implemented to construct, operate, and maintain water supply systems. The basic function of these water utilities is to obtain water from a source, treat the water so that its quality is acceptable, and deliver the desired quantity of water at the desired time. The analysis of a water supply system often is devoted to the evaluation of one or more of the five major systems components: the source, the water transmission, the treatment system, the storage system, or the distribution system. The storage system consists of both raw water storage and finished water storage.

The American Water Works Association (1974) defines a water distribution system as one "including all water utility components for the distribution of finished or potable water by means of gravity storage feed or pumps through distribution-pumping networks to customers or other users, including distribution-equalizing storage." The design or extension of a water distribution system generally involves large capital outlays as well as the continuing operation, maintenance, and repair costs. Because of the complexity of the problem arising from the large number of design components and their interaction, automated procedures that result in reliable, but minimal, cost designs are desired. Conventional design approaches consist of selecting a network configuration, pipe sizes, reservoir sizes and elevations, and facilities. This process is usually a trial-and-error procedure that attempts to find a design representing a low-cost solution. There is no guarantee that the resulting distribution system is a minimum-cost solution, nor is the reliability of the designed system generally measured (Mays, 1989).

To ensure delivery of finished water to the user, the water distribution system must be designed to accommodate a range of expected emergency loading conditions. These emergency conditions generally are classified into three groups: broken pipes, fire demands, and pump and power outages. Each condition must be examined with an emphasis on describing its impact on the system, developing relevant measures of system performance, and designing into the system the capacity required to handle emergency conditions with an acceptable measure of reliability.

A review of the literature (Mays, 1989, and Sec. 18.4) reveals that no universally acceptable definition or measure of the reliability of water distribution systems is currently available. This will become more evident throughout this chapter as the various co-authors have presented slightly different definitions of reliability in the different sections. *Reliability* is usually defined as the probability that a system will perform its mission within specified limits for a given period of time in a specified environment. For a large system, with many interactive subsystems (such as a water distribution system), it is extremely difficult to compute the mathematical reliability analytically. Accurate calculation of a mathematical reliability requires knowledge of the precise reliability of the basic subsystems or components and the impact on accomplishing the mission caused by the set of all possible subsystem (component) failures.

Many researchers, municipal engineers, urban planners, institutes, government agencies, and so forth have discussed the need to develop explicit measures and methodologies to evaluate the reliability of water distribution systems and their performance under emergency loading conditions. Some researchers have proposed candidate approaches using such concepts as reliability factors, economic loss functions, and forced redundancy in the designs. All these approaches have limitations in the formulation of problems and the techniques for solutions. Some investigators discuss the need to incorporate measures of reliability explicitly into optimization models to predict a systems operation under emergency loading conditions. At present, however, no "optimization-reliability" evaluation or design technique with general application has been developed.

The reliable delivery of water to the user requires the water distribution system to be designed to handle a range of expected emergency loading conditions. These emergency conditions can be classified: as (1) fire demands, (2) broken links, (3) pump failures, (4) power outages, (5) control valve failures; and (6) insufficient storage capability. A general application methodology that considers the aspects of minimum cost and reliability must consider each of these emergency conditions. These conditions should be examined within the methodologies to (1) describe their importance to the system; (2) develop relevant measures of system performance, and (3) design the capacity for the emergency loading conditions into the system with an acceptable measure of reliability.

18.1.2 Definitions of Distribution System Repairs

The American Water Works Association Research Foundation (AWWARF), in conjunction with the U. S. Environmental Protection Agency, developed a guidance manual titled *Water Main Evaluation for Rehabilitation/Replacement* (O'Day et al., 1986). Portions of that document will be cited throughout this section. One aspect of the *AWWARF manual* was the clarification of terminology regarding the reporting of leaks and breaks.

The lack of consistency between how utilities define leaks and breaks presents difficulties when trying to compare the conditions of distribution systems. The general term *leak* can include both structural failures as well as water loss, which occurs from improperly sealed joints, defective service taps, and holes caused by corrosion. *Breaks*, on the other hand, imply a structural failure of the water main, which results in either a crack or a complete severance of the main. Breaks cause leakage and water loss; therefore, it can be argued that breaks are a special form of leaks.

Distribution system repairs can be classified by the component that was repaired (main, service, hydrant), the nature of the leak (joint leak, break), and the type of repair. Distribution repair records should identify the specific component repaired. Distribution system components include the following (O'Day et al., 1986):

Main	Hydrant	Line Valve	Service
Barrel	Branch Line	Bolts	Tap
Joint	Valve	Bonnet	Pipe
Plug	Cap barrel	Stem	Pipe
		Curb stop	

A set of definitions was developed to help standardize the reporting and logging of distribution repairs. These definitions are presented below (O'Day, et al., 1986):

Leak repairs. The term leak repair refers to all actions taken to repair leaks in water mains, line valves, hydrant branches, and service lines.

Main leaks. Water main leaks include all problems that lead to leakage of water from the main. These include joint leaks, holes, circumferential breaks, longitudinal breaks, defective tapes, and split bells. Leaks related to hydrant service lines, and valves are not main leaks.

Joint leaks. Joint leaks represent a loss of water from the joint between adjacent sections of main. This is not a structural problem; it is a separation of sections of main caused by expansion and contraction, settlement, or movement or movement of joint materials because of pressure or pipe deflection.

Main breaks. Main breaks represent the structural failure of the barrel or bell of the pipe caused by excessive loads, undermining of bedding, contact with other structures, corrosion, or a combination of these conditions. Types of main breaks include (1) circumferential, (2) longitudinal, and (3) split bells, including bell failures from sulfur compound joint materials.

Hydrant leak repairs. All actions taken to repair hydrant branch lines, hydrant valves, and hydrant barrels are termed hydrant leak repairs.

Service leak repairs. All actions taken to repair leaking taps, corporation stops, service pipes, and curb stops are termed service leak repairs.

Valve leak repairs. All actions taken to repair leaks in valve flanges, bonnet, or body are termed valve leak repairs.

18.1.3 Failure Modes

Although water distribution networks have many features in common with other types of networks (Templeman and Yates, 1984), there are a number of significant differences between them when they are examined from a failure-based point of view. The differences arise primarily as a result of the ways in which water distribution networks can "fail." In its most basic sense, *failure* of water distribution networks can be defined as the pressure, flow, or both falling below specified values at one or more nodes within the network.

Under such a definition, there are two major modes of failure of water distribution networks: performance failure (i.e., demand on the system being greater than the design value) and component (mechanical) failure (the failure of individual network components: e.g., pipes, pumps, and valves). Both modes of failure have probabilistic bases that must be incorporated into any reliability analyses of networks.

18.1.3.1 Performance failure. In the first case, the network "fails" in a traditional sense of the load exceeding the design capacity. However this is not necessarily catastrophic, as might be the case of failure of a structural network, such as a truss. Instead, the pressure at one or more nodes may drop below "minimum required," the amount of flow available at the node or nodes may fall below the required level, or both. The implications of such a "failure" are difficult to define because they may constitute only a decreased level of services to domestic households over a relatively brief period of time: for example, 6 h. On the other hand, the failure may result in a shortage of water supply during a major fire, thus causing increased monetary damages and possible loss of life.

The probabilistic aspect of failure in the first case arises from the probability distribution of the loading condition under which the distribution network is expected to operate. In most cases, the probability distributions of the demands, which the network must fulfill, are reasonably well known. Choice of a design value for the demands on the network—for example, 10-year, 6-h maximum flow—is not a deterministic decision and implies acceptance of a probability of the network being unable to perform to specifications. What has been missing in network analysis to date is an explicit recognition of the probabilistic processes.

The implication of failure as a result of hydraulic loads being greater than design loads are similar to those resulting from component failure; flows or supply pressures can fall below the desired levels. The means of improving the reliability of a water distribution system, as measured by the probability of either failure occurring, also are similar. Obviously, larger component sizes will provide greater overall flow capacity in the network and therefore reduce the probability that actual demands exceed the design values. In addition, previous work has shown that pipe breakage rates in existing distribution networks are strongly correlated to diameter (e.g., Ciottoni, 1983; Kettler and Goulter,1985; Mays,1989, O'Day,1982,1983; and O'Day et al.,1980. Larger-diameter pipes break less often than smaller-diameter pipes. Thus, selection of larger-diameter pipes can reduce the probability of component failure while simultaneously improving the flow-exceedance -based measure of reliability.

It is important to note, however, that good performance according to a reliability measure based on one mode of failure does not necessarily mean good performance according to the other mode. For example, consider a large pipe near the source of supply for a particular network. Since this pipe is near the source, a larger percentage of the flow passing to more "distant" parts of the network will flow through it, and a large-diameter, large–capacity pipe will be required. This large-diameter pipe will have a low frequency of breakage and therefore would indicate good reliability according to component failure probabilities. The probability that the flow that same pipe is required to carry will be greater than its design capacity, however, may be large. If this is the case, its "flow exceedance" reliability is high. Improvements to both measures of reliability may still be achieved through the selection of larger-diameter pipes.

The improvements to network reliability that are achieved through specification of large pipes also can be achieved through selection of larger capacities for the other important components of a network. By the provision of more pumps, pump-station failure is reduced. The additional pumps also provide additional capacity for the pump station, thereby reducing the probability that demands on the station exceed the design capacity. A similar argument can be made for facilities on the network.

18.1.3.2 Component (mechanical) failure. The probability of component failure can be derived from historical failure records and can be modeled using an appropriate probability distribution (see Sec. 18.3 for time-to-failure analysis). Goulter and Coals (1986) proposed a Poisson probability distribution with parameter λ breaks/km/yr for pipe breakage. The probability of demands being greater than design values also can be derived from historical data and modeled appropriately. The most difficult problem, however, is that even knowing the appropriate probability of the two modes of failure, it is difficult to define reliability explicitly. Reduction in either the probability of component failure or the probability of demands being greater than design capacity obviously improves system reliability. The question is, by how much? If a pipe (or any other component) fails, what is the effect on the remainder of the network? Obviously, the more crucial the pipe in question is to the network, the more widespread and serious is the effect. The actual quantification of the effect is difficult to assess without extensive simulation of network operation. Similarly, the "network-wide" implication of a particular demand being greater than the design value for that area is difficult to assess without considerable effort. Furthermore, if network reliability, as defined from measures based on either mode of failure is unsatisfactory, what is the best way to improve it?

18.1.4 Reliability: Indexes and Approaches

The traditional engineering definition of reliability is the ability of a system to meet demand. Some authors prefer to label this concept "availability," while reserving the term "reliability" for the length of time that a system can be expected to perform without failure. The term *reliability*, in its general sense, is defined as any measure of the system's ability to satisfy the requirements placed on it.

A number of different measures of reliability indexes have been proposed, but no single one is universally accepted. Each can be useful in reliability analyses, depending on the purpose of the analysis:

1. Reliability is often defined as the ability of a system to meet demand under a defined set of contingencies. For example, a system can be judged to be reliable if it can satisfy demand under the drought of record or despite the failure of a key piece of equipment. Early design criteria were based on such rules of thumb instead of quantitative indexes, such as those below. Contingency criteria suffer from arbitrariness in choice of contingency and from inconsistency because different designs able to withstand the same contingency can have different probabilities of failure.

2. Availability, which equals the probability at any given moment that the system will be found in a state such that demand does not exceed available supply or capacity or that operating conditions are not otherwise unsatisfactory (e.g., low water pressure). *Average availability* is the mean probability over a period of time of being found in such a state. In this and the next section, the term availability will be used in the sense of average availability. Probability of system failure is sometimes defined as one minus availability.
3. *Severity indexes* describe the size of failures. In water supply studies, reliability has been defined as the ratio of available annual supply to demand.
4. *Frequency and duration indexes* indicate how often failures of a given severity occur and how long they last. Such indexes have found increasing use in electric-utility planning studies.
5. *Economic indexes* measure the economic consequences of shortages, also referred to as *vulnerability*. Letting S_u represent water supply and Q the quantity of water demanded in units of m³/h, these indexes can be phrased as follows:

   a. *Contingency.* A system is reliable if $P(S_u < Q + \text{contingency}) = 0$.

   b. *Availability.* The availability of the system is $P(S_u > O)$.

   c. *Probability and severity of failure.* The probability of a failure can be defined as $P(S_u < Q - s)$ at a particular moment or over a particular period, where s represents some predefined level of a failure's severity.

   d. *Frequency and duration.* The *expected frequency* $E(F)$ (in units of $1/T$) of a failure event of at least severity s in m³/h is related to the expected duration $E(D)$ by the equation $E(F)E(D) = P(S_u < Q - s)$.

   e. Economic consequences (measured in dollars) are related nonlinearly to the frequency, duration, and severity of events.

18.2 PRACTICAL ASPECTS OF PROVIDING RELIABILITY

18.2.1 Improving the Reliability of Water Distribution Systems

In some less-developed countries, providing water service on a part-time basis may be acceptable, but customers in developed countries expect an uninterrupted supply of water. This expectation of uninterrupted water increases the cost of the water distribution system and has an impact on the design, operation, and maintenance of water systems. Section 18.2.2 describes some facets of improving the reliability of water distribution (Walski, 1993a). These include a range of design considerations ranging from specification of adequate pipe material to use of transient analysis. Valving of the system into distribution segments is addressed in the following section. (A *segment* is the smallest portion of a distribution system that can be isolated by valving.)

18.2.1.1 Piping materials. The selection of pipe material can be complicated by the large number of materials available and the complexity of conditions at local sites. The key is to match the material with the laying condition and internal and external loads. With proper design, most pipe material will work in most conditions and the cost, including ease of maintenance and historical recording of the performance of that material in that utility, will determine the selection.

Bedding is the next consideration, and compliance with the manufacturer's directions usually will result in acceptable laying conditions. Depending on the soil type, it is usually justifiable to coat or wrap metal piping. Refer to Chap. 3 for further discussion of bedding.

Specifying the pressure rating of the pipe determines the thickness and thus the resistance of the pipe to breaks. In most cases, a pressure class of 350 results in a pipe with the ability to resist external loads and withstand internal pressure with an allowance for surge. However, in areas where the pipe will be subjected to extremely large loads and the impact of a failure will be great (e.g., adjacent to underground subway stations and transformers, under important intersections), the extra cost of specifying pipe with a greater wall thickness can be justified.

Preventing damage to pipe by accidental contact during excavation makes metal pipes more desirable, and the use of tracer wire is advisable for non metal pipes. Composite pipes (e.g., concrete cylinder, fiberglass) generally are more difficult to repair and tap than are pipes made from a homogeneous material, such as ductile iron or polyvinyl chloride.

18.2.1.2 Construction methods. The best designed water distribution system can be unreliable if the pipe is not installed properly or if record (as-built) drawings are not correct. Installation crews are paid according the amount of pipe laid, not based on pipe longevity, and have been known to shortcut pipe laying, which can result in long-term problems for the utility. Competent full-time inspection and pressure testing is necessary if a distribution system is to be reliable.

18.2.1.3 Pipe sizing. Pipe sizing affects the reliability of the distribution system in two ways. First, larger pipes, when located in loops, provide greater ability for the system to meet demands even when another pipe in the loop is out of service. Second, *beam breaks* (circumferential breaks) are the most common type of pipe failure. Because larger-diameter pipes have much greater beam strength than do smaller pipes, they will resist breaking to a greater extent. Increasing the diameter solely to increase structural strength is not justifiable economically.

18.2.1.4 Looped water distribution system. Eventually, all water distribution systems will have pipe breaks or other events requiring shutdown of the system. The number of customers affected will depend on the extent of valving in the system, which is covered in the next section. The ability to supply customers despite isolation of a portion of the system depends on the capacity of loops and parallel paths the water can take. If the system is grided well with adequately sized pipes, the loss of a single segment of the distribution system will not seriously affect most customers.

For small pipes in the distribution grid, the sizing of the pipes generally is controlled by fire-flow requirements, and the pipes are significantly oversized for average-day demands (velocity is extremely low). Loss of a single segment will go unnoticed outside of that segment (one reason why it is difficult to find closed valves in a distribution grid). It usually is safe to assume that if the system normally can handle fire requirements and is looped, loss of a single segment will not cause significant problems (except during fires). The probability of a fire occurring at the same time as an outage caused by a pipe break is extremely small.

In the case of large transmission mains and mains in rural systems without fire protection, velocities may not be low and loops may contain pipe with limited capacity; thus, the mere existence of loops does not guarantee sufficient capacity in the alternate path around the loop. Therefore, it is important during hydraulic design to test for the effect of losing individual segments of the distribution system. As will be shown in the next section, it is important to consider the location of valves when simulating the effect of losing a segment.

18.2.1.5 Emergency Storage. Depending on the size and layout of the distribution system, looping may not be the most efficient means of providing reliability. Distribution storage improves reliability in two ways. First, the existence of water in storage provides an alternate "source" of water in case a distribution segment is lost. Second, storage provides water needed during emergencies such as fires. Storage sizing and location is described in more detail in Chap. 10.

In large systems, terminal storage far from the source can reduce the cost of distribution mains significantly (both in terms and size and need for redundancy). Adequate storage means that mains serving an area need only meet peak day demands while shorter-term peaks can be met from water in storage. Storage also means that the principal main serving an area can be out of service for a short time without a serious interruption of service.

The reliability of tanks also raises some issues. Pressure zones with a single tank are considerably easier to operate than zones are with multiple tanks. However, tanks must be taken out of service for such processes as inspection, cleaning, and painting. In a pressure zone with a single tank, this means greater difficulty in maintaining pressures in the correct range and in meeting emergency peak demands when the tank is out of service.

18.2.1.6 Backup pumping and control valves. Pumps and control valves (e.g., pressure-reducing valves, pressure-sustaining valves) are two types of equipment that require routine maintenance. Therefore, it is crucial for spare valves and pumps to be available in each pressure zone.

Two valves should serve each pressure zone. Because most control valves require no power, there is some advantage to placing them at different points between the source and the pressure zone being served so that independent paths to the pressure zone are available.

Pumping stations, on the other hand, have much larger requirements in terms of structures, controls, telemetry, and power supply. Therefore, more than one pump usually is placed in each pumping station. The usual rule for the number of pump units is that the station must be able to meet design flows when the largest unit is out of service. In most instances, application of this rule is fairly straightforward. However, consider a small pressure zone (say less than 20 homes) with no elevated storage. A pump station serving this pressure zone might have two small pumps (roughly 40 gpm) and a fire pump at 750 gpm. The question then is whether there is a need for a spare fire pump. This is a judgment decision based on the likelihood that the fire pump will be needed while the single large pump is out of service.

18.2.1.7 Standby power. The next question regarding pumping station reliability is the need for a standby generator or a pump driven by an internal combustion engine. The need for a generator depends on the availability of storage downstream of the pump station, backup sources of water, and the reliability of the power supply to the station. If there is a backup source of water or if water can flow into the pressure zone by gravity from another source, then a standby generator usually is not needed.

An internal combustion engine, tied directly to a pump, is generally less expensive than a generator, but a generator provides the ability to use the pump station lighting, controls, and telemetry during a power outage and usually is preferred.

When limited storage is available downstream of the pump station, reliability can be provided by using a portable generator with a electrical transfer switch in the station or by using a portable water pump, provided adequate valving (or two hydrants, one on each side of the station) is available at the station.

18.2.1.8 Emergency controls. Pump stations should be designed so they can operate satisfactorily even if the pump controls fail. For example, variable-frequency drives (VFDs) are often used on pump stations delivering water to dead-end systems with no storage. If the VFD should fail, the pumps can easily over-pressurize the pressure zone, damaging the distribution system or customer's plumbing. If the VFD should fail, a pressure-relief valve should control pressure and a signal needs to be transmitted to the main control room or alarm system so that the problem can be corrected immediately.

Control equipment needs to be protected from electrical surges that can damage the controls without seriously affecting the pump motors. Conversely, control equipment should not be on the same circuit as the pumps so that a pump tripping a breaker or fuse will not cause the control or telemetry equipment to be placed out of service.

18.2.1.9 Emergency interconnections. Most water utilities prefer not to mix their water with that of neighboring utilities. However, interconnections with neighboring utilities are often the least expensive means of providing reliability. The interconnections should be installed before an emergency occurs and should be metered. The elevation of the hydraulic grade line in each system needs to be considered so that pumps or control valves can be sized and set properly.

When the connections are needed to meet peak demands or provide fire flow, the connection should allow flow automatically without the need for operator intervention. If the interconnection will be used only on rare occasions, having an operator open and activate the interconnection manually may be acceptable. If the distance between the last customer and the normally closed interconnection is fairly large, the quality of water in that pipe can deteriorate. Thus, it may be necessary to flush the line or run a blowoff periodically.

If the utility receiving water has the potential to have on adverse impact on service to the utility supplying water, some type of control valve (usually a pressure-sustaining valve) is required to maintain adequate pressure in the supply system.

18.2.1.10 Water distribution system modeling. Water distribution models are an important tool for analyzing the reliability of a water system and are covered in much greater detail later in this chapter. It is much more desirable to close pipes in a distribution system model and examine the effects than to close the real pipe and wait for customers to complain.

18.2.1.11 Transient analysis. Hydraulic transients (waterhammers) occur whenever water accelerates or decelerates; in most cases, the effects of these transients in the distribution system are negligible. However, in large transmission mains and in pumping stations with high velocities, transient analysis and control may be required. Chapter 6 addresses the topic of transient analysis.

18.2.1.12 Operational considerations. Although design and construction are the most effective ways of ensuring the reliability of water distribution systems, the system must be operated correctly as well. First, operators must monitor gauges and alarms. Minor problems can develop into major problems if the operator fails to respond correctly.

Second, operators must be trained to respond to emergencies, have adequate communications to call out the necessary repair and supervisory personnel, and make the required notifications to fire departments and the media in case of "boil water" advisories.
Third, valves and pumps must be operated in ways that minimize hydraulic transients and their effects. This usually involves the slow starting and stopping of equipment and the slow opening and closing of valves. It also involves monitoring the condition of surge-control equipment (e.g., water level in surge tanks) to make certain they are operable.

Fourth, water needs to be kept in tanks for emergencies. This does not mean that tanks should simply be kept full all the time but rather that water levels should fluctuate in the correct portion of the tanks (usually the upper half).

18.2.1.13 Maintenance considerations. Maintenance usually affects system reliability by both preventing component failures and by repairing any component failures quickly when they occur. Good training of maintenance employees, adequate inventory of repair parts and tools, and efficient mobilization practices in case of an emergency are at the center of maintaining of a reliable system.

Valve exercising is the single most important form of preventive maintenance for improving reliability. The next section discusses valving, but even the best laid out valves are not helpful if they are not exercised. Exercising valves keeps them functioning, familiarizes crews with their location, identifies inoperable valves, provides an opportunity to uncover paved-over valves and clean out valve boxes, and updates the utility's valve data-base.

18.2.2 Analyzing the Effect of Valving on System Reliability

18.2.2.1 Background. Given that water mains will fail and put some customers out of service, the key to improving the reliability of the distribution system is to install sufficient valves so that the outage can be limited to a small portion of the system. The more operable valves in a system, the greater its reliability. A system with no valves would have to be shut down completely at the source for any maintenance. Situations where large numbers of customers have been without water usually have been caused by a water main break and the inability of the utility to isolate the break to a small portion of the system because of inoperable valves.

Rules of thumb for locating isolating valves have evolved over the years. They include such considerations as placing a minimum of three valves at each cross-intersection and two valves at each T-intersection. The AWWA's *Introduction to Water Distribution* (1986) states that: "Isolating valves in the distribution system should be located less than 500 ft apart (150 m) in business districts and less than 800 ft (240 m) apart in other parts of the system. It is a good practice to have valves situated at the end of each block so that only one block will be without water during repair work."

The AWWA Manual M31 (1998) and the Ten States Standards (1992) contain similar wording.

The Ten States Standards state, "Where systems serve widely scattered customers and where future development is not expected, the valve spacing should not exceed one mile." Another rule of thumb is that no more than four valves should need to be operated to isolate a segment of a distribution system. Good practice also dictates that an isolating valve is needed for every hydrant lateral.

Most water systems are designed with adequate valving. However, as systems evolve, situations sometimes occur where the failure of a single pipe can have an adverse effect on service to a large number of customers, and a single valve could have prevented such impacts. An analysis of distribution segments can identify such problems.

18.2.2.2 Diagrams of distribution segments. Water distribution maps can be used to identify and analyze segments when the systems are fairly simple. However, when there are multiple pipes in a street, inadequacies in the valving can be masked by the system's complexity. This is especially true for transmission mains that traverse a neighborhood with few connections with the distribution grid. Walski (1993b, 1994) showed that for analyzing valving, it is better to view the distribution system as a series of distribution segments, represented by circles or ellipses, connected to one another by valves, represented by lines.

Figure 18.1A shows the piping and valves for a typical city block with the segments numbered and the valves identified by letters. Figure 18.1B shows the segment diagram for that same block, in which segments are represented by circles and valves are

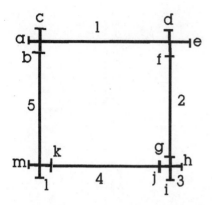
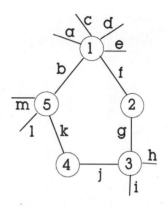

A. Distribution map showing piping and valves for a typical city block.

B. Segment diagram for block.

FIGURE 18.1 Example of creating a segment diagram from a distribution map.

represented by lines. Figure 18.1B shows that isolating segment 1 will be difficult because six valves must be operated successfully. The odds of that occurring are p^n, where p is the probability of operating any individual valve successfully and n is the number of valves to be operated. For example, if there is a 90 percent probability that any valve can be operated, the chance of operating all six is 53 percent ($0.90^6 = 0.53$).

Problems with inadequate valving usually occur along transmission mains and, more important, in places where they connect with the distribution grid. Transmission mains also are important because a pipe break there, combined with a failed valve, can expand the outage back to the source. Analyzing valving is illustrated best by some actual examples in the following sections.

18.2.2.3 Loops served from transmission mains. In this situation, segments 1, 2, and 3 are made up of a 12-in- (300-mm)- diameter pipe with a run of roughly 2500 ft (800 m) between the valves at each end of segment 2. This spacing of the valves was acceptable when segment 2 was installed to serve customers beyond segment 3. However, as the system grew, several subdivisions were installed along segment 2. Once this development occurred, a pipe break along segment 2 would result in a large number of customers being without water.

From the distribution maps, which were considerably more complicated than Figure 18.2A, it was not easy to see this problem. However, when the distribution segments were drawn, it became clear that at least three additional valves (between each set of T's) are needed in segment 2. So although the subdivision systems had looping for reliability, they did not have a reliable transmission main.

On a smaller scale, segment 4 also should be subdivided into more than one segment. This is not obvious from the distribution map, but the segment diagram shows that segment 5 also will be placed out of service if segment 4 must be shut down. There should be a valve in segment 4 between the two places where segments 4 and 5 tie together.

18.2.2.4 Emergency interconnections. In this case, utility A has a major user in segment 3, as shown in Fig. 18.3A. Because it was a long way back to the source (via segment 1), utility A installed two metered interconnections with neighboring utility B.

18.12 Chapter Eighteen

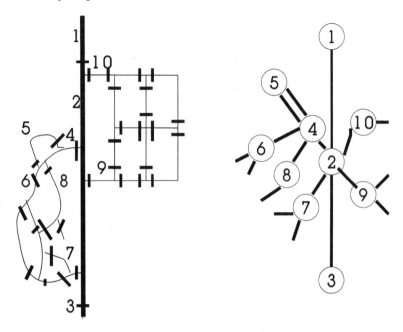

A. Distribution map B. Segment diagram
FIGURE 18.2 Inadequate valving in a transmission main.

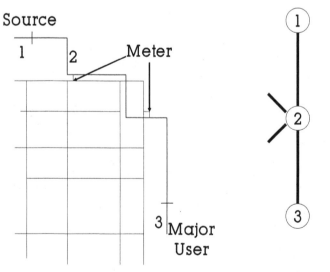

A. Distribution map B. Segment diagram
FIGURE 18.3 Interconnections tying into the same distribution segment.

This improved the reliability in terms of protecting the system from upstream outages in system A. However, a single pipe break in segment 2 would wipe out the flow from utility A's source and the two interconnections from utility B. Eventually, another metered interconnection with utility B was installed in segment 3 to correct this shortcoming.

18.2.2.5 Transmission lines connected to old systems. In Fig. 18.4A, a transmission main of 12-in-diameter pipe runs from the source to the storage tank through pipe segments 1, 2, 3, 4, 5, and 6. The weakness lies in the fact that the adjacent neighborhood (segments 10, 11, and 12) is a fairly large area served primarily by 4-in cast iron pipes laid in the 1890s. A break in one of these old pipes could take out the transmission main if a single valve is inoperable. In particular, segment 11 requires the operation of six valves to isolate it from the transmission main. If any one of these valves is inoperable, a major portion of this system must be shut down. The solution here would be to install a few isolating valves that would break segment 11 into smaller segments that could be isolated more easily.

18.2.2.6 Typical cross-intersections. Among the most typical situations involving placement of valves is the cross-intersection. Using four valves at a cross provides the highest level of protection from outages. However, usual practice is to have only three valves as is shown in Fig. 18.5. Given that the intersecting pipes are an 8 in (200 mm) and 6 in (150 mm) and the cost of valves is highly dependent on the diameter, one would want to use one 8–in and two 6-in valves.

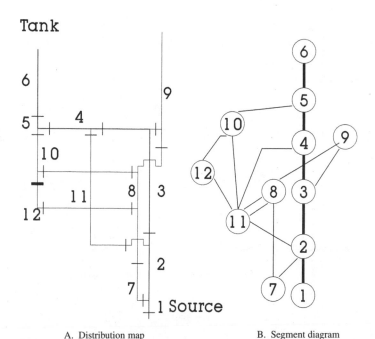

A. Distribution map   B. Segment diagram

FIGURE 18.4 Large segment with many valves needed for shutdown.

In most situations, one direction of flow is the primary direction, and the direction from which flow is occurring is labeled "source" in Fig. 18.5A and C. Of course, the system is assumed to be looped so that there are additional directions of flow through each segment. In the first alternative, the 8-in valve is placed on the source side of the intersection, yielding four segments as shown in Fig. 18.5B; in the second alternative, the 8-in valve is placed away from the source. The segment diagram shows that the second alternative is superior because a break in segment 1 or 3 can take out the entire intersection, whereas in the second alternative, only a break in segment 1 can impact service in the intersection. Similar logic can be applied to any situation where valve location is being determined.

18.2.2.7 Application of segments in valve locations and reliability evaluation. The above examples show that how drawing distribution system segments can help an engineer determine the best location for valves in a new system or evaluate situations where installation of new valves can greatly improve the reliability of an existing system.

Segments can be identified from distribution system maps by tracing out each segment, usually starting with the source, and marking each with a different colored pencil [or colored line type if the drawing is being done on computer-aided design (CAD) equipment]. Each segment is then labeled with a number. Each valve from the segment is identified with a number or letter, and a corresponding line is placed on the segment diagram.

Segments should be created throughout the system until one is far from the transmission mains and into the distribution grid. The decision regarding how far to go into the distribution grid depends on available resources and the level of reliability desired.

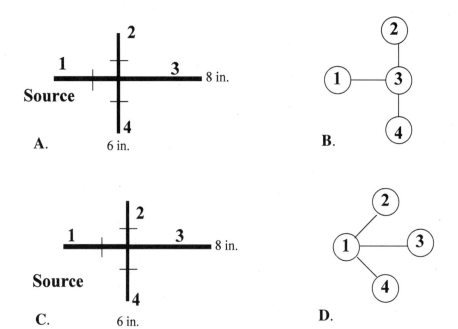

FIGURE 18.5 Valving at a typical cross-intersection.

In examining a segment diagram, one should look for segments that have a large number of valves attached to them. This indicates a difficult segment to isolate. One also should look for a segment through which virtually all the water for a given part of the system must pass. Such a segment will represent a major problem if it must be shut down.

When fire hydrants are being placed in the distribution system, it generally is better to place roughly one hydrant per segment rather than two hydrants on one segment and none on the next. In this way, taking a segment out of service will put only a single hydrant out of service.

In modeling the effect of a pipe break on distribution system hydraulics, it is important not only to take a single pipe out of service but to remove the entire distribution system segment from the model. This means that one or more pipes and junction nodes may have to be removed to simulate the shutdown correctly to repair the break.

If, after the segment diagram is developed and a particular valve is determined to be inoperable, the two segments that the valve connected should be blended into a single segment. This is an easy way to examine the impact of a failed valve.

18.3 COMPONENT RELIABILITY ANALYSIS

A system or its components can be treated as a black box or a lumped-parameter system, and the performances of the components of the components are observed over time. This reduces the reliability analysis to a one-dimensional problem involving time as the only variable. In such cases, the *time-to-failure (TTF)* of a system or a component of the system is the random variable. It should be pointed out that the term time could be used in a more general sense. In some situations, other physical scale measures, such as distance or length, may be appropriate for evaluation of system performance.

The TTF analysis is particularly suitable to assess the reliability of systems, repairable components, or both. For a system that is repairable after its failure, the time it would take to have it repaired to its operational status is uncertain; therefore, the *time-to-repair (TTR)* also is a random variable.

For a repairable system or component, its service life can be extended indefinitely if a repairable system available for service is greater than that of a nonrepairable system. This section focuses on characteristics of failure, repair, and availability of repairable systems by time-to-failure analysis.

18.3.1 Failure Density, Failure Rate, and Mean Time To Failure

Any system will fail eventually; it is simply a matter of time. Because of the presence of many uncertainties that affect the operation of a physical system, the time when the system fails to perform satisfactorily as intended is a random variable.

The *failure density function* is the probability distribution that governs the time when failure will occur. The failure density function serves as the common thread in the reliability assessments in TTF analyses. The reliability of a system or a component within a specified time interval $[o, t]$ can be expressed as

$$p_s(t) = P(\text{TTF} > t) = \int_o^\tau f(\tau)d\tau \qquad (18.1)$$

where TTF is random having $f(t)$ as the failure density function. The reliability $p_s(t)$ represents the probability that the system experiences no failure within $[o, t]$. The *probability of failure*, or *unreliability*, can be expressed as $p_f(t) = 1 - p_s(t)$. Note that unreliability $p_f(t)$ is the probability that a component or a system would experience its first failure within the time interval (o, t) as the component or system's age increases. Conversely, the failure density function can be obtained from reliability or unreliability as

$$f(t) = -\frac{d[p_s(t)]}{dt} = \frac{d[p_f(t)]}{dt} \tag{18.2}$$

The TTF is a continuous, nonnegative random variable by nature. Many distribution functions described in Sec. 18.3.1 are appropriate for modeling the stochastic nature of time-to-failure. Among them, the exponential distribution is perhaps the most widely used. The exponential distribution has been found, both phenomenologically and empirically, to describe adequately the time-to-failure distribution for components, equipment, and systems involving components with mixtures of life distributions.

The *failure rate is* defined as the number of failures occurring per unit time in a time interval $(t, + \Delta t)$ per unit of the remaining population at time t. The instantaneous failure rate $m(t)$ can be by

$$m(t) = \lim_{\Delta t \to 0} \left[\frac{N_F(\Delta t)/\Delta t}{N(t)} \right] = \frac{f(t)}{p_s(t)} \tag{18.3}$$

where $N_F(\Delta t)$ is the number of failures in a time interval $(t, t + \Delta t)$, and $N(t)$ is the number of failures from the beginning up to time t. This instantaneous failure rate also is called the *hazard function* or the *force of mortality function*.

The failure rate for many systems or components has a bathtub shape, as shown in Fig.18.6 in that three distinct life periods can be identified: early life (or infant mortality) period, useful life period, and wearout life period. In the early life period, quality failures and stress-related failures dominate, with little contribution from wearout failures. During the useful life period, all three types of failures contribute to the potential failure of the system or component, and the overall failure rate increases with age because wearout failures and stress-related failures are the main contributors and wearout becomes a more dominating factor in the failure of the system with age.

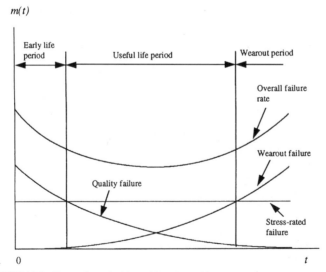

FIGURE 18.6 Shape of a typical hazard function and its components. (From Tung, 1996)

The reliability can be directly computed from the failure rate as

$$p_s(t) = \exp\left(\int_0^t m(\tau)d\tau\right) \quad (18.4)$$

Substituting Eq. (18.4) into Eq.(18.3), the failure density function $f(t)$ can be expressed in terms of failure rate as

$$f(t) = m(t)\left[\exp\int_0^t m(\tau)d\tau\right] \quad (18.5)$$

In general, the reliability of a system or a component depends strongly on its age. This can be expressed mathematically by the condition probability as

$$p_s(\xi|t) = \frac{p_s(t+\xi)}{p_s(t)} \quad (18.6)$$

where t is the age of system, the system not having failed up to that point, and $P_s\xi|t$ is the reliability over a new mission period, having operated successfully over a period of t.

A commonly used reliability measure of system performance is the *mean-time-to-failure (MTTF)*, which is the expected time-to-failure. The MTTF can be defined mathematically as

$$\text{MTTF} = E(\text{TTF}) = \int_0^\infty \tau f(\tau)d\tau \quad (18.7)$$

The MTTF for some failure density functions is given in Table (18.1)

For repairable water resources systems, such as pipe networks, pump stations, and storm runoff drainage structures, failed components within the system can be repaired or replaced so that the system can be put back into service. The time required to have the failed system repaired is uncertain; consequently, the total time required to restore the system from its failure to operational status is a random variable.

Like the TTF, the random TTR has the *repair density function g(t)*, which describes the random characteristics of the time required to repair a failed system when the failure occurs at time zero. The *repair probability G(t)* is the probability that the failed system can be restored within a given time period [0, t],

$$G(t) = P[\text{TTF} \leq t] = \int_0^t g(\tau)d\tau \quad (18.8)$$

The *repair rate r(t)*, similar to the failure rate, is the conditional probability that the system is repaired per unit time, given that the system failed at time zero and is still not repaired at time t. The quantity $r(t) dt$ is the probability that the system is repaired during the time interval $(t, t + dt)$ given that the system fails at time t. The relation between the repair density function, repair rate, and repair probability is

$$r(t) = \frac{g(t)}{1 - G(t)} \quad (18.9)$$

Given a repair rate r(t), the density function and the maintainability can be determined, respectively, as

$$g(t) = r(\tau)\exp\left(-\int_0^t r(\tau)d\tau\right) \quad (18.10)$$

and

$$G(t) = 1 - \exp\left(-\int_0^t r(\tau)d\tau\right) \quad (18.11)$$

TABLE 18.1 Mean-Time-to-Failure for Some Failure Density Functions

Distribution	PDF, $f(t)$	Reliability, $p_s(t)$	Failure, Rate $m(t)$	MTTF
Normal	$\dfrac{1}{\sqrt{2\pi}\,\sigma_T} \exp\left[-\dfrac{1}{2}\left(\dfrac{t-\mu_T}{\sigma_T}\right)^2\right]$	$\phi\left(\dfrac{t-\mu_T}{\sigma_T}\right)$	$\dfrac{f(t)}{\phi\left(\dfrac{t-\mu_T}{\sigma_T}\right)}$	μ_T
Lognormal	$\dfrac{1}{\sqrt{2\pi}\,\sigma\ln T}\exp\left[-\dfrac{1}{2}\left(\dfrac{\ln(t)-\mu_{\ln T}}{\sigma_{\ln T}}\right)^2\right]$	$\phi\left(\dfrac{\ln(t)-\mu_{\ln T}}{\sigma_{\ln T}}\right)$	$\dfrac{f(t)}{\sigma\left(\dfrac{\ln(t)-\mu_{\ln T}}{\sigma_{\ln T}}\right)}$	$\exp\left(\mu_{\ln T}+\dfrac{\sigma_{\ln T}^2}{2}\right)$
Exponential	$\beta e^{-\beta t}$	$e^{-\beta t}$	β	$\dfrac{1}{\beta}$
Rayleigh	$\dfrac{t}{\beta^2}\exp\left[-\dfrac{1}{2}\left(\dfrac{t}{\beta}\right)^2\right],\ \beta>0$	$\exp\left[-\dfrac{1}{2}\left(\dfrac{t}{\beta}\right)^2\right]$	$\dfrac{t}{\beta^2}$	$1.253\,\beta$
Gamma	$\dfrac{\beta}{\Gamma(\alpha)}(\beta t)^{\alpha-1}e^{-\beta t}$	$\int_t^\infty f(\tau)\,d\tau$	$\dfrac{f(t)}{p_s(t)}$	$\dfrac{\alpha}{\beta}$
Gumbel	$e^{\pm y - e^{\pm y}};\ y=\dfrac{t-t_0}{\beta}$	$1-e^{\pm e^{\pm y}}$	$\dfrac{f(t)}{p_s(t)}$	$x_0 \pm 0.577\,\beta$
Weibull	$\dfrac{\alpha}{\beta}\left(\dfrac{t-t_0}{\beta}\right)^{\alpha-1} e^{-\left(\dfrac{t-t_0}{\beta}\right)^a}$	$e^{\left(\dfrac{t-t_0}{\beta}\right)^a}$	$\dfrac{\alpha(t-t_0)^{\alpha-1}}{\beta^\alpha}$	$t_0+\beta\Gamma\left(1+\dfrac{1}{\alpha}\right)$
Uniform	$\dfrac{1}{b-a}$	$\dfrac{b-t}{b-a}$	$\dfrac{1}{b-t}$	$\dfrac{a+b}{2}$

The *mean-time-to-repair (MTTR)* is the expected value of time-to-repair of a failed system: that is,

$$\text{MTTR} = \int_0^t tg(t)dt \qquad (18.12)$$

The MTTR measures the elapsed time required to perform the maintenance operation and is used to estimate the downtime of a system. It also is a measure for maintainability of a system. Various types of maintainability measures are derivable from the repair density function.

The MTTR is a proper measure of the mean life span of a nonrepairable system. For a repairable system, a more representative indicator for the fail-repair cycle is the *mean-time-between-failure* (MTBF), which is the sum of MTTF and MTTR: that is,

$$\text{MTBF} = \text{MTTF} + \text{MTTR} \qquad (18.13)$$

The *mean-time-between-repair* (MTBR) is the expected value of the time between two consecutive repairs and is equal to the MTBF. Refer to Table 18.2.

18.3.2 Availability and Unavailability

A repairable system experiences a repetition of repair-to-failure and failure and failure-to-repair processes during its service life. Hence, the probability that a system is in operating condition at any given time t for a repairable system is different from that of a nonrepairable system. The term *availability* $A(t)$ generally is used for repairable systems to indicate the probability that the system is in operating condition at any given time t. On the other hand, reliability $p_s(t)$ is appropriate for nonrepairable systems, indicating the probability that the system has been in its operating state continuously, starting from time zero up to time t. Table 18.2 lists $P_s(t)$ for both repairable and nonrepairable systems.

Availability also can be interpreted as the percentage of time that the system is in operating condition within a specified time period. In general, the availability and reliability of a system satisfies the following inequality relationship:

$$0 \leq p_s(t) \leq A(t) \leq 1 \qquad (18.14)$$

with the equality holding for nonrepairable systems. The reliability of a system decreases monotonically to zero as the age of the system increases, whereas the availability decreases but converges to a positive probability. This is shown in Fig. 18.7.

The complement to availability is the *unavailability* $U(t)$, which is the probability that a system is in the failed condition at time t, given it is in operating condition at time zero. In other words, unavailability is the percentage of time that the system is not available for the intended service in time period $(0, t)$, given it is in operation at time zero. Availability, unavailability, and unreliability satisfy the following relationships:

$$A(t) + U(t) = 1.0 \qquad (18.15)$$

and

$$0 \leq U(t) \leq p_f(t) < 1 \qquad (18.16)$$

For a nonrepairable system, the unavailability is equal to the unreliability: that is, $U(t) = p_f(t)$.

TABLE 18.2 Summary of Constant–Rate Model

Repairable Systems	Nonrepairable Systems
Failure process	
$m(t) = \lambda$	$m(t) = \lambda$
$p_s(t) = e^{-\lambda t}$	$p_s(t) = e^{-\lambda t}$
$p_f(t) = 1 - e^{-\lambda t}$	$p_f(t) = 1 - e^{-\lambda t}$
$f(t) = \lambda e^{-\lambda t}$	$f(t) = \lambda e^{-\lambda t}$
MTTF = $1/\lambda$	MTTF = $1/\lambda$
Repair process	
$r(t) = \eta$	$r(t) = 0$
$G(t) = 1 - e^{-\eta t}$	$G(t) = 0$
$g(t) = \eta e^{-\eta t}$	$g(t) = 0$
MTTR = $1/\eta$	MTTR = ∞
Dynamic behavior of whole process	
$U(t) = \dfrac{\lambda}{\lambda+\eta}[1 - e^{-(\lambda+\eta)t}]$	$U(t) = 1 - e^{-\lambda t} = p_f(t)$
$A(t) = \dfrac{\eta}{\lambda+\eta} = \dfrac{\lambda}{\lambda+\eta}[1 - e^{-(\lambda+\eta)t}]$	$A(t) = e^{+\lambda t} = p_s(t)$
$w(t) = \dfrac{\lambda\eta}{\lambda+\eta} = \dfrac{\lambda^2}{\lambda+\eta}[1 - e^{-(\lambda+\eta)t}]$	$w(t) = f(t) = \lambda e^{-\lambda t}$
$\gamma(t) = \dfrac{\lambda\eta}{\lambda+\eta}[1 - e^{-(\lambda+\eta)t}]$	$\gamma(t) = 0$
$W(0,t) = \dfrac{\lambda\eta}{\lambda+\eta}t + \dfrac{\lambda^2}{(\lambda+\eta)^2}[1 - e^{-(\lambda+\eta)t}]$	$W(0,t) = p_f(t)$
$\Gamma(0,t) = \dfrac{\lambda\eta}{\lambda+\eta}t - \dfrac{\lambda\eta}{(\lambda+\eta)^2}[1 - e^{-(\lambda+\eta)t}]$	$\Gamma(0,t) = 0$
Stationary values of whole process	
$U(\infty) = \dfrac{\lambda}{\lambda+\eta} = \dfrac{\text{MTTR}}{\text{MTTF + MTTR}}$	$U(\infty) = 1$
$A(\infty) = \dfrac{\eta}{\lambda+\eta} = \dfrac{\text{MTTF}}{\text{MTTF + MTTR}}$	$A(\infty) = 0$
$w(\infty) = \dfrac{\lambda\eta}{\lambda+\eta} = \dfrac{1}{\text{MTTF + MTTR}}$	$w(\infty) = 0$
$\gamma(\infty) = \dfrac{\lambda\eta}{\lambda+\eta} = w(\infty)$	$\gamma(\infty) = 0$

Source: Adapted from Henley and Kumamoto (1981).

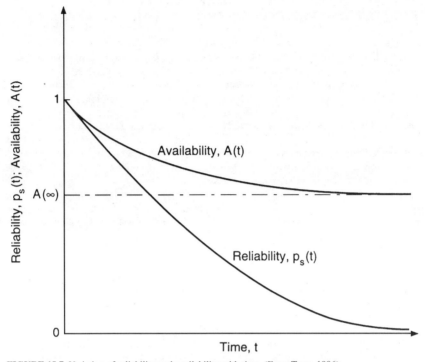

FIGURE 18.7 Variation of reliability and availability with time. (From Tung, 1996).

Determination of availability or unavailability of a system requires a full accounting of the failure and repair processes. The basic elements that describe processes are failure and repair processes. The basic elements that describe such processes are failure density function $f(t)$ and the repair density function $g(t)$. Consider a system with a constant failure rate λ and a constant repair rate η. The availability and unavailability, respectively, are

$$A(t) = \frac{\eta}{\lambda+\eta} + \frac{\lambda}{\lambda+\eta} e^{-(\lambda+\eta)t} \tag{18.17}$$

and

$$U(t) = \frac{\lambda}{\lambda+\eta} [1-e^{-(\lambda+\eta)t}] \tag{18.18}$$

As the time approaches infinity ($t \to \infty$), the system reaches its stationary condition. Then, the stationary availability and unavailability are

$$A(\infty) = \frac{\eta}{\lambda+\eta} = \frac{1/\lambda}{1/\lambda+1/\eta} = \frac{\text{MTTF}}{\text{MTTF}+\text{MTTR}} \tag{18.19}$$

and

$$U(\infty) = \frac{\lambda}{\lambda+\eta} = \frac{1/\eta}{1/\lambda+1/\eta} = \frac{\text{MTTR}}{\text{MTTF}+\text{MTTR}} \tag{18.20}$$

18.4 REVIEW OF MODELS FOR RELIABILITY OF WATER DISTRIBUTION SYSTEMS

The problem of defining and providing appropriate levels of reliability in water distribution systems has received a considerable degree of attention over the past two decades. This attention is expected to continue as the result of two different but related factors. The first is a growing emphasis on reliability of service in its own right. The second factor is the increasing levels of corporatization and privatization of water distribution systems occurring around the wold. As these systems become privatized, governments (e.g., municipalities, city councils, metropolitan water districts) are faced with the problem of specifying the service levels to which the water utilities must conform. These service specifications include both the definition of reliability as the quantity of water supplied in relation to the demand, the pressure at which that water is supplied, and, increasingly, the quality of the water delivered to the customer and the required levels of performance reliability for those definitions. Before proceeding with a consideration of the failure modes of water distribution systems and how those modes are captured in reliability analysis, it is necessary to define qualitatively what actually is meant by reliability of water distribution systems.

18.4.1 Reliability of a System Failure

Reliability of a water distribution system can be defined, using Goulter's (1995) approach, as " The ability of a water distribution system to meet the demands that are placed on it where such demands are specified in terms of (1) the flows to be supplied (total volume and flow rate); and (2) the range of pressures at which those flows must be provided." A useful extension or interpretation of this definition given by Cullinane et al. (1992) is "the ability of the system to provide service with an acceptable level of interruption in spite of abnormal conditions."

A key feature of Cullinane et al.'s definition is that it implies, or introduces, the concept of both the period over which the system is unable to meet demands and the particular point in time—that is, circumstances that caused the system to be unable to supply demands—as being important determinants in defining and calculating reliability. This interpretation of reliability also recognizes that "system failure" may occur either through the flows associated with the demands not being supplied or the flows associated with the demands being supplied but at pressures lower than the minimum acceptable or minimum specified for the particular circumstances. Equally important is that this definition also recognizes that failures should occur only in association with abnormal conditions, for example, component failure and or abnormally high demands or both.

The range of combinations of ways in which failure can occur in a water distribution system constitutes one, and arguably the major, source of the many theoretical and practical difficulties which have been encountered in establishing suitable, that is, comprehensive and computationally tractable, measures of reliability that can be used in the practical design and operation of water distribution systems. It must be recognized, however, that reliability has been an explicit factor in the design and operation of water distribution networks for a considerable time, as demonstrated by the existence of looped networks, which are not the most cost-efficient solution for water supply networks. However, the presence of loops does add reliability and redundancy to the system by providing excess capacity and alternative paths for the supply of water in the face of failure of one system component. The question facing practitioners in the water distribution industry is how to make the reliability of supply quantitatively explicit. This quantitative specification of reliability can then be used by municipal authorities and the like to define compliance levels for the reliability to be provided by water utility operators, both private and public. Similarly, such a specification can be used as a point against which the operators can fine-tune and optimize the performance of the system.

18.4.2 Failure Modes

Failure of a water distribution system, as defined above, arises in two major ways. First, the actual demand Qa, imposed on the network can be greater than the demand Qd for which the network was designed or, if it is an older network, may be larger than current capacity Qc, which may have been reduced from the design value through deterioration of the network with age. Goulter (1995) called this type of "demand variation failure" or, more simply, "demand failure." It also has been defined as "hydraulic performance failure." However, categorizing this failure mechanism as hydraulic performance failure has the following shortcoming. All quantity-based failures (incapacity to deliver the required volume of flow or flow rate) of water distribution systems—other than those caused by shortage at the primary source of water or by quality problems in the treatment facilities—are caused or exacerbated in part by lack of hydraulic capacity to deliver the flow under the circumstances, including component failure s, prevailing at the time.

Consideration of demand variation failure is, in part, an explicit recognition that flow demands are stochastic and therefore are defined most correctly using a probability distribution rather than a single deterministic value. This stochasticity has been recognized in the choice of design flow rates for water distribution systems: for example, 20-year maximum average daily flow or 10-year maximum hourly flow. The use of design values in these terms acknowledges that demands larger than the design values do in fact occur and do so with a known probability. In acknowledging the probability of flow demands larger than design values, designers also are defining the probability that the system will fail when those demands occur.

Consider the 50-year maximum average daily average flow Q_{50}. A network designed on the basis of the 20-year maximum average daily flow Q_{20} is unlikely to be able to meet the pressure and flow-rate requirements for the Q_{50} scenario and would "fail." On the average, the Q_{50} condition occurs once every 50 years. Therefore, it has a known probability: that is, 0.02 probability, of occurring in any given year. The probabilities of all demand scenarios can be calculated in similar fashion, and the total probability of demands being greater than a particular design value can be determined. Complicating factors in such analyses are the relationships among the flows in different design specifications. For example, what is the correlation between design flows in design specifications include both the 10-year maximum average hourly flow and the 20-year maximum average daily flow, and how should the probabilities of both events be combined to give an overall estimate of the probability of system failure?

A further complication arises from the changes in demand over time. Design values are established on the basis of projected loads on the system. However, demands can increase beyond these values because of unanticipated increases in the population served or changes in lifestyle. In these cases the probability distribution of demand shown in Figure 18.7 will shift to the right, but with Qd remaining in the same position on the horizontal axis, thereby resulting in an increased probability of failure. Government authorities are increasingly examining and applying new ways to overcome these problems: for example, by voluntary conservation and legislated water-use reduction measures (Briassoulis, 1994; Cameron and Wright, 1990; Nieswiadomy,1992; Weber,1993; Wilchfort and Lund, 1997). Interestingly, many of these approaches have been formed around failure conditions arising from shortages, relative to the total demand, in the primary water supply rather than around problems in the distribution network per se. There is scope to apply similar approaches to the management and operation of water distribution systems where the overall supply is adequate but the distribution system, particularly the distribution network, is not capable hydraulically of meeting the increased demands.

An additional factor that has an impact on consideration of demand failure is that the hydraulic capacity of the network also tends to decrease with time as the network itself ages. This deterioration in hydraulic capacity does not change the characteristics of the demand, as shown in Figure 18.7. However, the location of Q_c, as it represents the hydraulic capacity of the network, moves to the left, with the result that the overall probability that the demand will exceed the hydraulic capacity of the system again increases.

The second type of major failure is associated with failure of components of the distribution system—for example, pipe, pump, treatment plant process, or Supervisory Control and Data Acquisition (SCADA) system—that make up the physical or operational infrastructure of the system. The inability of the system to meet pressure or flow requirements in this situation comes from the reduction in the hydraulic capacity of the network or control of the system when the system has failed and under repair. It should be noted that if sufficient redundancy (and, implicitly, reliability) has been designed into the system, failure of a component does not have to result in failure of the system. However, the potential of a component to reduce the system's performance to the point where a system failure occurs must be addressed explicitly in any reliability analysis of the system.

An important factor in the consideration of component failure in reliability analysis is that components fail with probabilities that are known, or at least are able to be estimated from historical or experimental data with reasonable confidence. Thus, the probability of system failure caused by failure of its components can be calculated. Hence, as with "demand failure" of a water distribution system, failure of the system because of component failure has a strong probabilistic basis; and the constituent probabilities can be determined relatively easily.

Reliability analysis of any system requires consideration of the probabilistic characteristics of the actions or events that have an impact on the system's performance and, in the extreme, cause the system to fail. As noted above, the two major contributors to failure of water supply systems can be defined in the necessary probabilistic terms. These two major modes of system failure have been reasonably handled independently. However, difficulties arise when the two failure modes are considered jointly in a single measure of system reliability. In attempting to understand (1) how the two failure modes interact, (2) how a measure or technique that addresses both failure modes adequately can be developed and used, and (3) how the acceptable levels of system performance should be defined according to that measure or technique, it is necessary to examine the implications of network failure—in other words, loss of service,—in more detail.

The previous discussion examined failure from the perspective that failure could arise from flow requirements not being supplied, being supplied at delivery pressures that were below the minimum acceptable level, or both. Failure conditions at a single demand point, (demand node), are relatively easy to define under these circumstances—there is inadequate flow or pressure or both at that node. The parameters that define severity of the failure also are encapsulated fairly easily in the magnitude of the deficit in performance —the lower the delivery pressure at the demand node, the greater the magnitude of the failure or, conversely, the larger the difference between the required flow and the delivered flow, the greater the magnitude of the deficit.

However, the problem is considerably more complex when the system as a whole is considered. A system may experience failure through either inadequate flow or pressure at a single demand node, or any combination of demand nodes, up to the point where the flow at every node is below the demand or where every node in the system is supplied at delivery pressures that are below the minimum allowable. The deficit at different demand nodes also may vary significantly, with one node having a major failure and all other nodes experiencing no or insignificant failure. Alternatively, all nodes may experience a fairly uniform minor failure, with no single node facing a major or catastrophic failure circumstance. This variation in scenarios, each of which represents a form of system failure, demonstrates the problem of defining the reliability of the system as a unit and distinguishing between (i.e., ranking) the severities of the different failure scenarios. Consider the following cases:

Case 1: The flow demands across the system are supplied at delivery pressures that are at or above the minimum 95 percent of the time, whereas 5 percent of the time the required flows can be supplied only at pressures that are below the minimum level.

Case 2: Ninety-five percent of the required flow demand is supplied at delivery pressures that are at or above the minimum level 100 percent of the time.

Case 3: One hundred percent of the required flow demand is supplied 100 percent of the time but at delivery pressures that are only 95 percent of the minimum acceptable level.

Case 4: One hundred percent of the flow demand and pressure demand is satisfied 100 percent of the time at 95 percent of the nodes and there is no supply at all at 5 percent of the nodes.

Which of these four cases represents the most reliable network? All have elements of what might be described as "a 95 percent level of reliability." However, although Case 4 would clearly be unacceptable, it is considerably more difficult to differentiate among, and rank, the other three cases. The differences between the cases also demonstrate that the proportion of time the system is unable to meet demands is only one aspect of a reliability analysis.

When considering the reliability of water distribution systems, it is not sufficient to consider only the probability of the event that gives rise to the failure condition and the duration of that event. The magnitude of the deviation from the minimum level (i.e., the magnitude of the failure), must be addressed. These concerns lead to the key factor in specifying reliability in quantitative terms. This quantitative measure might be specified as "the delivery pressure any demand node in the system over a specified period of time should not be below the minimum allowable for more than 24 continuous hours, or less than the minimum for 1 h 20 times per year, etc." Such specifications also could be captured in a quantitative measure stating that the total volume of water supplied to a consumer at below minimum delivery pressure—that is, the "deficit" in water supplied at acceptable pressure—should not be greater than a defined value. This approach incorporates the impact of the failure circumstance by explicitly recognizing that, for delivery pressures that are below the minimum acceptable by the same amount, a failure at a high—demand node has greater consequences than does a failure at a demand with a smaller demand.

However, the deficit in water supplied at adequate pressures also is, not in itself, able to differentiate between the same volume of deficit supplied at delivery pressures below the minimum acceptable levels by different amounts. Delivery pressure, or more correctly, the amount by which the delivery pressures are below the minimum acceptable level, is important, and becomes increasingly so as the delivery pressure deviates further from the minimum acceptable level. In other words, the magnitude (size of the deviation from the minimum value of the parameter under which failure is defined) as well as the extent of the failure (duration and amount of flow demand not supplied under acceptable circumstances) are important factors in defining and considering reliability in water distribution systems. An additional complexity in analyses of this type is that flow and pressure in a water distribution network are highly related.

Before proceeding with the discussion of how these issues can be addressed from the perspective of the different failure modes, it is useful to note that the discussion to this point has emphasized the reliability aspects of the quantity, as opposed to the quality, of supply. Similar issues relate to defining and specifying reliability performance levels in relation to the quality of supply. Unacceptably low water-quality levels can arise from component failure in the treatment plant, decreased quality of the primary water supply (equivalent to drought caused by lack of water of adequate quality rather lack of water), contamination occurring within the distribution network after the water has left the treatment facility (synonymous with component failure), and increased requirements for the quality of supplied water (equivalent to increased demands). The following sections will address both the quantity and quality aspects of reliability analysis.

18.4.3 Approaches to the Assessment of Reliability

Consider the instance in which a reliability measure is based on deficits in volumes of water supplied. In recognition of the following (1) the need to incorporate both demand failure component failure, (2) the fact that demands are not actually located at nodes, as is normally assumed in network analysis and design, and (3) the probability nature of the events that give rise to failure—Bouchart and Goulter (1991, 1992) formulated the following measures for volume deficits:

1. Volume deficit arising from "demand variation failure."

$$ED_d^\eta = K \int_{Q_{de}^\eta}^{\infty} \left(Q^\eta - Q_{de}^\eta \right) p(Q^\eta) dQ^\eta \tag{18.21}$$

where Ed_d^η = expected volume of deficit of arising from demand at node η being larger than the design value Q_{de}^η at that node, Q^η = demand per unit time at node η, K = factor to expected flow rates to expected volumes, Q_{de}^η = design demand at node η, and $p(Q^\eta)$ = probability of demand Q_η at node η.

Volume deficit arising from "component (pipe) failure."

$$ED_f^s = q_d t_f E(NF) \tag{18.22}$$

where Ed_f^s = expected volume of deficit arising from a section of pipe in links failing and having to be isolated for repair, q_d = average demand per unit time in section of pipe isolated, t_f = average duration of repair, and $E(NF)$ = average number of failures per unit time of pipe section (equal to the length of pipe section between valves multiplied Times the number of breaks per unit length of pipe).

The expected deficit for the volume deficits resulting from component (pipe) failure and demand failure now have the same units. Hence, they can be added directly to give the total expected volume of deficit over the network that is,

$$TND = \sum_{\text{all links } s} ED_f^s + \sum_{\text{all nodes } \eta} ED_d^\eta \tag{18.23}$$

where TND = total expected volume of deficit due to both component failure and demand variation failure.

It should be recognized that the statement of reliability in Eq. (18.23), as used by Bouchart and Goulter (1991, 1992), does not address the impacts on the rest of the network caused by the reduction in the hydraulics of the network associated with removal of one component. For the same reason, it does examine the impacts of pump or valve failure (valve failure requires the valve to be repaired by isolating of the section of the network containing the valve and has some of the characteristics of pipe failure). The importance of this approach is that it recognizes both demand variation failure and component failure in a probabilistic sense. However, the technique is concerned only with purely volumetric deficit and does not take into consideration the ability of a network to meet flow demands at reduced pressures.

One approach to the measurement of reliability does take account of the pressure of delivery formulated and applied by Cullinane (1986) and Cullinane et al. (1992). The approach is based on the concept of *availability*, defined as the percentage of time the pressures in the network are below some preset level. In the case of a node, availability represents the percentage of time the pressure at the node is less than some predetermined level: that is,

$$A_j = \sum_{i=1}^{M} \frac{A_i t_i}{T} \qquad (18.24)$$

where, A_j = nodal availability at node j, A_i = availability during time period i, t_i = length of time period i, M = number of time period, i and T = time of simulation or time horizon for analysis of reliability.

The *network availability*, as opposed to nodal availability, can then be defined as

$$A = \sum_{j=1}^{N} \frac{A_j}{N} \qquad (18.25)$$

where, A = network availability, N = total number of nodes.

This measure is capable of considering both component (mechanical) and demand variation failure. The technique can also be extend to include the probabilistic aspects of mechanical and demand variation failure though the following expression:

$$AE_{jl} = R_{jl} A_{jl} + Q_{jl} A_j \qquad (18.26)$$

where, AE_{jl} = expected value of availability at node j, considering component l, R_{jl} = probability that component l is operational, Q_{jl} = probability that component l is not operational, A_{jl} = availability of node j with component l operational, and A_j = availability of node j with component l nonoperational.

An intriguing feature of the discussion by Cullinane et al. (1992) of this measure was their observation that "the judgement and experience of the modeler-designer is critical in system evaluation." The importance and relevance of this observation is discussed below in relation to simulation models for the assessment of network reliability.

It should be noted that demand variation failure is incorporated in Eq. (18.26) through the R_{jl} term, in which failure to meet the predetermined pressure can only occur through condition arising from demand failure. On the other hand, both component failure and demand failure are recognized in the $Q_j A_j$ term: component failure specifically through the Q_{jl} term, with demand the failure condition imposed through the A_j term. However, Eq. (18.26) does not explicitly consider the probabilistic aspects of "demand failure." Consideration of the probabilistic aspects can be obtained by modifying.

$$AE_{ji}^T = \int_0^\infty (R_{jl} A_{jl} + Q_{jl} A_j) p(Q^\eta) dQ^\eta \qquad (18.27)$$

where AE_{ji}^T = expected availability at node j considering demand variation and mechanical failure. All other terms as described above. Although Eq. (18.27) recognizes deficit in pressure as a factor to be addressed in defining network failure, the measure is not able to distinguish between failure conditions involving a large number of failures, each of which result in relatively small decreases in pressure below the minimum acceptable, and failure conditions involving a small number of failures, each of which cause significant decrease in pressure below the minimum acceptable. For example, a pipe failure that results in a reduction in delivery pressure heads at a node from 30 to 10 m is considerably more serious than a pipe failure of the same duration that results in reduction in delivery pressure head at the same node from 30 m to 29 m. Equation (18.27) cannot distinguish between these failures.

One approach to the problem is to use a measure of the following general form:

$$ND = \sum_{j=1}^{n} q^j H_{min} - \sum_{j=1}^{n} q_a^j H_{aj} \qquad (18.28)$$

and

$$= Q_T^* H_{min} - \sum_{j=1}^{n} q_a^j H_{aj} \qquad (18.29)$$

where, q^j = demand at node j, h = total number of nodes, Q_t^* = total network demand, q_a^j = actual flows supplied at node j, and H_{aj} = actual delivery pressure at node j, if $H_{aj} > H_{min}$ set $H_{aj} = H_{min}$, and ND = network deficit.

Note that this measure does not consider the stochastic nature of the demands and applies only for known deterministic nodal demands. However, the primary strength of this measure lies, in its ability to recognize explicitly the relationship between flows and pressures. If all flow demands are met, then any deficit is solely the result of deficits in supply pressure. If all demand pressures are satisfactory, any deficit is because the supply is not meeting the flow demand. In both cases, the size of the "deficit" increases as the amount by which the flow or pressure requirements are not met increases. However, if both pressure requirements are not met simultaneously— i.e., flows less than flow demands are supplied and these flows are only able to be supplied at pressures below the minimum acceptable, —the products of the q and H terms in Eq. (18.29) will be even less than if only one aspect of demand was not met. In this sense, the expression is very comprehensive in its interpretation of reliability.

However, the units of the measure are volume-pressure (i.e., m³/h · m) and, as such, do not directly affect reliability as engineers tend to know or assess it. For this reason, the expression should be considered as a heuristic measure of reliability, wherein it is known that if the deficit determined by the measure decreases the reliability of the system in improving. In addition, this expression does not consider that the duration or probabilities of the events giving rise to pressure or supplied flow rates are less than the requirements. These failures can be incorporated in the same fashion as they were for the "expected availability measure" of Eq. (18.27) by replacing the Q_j and A_{jl}, terms in that equation by elements from the right-side of Eq. (18.29) disaggregated by node, In other words,

$$ED_{jl} = R_{jl} [q^j H_{min}] + Q_{jl} [q_a^j H_{aj}] \qquad (18.30)$$

with the total network deficit given by

$$TND = \sum_{j=1}^{N} ED_{jl} \qquad (18.31)$$

where ED_{jl} = expected "deficit" at node j measured by the product of flow and pressure deficits and TND = total expected network deficit as measured by the product of flow and pressure deficits and without recognition of the stochastic nature of the demands.

Incorporation of demand variation can be achieved for each node by the following expression:

$$ED_{jl}^T = \int_0^\infty ED_{jl} p(Q^\eta) dQ^\eta \qquad (18.32)$$

Even when these probabilistic aspects have been incorporated in this manner, the measure gives only the average condition; it does not provide any indication of the extent to which the deficit varies around this mean: again, it does not indicate whether the mean is caused by a small number of extremely large and possibly unacceptable deficits or a large number of "acceptable deficits." Such issues are important in addressing reliability and are discussed more fully in the section on simulation models for the reliability analysis (Sec. 18.4.4.1).

This issue aside, the above discussion might indicate that the problem of assessing and measuring reliability in a water distribution network has been essentially solved. However, this is clearly not the case. It is generally accepted in the literature, (e.g., Cullinane et al., 1992, Goulter (1992a,b) that there is currently no method or measure for the assessment of reliability that is both comprehensive in its interpretation of reliability and computationally practical. The following sections discuss the simulation and analytical models currently available for assessing of reliability in water distribution networks and highlight the strengths and weaknesses of the approaches.

18.4.4 Models and Techniques for Assessing Network Reliability

The previous section examined the issues that should be considered in developing measures for assessing the reliability of water distribution networks. However, it did not address, how such measures would be calculated. The problem of computational practicality, mentioned above in relation to the lack of an acceptable measure for assessment of reliability, are in this calculation step.

Two approaches are available for the assessment of reliability.

1) Simulation approaches wherein the network is evaluated is under a range of scenarios formulated either on an individual case-by-case basis, or from a time series of "events" generated from the underlying distribution of conditions which define the requirements of the network

2) Analytical approaches wherein a closed form solution for the reliability is derived directly from the parameters which define the loads, (demand) on the network and from the ability of the network to meet those demands.

Two basic issues must be recognized in both approaches: one, hydraulic performance, as indicated by factors, such as flow and pressure head requirements on the network, pipe sizes, flow rates in the pipes, hydraulic gradients in the pipes, and pressures at the nodes and in the pipes; and, network configuration, as indicated by the level of connectivity between nodes, particularly demand nodes and sources, the shape of the network, (e.g., number of loops), and the existence or otherwise of branches. Network configuration and hydraulic performance are both important to simulation and analytical approaches to reliability. However, network configuration has received greater attention in the analytical approaches to reliability, whereas hydraulic performance, and its calculation has been the dominant feature of simulation.

The following sections examine and evaluate simulation and analytical approaches to assessment of reliability.

18.4.4.1 Simulation models. As noted above, determination of reliability by simulation models is usually undertaken on a case-by-case or scenario basis. The impacts of component failure are always assessed on a case-by-case or scenario basis in simulation models. Simulation models for assessment of reliability, for completeness, will still require consideration of the impacts of demand variation failure. If reliability is to be assessed in relation to expected volume deficit, the impacts of demand variation failure will, be determined analytically using Eq. (18.21). Use of other, more comprehensive, measures of reliability, such as availability [Eq. (18.26)] or the heuristic given in Eq. (18.29), will require case-by-case or scenario analysis for both demand variation and component failure.

In the case-by-case approach, a series of cases, i.e., set of demands or demand patterns and network configurations, are defined.The network is then modeled or simulated for each case to determine the pressures and flows that occur in the system as a result of those conditions. The performance of the network in relation to its ability to meet the demands over the complete set of cases can then be evaluated.

The demands used in these analyses can be any combination of demands, such as fire flows, that the network will be expected to supply. The component failure aspects of network performance in the simulation are handled through modifications to the network configuration, wherein the failure of any combination of network components, such as pumps, pipes and, valves, can be assessed by removing the relevant element from the network before hydraulic simulation. (When assessing the impacts of demand variation, it generally is not necessary to include component failure when determining the pressures and flows in the network arising from abnormal demands greater than the network design capacity. However, if the likelihood of a component failure during these periods of abnormal demand is believed to be sufficiently great, and the impacts of these joint occurrences are of sufficient magnitude, then the joint consideration of component failure and demand variation can be considered.)

The determination of the actual combination of load pattern and component failure to be included in each case, and the number of cases for which network performance is to be determined, presents some challenges. Ideally, the component failure considered in conjunction with a particular load or demand pattern will be the one most critical to network performance: a type of worst-case scenario. The critical component in this approach is that whose failure would be most problematic in relation to the ability of the network to meet the specified demand. Hence, the critical component for one load pattern can differ from that for another load pattern. For example, the critical component for a fire in an area will be different from the critical component for a fire in another area.

Network performance typically is assessed in this approach when only one component has failed at a time. This limitation to failure of a single component is not inconsistent with the approach for assessing the reliability of electrical distribution grids and is known as $k-1$ analysis, where $k =$ number of components in the network. The reason for restricting the analysis to the failure of one component at a time arises from the fact that the probability of two or more simultaneous failures (i.e., one component fails before another is repaired) is generally too small to justify consideration. An exception to this rule of thumb occurs when the impact of the failure of two components is so catastrophic that such an occurrence must be considered. Another important and significant exception to this general guideline is analysis of water distribution systems under earthquake conditions, when many components may fail simultaneously.

Nevertheless, even when of the analysis is restricted to the be failure of a single component, many combinations of demand pattern and component failure still must examined to obtain a comprehensive picture of network performance. Also recall that the network has to be simulated hydraulically for each of these cases, to determine the associated flows and pressures. In reality, a design engineer defines the subset of these combinations to be examined. The decision regarding which combinations are included in this subset is based on the designer's experience, perhaps guided by standards or municipal regulations.

Definition of the subset of combinations to be examined in this manner does, however, encounter a number of problems. The first, and perhaps most important, issue is whether the subset of combinations contains all the critical combinations and, in particular, whether it contains the worst—case combination. When dealing an experienced design engineer, it is reasonable to assume that all the extreme combinations have been defined and examined.

Another issue relates to specification of the reliability when deficits occur. If the network is satisfactory under all cases or scenarios i.e., the network is able to meet the demand in pressure and flow rate terms, under the various failure conditions, then the network is reliable (assuming that all critical cases have been addressed). In this case, there is no need to calculate expected deficits resulting from component failure because no deficits occur. However, consider the case when all deficits are not able to be met. The network still can be considered to be adequate: i.e., the deficits and the associated probabilities of occurrence may be acceptable and require no further evaluation. On the other hand, knowing the extent to which demand is not met, and the probability of the associated events, may not be sufficient. The expected deficit, or at least the distribution of deficits, may need to be evaluated. If some of the critical combinations for which simulation was performed result in deficits, it is likely that the other combinations of demand and component failure not considered in the original simulation subset also will result in deficits. Such deficits may not be as large as those associated with the critical conditions, but they do exist. An expected deficit component failure that is calculated only on the basis of the combinations contained in the simulation subset will not provide a true picture of either the expected deficit or the distribution of deficits. Calculation of the complete set of deficits for the purpose of defining the distribution of deficits leads adequately to the potential to require hydraulic simulation; in turn this may require evaluation of an extremely large set of combinations and an associated large computational effort. It should be noted that this computation issue is not restricted to simulation approaches; it also is an especially difficult problem for analytical approaches to reliability assessment. At this point, there appears to be no easy solution to the problem other than that, the increasing power of computers is reducing the magnitude of the problem.

One potential solution, however, may lie in work by Jowitt and Xu (1993), who proposed a method in which the distribution of flow in a network following component failure (and subsequent removal of that component from the network for repair) is predicted a priori using a microflow analysis without the requirement for a full hydraulic simulation. The authors showed that the accuracy of the microflow- based predictions of failed nodes varied from 72 percent for head definitions of node failure to 47 percent for demand-based definitions of node failure. The performance of the method with respect to the head-based definitions of failure suggests that it warrants further refinement and investigation across a wider range of test networks as a way of reducing some of the computational burden of case-by-case simulation methods for reliability assessment. This approach can be limiting because a fire or pipe break dramatically changes the microflow pattern, in which case a full hydraulic simulation is needed.

The other major simulation approach to assessment of reliability is based on the generation and simulation of a "time series" of scenarios based on the underlying probability distribution for pipe and pump failures and repair times. The hydraulic performance of the network is analyzed under the conditions contained in that time series. A number of reliability measures reflecting a wide range of perspectives on the reliability performance of the network can then be derived from the hydraulic outcomes arising from the time series. This approach assumes that the time of scenarios generated from the underlying probability distributions adequately represents the range of conditions to which the network will be subjected.

Wagner et al. (1988b) list the following 20 different measures of reliability, which were able to be obtained from their use of this type of simulation analysis:

Event-related

Type of event (failure or repair)

Interfailure time and repair duration

Total number of "events" in the simulation period System status during each event (i.e., normal, reduced service, or failure)

Node-Related

Total demand during the simulation period

Shortfall (total unmet demand)

Average head

Number of reduced service events

Duration of reduced service events

Number of failure events

Duration of failure events

Link-related

Number of pipe failures

Percentage of time of failure time for each pump

Percentage of failure time for each pipe

Number of pump failures

Total duration of failure time for each pump

System-related

Total system consumption

Total number of breaks

Maximum number of breaks per event

It is interesting that the predominantly analytical approach to reliability reported by Duan and Mays (1990) was able to derive eight different reliability measures because of incorporation in to the approach of a number of simulation features similar to those used by Biem and Hobbs (1988), Hobbs and Biem (1988), and Wagner et al. (1988a), Bao and Mays (1990) also used a Monte Carlo simulation approach to the measurement of reliability. In their case, however, the reliability issue focused on "hydraulic reliability" in that the time series of scenarios was generated by modeling the probability distribution of demand, pressure head, and pipe roughness. A complete analysis of network reliability using the time series-based simulation approach would require both the mechanical-failure aspects of Wagner et al. (1988) and the flow-based aspects of the work by Bao and Mays (1990). To the author's knowledge, this comprehensive analysis has yet to be undertaken.

Germanopoulis et al. (1986) proposed a similar approach to reliability based on extended simulation incorporating the probability distribution of pipe failures. An important feature of this work was the consideration of conditions between the failure event and the time full service was reestablished. This consideration is important in reliability analysis of water distribution network because it recognizes the ability of the system operator to manage the system in the face of a mechanical failure to minimize the impacts of the failure and thereby contribute to improved reliability.

An important feature of these simulation approaches is the need to generate the time series and to model and simulate the hydraulic performance of the network for each case or condition generated in the time series. This requirement obviously is a serious disadvantage of this type of approach since it imposes significant computational requirements. However, the problem is similar to that encountered with the case-by-case approach to simulation analysis of network reliability discussed above. It should be noted, however, that, although being more computationaly intensive than the case-by-case approach, the time-series approach generally gives a more complete description of how the network performs by considering a wider range of cases.

The important feature of all simulation approaches is that they permit the use of any reliability measure that can be derived from the hydraulic performance of the network. It is this ability that distinguishes these approaches from analytical approaches and represents their greatest advantage over analytical techniques.

18.4.4.2 Analytical approaches. Analytical approaches are distinguished by techniques that integrate all the parameters related to the network and its loads— for example, network layout and characteristics of loads— into single or multiple measures of reliability without the requirement of having to undertake complete simulation of the network. An important feature of analytical techniques is an increasing use of graph theory to describe the underlying performance of the network in terms of its shape and the connectivity between nodes, particularly between source nodes and demand nodes. This increasing use of graph theory is consistent with the prediction of Goulter (1992b), who indicated that analysis of hydraulic performance and incorporation of that performance into reliability approaches had reached a plateau and that further development in the area of reliability analysis would be the result of incorporating concepts derived from graph theory.

The features of the theory used most commonly in reliability analyses of water distribution networks are (1) *reachability*, the connection of a specific demand node to a source node; (2) *connectivity*, the connection of every demand node to at least one source; and (3) *cut-set*, a set of links that when removed from a network, completely disconnects one or more nodes in the remainder of the network

It should be recognized that connection of a node is a necessary, but not sufficient, condition for the node to meet its demands. If the connection between the source is constituted by a pipe that is too small or if the pressure in the system as a whole is low, that node may not receive any water even though it is connected to the source. The development of an effective means of integrating these hydraulic performance aspects with features of graph theory associated with the layout of the network is a major problem in analytical techniques for assessment of network reliability. It is important to emphasize here the underlying principle of the use of analytical techniques to assess the reliability of a of distribution network. Analytical techniques also are used to assess the reliability performance of the method on the basis of an analysis of the network's fundamental parameters (e.g., number of links connected to each node, size of pipes, location of nodes, demands), rather than an analysis based on an investigation of a range of scenarios.

The principles of graph theory have been applied extensively in the analysis of electrical distribution networks, communication networks and so on. Some of the first formal considerations of their use for reliability assessment of water distribution networks were reported by Goulter (1988) and Jacobs and Goulter (1988, 1989). Since that time, graph theory approaches have been used increasingly in analytical techniques for assessing the reliability of water distribution networks.

Wagner et al. (1988a) examined a series of analytical techniques based on the concepts of reachability and connectivity and observed that although the particular techniques proposed for assessment of probability of reachability and connectivity were effective for some networks, significant computational problems were encountered when the techniques were applied to another network that was generally accepted in the literature as a benchmark for analysis.

Goulter and Jacobs (1989) noted similar problems in simple extensions of the networks for which the methods of Wagner et al. (1988a) worked well. The computational difficulties encountered in applying these and other analytical techniques to the assessment of reliability are symptomatic of the fact that the problem is NP-hard: that is, the computation effort increases exponentially with increases in network size (Jacobs and Goulter, 1989). Furthermore, the assumptions or requirements that are imposed on the networks in order for them to be evaluated by analytical techniques generally are rigorous.

Therefore, the results do not provide an exact interpretation of reliability. In addition, the techniques and results cannot be generalized from one network situation to another.

Nevertheless, the use of graph theory principles, either in a fundamentally theoretically correct form or in some modified form, in analytical techniques for assessing reliability of water distribution networks can continue to increase. The techniques that use graph theory in some reliability assessments are identified in Table 18.3 which lists (in chronological order) and summarizes the analytical and simulation techniques available to assess the reliability of water distribution networks. Of particular relevance to the above discussion of analytical techniques and their associated computational requirements is the work of Su et al. (1987). The computation problems associated with the use of these techniques and the difficulties associated with generalizing the results can be seen in the increase in computational time for Su et al's model (1987) on a Dual Cyber machine from 1157 s for a 14-link, one-loop network to 200.5 min for a 17-link, three- loop system based on the 14-link, one-loop network. Not withstanding these computational problems, an important feature of Su et al.'s model (1987) was the use of a modified aspect of the cut-set approach. Although cut-set approaches had been used either explicitly, (e.g., Mays et al., 1986; Quimpo and Shamsi, 1987; Shamir and Howard, 1985; Shamsi, 1990; Wagner et al., 1988a) or implicitly ("Probability of Node Isolation" in Goulter and Coals, study (1986). Su et al. (1987)) modified the strict definition in graph theory of set links, the removal of which disconnected at least one demand node from the rest of the network to a definition of set links which, when removed, caused the network to be unable to meet the demand. Defined in terms of flow rates and delivery pressures at one or more modes, it is of integration of graph theory and hydraulic performance that probably holds the most promise for future computational effectiveness and appropriate comprehensiveness for assessment of reliability in water distribution networks (Goulter, 1992b).

Other approaches to analytical assessment of reliability have attempted to address the computation problem by "simplifying" the network using other principles in graph theory principles. Two techniques have received special attention in this regard: namely, series reduction algorithms and parallel reduction algorithms. These techniques have been applied by Awumah and Goulter (1992), Wagner et al. (1988a), Wu and Quimpo (1992), and Wu et al. (1993). The underlying feature of these techniques is a reduction in the complexity of the network,—e.g. a reduction in the number of links which must be considered—by "lumping" links and nodes together without redefining the underlying determinants of reliability. These techniques have achieved some success in reducing the computational requirement while still maintaining an appropriate level of comprehensiveness in their interpretation of reliability. However, the techniques are still hampered by a significant computational burden (Wu et al., 1993) and are not always effective even for relatively simple networks, as Goulter and Jacobs (1989) pointed out in their discussion of Wagner et al. (1988a).

An interesting response to this computational burden was the study by Jacobs and Goulter (1991). They attempted to develop a technique based on cut-set theory whereby the reliability of a network, defined as the probability that each node was connected to a source, could be estimated within a known or specified accuracy. However, the technique did not address the hydraulic capacity of the links connecting a node to the network, this neglecting the fact that a network might be reliable on a connectivity basis alone, whereas in terms of hydraulic performance, it might not meet the demands and therefore be in a failed state. Other follow-up work by Jacobs (1992) extended the analysis to the consideration of the hydraulic of the network when links were removed, but the extension required extensive hydraulic simulation of the networks with an associated large computational burden.

TABLE 18.3 Summary of Major Simulation and Analytical Approaches to Assessment of Reliability in Water Distribution Networks

Study	Approach	Simulation or Analytical	Issues Addressed	Remarks
Rowell and Barnes (1982)	Minimum-cost branched network with cross-connections	S	Design branched system—add cross-connections (with a branch failed) to meet demands	Some hydraulic inconsistencies
Morgan and Goulter (1985)	Minimum-cost design model for looped systems	S	Designed for a range of combinations of critical flows (fire flows) and pipe failure	Judgement of engineer needed to define the worst location of a broken pipe for each fire flow
Kettler and Goulter (1983)	Minimum-cost design model with constraints on the probability of a pipe failing	A	"Reliability" constrained-probability of pipe breakage ≤ acceptable level "removed"	No concern with, or recognition of, performance of network with the pipe
Goulter and Coals (1986)	Minimum-cost design model under constraints on probability of node isolation	A	Probability of a node being disconnected from the network must be < an acceptable value — if unacceptable which link should be improved? be able to meet the demand	Does not address the problem of "partial supply — remaining connections to the network with links after one (or two) links(s) fail may not
Su et al. (1987)	Minimum-cost design model with restrictions on the probability of "minimum cut-sets"	S → A	Examines the impacts of removal of one (and two) links on the ability of network to meet demands in the network — uses probability of pipe breakage Graph theory	Computationally intensive! 14 links, 1 loop: 1157 s 17 links, 3 loop: 12030 s (200.5 min on a Dual Cyber machine)
Germanopoulos et al. (1986)	Assessing reliability of supply and level of service	S	Network performance in failure/post failure Simulation of failure occurrences and repair times	Examines management of a failure to minimize impacts
Wagner et al. (1988a)	Reliability analysis-analytical	A	Reachability and connectivity Series and parallel reductions to get trees Probability of sufficient supply as a reliability measure	Not general for all networks Need for supporting simulation User graph theory

Reliability Analysis for Design **18.35**

TABLE 18.3 (Continued)

Study	Approach	Simulation or Analytical	Issues Addressed	Remarks
Wagner et al. (1988b)	Reliability analysis — simulation	S	Models failures of the components Models repair times for failure Looks at a range of reliability measures	Time-consuming Hard to generalize the results
Lansey et al. (1989)	Minimum-cost design model chance constrained on probability of meeting demands	A	Uncertainties in: Future demands Pressure requirements Pipe roughness	Cost-reliability function is convex—the cost of achieving improvements in reliability is greater at higher levels of reliability
Bao and Mays (1990)	Reliability of water distribution systems	S	Distribution of operating Scenarios from Monte Carlo simulation Probability of head being larger minimum required	Difficulty in distinguishing between critical and noncritical events
Goulter and Bouchart (1990)	Minimum-cost design model with reliability constraints on node performance	A	Probability distribution of demand at each node Probability of node isolation mechanical failure and failure to meet demand	Emphasized the distribution of demand at each node particularly with respect to extreme events (fire flows) Used a "heuristic" combination of
Duan and Mays (1990)	Reliability analysis of pumping systems Frequency/duration analysis	A	Mechanical failure and hydraulic failure of pumps not networks Eight parameters related to reliability, failure Probability, failure frequency, cycle time, and expected duration of failure, expected unserved demand of a failure, expected number of failure, expected total duration of failures, and total expected unserved demand	Does not consider nodal reliability Does not consider mechanical failure of network
Duan et al. (1990)	Optimal reliability–based design of pumping and distribution systems	A	Extension of work of Duan and Mays (1990) into the design of the distribution network	

Reliability Analysis for Design **18.37**

TABLE 18.3 (Continued)

Study	Approach	Simulation or Analytical	Issues Addressed	Remarks
Kessler et al. (1990)	Least cost improvements in network reliability	S	Topologic redundancy from alternative trees in the network Level-one redundancy Different levels of acceptable service under component failure	Not probability based Avoids need for cut-set calculations Use graph theory
Fujiwara and De Silva (1990)	Reliability-based optimal design of water distribution networks	A	Ratio of expected maximum demand to total water demanded	
Fujiwara and Tung (1991)	Improving reliability through increasing pipe size	A		
Awumah et al. (1990) Awumah et al. (1991)	Entropy-based measures of network redundancy	A	Associates reliability with redundancy	Heuristic or surrogate measure Employs aspects of graph theory
Bouchart and Goulter (1991), (1992)	Improving reliability through valve location	A	Demands are not located at nodes Variation in demand at node Mechanical failure of links (nodes) Expected volume of deficit	Recognizes the option, through valve location, of managing the reliability of network to minimize the impact of component failure
Quimpo and Shamsi (1991)	Reliability-based distribution system maintenance	S	Component reliability Enumeration of cut-sets and path sets Nodal pair reliability Lumped systems Computional problem in identifying	Uses graph theory Does not recognize that existence of a path. Does not guarantee demand can be met links to "improve" to obtain the greatest increase in reliability
Jacobs and Goulter (1991)	Estimation of network reliability—cut-set approaches	A → S	Probability of isolation Node Groups of nodes Probability of m links failing simultaneously Probability of m simultaneous link failures causing network failures	Does not consider capacity issues Uses graph theory

TABLE 18.3 *(Continued)*

Study	Approach	Simulation or Analytical	Issues Addressed	Remarks
Cullinane et al. (1992)	Minimum cost model with availability constraints	S/A	Considers pipes, tanks, and pumps Repair time for failures	Examines probability of failure and impacts of failure on pressure Recognizes "fuzziness" in level of acceptable pressure Results are consistent with engineering rule of thumb
Wu et al. (1993)	Capacity-weighted reliability	A	Connectivity based (nodal pair reliability) Includes capacity of links Reduces system by block reduction and path set methods	Recognizes practical contributions to meeting demands Weighted nodal pair reliability—importance of link in supplying a demand Computational problems Heuristic reliability index Users graph theory
Park and Liebman (1993)	Redundancy-constrained minimum cost model	S	Expected shortage due to pipe failure Based on geometry of the network Reduces system by block reduction and set methods	Computationally demanding Recognizes that network "operator" can adjust system in the face of a component failure to minimize impacts
Jowitt and Xu (1993)	Predicting pipe failure effects on service	S	Failure of pipes Simplified prediction of nodal conditions — network performance under failed pipes Expected shortfalls at nodes	Proposes a new methodology for allocating flows and identifying pressure and demand-based failure without the need for a full hydraulic simulation
Gupta and Bhave (1994)	Reliability analysis considering nodal demands and heads simultaneously	S	Failure of pipes and pumps node reliability network reliability	Comments about aggregation of demand at nodes Comments about isolation of breaks by values Reliability index Recognizes unacceptable reliability of 100% of demands 100% of time at 90% of nodes No supply at 10% of nodes

18.4.4.3 Heuristic techniques. A recent development in measures of reliability in water distribution networks has been the use of heuristic measures to evaluate reliability and to guide the designer in improving the reliability of networks. *Heuristic measures* in this context are defined as measures that do not reflect reliability precisely (e.g., their units may not have any real meaning regarding traditional quantitative aspects of reliability) but do reflect reliability in that changes to their values reflect changes in reliability. These heuristic measures of reliability have received particular attention in optimization approaches to the design of water distribution networks.

The first uses of such heuristic measures were reported by Awumah and Goulter (1992), Awumah et al. (1990, 1991) and Lansey et al. (1992). Awumah and Goulter (1992) used a redundancy measure from entropy theory. The underlying principle in the redundancy measure is that entropy theory is able to measure flexibility and diversity in the system and therefore reflect the redundancy of the network as it contributes to reliability. The measure is able to be stated in simple mathematical terms and, as such, was to be used in an optimization framework. Awumah et al. (1991) used the measure in an optimization application and showed that it was capable of producing network designs that were comparable in reliability to networks designed by simulation-based approaches with significantly larger computational effort.

Wu et al. (1993) also proposed a heuristic reliability index that recognized the partial contribution of links to meeting demands. This recognition was achieved by weighting a connectivity-based reliability measure for a particular path by the importance of that path with regard to the ratio of the total flow in the path to the total demand of the node for the reliability being assessed. The index did not reflect conditions in the network precisely and had the same computational problems normally associated with analytical approaches to reliability based on computation. However, the approach is important in that it represents an attempt to integrate graph theoretic and hydraulic performance issues into a single reliability measure.

The measure of reliability used by Goulter and Bouchart (1990) also fits within this heuristic category because it combines, by direct multiplication, the probability of "no node isolation" with the probability of "no demand failure." Recall that the reliability expression in Eq. (18.29) also is a heuristic because its meaning and units do not conform to normal quantitative measures of reliability However, as noted in the discussion of that equation, it does provide a reliability comprehensive statement of deficits and their magnitudes. Changes in the value of the measure also reflect changes in the reliability in the network.

Earlier, it was asserted that future developments in the assessment of reliability in water distribution networks were likely to be based on further integration of hydraulic theory and graph theory. The strategy of using heuristics as a means of combining the principles in both theories in a computationally sensible manner, while maintaining a comprehensive and reasonably realistic statement of reliability, also should be recognized as a useful approach. This approach is not just for improving current approaches to assessment of reliability; it also is for incorporating reliability into optimization models more easily.

18.4.4.4 Redundancy based measures. Another recent development in the consideration and assessment of reliability in water distribution networks has been an increase in the use of redundancy, as opposed to reliability in its purest sense. *Redundancy* in this context is defined as the existence of alternative pathways, or capacity in excess of that required in normal operating conditions, which are able to be used when one or more components used in the normal operating condition fail. This renewed interest in redundancy has arisen from the realization that the reliability of a network, to a large extent, is determined by the layout or shape of the network and the redundancy inherent in that layout. The work by Morgan and Goulter (1985) and Rowell and Barnes (1982), which focused on providing alternative paths of sufficient capacity in the network should a pipe fail, is an early example of the explicit consideration of redundancy. In another early work, Goulter (1988) examined redundancy closely from the perspective of clusters and interconnectivity derived from graph theory considerations. The studies of Awumah and Goulter (1991) and Awumah et al. (1990, 1991), cited above with respect to heuristic techniques also are examples of explicit consideration of redundancy. Kessler et al., (1990) and Ormsbee and Kessler (1990) used graph theory directly when considering redundancy by identifying independent spanning trees that could be available to supply a node should any link in the other tree fail. An important feature of this approach was recognition that a lower level of service-for example, when 80 percent or less of the demand is supplied-might be acceptable while the failed link was being repaired. These two studies also acknowledged that the acceptable level of reduced service could be part of the design criteria.

More recently, Park and Liebman (1993) proposed what they termed a redundancy-constrained minimum-cost model for pipe network design. The approach attempts to qualify redundancy through a surrogate measure based on the expected shortage caused by failure of individual pipes. As such, this approach fits more easily into the category of simulation methods for consideration of reliability. It is discussed here, however, because of its explicit recognition of redundancy as an underpinning element of reliability.

18.4.5 Overview of Reliability Measures

In addition to providing a summary of the most well-known and commonly used analytical and simulation techniques for reliability assessment, Table 18.3 provides the opportunity to highlight the major steps and break-through in the development of reliability measures for water distribution networks.

The first article in the table (Rowell and Barnes, 1982) was the first optimization model for design of water distribution networks to recognize explicitly the capacity required of interconnecting links in looped networks in light of the failure of other links in the system. Previous optimization models, such as those of Alperovits and Shamir (1977) and Quindry et al. (1981), considered capacity in the interconnections in looped networks, but only by placing constraints on the minimum sizes of pipe values rather than by examining what capacity would be required in those links.

Morgan and Goulter (1985) made the next major step by developing a technique that was able to examine simultaneously a range of combinations of critical flows (fire flows) and failed links in the design of a looped network. However, similar to simulation approaches, their method required engineering judgment to define the various fire flows and failed link conditions.

The first explicit consideration of probabilistic issues in the reliability of networks was reported by Kettler and Goulter (1983), who, in an optimization model for design of networks, constrained the possibility of breakage of the pipes in each link below some acceptable level. Although Goulter and Coals (1986) also considered the probability of pipe breakage in their model for reliability assessment of pipe network, they were among the first to consider graph theory in the analysis (in their case, it was implicitly through examining the probability of node isolation). Shamir and Howard (1985) reported other early uses of graph theory concepts (such as cut-sets). Mays et al. (1986) and Su et al. (1987) combined those graph theory concepts with the probability of occurrence of the events associated with the concepts that would cause failure of the network under a definition based on graph theory. Wagner et al. (1988a) were first to propose series and parallel reductions of looped networks to obtain simplified "tree" representations of the networks that were more amenable to reliability analysis.

Around the same time, Germanopoulis et al. (1986) began to examine both techniques for assessing the reliability of supply and level of service and the way a network might be operated between the time of component failure and the time when full service was restored to improve reliability by minimizing the impact of the failure. Germanopulis et al. (1986), followed by Biem and Hobbs (1988) and Hobbs and Biem (1988), also introduced time series simulation of the scenarios under which the networks were expected to operate.

Most of the discussion on reliability analysis up to that point had concentrated on reliability aspects related to component failure rather than on demand variation failure. Lansey et al. (1989) explicitly introduced consideration of variation in the demand into reliability calculations. Duan and Mays (1990) subsequently introduced the performance of pumps into the consideration of reliability, and Bouchart and Goulter (1991, 1992) proposed consideration of the fact that demands are not actually located at the nodes but are only placed these for analytical purposes. Bouchart and Goulter (1991, 1992) also examined the impact of the placement and operation of valves on the reliable performance of the network.

Setting aside heuristic and redundancy-based measures at this time because of their low level of development and use compared with techniques that can be categorized as analytical or simulation, it is possible to establish some general conclusions about the strengths and weaknesses of analytical and simulation methods. These approaches have common as well as different strengths and weaknesses, as summarized in Table 18.4. A review of Table 18.4 indicates that analytical and simulation techniques should be partners rather than alternatives when assessing and considering of improvement in the reliability of water distribution networks. Wagner et al. (1988a, 1988b) argued for the integration of the two types of techniques and asserted that analytical techniques are best used for screening the system: for example, to determine the level of connectivity and reachability in the network. A more elaborate analysis using simulation techniques can then be undertaken on a subset of conditions or scenarios to determine other features related to reliability, such as whether the connections have enough capacity to meet the demand.

Notwithstanding the accepted practice of integrating analytical and simulation techniques as the most reliable means of performing reliability analyses of water distribution systems, among the most interesting, and potentially most useful, new approaches for comprehensive consideration of reliability in water distribution networks has been proposed by Xu and Goulter (1998). This analytical approach can consider uncertainties in a wide range of factors that affect the reliability of water distribution

TABLE 18.4 Strengths and Weaknesses of Analytical and Simulation Techniques for Assessing of Reliability in Water Distribution Networks

	Weaknesses		Strengths	
	Analytical	*Simulation*	*Analytical*	*Simulation*
	Simplistic interpretation of reliability, e.g., connectivity vs hydraulic performance Computationally Intensive	Only evaluates "sample conditions" identified for consideration Computationally Intensive	Considers the complete network rather than "samples"	Can generate or consider a broader range of reliability measures Realistic interpretation of reliability

systems. A key feature of the method is the use of a linear approximation in the hydraulic simulation component. This linear approximation provides accurate results under a wide range of conditions and enables the direct inclusion in the reliability analysis of uncertainties concerning the capacities of the pipes, the nodal demands, and the reservoir storage tank levels. The impacts of these uncertainties are then combined with the reliability implications of component failure to give good estimates of nodal and overall system reliability under commonly accepted definitions of reliability.

Fujiwara and Li (1998) used reliability analysis and optimization to distribute the impacts of system failure. This approach considers relationship between delivery pressure and flow, allows system operators to redistribute flows to minimize the worst impacts of system failure, and does not consider probabilities. Xu and Goulter (1998) used reliability analysis to consider the probability of component failure together with uncertainties in nodal demands, pipe capacities, and reservoir levels. Their approach integrates component failure and variations in demand, uses a probabilistic hydraulic model to reduce computational requirements, and considers other uncertainties contributing to network reliability, such as reservoir levels and hydraulic (pipe) capacities.

18.4.6 Observations

Section 18.4 has reviewed the methods available for reliability analysis of water distribution networks from the perspective of (1) the issues that must be considered in assessing water distribution networks, (2) the type of mathematical techniques and expressions that have been, or should be, used for assessment of reliability, (3) the approaches (particularly simulation and analytical techniques) currently available to calculate using these techniques and expressions, (4) the relative strengths and weaknesses of the current methods or measures for assessing reliability, and (5) potential future directions for improved methods or measures for the assessment of reliability. The need to consider the reliability aspects of both component failure and demand that exceeds network capacity was highlighted together with the need to consider the probability aspects of both types of system failure.

A number of previous reviews of water-distribution network designs highlighted the fact that was not a measure for reliability analysis of water distribution networks that was both comprehensive in its interpretation of reliability and computationally practical. This situation remains, although marginal improvements in the techniques available to assess the reliability of water distribution systems and significant improvements in computing power have diminished the problem somewhat. Interestingly, the greatest advances in the measures available to evaluate reliability have arisen from increased use of graph theory in the measures, perhaps indicating that evaluation of the hydraulic aspects of reliability is approaching the maximum achievable level of comprehensiveness and efficiency.

An important feature of our current ability to assess reliability is that despite their relative strengths and weaknesses, both simulation and analytical methods should be used to assess reliability. Analytical techniques should be used in the initial screening and simulation techniques should be used for more detailed analyses. The primary reason for recommending one technique over the other lies, on the one hand, on the stringent assumptions required for analytical analysis of reliability and, on the other hand, on the ability of simulation techniques to "calculate" a wide range of reliable measures covering all aspects of reliable performance of a network.

Looking at the future of reliability analysis, a number of recommendations can be made. The ability of system operators to mitigate the effect of a component failure, abnormal demand, or both is increasingly recognized. Future measures of reliability must take this flexibility of network operation into consideration in evaluation of a network's reliability. Heuristic techniques that do not yield exact or typical quantitative statements regarding reliability also warrant further examination. Such techniques will have value if they can be useful for assessing relative reliability and for guiding the improvement of reliability.

Finally, redundancy, as an underlining feature of reliability, should receive additional attention, perhaps in conjunction with the investigation and development of improved heuristic measures of reliability. Implicit with the recommendation to consider redundancy-derived measures is the emphasis on graph theory as a means of considering the impacts of a network's shape or layout on the level of reliability that can be achieved.

For discussions about the future of reliability analysis, we recommend the following works: Bouchart and Goulter (1991), Briassoulis (1994), Cameron and Wright (1990), Cullinane (1986), Cullinane et al. (1992), Goulter (1995), Nieswiadomy (1992), Weber (1993), and Wilchfort and Lund (1997).

18.5 MEASURE OF LINK IMPORTANCE

The *vulnerability* of a network to a mechanical failure (e.g., a burst pipe) can be defined as the magnitude of the shortfall in the water supplied to consumers as a result of the failure. Although most networks exhibit a degree of resilience, whereby flows are redistributed within the network to compensate for the loss of the failed component, the continued and full delivery of all water demands may not be possible, depending on which specific component has failed.

For example, a burst pipe in link 1 of the network depicted in Fig. 18.8 will result in a complete failure of the system, with all consumers experiencing interruption of supplies while the pipe is repaired. By comparison, a burst pipe in link 9 of the same network will

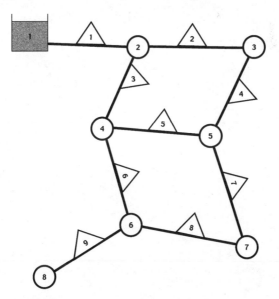

FIGURE 18.8 Sample distribution network.

result in a localized failure; thus, only a relatively small number of consumers will be affected (unless demand node 8 corresponds to a substantial portion of the consumers serviced by the network). This suggests that link 1 is far more important than link 9 because the consequences of a failure in link 1 are significantly greater.

Bouchart and Goulter (1989) defined the *link importance* (LI_i) measure for a given link i as the proportion of the demand in the network as a whole that is not supplied when the link in question has failed.

The link importance of a given link i is calculated as

$$LI_i = \frac{D - F_i}{D} 100\% \tag{18.33}$$

where LI_i = link importance measure of link i (%), D = total demand within the network (L/s), and F_i = total demand that can be supplied when link i fails (L/s). The link-importance measure is an indicator of the relative value of a link to the network as a whole. Once the link-importance measures of all links are determined, one can identify the links in which failure results in a major collapse in service. When combined with a reliability analysis of the entire network and of individual components within the network, the link-importance (LI) values highlight the pipe segments that are crucial to the conveyance of water throughout the network. For example, an analysis of the mechanical reliability of link 1 in Fig. 18.8 might reveal that its probability of failure is low. However, the consequences of such a failure are extremely high, because all consumers are affected. Conversely, failure of some links in the network results in only a minor reduction in the total water supplied; consequently, those links are less important to the network.

Consider the branched water distribution network depicted in Fig. 18.9. Tables 18.5 and 18.6 contain the relevant data needed to solve for the network flows and pressures. Links 5 and 8 are the additional links required to transform the branched network depicted in Fig. 18.9 into the network shown in Fig. 18.8. The relative importance of each pipe link

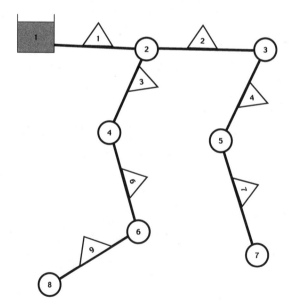

FIGURE 18.9 Branched networks.

TABLE 18.5 Node Data

Node	Elevation (m)	Demand A (L/s)	Demand B (L/s)
1	60	–	–
2	0	3	2
3	0	15	8
4	0	8	4
5	0	23	11
6	0	17	9
7	0	5	20
8	0	7	3

TABLE 18.6 Link Data

Link	Length (m)	Diameter (mm)	Roughness Coefficient
1	1000	400	100
2	1000	300	100
3	1000	400	100
4	1000	200	100
5	1000	300	100
6	1000	250	100
7	1000	200	100
8	1000	200	100
9	1000	200	100

in the branched network can be derived intuitively without recourse to the LI measure proposed by Bouchart and Goulter (1989). Links closer to the source are undoubtedly more important than are links further downstream because the failure of such upstream links prevent a greater proportion of the demand from being met. Table 18.7 presents the LI values for the pipe links in the branched network based on the two demand loading conditions defined in Table 18.5.

The LI values for the branched network exhibit the expected pattern, with links near the source being relatively more important than are those at the ends of the branches. It also is important to note that the LI measure allows for direct comparisons between different loading conditions because it is based on the proportion (or percentage) of the demand that cannot be supplied rather than on the absolute shortfall in supply. Such comparisons would be more difficult to undertake using the absolute value of the shortfall in supply because the total water demand on the system can be different for each loading case (e.g., total demand is 78 L/s under loading A and 57 L/s under loading B). As defined,

TABLE 18.7 Link Important Measures for the Branched Network

Link	Link Importance (%)	
	Loading A	Loading B
1	100	100
2	55	68
3	41	28
4	36	54
6	31	21
7	6	35
9	9	5

the LI measures for different loading scenarios can be compared to obtain the range of LI measures for all links in the network. Finally, the results in Table 18.7 demonstrate the potential importance of links located at the extremes of the network. If the demand loading B represents an infrequent event (e.g., a fire at node 7), then the relatively high LI value for link 7 might be disregarded within the overall risk analysis. However, this LI value of 35 percent becomes far more relevant if the loading condition B occurs frequently. In this latter scenario, the determination of the LI values suggests that link 7 is a vital component of the network (much more so than link 9, the LI values of which are below 10 percent).

Determination of the LI measures for branched networks provides only limited additional insights into the reliability—or more correctly, the vulnerability—of the network with regard to mechanical failures in pipe links. The reason for this limited scope is the inherent lack of resilience in branched networks, which by definition lack the alternative flow paths needed to redistribute the flows within the network to compensate for the loss of a failed component. On the other hand, looped networks contain the redundant flow paths necessary to reduce the impact of mechanical failures on levels of service.

When links 5 and 8 are added to the network depicted in Fig. 18.9, the resulting network (see Fig. 18.8) becomes resilient to all mechanical failures other than a failure in links 1 and 9. This ability to cope with mechanical failures is reflected in the LI values presented in Table 18.8. In the two scenarios considered, the full delivery of water to meet demand is maintained at pressures well above any stipulated level of service (e.g., 20 m). This can be verified using any commercially available hydraulic network solver. The significance of these results is that when undertaking a reliability analysis of this network, the mechanical reliability of link 1 is for more significant than is the reliability of the network as a whole.

The two sample networks considered above fail to demonstrate the full benefits of the link importance measure. In the case of the branched network, it is possible to ascertain intuitively the importance of a link by its location relative to the source. Furthermore, the results obtained for the looped network might suggest that the LI measure simply highlights all the branches within a water distribution network (link 1 is the only connection between the source and the rest of the network). Although these branches no doubt will always represent critical connections within a network, all other links in this case are unimportant because the network exhibits a high degree of redundancy.

TABLE 18.8 Link Important Measures for the Looped Network

Link	Link Importance (%)	
	Loading A	Loading B
1	100	100
2	0	0
3	0	0
4	0	0
6	0	0
7	0	0
9	9	5

Consider a slight modification to the looped distribution network depicted in Fig. 18.8: namely, a reduction in the pipe diameter of link 2 from 300 to 200 mm. For this new network layout, the flow patterns under both loading conditions result in pressure heads well above the minimum required level of 20 m. However, the network is less resilient in times of failure, as evidenced by the *LI* values in Table 18.9.

The first observation that can be made from the results in Table 18.9 is that defining the *LI* measure as a percentage rather than as the absolute shortfall in supply does not completely subsume the magnitude of the demand loading on the network. The network is more capable of "absorbing" the loss of link 3 when the total demand is low (loading case B) compared to when it is high (loading case A). Second, the loss of capacity caused by the use of a smaller pipe in link 2 has made the network more vulnerable to mechanical failures in link 3. There is now a 30 percent shortfall in supply when link 3 fails (based on loading condition A).

The final consideration in the use of the *LI* measure is the method used to obtain the F_i values. Standard, commercially available hydraulic network solvers are demand-driven. For a specified demand loading on the network, the hydraulic network solver determines the flows in each pipe segment and the corresponding nodal pressure heads. The water demands are satisfied regardless of the actual pressures at the nodes. Failures of the

TABLE 18.9 Link Important Measures for the Modified Looped Network

Link	Link Importance (%)	
	Loading A	Loading B
1	100	100
2	0	0
3	30	9
4	0	0
6	0	0
7	0	0
9	9	5

system are therefore not defined by an inability to supply the water demands (since the solution algorithm assumed that the demands are met) but by pressure heads at the nodes being below a specified minimum. The reality of the situation is that as the pressure in the network drops below some threshold level, the volume of water actually delivered to consumers begins to fall short of the total demand. For example, this pressure threshold usually is assumed to be below the minimum of 20 m specified in network studies. The relevance of this relationship between pressure and water actually supplied is that during a link failure, the pressures may fall below the threshold value, thereby reducing the total water supplied. This situation occurs when link 3 fails in the above network because the remaining links are unable to supply the full demand. Unfortunately, the extent of this shortfall cannot be obtained from the majority of commercially available hydraulic network solvers. Instead, the failure is reflected by either low or negative pressures. Therefore, an alternative method is needed to obtain the required F_i values.

The first approach to obtaining the required F_i values is to reduce the demands progressively at nodes experiencing low pressures until their associated pressures (as calculated using a standard hydraulic network solver) rise above the specified minimum (e.g., 20 m). The use of this minimum pressure ensures that the calculated shortfall in supply represents the upper bound; in fact, this shortfall might be slightly less by accepting slightly lower pressures, but the calculated shortfall does represent the worst case. A second approach would be to set a key downstream node to its minimum acceptable head.

An alternative approach to obtaining the required F_i values is to formulate a nonlinear optimization model, the objective of which is to maximize the total demand supplied to the network. Although this approach involves the added complexity of formulating and solving this optimization model, it does guarantee the identification of the maximum volume of water that can be supplied, subject to meeting the minimum pressure requirements. For a pipe failure in link i the model is

$$\text{Maximize } F_i = \sum_{m=1}^{ND} d_m \qquad (18.34)$$

subject to

$$d_m \leq D_m \qquad (18.35)$$

$$\sum_{k \in IN_m} Q_k - \sum_{k \in OUF_m} Q_k - d_m = 0 \qquad (18.36)$$

$$\sum_{j \in P_m} \Delta H_j \leq H_s - H_m = 0 \qquad (18.37)$$

and

$$\sum_{j \in LOOP_m} \Delta H_j = 0 \qquad (18.38)$$

where d_m = demand actually supplied to node m, D_m = demand at node m, ND = total number of nodes in the network, Q_j = flow in link j, IN_m = set of links directly connected to node m and that carry water to node m (excludes link i), OUT_m = set of links directly connected to node m and that carry water away from node m (excludes link i), Hj = headloss caused by friction in link j, H_s = head at source, H_m = minimum allowable head at node m, P_m = set of links included in the selected path from the source to node m (excludes link i), and LOOP = set of links included in a given loop (excludes link i).

This model is formulated and solved for each possible link failure. The resulting F_i values are then used to calculate the link-importance values.

REFERENCES

Alperovits, E., and Shamir, U., "Design of Optimal Water Distribution Systems," *Water Resources Research*, 13: 885–900, 1997.

AWWA, "Water-Distribution Research & Applied Development Needs," *Journal of the American Water Works Association,* pp. 385–390, June 1974.

AWWA, *Introduction to Water Distribution*, American Water Works Association, Denver, CO, 1986.

AWWA, *Distribution Requirements for Fire Protection*, American Water Works Association, Denver, CO, 1988.

Awumah, K., I., Goulter, and S. Bhatt, "Assessment of Reliability in Water Distribution Networks Using Entropy Based Measures," *Stochastic Hydrology and Hydraulics*, 4:325–326, 1990.

Awumah, K., and I. Goulter, "Maximizing Entropy-Defined Reliability of Water Distribution Networks," *Engineering Optimization:* 20(11): 57–80, 1992.

Awumah, K., I., Goulter, and S. Bhatt, "Entropy-Based Redundancy Measures in Water Distribution Network Design," *ASCE Journal of Hydraulic Engineering*, 117: 595–614, 1991.

Bao, Y., and L., Mays, "Model for Water Distribution System Reliability," *ASCE Journal of Hydraulic Engineering*, 116: 1119–1137, 1990.

Biem, G., and G. Hobbs, "Analytical Simulation of Water System Capacity Reliability. 2: A Markov-chain Approach and Verification of Models," *Water Resources Research*, 24: 1445–1458,1988.

Bouchart, F., and I. C. Goulter, "Implications of Pipe Failure on the Hydraulic Performance of Looped Water Distribution Networks," *Proceedings of the First Caribbean Conference on Fluid Dynamics*, St. Augustine, Trinidad, 291–297, 1989.

Bouchart, F., and I. Goulter, "Reliability Improvements in Design of Water Distribution Networks Recognizing Valve Location," *Water Resources Research*, 27: 3029–3040, 1991.

Bouchart F. and I. Goulter, "Selecting Valve Location to Optimize Water Distribution Network Reliability," in J-T Kuo and G-F. Lin, eds., *Proceedings of the 6th IAHR International Symposium on Stochastic Hydraulics*, J-T. Taipei, Taiwan, May 18–20, 155–162, 1992.

Briassoulis, H., "Effectiveness of Water Conservation Measures in Greater Athens Area," *Journal of the American Water Works Association*, 86: 764–778, 1994.

Cameron, T., and M. Wright, "Determinants of Household Water Conservation Retrofit Activity-A Discrete Choice Model Using Survey Data," *Water Resources Research*, 26: 179–188. 1990.

Ciottoni, A. S, "Computerized Data Management in Determining Causes of Water Main Failure," *Proceedings of the 12th International Symposium on Urban Hydrology, Hydraulics and Sediment Control*, University of Kentucky, Lexington, 323–329, 1983.

Cullinane, M. J., "Hydraulic Reliability of Urban Water Distribution Systems," in M. Karamouz et al., eds., *Proceedings of ASCE Conference, Water Forum '86: World Issues in Evolution*, Long Beach, CA, 1986, pp. 1264–1271.

Cullinane, M., K. Lansey and L. Mays, "Optimization-Availability Based Design of Water-Distribution Networks." *ASCE Journal of Hydraulic Engineering*, 118: 420–441, 1992.

Deb, A. K., C. Zernentsch, and J. McElheny, "Need for Rehabilitation of Nation's Aged Water System: A Case Study," presented at the AWWA Annual Conference, Miami, FL, May 16–20, 1982.

Duan, N., and L. Mays, ""Reliability Analysis of Pumping Systems," *ASCE Journal of Hydraulic Engineering*, 116: 230–248, 1990.

Duan, N., L. Mays and K. Lansey, "Optimal–Based Design of Pumping and Distribution Systems," *ASCE Journal of Hydraulic Engineering*, 116: 268, 1990.

Fujiwara, O., and A. De Silva, "Algorithm for Reliability Based Optimal Design of Water Networks," *ASCE Journal of Environmental Engineering*, 116: 575–587, 1990.

Fujiwara, O., and J. Li, "Reliability Analysis of Water Distribution Networks in Consideration of Equity, Redistribution, and Pressure Dependent Demand," *Water Resources Research*, 34: 1843–1850, 1998.

Fujiwara, O., and H. Tung, "Reliability Improvement for Water Distribution Networks Through Increasing Pipe Size," *Water Resources Research*, 27: 1395–1402, 1991.

Germanopoulis, G., P. Jowitt, and J. Lumbers, "Assessing the Reliability of Supply and Level of Service for Water Distribution Systems," *Proceedings, ICE*, Part 1, 80(Apr.): 413–428, 1986.

Goulter, I. C., "Measures of Internal Redundancy in Water Distribution Network Layouts," *Journal of Information and Optimization Science*, 9: 363–390, 1988.

Goulter, I. C., "Systems Analysis in Water Distribution Network Design: from Theory to Practice," *ASCE Journal of Water Resources Planning and Management*, 118: 238–248, 1992a.

Goulter, I. C, "Modern Concepts of a Water Distribution System: Policies for Improvement of Networks with Shortcomings," in E. Cabrera and F. Martinez, eds., *Water Supply Systems—State of the Art and Future Trends*, Computational Mechanics Publications, Southampton, UK, 119-138, 1992b.

Goulter, I., "Analytical and Simulation Models for Reliability Analysis in Water Distribution Systems," in E. Cabrera and A. Vela, eds., *Improving Efficiency and Reliability in Water Distribution Systems*, Kluwer Academic Publishers, Dordrecht, The Netherlands, 235-266, 1995.

Goulter, I. C., and F. Bouchart, "Joint Consideration of Pipe Breakage and Pipe Flow Probabilities," Proceedings of the ASCE 1987 National Conference on Hydraulic Engineering, Williamsburg, VA, August 3–5, 1987.

Goulter, I., and F. Bouchart, "Reliability-Constrained Pipe Network Model," *ASCE Journal of Hydraulic Engineering*, 116: 211–229, 1990.

Goulter, I., and A. Coals, "Quantitative Approaches to Reliability Assessment in Pipe Networks," *ASCE Journal of Transportation Engineering*, 112: 287–301, 1986.

Goulter, I., and P. Jacobs, "Discussion of 'Water Distribution' Reliability: Analytical Method by Wagner, Shamir and Marks," *ASCE Journal of Water Resource Planning and Management*, 115: 709–711, 1981.

Great Lakes and Upper Mississippi Board of State Public Health and Environmental Managers, Ten States Standards, Recommended Standards for Water Works, Albany, NY, 1992.

Gupta, R., and P. Bhave, "Reliability Analysis of Water Distribution Systems," *ASCE Journal of Hydraulic Engineering*, 120: 447–460, 1994.

Henley, E.J. and H. Kumamoto, *Reliability Engineering and Risk Assessment*, Prentice-Hall, Englewood Clifts, N.J., 1981.

Hobbs, B., and G. Biem, "Analytical Simulation of Water System Capacity Reliability. L: Modified Frequency-Duration Analysis," *Water Resources Research*, 24: 1431–1444, 1988.

Jacobs, P., "A Removal Set Based Approach to Urban Water Supply Distribution Network Reliability," Ph.D. Thesis, University of Manitoba, Winnipeg, 1992.

Jacobs, P., and I. Goulter, "Evaluation of Methods for Decomposition of Water Distribution Networks for Reliability Analysis," *Civil Engineering Systems*, 5: 58–64, 1988.

Jacobs, P., and I. Goulter, "Optimization of Redundancy in Water Distribution Networks Using Graph Theoretic Principles," *Engineering Optimization* 15: 71–82, 1989.

Jacobs, P., and I. Goulter, "Estimation of Maximum Cut-Set Size for Water Network Failure," *ASCE Journal of Water Resources Planning and Management*, 117: 588–605, 1991.

Jowitt, P., and C. Xu, (1993). "Predicting Pipe Failure Effects in Water Distribution Networks." *ASCE, Journal of Water Resources Planning and Management*, 119(1), 18-31, 1993.

Kessler, A., L. Ormsbee And U. Shamir, "A Methodology for Least-Cost Design of Invulnerable Water Distribution Networks," *Civil Engineering Systems*, 1(1): 20–28, 1990.

Kettler, A. J., and I. C. Goulter, "Reliability Consideration in the Least-Cost Design of Looped Water Distribution Networks," Proceedings of the 12th International Conference on Urban Hydrology, Hydraulics, *and Sediment Control*, University of Kentucky, Lexington, 305–312, 1983.

Kettler, A. J., and I. C. Goulter, "Analysis of Pipe Breakage in Urban Water Distribution Networks," *Canadian Journal of Civil Engineering*, 12: 286–293, 1985.

Lansey, K., N. Duan, L. Mays, and Y-K. Tung, "Water Distribution Design Under Uncertainties," *ASCE Journal of Water Resources Planning and Management*, 115 : 630–645, 1989.

Lansey, K., K. Awumah, Q. Zhong, and I Goulter, "A Supply Based Reliability Measure for Water Distribution System," in J-T. Kuo and G-F Lin, Eds., IAHR International Symposium on Stochastic Hydraulic, Taipei, Taiwan, May 18–20, 171–178, 1992.

Mays, L. W., ed., *Reliability Analysis of Water Distribution Systems,* American Society of Civil Engineers, New York, 1989.

Mays, L., N. Duan, and Y. Su, "Modeling Reliability in Water Distribution Network Design." In M. Karamouz et al. eds., Proceedings of the Speciality Conference Water Forum-86: World Water Issues in Evolution, American Society of Civil Engineers, New York, 1272–1279, 1986.

Morgan, D., and I. Goulter, "Optimal Urban Distribution Design," *Water Resources Research*, 21: 642–652, 1985.

Nieswiadomy, M., "Estimating Urban Residential Water Demand—Effects of Price Structure, Conservation, and Education," *Water Resources Research*, 28: 604–615, 1992.

Ormsbee, L., and A. Kessler, "Optimal Upgrading of Hydraulic—Network Reliability," *ASCE Journal of Water Resources Planning and Management*, 116: 784–802, 1990.

O'Day, D. K., "Organizing and Analyzing Leak and Break Data for Making Main Replacement Decisions," *Journal American Water Works Association*, 74 (11): 584–596, 1982.

O'Day, D. K., "Analyzing Infrastructure Condition—A Practical Approach," *Civil Engineering*, 53 (4): 39–42, 1983.

O'Day, D. K., C. M. Fox, and G. M. Hugnet, "Aging Urban Water Systems: A Computerized Case Study," *Public Works*, 3 (8): 61–111, 1980.

O'Day, D. Kelly and L. A. Newman, "Assessing Infrastructure Needs: The State of the Art", in R. Hanson, ed., *Perspectives on Urban Infrastructure,* National Academy Press, Washington, DC, 72–75, 1984.

O'Day, K. D., R. Weiss, S. Chiaveri, and D. Blair, *Water Main Evaluation for Rehabilitation/Replacement,* American Water Works Association Research Foundation, Denver, CO, and U. S. Environmental Protection Agency, Cincinnati, OH, 1986.

Park, H., and J. Liebman, "Redundancy-Constrained Minimum Design of Water-Distribution Nets," *ASCE Journal of Water Resources Planning and Management*, 1991(1): 83–98, 1993.

Quimpo, R., and U. Shamsi, "Network Analysis for Supply Reliability Determination," in R. Ragan, Ed., Proceedings of the National Conference on Hydraulic Engineering American Society of Civil Engineers, New York, 716–721, 1987.

Quimpo, R. J, and U. Shamsi, "Reliability-Based Distribution System Maintenance," *ASCE Journal of Water Resources Planning and Management*, 117: 321–339, 1991.

Quindry, G., E. Brill, J. Liebman, and A. Robinson, "Comments on 'Design of Optimal Water Distribution Systems by Alperovits and Shamir," *Water Resources Research*, 15: 1651–1656, 1979.

Quindry, G., E. Brill, and J. Liebman, "Optimization of Looped Water Distribution Systems," *ASCE Journal of the Environmental Engineering Division*, 107: 665–679, 1981.

Rowell, W., and J. Barnes, "Obtaining Layout of Water Distribution Systems," *ASCE Journal of the Hydraulic Engineering Division*, 108: 137–148, 1982.

Shamir, U., and C. Howard, "Reliability and Risk Assessment for Water Supply Systems," in H. Torno, Ed., Proceedings of the Specialty Conference on Computer Applications in Water Resources, American Society of Civil Engineers, New York, 1218–1228, 1985.

Shamsi, U., "Computerized Evaluation of Water Supply Reliability," *IEEE Transactions on Reliability*, 39(1): 35–41, 1990.

Su, Y., L. Mays, N. Duan, and K. Lansey, "Reliability Based Optimization for Water Distribution Systems," *ASCE Journal of Hydraulic Engineering*, 113: 589–596, 1987.

Templeman, A., "Discussion of 'Optimization of Looped Water Distribution Systems' by Quindry et al.," *ASCE Journal of the Environmental Division*, 108: 599–602, 1982.

Templeman, A. B., and D. F. Yates, "Mathematical Similarities in Engineering Network Analysis," *Civil Engineering Systems*, 104 (1), 1984.

Tung, Y-K., "Uncertainty and Reliability Analysis, in *Water Resources Handbook,* ed. by L.W. Mays, McGraw-Hill, 7.1-7.65, 1996.

Tung, Y-K., "Model for Optimal Risk-Based Water Distribution Network Design," in M. Karamouz, et al., eds., Proceedings of the ASCE Conference: Water Forum '86—World Water Issues in Evolution, Long Beach, CA, 1280–1286, 1986.

Wagner, J., U. Shamir, and D. Marks, "Water Distribution Reliability: Analytical Methods," *ASCE Journal of Water Resources Planning and Management,* 114: 253–275, 1988a.

Wagner, J., U. Shamir, and D. Marks, "Water Distribution Reliability Simulation Methods," *ASCE Journal of Water Resources Planning and Management*, 114: 276–293, 1988b.

Walski, T. M., "Practical Aspects of Providing Reliability in Water Distribution Systems," *Reliability Engineering and System Safety*, 42: 13–19, 1993a.

Walski, T. M., "Water Distribution Valve Topology for Reliability Analysis*,*" *Reliability Analysis* and *System Safety*, 42: 21–28, 1993b.

Walski, T. M., "Valves and Water Distribution System Reliability," *Proceedings of the* AWWA Annual Convention, New York, 1994.

Walski, T. M., and A. Pelliccia, "Economic Analysis at Water Main Break," *Journal of the American Water Works Association*, 140–147, March 1982.

Weber, J., "Integrating Conservation Targets into Water Demand Projections," *Journal of the American Water Works Association*, 85: 63–70, 1993.

Wilchfort, O., and J. Lund, "Shortage Management Modeling for Urban Water Supply System," *ASCE Journal of Water Resources Planning and Management*, 123(4): 250–258, 1997.

Wu, S-J., and R. Quimpo, "Predictive Model and Reliability Analysis for Water Distribution Systems" in Proceedings of the J.-T. Kuo and G-F. Lin, eds., 6th IAHR International Symposium on Stochastic Hydraulics, Eds., Taipei, Taiwan, May 18–20, 163–170, 1992.

Wu, S-J., Y-H. Yoon, and R. Quimpo, "Capacity-Weighted Water Distribution System Reliability," *Reliability Engineering and System Safety*, (42): 39–46, 1993.

Xu, C., and I. Goulter, "Probabilistic Model for Water Distribution Reliability," *ASCE Journal of Water Resources Planning and Management*, 124: 218–228, 1998.

INDEX

Index note: The f or t following a page number refers to a figure or table, respectively on that page.

Acoustic (sonic) velocity, 6.2
Advective transport, 9.2
   In pipes, 9.17
Aerobic zone, 9.2
Affinity laws, 5.14
Alkalinity, 9.2, 9.8
Altitude valves, 10.18
Anaerobic zone, 9.2
Ancient urban water supply, 1.3 – 1.9
Apodyterium, 1.6
Aqueducts,
   Ancient Rome, 1.8f
   Segovia, 1.9f
Atmospheric vacuum breakers, 8.6
Automated control, 15.3
   Computer-assisted control, 16.5f
   Computer-directed control, 16.5f
   Computer-monitored control, 16.4f
Availability, 18.6, 18.19
   Expected availability, 18.26
   Network availability, 18.26
   Stationary availability, 18.20
   Stationary unavailability, 18.20
   Unavailability, 18.19

Backflow, 8.5
Backflow prevention, 8.6 – 8.8
Backflow preventors, 8.6
   Application of, 8.7 – 8.8
   Types of, 8.6 – 8.7
      Air gap, 8.6
      Atmospheric vacuum breakers, 8.6
      Barometric loops, 8.6
      Dual check valves, 8.7
      Double-check valves, 8.6
      Pressure vacuum breakers, 8.6
      Reduced-pressure backflow
         preventors, 8.6
      Single check valves, 8.7
Backpressure, 8.5
Backsiphonage, 8.5
Bacterial growth, 11.5
Bernoulli equation, 2.13

Biofilms, 9.2, 9.9
   Composition of, 9.9
   Control of, 9.10
   Origins of, 9.9
   Significance of, 9.10
   Treatment of, 9.10
Biological corrosion, 9.7
Booster disinfection, 9.5, 16.32 – 16.41
   Optimal scheduling of dosages,
      16.32 – 16.41
Booster pumps, 15.9
Boundary layer theory, 2.14
Branching pipe system, 4.7
Bulk modulus of elasticity, 2.36
Bulk reactions, 9.2, 9.17

Calcium hypochlorite, 8.2
Caldarium, 1.6
Calibration, 12.8, 14.1 – 14.23
   Basic steps, 14.5 f
   Data collection, 14.11 – 14.13
      Fire-flow tests, 14.12
      Telemetry data, 14.13
      Water-quality data, 14.13
   Demand-load simplification, 14.2f
   Evaluation of results, 14.13
   Fire-flow tests, 14.12
   Macro-level calibration, 14.15
   Micro-level calibration, 14.15 – 14.17
      Analytical approaches, 14.16
      Optimization approaches, 14.17
      Simulation approaches, 14.16
   Model parameter estimates, 14.3 – 14.11
      Data collection, 14.11 – 14.13
      Nodal demands, 14.9 – 14.11
      Pipe roughness, 14.3 – 14.9
   Network characterization, 14.1
   Network data requirements, 14.1
   Node-link characterization, 14.2f
   Sensitivity analysis, 14.15
Castella, 1.7
Castellum, 1.7
Cavitation, 5.22

I.2 Index

Check valves for backflow prevention, 8.6, 8.7
    Dual check valves, 8.7
    Double-check valves, 8.6
    Single check valves, 8.7
Chlorine residual, 13,21
    First-order decay, 13.21
Clearwell, 11.2
Clearwell storage, 10.7
Colebrook-White equation, 2.17
Coliform bacteria, 9.2
Component reliability analysis, 18.15 – 18.20
    Availability, 18.19
    Failure density, 18.15
    Failure rate, 18.15
    Hazard function, 18.16
    Mean-time-between-failure (MTBF), 18.19
    Mean-time-to-failure (MTTF), 18.17, 18.18t
    Mean-time-to-repair (MTTR), 18.19
    Repair probability, 18.17
    Repair rate, 18.17
    Stationary availability, 18.20
    Stationary unavailability, 18.20
    Unavailability, 18.19
    Unreliability, 18.15
Components, 1.19f
Compound meter, 1.29f
Computational fluid dynamics, 11.25
    Application of models, 11.27
    Commercial packages, 11.27t
    Mathematical formulation of, 11.26
    Mesh for cylindrical tanks, 11.26f
Computer models, see water distribution system modeling
Conservation equations, 7.3
Continuity, see conservation laws
Continuously stirred tank reactor (CSTR), 11.28
Conservation laws of, 2.6 – 2.7, 4,3, 4.11, 4.24
    Chemical species, 2.7
    Energy, 4.3, 4.11
    Mass, 4.3, 4.11
    Molecular species, 2.7
    Momentum, 4.24
Control volume, 2.7
Corrosion,
    Bimetallic connections, 17.11
    Biological corrosion, 9.7
    Control of, 9.8
    External soil corrosion, 17.10 – 17.11
    Factors affecting, 9.7
    Indicators of, 9.8
    Internal corrosion, 17.11
    Maintenance, 17.2
    Pipe breaks, 17.10
    Pitting, 9.7
    Stray current corrosion, 17.11
    Tuberculation, 9.7
    Types of, 9.7
    Uniform corrosion, 9.7
Cross-connection, 8.5
Cross-connection control, 8.5
Cutoff head, 4.4

Darcy-Weisbach equation, 2.16, 4.9, 5.12
Dead storage, 10.17 – 10.18
Densimetric Froude number, 11.11
Design optimization, see optimization of design
Detention time, 11.2, 11.12
Discrete volume method (DVM), 9.19
Disinfectant loss, 9.3 – 9.6
    Mitigation of, 9.5
    Rates of, 9.5
Disinfectant residual, 11.2 – 11.3
Disinfection, 9.3
Disinfection by-products (DBP), 9.2
    Formation of, 11.3
    Growth of, 9.6
Disinfection methods, 9.4
    Booster disinfection, 9.5
    Post disinfection, 9.4
    Primary disinfection, 9.4
    Secondary disinfection, 9.4
Disinfection of new water mains, 8.1 – 8.5
    Chemicals, 8.2
    Need for, 8.2
    Procedures, 8.2 – 8.3
        Continuous method, 8.3
        Slug method, 8.3
        Tablet method, 8.2
    Reducing agents, 8.4
        Dose, 8.4t
    Testing new mains, 8.3
Disinfection of storage tanks, 8.4 – 8.5
    Procedures for filling tanks, 8.4
Dispersive transport, 9.2
Dissolved oxygen, 9.8

Effective storage, 10.7f
Emergency storage, 10.16
Energy gradeline, 2.14, 2.22f
Entrapped air, 6.2
EPANET, 12.13 – 12.20
    Background of, 12.13
    Data flow diagram, 12.18f
    Input data file, 12.19
    Program features, 12.14
    Programmer's toolkit, 12.20
    Solver module, 12.17 – 12.18
    User interface, 12.15, 12.15f, 12.16f, 12.17f

Equalization, 10.2
Equalization storage, 10.12 – 10.14
Eulerian approach, 9.3
Euler number, 6.10
Event-driven method (EDM), 9.19
Extended period simulation, 4.12, 4.23

Failure density, 18.15
Failure rate, 18.15
Finite difference method (FDM), 9.19
Fire-flow tests, 14.12
Fire storage, 10.14 – 10.16, 10.15f
First-order decay, 13.21
First-order reaction, 9.3
Fistulae, 1.7
Fixed grade node, 4.2
Floating-on-the-system, 10.4
Flow control, 2.1
Flushing of distribution systems, 8.8 – 8.10
    Alternating of disinfectants, 8.9
    Directional flushing, 8.9
    Procedures, 8.8 – 8.9
Frigidarium, 1.6
Froude number, 11.23

Geographic information systems (GIS), 17.22
Gradient algorithm, 4.20

Hardness, 9.8
Haloacetic acids (HAA's), 11.3
Hardy Cross, 1.16
Hardy Cross method, 1.16, 4.11
Hazard function, 18.16
Hazen-Williams equation, 2.18, 4.6, 5.9, 5.10, 17.19
    Coefficients for, 5.11t, 14.4t
Heterotrophic bacteria, 9.3
High-service pump stations, 15.9
History,
    Ancient systems, 1.3 – 1.8
    Chronology of knowledge, 1.3t
    Early pipe flow formulas, 1.16, 1.17t
    Modeling, 39, 12.2 – 12.3, 13.4 –13.5
    U. S. 19th Century, 1.9 – 1.16
Hydraulic carrying capacity, 17.16 – 17.21
Hydrant flow test, 17.17f
Hydrant leak repairs, 18.3
Hydraulic gradeline, 2.14, 2.22f, 2.28f
Hydraulic head, 4.2
Hydrogen sulfide, 11.5
Hydropneumatic tanks, 10.6 – 10.7

Incipient cavitation, 5.23
Internal corrosion, 9.6
Iron buildup, 11.4

Jet mixing, 11.8 – 11.9, 11.11
Joint leaks, 18.3
Joukowsky equation, 6.2
Joukousky head rise, 6.3

KYPIPE, 4.18

Lagrangian approach, 9.3
Laminar flow, 2.16
Leachate, 11,5
Leak detection, 17.5
    Contact point sonic leak detection, 11.6f
    Ground sonic leak detection, 17.6f
    Leak correlator equipment, 17.6f
Leak repairs, 18.3
Linear theory method, 4.17
Liquid chlorine, 8.2
Liquid column separation, 6.2
Liquid-level switches, 15.6
Loop equations, 4.11

Main breaks, 18.3
Main leaks, 18.3
Maintenance information systems, 17.21 – 17.23
    Geographic information systems (GIS), 17.22
    SCADA, 17.23
    System database, 17.22
    System mapping, 17.22
Maintenance management systems, 17.23
Maintenance problems, 17.1 – 17.2
    Corrosion, 17.2
    Normal wear, 17.1
    Unforeseen loads, 17.2
Manganese buildup, 11.4
Man-machine interface (MMI), 15.8
Manning's equation, 5.11
Mean-time-between-failure (MTBF), 18.19
Mean-time-to-failure (MTTF), 18.17, 18.18t
Mean-time-to-repair (MTTR), 18.19
Mixed flow, 11.8f, 11.31
Mixing in,
    Pipe junctions, 9.17
    Storage facilities, 9.17, 11.8 – 11.12
        Ideal flow regimes, 11.8f
        Jet mixing, 11.8 – 11.9, 11.10f
        Mixed flow, 11.8f, 11.31
        Mixing times, 11.9
        Mixing time formulations, 11.11t
        Plug flow, 11.8f, 11.32 – 11.33
        Stratification, 11.10 – 11.11
Model calibration, 9.21
    Process, 9.22

Modified linear theory Newton method, 4.18
Monitoring, 11.12 – 11.17, 15.3, 15.5 – 15.9
  Biofilm monitoring, 11.12
  Frequency, 11.18
  Nitrification monitoring, 11.13
  Performance with SCADA, 15.5 – 15.9
  Routine monitoring, 11.12
  Sediment monitoring, 11.13
  Sediment monitoring parameter, 11.16t
  Temperature monitoring, 11.20

Net positive suction head, 5.20 – 5.24
  Available, 5.20
  Cavitation,, 5.22
  Criteria from pump tests, 5.22f
  Incipient cavitation, 5.23
  Major considerations, 5.22
  Required, 5.21
  Safety factor considerations, 5.22
  Suppress visible cavitation, 5.23f
Network availability, 18.26
Network reliability, 18.29 – 18.40
  Connectivity, 18.33
  Cut-set, 18.33
  Heuristic techniques, 18.39
  Reachability, 18.33
  Redundancy-based measures, 18.35
  Simulation, 18.29
  Summary of approaches, 18.35t
Newton-Rhapson method, 4.8
Newton's second law, 2.8
Nitrification, 11.6
Node equations, 4.18
No-slip condition, 2.14

Odor, 11.3
Operation,
  Emergency operation, 15.4
  Operating criteria, 15.3 – 15.4
  Operating indices, 15.2 – 15.3
  Water quality considerations, 15.4
Opportunistic pathogens, 9.3
Optimal control systems, 1.30f
Optimal operation,
  Booster disinfection scheduling, 16.32 – 16.41
    Maximum set-covering problem, 16.37 – 16.39
    Mixed integer linear programming problem, 16.35 – 16.37
    Optimal location of booster stations, 16.37 – 16.39
  Formulation to minimize energy costs, 16.3 – 16.16

Energy management, 16.3
Management models, 16.3 – 16.9
Management strategies, 16.3
Optimization models, 16.9 – 16.16, 16.11t
Formulation to satisfy water quality, 16.16 – 16.19
Mathematical programming approach, 16.19 – 16.20
Simulated annealing approach, 16.22 – 16.26
Solution methods, 16.19 – 16.31
Optimization of design,
  Applications, 7.9 – 7.12
  Branched systems, 7.4 – 7.5
  Linearization, 7.5 – 7.7
  Looped systems, 7.5 – 7.7
  Mathematical formulations, 7.2
  New York City water supply problem, 7.5 – 7.12
  Nonlinear programming, 7.7 – 7.8
  Optimization methods, 7.4 – 7.9
  Optimization problems, 7.1 – 7.2
  Optimization-simulation model link, 7.7f
  Stochastic search techniques, 7.8 7.9

Parallel pipe systems, 4.5
PH, 9.8, 11.4
Piezometric head, 4.1
Pigs, 17.20f
Pipe breaks, 17.10 – 17.16
  Characteristics of, 17.5t
  Corrosion, 17.10 – 17.11
  External loads, 17.11 –17.12
  Poor tapping, 17.13
  Pressure-related breaks, 17.13
  Repair, 17.14
  Replacement, 17.14
Pipe flow, 2.18 – 2.19
  Extended period simulation, 4.12
  Flow in branching systems, 4.7
  Flow in parallel, 2.18f, 4.5
  Flow in series, 2.18f, 4.5
  Local losses, 2.21,4.5
  Networks, 4.11
    Gradient method, 4.20
    Hardy Cross method, 4.11
    Linear theory method, 4.17
    Newton-Rhapson method, 4.18
    Unsteady flow, 4.23
Pipe leaks, 17.5 – 17.7
  Characteristics of, 17.5t
Pipeline models,
  Modeling approach, 2.4

Pipeline models (Cont.)
  Numerical models, 2.2 – 2.4
    Operational models, 2.4
    Planning models, 2.4
Pipe material characteristics, 6.3
  Acoustic velocity, 6.4f
  Hoop stress, 6.3
  Physical properties, 6.3
  Poisson ratio, 6.3
  Young's modulus, 6.3
Pipe rehabilitation, 17.20 – 17.21
  Costs, 17.21
  Evolution of, 17.21
  Technology of, 17.20 – 17.21
Pipe systems, 4.5 – 4.11
  Branching systems, 4.7
  Networks, 4.11
  Parallel systems, 4.5
  Series systems, 4.5
Pipe wall reactions, 9.3, 9.12 – 9.18
Piscinae, 1.7
Pitting, 9.7
Plug flow, 11.8f, 11.32 – 11.33
Post disinfection, 9.4
Pressure problems, 17.16 – 17.20
  Correction of, 17.17 – 17.20
  Diagnosis of, 17.16 – 17.17
Pressure-reducing valves, 1.25f, 1.26f
Pressure-regulating valves, 4.4
Pressure-sustaining valves, 1.25f, 1.26f
Pressure transducer, 15.7f
Pressurized pipeline, 2.1
Pressurized storage tanks, 10.6 – 10.7
Primary disinfection, 9.4
Programmable logic controller, (PLC), 15.6
Pump definitions, 5.2 – 5.6
  Allowable operating range, 5.2
  Atmospheric head, 5.2
  Centrifugal point, 5.2
  Condition points, 5.2
    Best efficiency points, 5.2
    Normal condition point, 5.2
    Rated condition point, 5.2
    Specified point, 5.2
  Cutoff head, 4.4
  Datum, 5.2
  Elevation head, 5.2
  Friction head, 5.2
  Gauge head, 5.2
  Head, 5.2
  High-energy pump, 5.2
  Impeller balancing, 5.2
    Single-plane balancing, 5.2
    Static balancing, 5.2
    Two-phase balancing, 5.3
  Overall efficiency, 5.2
  Power, 5.3
    Brake horsepower, 5.3
    Pump efficiency, 5.4
    Pump input power, 5.3
    Pump output power, 5.3 – 5.4
    Water horsepower, 5.3
  Pump pressures, 5.4
    Field test pressure, 5.4
    Maximum allowable casing working pressure, 5.4
    Maximum suction pressure, 5.4
    Shutoff, 5.4
    Speed, 5.4
  Suction conditions, 5.4 – 5.6
    Maximum suction pressure, 5.4
    Net positive suction head available, 5.4
    Net positive suction head required, 5.5
    Static suction lift, 5.5
    Total discharge head, 5.5
    Total head, 5.5
    Total suction head, closed suction tests, 5.4
    Total suction head, open suction, 5.4
    Total dynamic head, 5.4
  Velocity head, 5.4
Pump characteristics, 6.9
  Abnormal (four quadrant) characteristics, 6.11
  Corrected pump curves, 5.24 – 5.28
    Example calculation, 5.24 – 5.28
    Minor losses, 5.24t
  Design point, 6.11
  Duty, 6.11
  Head and torque characteristics, radial flow pump,
    Four quadrant, 6.17f
    Negative rotation, 6.16f
    Positive rotation, 6.15f
    Suter diagram, 6.17f
  Karman-Knapp circle diagram, 6.14
  Nameplate, 6.11
  Rated conditions, 6.11
Pump hydraulics, 5.9 – 5.20
  Hydraulics of valves, 5.13
  Impeller wear, effect of, 5.16f
  Operation, 5.14
    Affinity laws, 5.14
    In parralel, 5.14, 5.17f
    Homologous laws, 5.14
  Pipeline hydraulics, 5.9 – 5.13

I.6 Index

Pipeline hydraulics (Cont.)
    Colebrook-White equation, 2.17
    Darcy-Weisbach equation, 2.16,
        4.9, 5.12
    Hazen-Williams equation, 2.18, 4.6,
        5.9, 5.10
    Manning's equation, 5.11
    Roughness factors, 5.12
    Pump performance curves, 4.3, 5.8f, 5.9
    Pump specific speed,
        Equation for, 5.16
        Suction specific speed, 5.18
    System curves, 5.9, 5.14, 5.15f
    Variable speed pumps, 5.14
        Typical discharge curves, 5.18f
Pump performance
    Homologous (affinity) laws, 5.14,
        6.10 – 6.11
    Dynamic similarity, 6.10
    Kinematic similarity, 6.10
    Power coefficient, 6.1
    Torque coefficient, 6.11
    Performance parameters, 6.9 – 6.10
    Flow coefficient, 6.10
    Flowrate, 6.10
    Head coefficient, 6.10
    Impeller diameter, 6.9
    Power coefficient, 6.10
    Rotational speed, 6.9
    Total dynamic head, 6.10
Pump operation for transients, 6.11 – 6.14
    Abnormal (four quadrant), 6.11 – 6.14
    Zones of possible pump operation, 6.13f
Pump selection, hydraulic considerations,
    5.28 – 5.29
    Flow range of centrifugal pumps, 5.28
    Steps, 5.30 – 5.32
Pump specific speed,
    Equation for, 5.16
    Suction specific speed, 5.18
Pump standards, 5.1
Pump station design, 5.32 – 5.36
    Hydraulic transients, 5.37
        Pipe material selection, 5.37
        Valve selection, 5.37
    Operating problems, 5.34
    Piping, 5.34
    Piping design, 5.35
        Criteria, 5.36
        Pressure design, 5.35
        Vacuum condition, 5.36
    Pump operating ranges, 5.32t, 5.32f
Pump surge protection devices, 6.18 – 6.24
Pump surge control devices, 6.18 – 6.24

Quasi-steady flow, 2.12, 2.31 – 2.32

Rehabilitation problems, 17.1 – 17.2
Reliability, 18.5, 18.20
    Effect of valving, 18.10 – 18.15
    Improvement in system, 18.6 – 18.10
Reliability analysis,
    Component reliability analysis,
        18.15 – 18.20
    Availability, 18.19
    Failure density, 18.15
    Failure rate, 18.15
    Hazard function, 18.16
    Mean-time-between-failure
        (MTBF), 18.19
    Mean-time-to-failure (MTTF),
        18.17, 18.18t
    Mean-time-to-repair (MTTR), 18.19
    Repair probability, 18.17
    Repair rate, 18.17
    Stationary availability, 18.20
    Stationary unavailability, 18.20
    Unavailability, 18.19
    Unreliability, 18.15
Reliability assessment,
    Approaches, 18.25 – 18.40
    Availability, 18.26
    Component (pipe) failure, 18.25
    Demand variation failure, 18.25
    Expected availability, 18.26
    Network availability, 18.26
    Network reliability, 18.29 – 18.40
        Connectivity, 18.33
        Cut-set, 18.33
        Heuristic techniques, 18.39
        Reachability, 18.33
        Redundancy-based measures, 18.35
        Simulation, 18.29
        Summary of approaches, 18.35t
    Total expected volume of deficit, 18.25
    Volume deficit, 18.25
Reliability indexes, 18.5 – 18.6
    Availability, 18.6
    Economic indexes, 18.6
    Frequency and duration indexes, 18.6
    Severity indexes, 18.6
Reliability measures, 18.32, 18.40 – 18.43
    Link importance, 18.43
    Overview of, 18.40 – 18.42
Remote terminal unit, (RTU), 15.6, 15.7f
Repairs, definitions of, 18.3 – 18.4
Repair probability, 18.17
Repair rate, 18.17

Reynold's number, 2.15, 11.23
Rigid column analysis, 6.7
Rivus, 1.7

Safe Water Drinking Act, 1.2, 4.26, 13.1 – 13.2
   Amendments, 1.2, 13.1
   Lead and copper rule, 13.2
   Maximum contaminant levels (MCL), 1.2
   Stage 1 disinfectants and disinfection by-products rule, 13.2
   Total coliform rule, 13.2
   Trihalomethane regulation, 13.2
Sanitation needs, 1.2t
SCADA, 1.27, 15.1, 15.5, 16.2, 17.23
Secondary disinfection, 9.4
Sediment buildup, 11.7
Series pipe systems, 4.5
Service leak repairs, 18.4
Sodium hypochlorite, 8.2
Sonic leak detection, 17.5
Specus, 1.7
Standpipe, 9.3, 11.2
Steady flow, 2.11, 2.13 – 2.31
   Quasi-steady flow, 2.12, 2.31 – 2.32
Storage tanks,
   Aesthetics of, 10.4
   Cathodic protection, 10.18
   Clearwell storage, 10.7
   Coatings, 10.18
   Dead storage, 10.17 – 10.18
   Design issues, 10.4 – 10.7
      Effective versus total storage, 10.6
      Floating-on-the-system, 10.4
      Floating versus pumped storage, 10.4 – 10.5
      Ground versus elevated tank, 10.5
      Pressurized tanks, 10.6 – 10.7
      Private versus utility owned tanks, 10.6
   Effective storage, 10.7f
   Effect on water quality, 10.3
   Emergency storage, 10.2, 10.16
   Energy consumption, 10.3
   Equalization, 10.2
   Equalization storage, 10.12 – 10.14
   Fire storage, 10.2
   Hydraulic gradeline for, 10.3f
   Hydropneumatic tanks, 10.6 – 10.7
   Multiple pressure-zone systems, 10.9
   Multiple tanks, 10.8
   Overflows, 10.18
   Pressure maintenance, 10.2
   Location, 10.7 – 10.9
   Tank levels, 10.9 – 10.11
   Overflow levels, 10.9
   Pressure zones, 10.10, 10.11f
   Service areas, 10.10
   Tank terminology, 10.5f
   Tank volume, 10.11 – 10.18
      Functional design, 10.12 – 10.18
      Staging requirements, 10.16 – 10.17
      Standards-driven sizing, 10.12
      Trade-offs, 10.11
   Vents, 10.18
Storage tank disinfection, 8.4 – 8.5
Storage tank inspection, 8.5
Storage, water quality of,
   Aging, 11.11 – 11.12
   Chemical problems, 11.2 – 11.5
   Design for, 11.30 – 11.33
      Flow regimes, 11.30
      Modes of operation, 11.30
      Water quality objectives, 11.30
   Inspection, 11.34 – 11.35
   Maintenance, 11.35
   Microbiological problems, 11.5 – 11.8
   Mixing, 11.8 – 11.12
   Modeling, 11.22
      Computational fluid dynamics, 11.25 – 11.27
      Scale models, 11.22 – 11.24
      Similitude, 11.23
   Monitoring, 11.12 – 11.17
      Biofilm monitoring, 11.12
      Frequency, 11.18
      Nitrification monitoring, 11.13
      Routine monitoring, 11.12
      Sediment monitoring, 11.13
      Sediment monitoring parameter, 11.16t
      Temperature monitoring, 11.20
   Operation, 11.30 11.33
   Sampling, 11.17 – 11.20
      Equipment, 11.17
      Methods, 11.17
   Water quality parameters,
      And associated regulation for storage facilities, 11.14t
      For finished water storage facilities, 11.15t
Stratification in reservoirs, 11.10 – 11.11, 11.33
Supervisory control and data acquisition (SCADA), 1.27, 15.1, 15.5 – 15.9, 17.23
   Alarm recording, 15.9
   Anatomy of, 15.6 – 15.9
   Elements of, 15.6f
   Flow measuring device, 15.7f
   Linking with models, 15.11 – 15.14

Supervisory control and data acquisition
   (SCADA) (Cont.)
      Databases, 15.14
      Data requirements, 15.12
      Link establishment, 15.13 – 15.14
      Open Data Base Connectivity (ODBC),
         15.14
   Liquid-level switches, 15.6
   Man-machine interface (MMI), 15.8
   Pressure transducer, 15.7f
   Programmable logic controller, (PLC), 15.6
   Remote terminal unit, (RTU), 15.6, 15.7f
   Schematic of, 15.8f
Surface water treatment rule, 9.3
Surge, 2.12
Surging, 6.2
System models, 11.2, 11.28
   Application of, 11.28 – 11.29
   Compartment models, 11.28
   Elemental system models, 11.28
      Behavior of, 11.29f

Taste, 11.3
Temperature modeling, 11.25
Ten State Standards, 18.10
Tepidarium, 1.6
THM, 9.6, 11.3
Time-driven method (TDM), 9.19
Total coliform rule, 13.2
Total dissolved solids, 9.8
Tracer chemical, 9.3
Tracers, 11.24
   Movement of tracer dyes, 11.25
Tractive force, 2.24
Transients, 2.11
   Bulk modulus of water, 2.36
   Causes, 2.34
   Joukowsky relations, 2.39
   Kinetic energy, 2.35
   Physical nature, 2.35
   Wavespeed, 2.38
   Valves, 2.40
Transitions, 2.23t
   Local loss coefficients, 2.23
Trihalomethane regulation, 13.2
Trihalomethanes, 9.6, 11.3
Tuberculation, 9.7, 17.18f
Tubercle, 9.3
Turbine meter, 1.27f, 1.28f
Turbulent flow, 2.15 – 2.16

Unaccounted-for water, 17.2 – 17.10
   Breaks, 17.5t
   Causes, of, 17.3 – 17.4
   Components of, 17.5
      Authorized unmetered uses, 17.8
      Blowoffs, 17.8
      Fire fighting, 17.7
      Flat-rate customers, 17.8
      Main flushing, 17.8
      Meter unmetered uses, 17.8
      Service pipe leakage, 17.7
      System pressure, 17.7
      Theft, 17.7
      Water main leakage, 17.5
   Indicators for, 17.3
   Leaks, 17.5t
Unavailability, 18.19
Uniform corrosion, 9.7
United Nations International Drinking water
   Supply and Sanitation Decade, 1.1
Unreliability, 18.15
Unsteady flow in pipe networks, 4,24 – 4.26
U.S. regulatory limits for finished water
   quality, 9.12t

Valve leak repairs, 18.4
Valves, 3.11 – 3.48
   Air release valves, 3.47
   Air vacuum valves, 3.47
   Altitude valves, 3.47
   Ball valves, 6.5f
   Blow-offs, 3.47
   Butterfly valves, 3.45, 6.5f
   Check valves for backflow prevention,
      8.6, 8.7
      Dual check valves, 8.7
      Double-check valves, 8.6
      Single check valves, 8.7
   Control valves, 3.46
   Flow control valves, 3.47, 6.8
   Gate valves, 3.44, 6.5f
   Geometric characteristics, 6.5, 6.8f
      Cross-sections of, 6.5f
   Globe valve, 6.5f
   Headlosses, 6.6
      Equation for, 6.7
   Hydraulic characteristics, 6.8f
   Isolation valves, 3.44
   Pressure-reducing valves, 1.25f, 1.26f, 3.46
   Pressure-relief valves, 3.47
   Pressure-sustaining valves, 1.25f,
      1.26f, 3.46
   Needle valve, 6.5f
Valve closure, 6.7
   Classification of, 6.9t

Index **I.9**

Valve operation, 6.8
Vamus, 1.7
Venturi condition, 2.22f
Vulnerability, 18.43

Water age, 9.3
Water distribution design in the U.S., 13.2 – 13.3
Water distribution hydraulics, pipe flow, 2.18 – 2.19
   Extended period simulation, 4.22
   Flow in parallel, 2.18f, 4.5
   Flow in series, 2.18f, 4.5
   Local losses, 2.21, 4.5
   Networks, 4.11
      Gradient method, 4.20
      Hardy Cross method, 4.11
      Linear theory method, 4.17
      Newton-Rhapson method, 4.18
      Unsteady flow, 4.23
Water distribution system control, 15.9 – 15.11
   Advanced control, 15.10
   Automatic control, 15.10
   Central databases, 15.15
   Centralized control, 15.11
   Control strategies, 15.10 – 15.11
   Data centric, 15.15f
   Local control, 15.11
   Supervisory control, 15.10
Water distribution system modeling, 3.9 – 3.12, 4.31 – 4.37, 13.4 – 13.6
   Application, 4.32
   Calibration, 12.8
   Calibration process, 4.32
   Computer models,
      History of, 12.2
      Uses of, 12.2
   Computer model internals, 12.8 – 12.13
      Extended period solver, 12.11 – 12.12
      Input process, 12.9
      Linear-equation solver, 12.11
      Hydraulic solution algorithm, 12.9
      Output processing, 12.12
      Topological processing, 12.9
      Water-quality algorithms, 12.12
   DWQM, 13.6
   Dynamic water quality models, 13.5 – 13.6
   EPANET, 4.33, 13.22
   Evolution of models, 13.4f, 13.22 – 13.23
   History of, 3.9, 12.2 – 12.3,13.4 – 13.5
   Model development, 3.10
   Model selection, 4.32
   Network representation, 12.3
   Junctions, 12.3
   Network components, 12.3
   Network skeletonization, 12.4
   Reservoir nodes, 12.3
   Operating characteristics, 12.7
   Problem definition,4.32
   Propagation of contaminants, 13.23 – 13.44
   Software packages, 3.10, 3.11t
   Steady-state water quality models, 13.5
   Verification, 4.32
   WADISO, 13.7
   Water quality modeling, 4.26 – 4.31
      Conservation of energy, 4.3
      Conservation of mass, 4.3
      Fixed grade node, 4.2
      Hydraulic head, 4.1
      Loading condition, 4.1
      Hardy-Cross method, 4.11
      Piezometric head, 4.1
      Pressure regulating valves, 4.4
      Pump curve, 4.4
Water distribution pipeline design,
   Pipeline design 3.34 – 3.44
      Flexible pipe, 3.37
      Internal pressures, 3.34
      Loads, 3.34
      Rigid pipes, 3.36
      Thrust restraints, 3.38
   Preliminary design, 3.12 – 3.13
      Alignment, 3.12
      Rights-of-way, 3.13
      Subsurface conflicts, 3.12
Water distribution piping materials, 3.13 – 3.34
   Asbestos-cement pipe (ACP), 3.31
   Ductile iron pipe (DIP), 3.13
   High-density polyethylene (HDPE) pipe, 3.29
   Polyvinyl chloride (PVC) pipe, 3.18
   Reinforced concrete pressure pipe (RCPP), 3.25
   Steel pipe, 3.20
Water distribution systems, 3.1
   Average day demand, 3.2
   Fire demands, 3.7
   Maximum day demand, 3.2
   Peaking coefficients, 3.8, 3.9t
   Peaking factors, 3.2
   Peak hour demand, 3.2
   Planning and design criteria, 3.4
   Service pressures, 3.8, 3.9t
   Storage, 3.7
      Emergency storage, 3.7
      Fire storage, 3.7

I.10 Index

Water distribution systems (Cont.)
  Operational storage, 3.7
  Supply, 3.7
  Water demands, 3.3t
Water duties, 3.2, 3.5t
Water hammer, 2.12, 6.1
  Case studies, 6.27 – 6.31
  Definition of, 6.2
  Entrapped air, 6.2
  Joukowsky equation, 6.2
  Joukowski head rise, 2.39, 6.3
  Liquid column separation, 6.2
  Mitigation of, 6.1
  Time constants, 6.24
    Elastic time constant, 6.24
    Flow time constant, 6.24
    Pump and motor inertia time constant, 6.24
    Surge tank oscillation inelastic time constant, 6.24
Water knowledge,, chronology of, 1.3t
Water properties, 2.5t
Water quality in networks, 13.3 – 13.4
  Transformation in bulk water phase, 13.3f
  Transformation in pipe wall, 13.4f
Water quality model calibration, 9.21 – 9.22
Water quality modeling, 9.4 – 9.22, 13.6 – 13.45
  Calibration, 9.21
  Current trends, 13.44 – 13.45
  Data requirements, 9.20 – 9.21
  Dynamic model solution methods,
    Discrete volume (DVM), 9.19
    Event-driven method (EDM), 9.19
    Finite difference method (FDM), 9.19
    Time-driven method (TDM), 9.19
  Early applications, 9.16, 13.6 – 13.22
    Cabool, Missouri, 13.22
    North Penn Study, 13.6 – 13.9
    South Central Connecticut Regional Water Authority, 13.9 – 13.21
  Evolution of, 13.22 – 13.23
  Governing equations, 9.16 – 9.18
  Properties of contaminants (case studies), 13.23 – 13.44
    Gideon, Missouri, 13.36 – 13.44
    North Marin Water District, 13.24 – 13.36
  Solution methods, 9.18 – 9.19
    Dynamic models, 9.18
    Steady-state models, 9.18 – 9.19
Water quality monitoring, 9.11 – 9.15
  Routine monitoring, 9.11
  Synoptic monitoring, 9.11
Water quality parameters, 9.13t
Water quality transformations,
  In distribution system, 9.4f
  In pipe wall, 9.4f
Water supply needs, 1.2t
Wavespeed, 2.37 – 2.38
Weber number, 11.23